AF528053

HORTICULTURE

HORTICULTURE

By
Mahmood N. Malik

2015
Biotech Books
New Delhi - 110 002

ISBN 978-81-7622-098-9

Published by : **BIOTECH BOOKS**
4762-63/23, Ansari Road, Darya Ganj,
New Delhi - 110 002
Phone: +91-011-23262132
e-mail: biotechbooks@yahoo.co.in

Printed at : **Chawla Offset Printers**
Delhi - 110 052

PRINTED IN INDIA

CONTENTS

FIGURES

TABLES

FOREWORD

The need for locally written textbooks on agriculture has been felt for a long time. This is because agriculture is a highly site–specific field, and students find it difficult to relate what is stated in textbooks written abroad to the widely different physical and socio–economic environment of Pakistan.

It became possible to meet this need with the obligation of USAID funds for this purpose under Project Implementation Letter No. 17 of 14 February 1990, largely due to the interest of Cordell Hatch, MART Project and Harry Dickherber, then of USAID, Islamabad. Strong interest and support were provided by former TIPAN Team Leaders Dr. Raymond G. Cragle and Dr. Gilbert H. Kroening, and by Abdur Rehman Khan, the former Vice Chancellor of the NWFP Agriculture University.

The project started with the constitution of a Task Force to plan the work. It comprised nominees of all the three agricultural universities, Pakistan Agricultural Research Council, National Book Foundation, USAID, and the TIPAN Project of the NWFP Agricultural University.

The first meeting of the Task Force was held on 25 March 1990. It identified the following topics and priorities for the textbooks to be written:

Textbook topic	Priority
Crop Production	1
Animal Production	1
Soil Science	1
Horticulture	1
Plant Breeding and Genetics	1
Extension Methods	1
Farm Management	2
Agricultural Entomology	2
Plant Pathology	2
Water Management	3
Range Management	3

It also identified managing authors and several contributing authors for the books at priority 1.

An authors' conference was held in Faisalabad on 6–8 May 1990, at which time panels of authors decided the contents of the books and named the most appropriate persons to write the various chapters. Dr. Everett Edington, Teaching Specialist, TIPAN Project, and Dr. J. Cordell Hatch,

Consultant MART Project, facilitated the conference. And thus the work got under way.

These textbooks have been written at the level of the beginning student of agriculture, and, as far as possible, cover indigenous research and experience.

The best part of the exercise has been the collaboration of over a hundred agricultural scientists from all over Pakistan.

S. Basit Ali Shah
Vice Chancellor
NWFP Agricultural
University

Irshad Ali
Vice Chancellor
Sindh Agricultural
University

Muhammad Rafiq Khan
Vice Chancellor
Agriculture University
Faisalabad

Oval Myers, Jr.
Team Leader
TIPAN

EDITOR'S PREFACE

Every effort has been made to present the ideas and interests of each of our authors as clearly and faithfully as possible, while at the same time adjusting the various formats of their separate articles into a consistent style suitable for a textbook. We have tried to keep the language as simple and straightforward as possible, consistent with accurate representation of the content. We hope that this will make the content of the book more accessible and useful to students.

Typographical conventions employed in this book are as follows. *Italic type* is used: (1) for botanical names at the genus and species level; (2) for non–English words (e.g. Urdu, Russian, Latin, Pashto) which are not normally used in ordinary English conversation or writing (e.g. *in situ*, or *floris*); (3) for the titles of books mentioned in the text; (4) to contrastively stress a particular word in a sentence (e.g. classified by *shape*, not function). Non–English words or botanical names which have become part of ordinary English usage in Pakistan or elsewhere are not italicized (e.g. berseem, dal, barani, rabi, kharif, chernozem, podzol, eucalyptus, delphinium, salvia). **Boldface type** is used to introduce an important new word or concept in the text.

Spelling generally follows the British tradition rather than the American. Individual author preference is respected, however, in this regard. Bibliographical conventions and abbreviations for standard units of measure are those specified in the fifth edition (1983) of the *CBE Style Manual*, published by the Council of Biology Editors, 9650 Rockville Pike, Bethesda, MD 20814, USA.

The editors take this opportunity to express their appreciation to the authors for the spirit of cooperation and enthusiasm which they have shown at all stages of our endeavor to produce the best possible locally–oriented textbook. The illustrations in the book owe a tremendous amount to the invaluable help of Professor Habib–ul–Rahman Mian of the Department of Plant Breeding and Genetics, NWFP Agricultural University with the coordination and execution of the artwork.

INTRODUCTION

Horticulture is a fascinating subject. Since it deals with beautiful flowers and foliage and delicious and nutritious fruits and vegetables, it can be both a profitable business and a rewarding hobby. Also, it improves the human environment, both physically and aesthetically. All these, and many more benefits of horticulture are covered by Iqrar Ahmad Khan in his Introduction to this volume. He ranges widely over the subject, touching several facets which are covered in greater detail in the subsequent chapters.

Altaf–ur–Rehman Rao has the difficult task of introducing the neophyte to the basic concepts and the vocabulary used in describing the structure of flowers, the processes of fertilization and seed formation, the development of fruit and their structure, and the several changes that occur during fruit development. Noor Badshah, in his concise chapter, briefly describes the processes of food manufacture, respiration, and carbon metabolism, and highlights their importance to plants as well as to humans and animals which feed on plants.

Abdul Fatah Baloch, in his chapter on phases of plant growth, describes the vegetative and floral growth and development of plants. His account of the effects of daylength, plant hormones, and temperature on plant growth and their applications in practical horticulture is especially interesting and instructive.

Musahib–ud–Din Khan and Habib–ul–Rahman launch into the difficult but very important subjects of the genetic basis of variability in crop species, the various methods of breeding plants, the state–of–the–art work of biotechnology and genetic engineering, and the breeding methods for some important fruits, vegetables, and ornamental crops.

Biological diversity has kept humans alive and well so far, and is humanity's insurance against starvation in the rapidly changing climatic conditions forecast for the future, which are caused and accelerated by environmental degradation. As the greenhouse effect takes hold, making earth's climate subject to violent fluctuations, biological diversity and the ability to make use of it for breeding better and more adapted crops will become increasingly important for the survival of the human race.

Saeed Ahmed's chapter on plant propagation takes us into the heart of horticulture, telling us how to propagate fruits, vegetables, and ornamentals by seed as well as by asexual methods. A short section deals with plant tissue culture, which is likely to attain considerable importance in plant propagation. Environmental conditions have a profound influence on which species of plants can be grown where, and at what times of the year. Iqrar Ahmad Khan's chapter on plant environments therefore deals with the effects of light, temperature, water, and air on plant growth.

In horticulture, we are mostly concerned with plants rooted in soil. Muhammad Ibrahim Chaudhary's chapter on soils and fertilizers is therefore a good introduction to the significance of soil factors in horticulture. After describing the important attributes of soils, Ibrahim also discusses organic matter and inorganic fertilizers which are so important in the sustainable development of horticulture.

Mahmood Niaz Malik's two chapters on the establishment of gardens and management practices form the centrepiece of this text. Malik draws upon his life–long experience of teaching horticulture in telling us how to select a suitable site for an orchard, how to prepare it, how to select the planting stock, how to plant, and how to manage the numerous aspects of successful horticulture—soils and their cultivation, water and nutrients, weed and pest management, and pruning and training. Only scrupulous attention to these details can ensure success in horticulture.

Insect pests and diseases are a bane of horticulture. Unless you can successfully manage these, you can forget about practicing horticulture. Ali A. Hashmi's chapter on insect pests and diseases is therefore crucial reading in this introductory volume on horticulture. Though you will have to take many more courses in both entomology and plant pathology to become adept in managing the pests and diseases of horticultural plants, it is well to get an early start in becoming familiar with the problems. Of particular interest in this discussion of diseases and insect pests are the various methods of cultural and biological control, and the message that chemical control should be used as a last resort, and even then in moderation, using only those pesticides which are the safest and the least persistent. This is an essential requirement for practicing sustainable horticulture which will not endanger human health.

As Wasim Farooqi points out in his chapter on post–harvest handling, 20–40 percent of our fruit and vegetable harvest is lost annually due to improper attention to the numerous details between the ripening of the produce and its consumption. This is a tremendous recurring annual loss which we can ill afford. All possible measures must therefore be taken to reduce it. An important measure is to educate the student of agriculture. Wasim sets himself this task and takes us through the factors predisposing horticultural crops to these losses, the processes that cause them, and the steps which must be taken to counter them.

The most important fruit crops of Pakistan are citrus, mango, guava, date palm, and banana. Muhammad Ibrahim Chaudhary deals with them in detail in his chapter on fruit crops. Topics covered for each include botany, soil and climate, propagation, and cultural practices. Short accounts are also given for several other fruits grown on a smaller scale—apple, pear, plum, and apricot.

Abdul Fatah Baloch follows with his chapter on vegetable crops, extending a similar treatment to the major vegetable crops of Pakistan.

The most important chapter of this book is on floriculture and landscape gardening by Daud Ahmad Khan. This is because it deals with the most beautiful organism on earth—flowers, and because landscaping is a potent device for making our cities more habitable. It deals with garden designs, lawns, flowers, trees, shrubs, and house plants.

The purpose of this book is to whet your appetite for learning about horticulture. To do so, you will have to spend several years and study several courses. But this is a good start.

Ghaus Mohammed Khattak
Coordinator

Mahmood Niaz Malik
Managing Author

TO THE STUDENT:

This textbook has been written by a team of agricultural researchers and teachers from various parts of Pakistan, each with his own particular expertise and focus of interest. In order to make its contents maximally accessible to you, its users, we have provided the following features.

- **Chapter outlines.** Each chapter is preceded by an outline of its main content areas. Read this first to get a broad idea of the structure of the chapter and the topics it covers.
- **Learning objectives.** At the beginning of each chapter is a list of learning objectives; that is, things which you should be able to do after studying the chapter. Read and think about these before you begin to study the chapter to clarify in your mind what you are going to learn about. Try to keep them in mind to guide and focus your reading of the chapter.
- **Study questions.** At the end of each chapter there is a set of questions. These are intended to re-focus your attention on the most important content of the chapter. They can be used for self-study or assigned by your teachers.
- **References.** A list of references follows each chapter. This list contains all works referred to in the text, as well as suggestions for further reading in some cases. This is a valuable tool for you to use in pursuing an idea beyond what is given in the textbook itself.
- **Abbreviations.** Abbreviations and acronyms used specifically in this book as well as those for standard units of measurements are listed in a separate section at the end of the book.
- **Glossary.** After the final chapter there is a glossary of important technical terms and words which the authors have felt may need to be precisely defined. Use this like a mini-dictionary to find the meaning of an unfamiliar or difficult word quickly.
- **Index.** At the end of the book there is a topic index. In it are listed words and phrases which refer to important concepts, institutions, or persons. If the topic or idea you are interested in is not listed in the table of contents, look for it in the index.

Outline of Chapter 1

1. INTRODUCTION

Iqrar Ahmad Khan[1]

LEARNING OBJECTIVES

After reading this chapter, a student should be able to:

- Explain the place of horticulture as a branch of agriculture
- Discuss the importance of horticulture in our lives and economy
- Summarize the history and development of horticulture
- Talk about the present status and future potential of this branch of agriculture in Pakistan
- Identify the factors affecting or limiting the growth of different branches of horticulture in Pakistan

1.1 Definition of horticulture

An important objective of this introductory chapter is to introduce the discipline of horticulture as it relates to agriculture in Pakistan. A brief history of the discipline, classification of horticultural plants and discussion of their importance, and a description of the production systems of various divisions of horticulture are presented.

The term *horticulture* is derived from two Latin words *hortus* 'garden' and *colere* 'to cultivate'. Halfacre and Barden (1979) state that the first known use of the term horticulture was in 1631. Bailey (1939) observed that horticulture is concerned with production within an enclosure. Gardens are distinguished from fields by the concept of **enclosure**. Janick (1986:1) defines horticulture as "that branch of agriculture concerned with intensively cultured plants used for food, for medicinal purposes, or for aesthetic gratification." Therefore, if agriculture is defined as the technology of raising plants and animals, then horticulture would be that part of plant agriculture which deals with garden crops (Fig. 1.1). **Agronomy** and **forestry** are the

---

[1]Associate Professor, Department of Horticulture, University of Agriculture, Faisalabad.

branches of agriculture covering field crops and forest trees, respectively. Horticulture may be considered as a parallel discipline to agriculture, since the Latin word *ager* means 'field', and agriculture is defined as the cultivation of field crops, whereas horticulture is the cultivation of garden crops. However, the concept of garden has evolved from that of an enclosed place to that of intensively cultivated land.

The differences among horticulture, agronomy, and forestry are mainly defined by custom. Horticultural produce usually has a high water content, is highly perishable, and is usually utilized fresh; whereas agronomic and forestry products are often utilized in the non–living state and usually contain high percentages of dry matter. Crops like potatoes and sweet potatoes may be considered agronomic crops when they are used as staple foods. For us in Pakistan, potatoes and sweet potatoes are vegetable crops. Trees like walnut, mango, and pine are forest crops when planted for wood, and horticultural crops when used as fruit and ornamental trees.

Horticultural crops often have high cash value and are intensively cultivated on relatively small areas. Thus melons can be either an agronomic or a horticultural crop, depending on whether they are grown extensively or intensively. The high cash value of horticultural crops justifies a large input of capital, labour, and technology per unit area of land. Sweet corn is different from maize only by a single gene, but this one gene increases its value enough to warrant the use of hybrid seed and intensive cultural methods. This, plus its use by humans, changes its category from an agronomic crop to a horticultural crop.

Figure 1.1. An aerial view of a young citrus grove lined with date palms.

Customs change, and intensity of cultivation in small areas is not necessarily the defining characteristic of horticulture. The input per unit area for agronomic crops like cotton and sugarcane is also very high and may be more than for some horticultural crops. Thus we may need to identify other common characteristics among horticultural products. One such characteristic is that horticultural products are sources of aesthetic pleasure, i.e. beauty and pleasant flavor. Some fruits and vegetables like cashews and garlic are highly nutritious but they are eaten primarily as a source of flavor and pleasure. Thus horticultural foods may be distinguished from agronomic food products by their aesthetic as well as their food value.

Figure 1.2. Roses are one of the most common and beautiful flowers. By–products like *attar* and *gulkand* are also made from them.

Horticulture deals with a large number of plant species. Traditionally, it includes fruits, vegetables, and ornamental plants. Even medicinal plants, beverage plants (tea, coffee), and spices are considered horticultural crops.

Horticulture is an art as well as a science. It deals with a combination of the botanical and agricultural aspects of plants. Thus, one may define horticulture as the culture and biology of garden crops, including both the aesthetic and the scientific dimensions. Basic principles of physics, chemistry, and biology are used by horticulturists to understand and manipulate plant life. Biotechnology is now finding direct applications in horticulture.

1.2 Divisions of horticulture

There are three basic divisions of horticulture: **pomology**, or fruit production; **olericulture**, or vegetable production; and **floriculture**, or flower production. Landscape horticulture and designing have also emerged as distinct branches of horticulture. Floriculture and landscaping are treated together as one branch of horticulture, also known as ornamental horticulture. From the commercial point of view, production of nursery plants or nursery culture is also a viable branch of horticulture. There are other commercial branches like seed production and marketing, greenhouse crops, pharmaceutical crops, processing and storage, along with many support industries.

1.2.1 Pomology

The study of fruits is called **pomology**. Botanically, a fruit is a ripened ovary. The horticultural definition of fruit includes other floral parts as well, but here, fruit is the edible, fleshy or dry portion of a plant whose development is closely associated with the flower (Figs. 1.3 and 1.4).

The commercial production of fruits is known as **orcharding**. It is typically based on long–lived perennials, many of which do not bear fruit until several years after they are planted (Figs. 1.5 and 1.6). Grape plantations are called vineyards, and the cultivation of grapes is called **viticulture**. Similarly, citrus orchards are typically called citrus **groves** and the cultivation of citrus is known as **citriculture**.

Figure 1.3. A mango tree at the flowering stage. Mango squash and *acchar* are common ways of preserving the fruit.

Figure 1.4. A heavily bearing date palm a few weeks after pollination. The dioecious date palm is grown in all provinces of Pakistan, with Balochistan the leading producer.

Figure 1.5. A pear plantation intercropped with *motia*. Pears are commercially grown in the Peshawar Valley of the NWFP Province.

Orcharding and other types of fruit growing require high capital investment for years on a fixed site, without immediate return. In Pakistan, most of these operations are carried out manually and year–round, and both skilled and unskilled labour is required. Since it is costly to get into or out of the fruit growing business, selection of sites, species, and cultivars should be made very carefully.

Figure 1.6. The papaya is an important fruit of Sindh. Two plants are seen standing side-by-side; only the female plant is bearing.

All major fruits are clonally propagated. In most cases, the commercial varieties are used as scion, and the rootstock is of a different species or variety. Farm operations like pruning and training of trees and thinning of fruits are unique to pomology. With the exception of a few nuts, fruits are highly perishable. Post–harvest handling of fruits is itself a specialized discipline, dealing with grading, packing, storing, processing, and shipping operations. Similarly, marketing of fruits requires special attention to take advantage of seasonal markets and to avoid losses due to perishability.

World production (by weight) of fruits is close to that of staple food crops. The greatest production is of grapes, followed by citrus fruits, bananas, and apples. In Pakistan, thanks to climatic and soil diversity, about 30 different fruits are grown.

Among the fruits grown in Pakistan, citrus ranks first, with 95 percent of its area and production concentrated in the plains of the Punjab. Mango is next to citrus, and is mostly grown in the southern districts of Punjab and Sindh. Banana, guava, date palm, and apple are next in rank. Significant quantities of stone fruits (drupes—peaches, plums, apricots, and almonds) are also produced. There are several other important fruits grown in different parts of this country (Table 1.1 and Fig. 1.7).

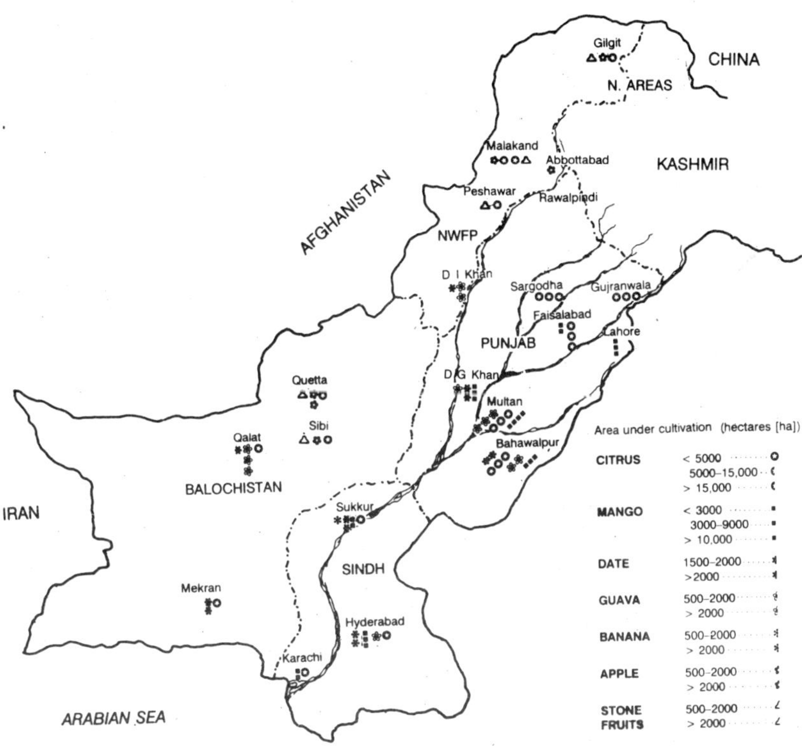

Figure 1.7 Fruit map of Pakistan (Drawn by Habib ul Rahman Mian)

1.2.2 Olericulture

The study of vegetable production is called **olericulture**, derived from *oleris* 'an herb'. A **vegetable** may be defined as the edible portion of an herbaceous plant used fresh or processed. The edible portion may be the fruit, foliage, tuber, root, or any other plant part. The edible portion of the tomato is a fruit; the potato is a tuber (stem); the sweet potato is a root; peas are seeds; and lettuce is the leaf. Potatoes and other vegetables together produce more food (by weight) than any other food crop in the world. In Pakistan, about 50 different vegetables are grown, in two growing seasons (Table 1.2). As compared to orcharding, the vegetable industry is characterized by its flexibility. Because most vegetables are grown as annuals, shifts in cultivars and crops can be readily made. There are three main categories of vegetable production: home gardening, market gardening, and truck gardening. In addition, there are several small, specialized production types including vegetable forcing, production for processing, seed production, and mushroom culture.

Home gardening. Home gardening is the growing of vegetables for home use. It is the oldest form of vegetable production, and is still the most important source of vegetables in our rural households. Besides being very economical, this is the only method of assuring fresh produce, and the choice of crops can be made according to the likings of the family members. If intensive methods are followed, from an area of 10–20 marlas sufficient vegetables and some fruits can be grown for an average family, and at times, additional income can be earned by selling the excess. Unfortunately, home gardening is disappearing from our villages, and rural people are also becoming dependent on market–supplied vegetables. This trend is unhealthy and must be reversed by popularizing home gardening through proper education and easy availability of supplies, especially seeds.

In the western world, organic farming—farming without the use of chemical fertilizers or pesticides—has emerged as an important form of home gardening. People have realized that the produce sold in the market carries a heavy load of chemicals which may be injurious to their health. The only way to get fresh and uncontaminated vegetables is through organic home gardening.

Market gardening. Near large centers of population, many kinds of perishable vegetables are grown for sale. The produce from this kind of vegetable culture is sold in local markets which are usually within a few miles from cities. This type of **market gardening** has developed to meet the requirements of people with no land, time, or interest to grow their own vegetables. Improved roads and transport facilities have made it possible for market gardeners to serve distant areas. Because of the high value of land in the immediate vicinity of cities, and the availability of road transport, market gardening is now also practiced at places far away from the centers of consumption. Because of intense competition among producers, specialized production of particular crops has begun to be practiced, with emphasis on grading and preservation of freshness and appearance of produce.

Market gardens in the immediate vicinity of big cities utilize city sewer water as a source of fertilizer and irrigation. Often, heavy use of chemicals is made to control the spread of insects and diseases carried in city refuse. Vegetables produced on land fertilized with sewage may have a heavy load of chemicals and amoebic organisms which are a health hazard to humans.

Market gardeners typically face glut periods when there is surplus produce in the market with very little return to the growers. By choosing the time of sowing carefully, and selection of the right kind of varieties, it is possible for progressive growers to avoid glut seasons, and the wastage of produce and loss of income they entail.

Truck gardening. Truck gardening or farming is the production of vegetables in relatively large quantities for distant markets. Special crops are selected and grown in different agro-ecological regions. The usual choices are the less perishable or non-perishable crops. Production of potatoes in Okara District, melons and peas in Gujranwala District, winter muskmelons in lower Sindh, and a late summer crop of onions in Balochistan and Swat are a few examples of truck farming. Truck farming is less intensive than market gardening. Also, the grower can usually wait from a few days to weeks between harvesting and marketing. The marketing is done in wholesale markets. With improvements in means of communication and transport—roads, trucks, and refrigerated transportation—and information on distant markets, the outlook of truck farming has changed. Now, most of those perishable vegetables which were formerly grown in market gardens can be grown on the truck farming pattern. Tomatoes, even though highly perishable, are being cultivated in frost-free areas during winter and shipped to distant markets.

Vegetable forcing. Production of vegetables out of their normal season of outdoor production is known as **vegetable forcing**. There is demand for out-of-season produce and the consumer is willing to pay extra. The most common form is the early or late production of summer vegetables. A usual limitation on the production of summer vegetables like tomatoes, cucumbers, eggplant, bell peppers, melons, and gourds is frost. In developed countries, such crops are grown in greenhouses, making it the most intensive type of cultivation. Commercial production of greenhouse crops in our country is not feasible. Nature has blessed us with a wide range of agro-ecological climates. There are several frost-free pockets in Punjab and NWFP, and summer vegetables can be grown out-of-season in the entire lower Sindh. Parts of Malakand and the *katcha* area of Khushab District are big producers of tomatoes. Most summer vegetables are shipped from lower

Sindh to other parts of the country. Traditionally, market gardeners produce early vegetables on a small scale, providing protection from frost by covering the germinated seedlings, using *sarkanda* or branches of trees. The use of plastics has gained importance in recent years. Low plastic tunnels are erected on field beds for crops like vegetable marrow. High plastic tunnels are used for crops like cucumbers, tomatoes, bottle gourds, peppers, and eggplant. In the west, plastic tunnels are used as a low–cost alternative to the greenhouse.

Vegetable production for processing. Processing refers to ways and means by which vegetables or their products are preserved for future use instead of fresh consumption. This sector of the vegetable industry has not developed properly in this country. In many parts of the world, this is an important area and vegetables are grown specifically for processing. Vegetable processing can utilize the excess produce during the glut season, stabilize prices and availability of vegetables over prolonged periods, and fulfill emergency needs. Important types of vegetable processing include canning, freezing, dehydration, and making industrial products like tomato ketchup, potato chips, and juices. Peas, beans, and mushrooms can be canned. Vegetables like spinach, okra, peas, and sweet corn can be frozen. Onions, garlic, and many root vegetables can be dehydrated and stored for future use.

Vegetable seed production is a small component of the vegetable growing industry but is considered as an important and highly specialized area. Similarly, **mushroom culture** is a branch of olericulture which has the potential for expansion in some areas.

1.2.3 Floriculture and ornamental horticulture

This branch of horticulture deals with plants and their layout for beautification of the environment. A tremendous number of plants are classed as ornamentals based on their decorative value and personal choices. There are seasonal flowering plants, foliage plants, lawn grasses, evergreens, and deciduous shrubs and trees. Fruit trees grown in home gardens can also be selected for ornamental purposes. Selection of plants according to the physical situation of a house, public or private building, road, public facility like an airport, or recreational place like a park is known as landscape designing.

Ornamental horticulture can play an important role in modification of domestic and urban environments and pollution control. Growing awareness of the importance of plants in the environment has increased the appreciation of the public and policy makers for ornamental horticulture. The nursery business has expanded in recent years to meet increasing demands for ornamental plants. Producers and wholesalers are located both in the vicinity of large cities and in more remote areas which supply plant materials to retailers in the cities. Pattoki in District Kasur is a major center of the nursery business. A variety of plant materials are imported for ornamental purposes. Also, significant quantities of seeds of flowering plants are imported every year. Ornamental plants are also grown for the extraction of by-products like essential oils and aromatic compounds, for sale as cut flowers, and for medicinal purposes. There is considerable scope for the export of cut flowers. To enhance the scope of ornamental horticulture, annual flower shows and design competitions are held in all the big cities. A selected list of common ornamental plants is found in Chapter 15.

1.3 Classification

1.3.1 Horticultural classification

Horticultural classification is a logically conceived system of description, nomenclature, and identification of plants. There can be several ways to classify plants, like growth habit, life span, temperature relations, uses, morphology, and cultural requirements. However, the botanical classification system remains the basis of plant nomenclature and identification.

On the basis of growth habit and physiological characteristics, horticultural plants can be classed as succulent, herbaceous, or woody. In floriculture, the term **succulents** is usually used for foliage plants with extremely tender and watery stems and leaves (See Chapter 15). **Herb** is a term used for self-supporting succulents. Herbaceous plants usually have tender stems, either drooping or self-supporting. Most vegetables and many floral and ornamental plants are herbaceous in their growth habit. Self-supporting woody plants are known as shrubs or trees. The distinction between shrubs and trees is made arbitrarily on the basis of number of stems and plant height. Trees are characterized by a single central stem, and shrubs have more than one to many stems. Trees are usually taller than shrubs, but the distinction between trees and shrubs may be obscured by growth environments and training practices. Plants with climbing or trailing stems can be woody or non-woody. A climbing plant with a non-woody stem is known as a **vine**, whereas a woody plant with a climbing growth habit is called a **liana** (Janick 1986:29-30).

Some plants shed their leaves during winter; these are referred to as **deciduous**. Plants with persistent leaves are called **evergreen**. Most deciduous plants are native to temperate climates, and, with the exception of pines, evergreens are considered tropical in origin.

Plants are annuals, biennials, and perennials, according to their life span. An **annual plant** completes its entire life cycle in a single growing season. Many vegetables and seasonal flowering herbs are annuals. The

biennial plant completes its life cycle in two growing seasons. The first season's growth is entirely vegetative, and is typically characterized by short, low internodes called rosettes. In the second season, it **bolts**, i.e. sends up a flowering stalk with extended internodes bearing flower and fruits. Root vegetables like carrots and beets; leafy vegetables like lettuce and cabbage; and others like onions are biennial in nature. Climate is the critical factor in determining the life span of plants. Annuals and biennials can vary in their life spans in different climates. Most of the above named biennial vegetables are harvested for consumption after the first season of growth and are thus treated as annuals for cultivation purposes.

Perennial plants grow for years, and most of them are woody. Their growth can be divided into juvenile and mature phases. During the first phase, the plants grow vegetatively for several years. In the second phase, reproduction starts, and vegetative and reproductive growth are concurrent. All of our fruit trees and ornamental shrubs and trees are perennial, and some herbaceous plants are also perennial. For example, asparagus, potatoes, and many bulbs are perennial. The above–ground parts of such plants are killed in winter while the underground storage structures, like the tubers of potatoes, survive. Some plants like tomato and eggplant are perennial in tropical climates and annual in temperate zones because of winter kill.

The flowering habits of plants are very important in horticulture. Flowers can be described according to their functional parts as **hermaphrodite**, **perfect**, **complete**, or **incomplete**. Plants with flowers having both male and female sex organs in the same flower, like peas, are hermaphroditic. Plants with flowers of only one sex are dioecious (date palm, papaya, spinach, asparagus). Another category of plants bear two or three kinds of flowers, male, female, and sometimes hermaphroditic flowers as well. Plants with separate flowers of a single sex are called **monoecious** (e.g. cucurbits).

Plants can also be classified according to their temperature relations, i.e temperature requirements for growth and tolerance to low temperature. Vegetables are grouped as summer (warm season) and winter (cool season) vegetables on the basis of their growing season temperatures. A winter vegetable in the plains of Punjab can be successfully grown during the summer at higher attitudes. Most summer vegetables are grown all year round in the frost–free areas of lower Sindh. Cool–season crops require that the maximum temperature not exceed 80–85°F and the minimum not fall below 35-40°F. The optimum for most such crops is 65°F. For warm season crops, the optimum is above 80°F. They are usually frost–sensitive and cannot grow if the minimum temperature falls below 50°F. Fruits are categorized as temperate, subtropical, and tropical. Temperate fruits are mostly deciduous and require a certain amount of chilling to flower. Apples, pears, peaches, plums, almonds, and apricots are important temperate fruits. Subtropical and tropical fruits are native to warm climates; they are frost–

sensitive and unless hardened properly in the late summer, the foliage can be destroyed by frost. Subtropical plants like citrus, guava, and grapes can tolerate frost. Tropical plants like banana, papaya, and mango are much more sensitive to low temperature exposures. Horticulturists speak of tender and hardy plants according to their ability to withstand low winter temperatures. Hardy plants are resistant to frost and tender ones are not.

Fruits borne on low-growing plants like shrubs and vines, e.g. grapes, falsa, and strawberry, are known as **small or soft fruits. Nuts** are edible seeds like almonds, walnuts, pecans, etc. **Fleshy fruits** are those whose edible part is the soft flesh as opposed to the seed. Systematic pomology is a branch of pomology which describes fruits according to their structural development. **Pomes** (apple, pear, quince) are false fruits in which the edible part is the thalamus. **Berries** (grape, banana, citrus) are true fruits developed from the ovary walls. **Stone fruits,** also known as **drupes**, are also true fruits (peach, apricot, plum, cherry). An **aggregate fruit** is derived from a flower having many pistils on a common receptacle. The individual fruits of the aggregate may be drupes, as in blackberries, or achenes, as in strawberries. **Multiple fruit** is a name for fruits derived from many separate but closely clustered flowers. Common examples are pineapple, fig, and mulberry.

Vegetables may be grouped according to their edible portions and cultural requirements. Root crops are vegetables with underground edible parts like carrots, radishes, turnips, potatoes, and sweet potatoes. In a strict botanical sense, the potato is not a root, but all the above require similar methods of sowing and other cultural practices. Leafy vegetables like lettuce, cabbage, and spinach have edible leaves. Vine crops include most of the cucurbits. Other important groups are **solanaceous fruits** (tomatoes, peppers, eggplant) and flower crops like cauliflower and broccoli.

Plants used as ornamentals are commonly separated into flowering or landscape plants. Flowering plants include seasonal annuals, perennials, and bulbs with underground storage organs. Landscape plants are usually foliage plants varying from ground covers or lawn grasses to hedges, and trees and shrubs. Indoor plants are ornamental plants with persistent evergreen foliage characters.

Other horticultural plant categories which do not fit into the popular horticultural classification of fruits, vegetables, and ornamentals are beverage plants which are used for their flavours, and plants which produce aromatic or fragrant products. There are industrial crops like jojoba (an oilseed), rubber, resin crops, and many drug and medicinal plants.

1.3.2 Botanical classification

Scientific plant classification is based on the phylogenetic relationships of organisms. The science of classification is known as **taxonomy**, derived from the Greek word *taxon*, meaning group or category. All forms of life are related, with lower organisms the progenitors of higher organisms. The similarities and differences among the organisms form the basis for classification. In the mid-18th century Linnaeus recognized the value of using the morphology of the sexual or reproductive parts as a basis for taxonomy. These plant organs are less influenced by environmental conditions, and similarities and differences in floral organs have become basic to the classification of higher plants.

The plant kingdom consists of about a dozen major phyla or divisions. The most highly evolved or advanced division is known as Tracheophyta. A vascular or tracheary system is the common feature of all higher plants, hence the name Tracheophyta. Almost all horticultural plants excepting mushrooms belong to this division. The division Tracheophyta is divided into several **classes**. Horticulturally important ones are Filicinae (ferns), Gymnospermae (cycads, conifers), and Angiospermae (flowering plants). Each class is subdivided into **orders**, and the orders are further divided into **families**. Each family comprises a number of genera (pl. of **genus**). Within each genus, there are various **species**. An identical group of individuals within a species is called a **variety/strain**. All categories need not be used, but the sequence is important. Sometimes intermediate subdivisions are named with the prefix *sub-*. The categories from kingdom to family are called the major taxa, and those below the level of family are called the minor taxa. The magnitude of genetic diversity decreases in order from genus to variety and to individual (Janick 1986:40-43).

The **gymnosperms** are a small group of about 700 living species. Most of them are evergreen trees with needle-shaped leaves, belonging to the temperate zones. The name is derived from the characteristic of their seeds, which are borne 'naked', as opposed to angiosperms which bear enclosed seeds. Gymnosperms are sources of timber, wood pulp, turpentine, resin, edible seeds, and many highly valued ornamental plants (Janick 1986:43).

The **angiosperms** are the largest group of higher plants, with more than 250,000 species. Characteristically, their seeds are enclosed in the fruit, and the leaves are broad (not needles). They are the primary sources of food, fiber, and shelter. The class Angiospermae is divided into two subclasses, the Dicotyledonae (dicots) and the Monocotyledonae (monocots). The dicots have two cotyledons (seed leaves), flower parts in fours or fives or multiples of these numbers, reticulate leaf venation, and the presence of vascular cambium. The monocots have a single cotyledon, flower parts in threes or multiples thereof, parallel leaf venation, and lack of vascular cambium. There are about 200,000 dicotyledonous plant species, and approximately 50,000 monocots. Within each class there are several orders and families (Janick 1986:43).

The identification and description of plant materials is based on two–part Latin names which refer to the genus and the species. The genus is named first, followed by the species. The genus begins with a capital letter and the species name with a small letter. When printed, both names are italicized. The name of the person who first named and described the species (or abbreviation of the name) follows the binomial designation. Botanical and English or common names of important fruits and vegetables are listed in Tables 1.1 and 1.2. Chapter 15 contains a similar list of ornamental species.

The genus is a group of species which have many common morphological, genetic, and cytological features. The members of a genus can sometimes cross among themselves, but can never cross with the plants of any other genus. Thus, a species is made up of plants having morphological similarities and producing like progeny. A species can be considered as a normally exclusive interbreeding population.

A sub–classification of the traditional species is the **variety**. When a population of plants within a species differs in appearance from the original members of the species, it may be called a variety. A variety is named by adding a third name after the species name. It should not be confused with the cultivated variety, or **cultivar**. The term *cultivar* refers to a named group of plants within a cultivated species which maintain their identity when propagated either sexually or asexually. In horticulture, clonal cultivars which are propagated vegetatively (asexually) are very important. Variants among clones are called **strains** or **sports**. Sexually propagated cultivars fall into three groups: (1) pure–line or self–pollinated cultivars, (2) open–pollinated or cross–pollinated cultivars, and (3) hybrid cultivars. Hybrids are developed by crossing divergent parents, and do not breed true by sexual propagation (Janick 1986:50).

1.4 Significance of horticulture

World production of fruits and vegetables is in millions of tonnes (Table 1.3). The horticulture industry provides many essential components of our daily diet (Tables 1.4, 1.5, and 1.6), and meets the individual's aësthetic needs. It is a profession for many researchers and teachers and an occupation or vocation for others working in the production phases. It is a business for merchants and a source of exercise and a small income for amateurs. Horticulture promotes the physical and mental health and economic prosperity of individuals and nations.

1.4.1 Dietary Importance

A complete food must contain carbohydrates, proteins, fats, vitamins, minerals, and roughage or fiber. Horticultural foods are an excellent source of all the essential components of the human diet (Tables 1.4, 1.5, 1.6). Approximately 30 percent of the food consumed in the world is produced in the horticulture sector. All fruits and vegetables have some quantity of digestible carbohydrates and other food components in varying proportions. A survey of the composition of important fruits and vegetables is given in Tables 1.5 and 1.6. Potatoes and sweet potatoes are especially high in starch, containing 19 percent and 27 percent starch, respectively. These vegetables, and fruits like bananas, have the potential to supplement our future energy (calorie) requirements. On a per-acre basis, potatoes and bananas produce more calories than wheat. Potatoes are the fourth ranking food crop in the world, and increased potato cultivation could reduce the pressure on grain production. Peaches, beans, sweet corn, and pecans are rich in protein. Avocados, olives, and most nuts contain a very high percentage of fat. Spices and beverages add pleasure to eating. Importantly, fruits and vegetables provide the vitamins and minerals which are lacking in the staple foods.

Varied colours, texture, and flavours create interest in eating. Fiber or roughage is deficient in our diets, despite heavy intake of cereals. Meat and milk do not contribute fiber. Very high fiber content is found in vegetables and in some fruits. Leafy vegetables like celery, cabbage, lettuce, and others with high cellulose content add bulk to food. In the stomach, this roughage helps motility and neutralizes acidity created by the digestion of proteins. For a balanced diet, WHO recommends 450 grams of vegetables and fruits every day. In Pakistan, however, less than 200 grams per day is available.

Deficiencies of vitamins and minerals cause chronic diseases. Vitamin A deficiency causes skin diseases, night blindness, and kidney stones. It is synthesized in animals and humans from its precursors or provitamins called **carotenoids**, which are richly provided by orange-coloured and dark-green fruits and vegetables. The red pigment in 'red-blood' oranges is an anthocyanin, while the red pigment in some grapefruit is carotene. Shortage of ascorbic acid or vitamin C is responsible for hemorrhaging and swollen gums, and lowered body resistance against infection. Citrus, ber, guavas, tomatoes, and melons are rich in ascorbic acid. Vitamin E-deficient adults show symptoms of sterility. Onions, lettuce, oranges, bananas, and avocados are rich in vitamin E. The protective effects of three antioxidant vitamins, beta carotene (the plant provitamin A), ascorbic acid, and vitamin E are well known. Low intake of these vitamins may lead to increased risk of coronary heart diseases and cancer of the lungs, breast, and prostate.

Deficiencies of B-vitamins can cause beriberi, loss of sensation, and heart enlargement. Thiamine, riboflavin, and nicotinic acid are important B-vitamins present in appreciable quantities in many fruits and vegetables. Vitamin D helps the body to utilize calcium. Children whose bodies are deficient in vitamin D develop rickets, a softening and curvature of the bones. Vitamin D is synthesized by the body in the presence of sunlight, and is also available in green leafy vegetables. A blood–clotting factor known as Vitamin F is present in spinach and other leaves. People whose diet contains insufficient folic acid are more likely in old age to develop cataracts, a form of impaired vision in which the lens of the eye becomes opaque. Fresh cooked spinach, white beans, avocados, asparagus, and turnip greens are good sources of folic acid.

Minerals play an important role in the development of the human body as constituent parts of tissue, and regulate metabolic activities as catalysts and co–factors. Important minerals provided by fruits and vegetables are Ca, Na, P, Co, Cu, Mg, Mn, Fe, and I. Deficiency of Ca affects the contractibility of muscles, blood coagulation, the bones, and heartbeat. Phosphorus is an important component of nucleic acids and plays an integral role in energy transformation. Iron is the oxygen carrier, and its deficiency causes anemia. Iodine is needed for normal functioning of the thyroid gland. Fruit juices, seed pods, leafy vegetables, apples, apricots, plums, dates, olives, and peaches are rich in minerals. Large amounts of sugar, organic acids, enzymes, and many pigments are important constituents of fruits. Because of its high pectin content, the albedo (white portion) of citrus peel is used for commercial pectin production and marmalades. Diced citrus peel is also used in cakes and cookies.

The nutritional value of fruits and vegetables varies with their culture conditions, stage of consumption, and treatment in the field and after harvesting. A high level of nitrogen in the soil lowers the Ca content of vegetables and affects the composition of proteins. Boron deficiency reduces the carotene and riboflavin content of tomatoes. The amounts of rainfall and sunlight received by a crop influence its ascorbic acid content. The gap between supply and demand is further widened by high post–harvest losses due to the lack of a well–developed processing industry. Vitamin losses during shipping and storage are common. Storage of crates or bags at high temperatures, high respiration levels, and loss of water lead to loss of vitamins. Water loss by ruptures takes away the water–soluble vitamins. Oxidation affects thiamine content, but vitamin A, niacin, and riboflavin are stable.

Cooking methods also play a role in the ultimate vitamin content of the food consumed. Loss of vitamins occurs due to heat and leaching. The best way to eat vegetables is fresh, or preferably raw. The quality of fruit is also subject to change during storage. Temperatures below 56°F cause brown color to develop on the skin of bananas. Citrus fruits need storage above refrigeration temperature. Chilling injury occurs in apples stored at low temperature. Dry fruits must be stored at room temperature (72°F). The vitamin content of processed fruits and vegetables is invariably altered, and often manufacturers of processed food sell fortified products to raise the vitamin and mineral content.

1.4.2 Economic position

In Pakistan there is greater demand for fruits and vegetables than can be met by existing production. Formal data on the volume of business in the area of ornamental horticulture is not available. In general, however, our supply of all horticultural products falls short of the amounts needed to meet minimum requirements. In the USA, 40 percent of the food weight consumed consists of horticultural products, whereas the percentage of horticultural products consumed in Pakistan is negligible. It is clear that concerted efforts should be made in the country to increase production of horticultural products. Higher fruit and vegetable production would improve the nutritional status of the people of Pakistan. Also, more fruits and vegetables could be exported to respond to the demand abroad. There is clear scope for the export of our produce (Table 1.8).

Analysis of farmers' costs of production and returns (Table 1.7) shows that growers of fruits and vegetables regularly earn higher profits than growers of agronomic crops. High investment in the various phases of production, handling, processing, and marketing of these crops involves a higher flow of capital. Thus, cultivation of horticultural crops brings about accelerated economic activity in the agricultural sector. The expansion of the horticulture industry will create more infrastructure, more job opportunities, and better returns to farmers. Export of horticultural produce will bring foreign exchange into the country. Fruit and ornamental trees planted today are an investment in the future. They will be a source of timber and firewood. Raw materials are provided for many industries like paper, perfumes, feed, fertilizer, furniture, and various other articles of daily use.

1.4.3 Aesthetic value

The beauty of plants and the pleasure received are not tangible quantities that can be measured or weighed (Figs. 1.8 and 1.9). These are value judgements which vary with persons, places, traditions, and cultures. People of different heritages will have quite different opinions about what is beautiful and what is ugly. In horticulture, the elements of plant beauty are combined to enhance their utility for human use. Whether horticultural plants are encountered as foods, as desserts, or in a community park, their aesthetic value always takes precedence over economics. The aesthetic value of horticulture has been used to promote mental health and a mode of relaxation. Horticultural therapy is now a well-recognized field of medical science.

1.4.4 Role in the environment

Figure 1.8. Bougainvillea, a common creeper planted along walls and pillars.

Plants serve as lungs in cleaning our environment by regulating the CO_2 content of the air. Air pollution and general degradation of the environment due to industrialization can best be controlled by the use of proper plant materials. Plants prevent soil erosion, and hardy trees like ber, guava, date palm, and pomegranate can be planted to reclaim waterlogged soils. Trees act as a barrier to reduce the velocity of winds. The presence of vegetation results in mild climates at both the micro and macro levels.

1.4.5 Medicinal plants

Many plants with medicinal value are cultivated in the horticulture sector. Pharmacology, the science of drugs and medicines, is dependent on these plants. The cultivation and maintenance of such plants has its own aesthetic and economic importance. Falsa, sweet lime, and jaman are known for their cooling effects. Grapefruit is recommended for dieting patients. Bitter gourd and jaman are also considered to help diabetics.

1.5 History and development of horticulture

1.5.1 Overview

Cultivated fruits, vegetables, and many ornamental plants have been selected and domesticated by man during the development of civilization. Agriculture and civilization have common roots. Based on anthropological, archeological,

Figure 1.9. Canna, a common flowering bulb in Pakistan.

and paleo–botanical information, the history of civilization and agriculture can be divided into four periods: Paleolithic, Neolithic, historical, and modern times. During the early stone age or the Paleolithic time (more than 50,000 years ago), man lived in caves and ate from the wild. The earliest forms of food were perhaps fruits and herbs, which were the easiest to collect and eat. This was a period of great uncertainty, and man faced starvation due to scarcity of edible wild plants, lack of ability to preserve gathered food, and seasonal migrations. Lack of hunting activity due to unfavourable weather or shortage of game also endangered human existence. In the later Paleolithic and the onset of the Neolithic period (around 11,000 years ago), people knew how to make fire, and had learned to make better hunting implements by grinding stone.

As human population increased and permanent settlements were established, a transition from hunting and gathering to village life and agriculture took place. Man started to collect and sow the seeds of edible plants. Well–watered and fertile land was required. There is evidence that the cultivation of plants started along the banks of the rivers Tigris and Euphrates about 9000–11,000 years ago, during the Neolithic period. Cultivation along the Indus and Nile probably came later. Annual vegetables and cereals must have been the first cultivated plants. Fruits were certainly not among the first to be cultivated because of the long time required by the fruit trees to reach bearing age. Peas, beans, and cucurbits were probably the earliest cultivated plants.

1.5.2 Domestication of plants

Most cultivated plants were domesticated before the beginning of the historical period (2500–3000 years ago), but the search for more edible plants continues. Sugar beets were domesticated in the 18th century, rubber and some oil palms in the 19th century, and modern strawberries and Macadamia nuts in the 20th century. The kiwi fruit of New Zealand is a newly domesticated fruit.

At the beginning of civilization, man erected walls and constructed shelters. While learning to grow plants, he realized that some plants require more care than others. Cereal grains could be grown in the open without much protection, but many vegetables and herbs needed constant care. Thus he decided to cultivate vegetables and other precious plants in enclosures and within walls.

During the process of domestication, cultivated plants underwent substantial changes. Unconsciously, man selected among edible plants for better taste, colour, flavour, and yield. This directed form of evolution induced recognizable differences between the cultivated plants and their wild relatives. Darwin in *The Origin of Species by Means of Natural Selection* presented explanations for evolution, and Mendel's laws of inheritance suggested the genetic basis for variation upon which natural selection acted. Modern plant breeders have combined the principles of evolution and domestication to evolve new and superior varieties of fruits, vegetables, and ornamentals.

The history of plant domestication is closely associated with geography. Nine centres of origin or diversity of crops are recognized: Chinese, Indian and Indo–Malayan, Central Asiatic, Near Eastern, Mediterranean, Abyssinian, South Mexican and Central American, South American, and Australian. Fig. 5.3 (Chapter 5) shows the origins of some important horticultural crops.

The Fertile Crescent, which curves from the Nile Valley northeast into the Near East, Turkey and Central Asia, and southward through the valleys of the Tigris and Euphrates, is considered to be the oldest and the richest agricultural region. For centuries, there was an exchange of domesticated plants between the Fertile Crescent and Indo–China in the east and Europe in the west through well–traveled trade routes.

The introduction and adoption of plants from their places of origin to new environments has played a significant role in the success of many edible crops. Citrus, for example, has its origin in the Indo–Chinese regions. From India, the citron was carried to Europe by Alexander. In the Near East, it was an important part of the Jewish diet and was traded into the Roman Empire. When the Romans established sea routes to India with the cooperation of Arab merchants, lemons, limes, and oranges were important trade items. With the rise of Islam, citrus fruits were brought to North Africa and Spain. When the Portuguese rounded South Africa on the way to India, citrus also found its way into South Africa. From Europe, the seeds of citrus fruits were taken to the New World by Columbus.

Persia (Iran) had a lasting impact on the art of gardening. The formal style of gardens, which has characterized the whole Islamic world for more than a thousand years, started in Persia. In ancient Persia, walls of gardens were hung with grapevines and climbers; fruit trees including peach, apple, cherry, banana, date, fig and olive, were cultivated. The Persians also grew flowers such as poppies, lilies, chrysanthemums, narcissus, and roses. They created canals in the gardens. The Persian ideas of formal gardens are found in Persian writings, paintings, and woven carpets. Alexander took the ideas of Persian garden design to Greece. Persian arts were also adopted by the neighbouring Indians. Water gardens were common in ancient India in the form of water ponds developed in community parks at temples. The Indo-Persian art of water gardens was adopted in Moorish gardens with Muslim influence in North Africa and Spain. From there, the Spanish style was passed to South America. The Moorish tradition of water gardens is still very much alive in modern house design in the form of fountains, waterfalls, and swimming pools.

Names of many fruits are mentioned in the holy books. In the Holy Quran, dates, grapes, figs, and olives are repeatedly named, reflecting a very ancient origin for these fruits.

The great philosophers of Greece wrote extensively on edible plants around 300 B.C. Citron brought by Alexander from India was discussed in these writings. The Greeks made little advance in practical farming, but the Romans were deeply interested in practical horticulture and agriculture. They liked flowers and ornamental plants, and refined the existing horticultural techniques. They grew vegetables in their country homes. Salad crops like cabbage were their favourite vegetables; lettuce, carrots, parsley, fennel, and melons were also cultivated. The Romans introduced many vegetables and fruits into Europe along with the knowledge of horticulture. Significant advances in the selection of edible plants and landscape architecture occurred again during the Renaissance and afterwards.

Concurrent with the Roman and Persian influences was the impact of Islam. With the dawn of Islam, Muslim conquerors from the Arabian peninsula went east, west, and north, accelerating the spread of many edible plants in different parts of the world. The Chinese also had a great impact on the early history and spread of edible plants; references on citrus fruits are found in ancient Chinese manuscripts.

The discovery of the New World by Columbus in 1492 began a new era in the development of agriculture. Important plant introductions and exchanges started between the new and the old world, which are still continuing. Important horticultural crops of the New World are maize, potato, tomato, sweet potato, squashes, pumpkins, beans, cranberry, avocado, cashew, pecan, pineapple, chocolate, vanilla, chili, quinine, cocaine, and tobacco. Mutual exchange of many horticultural crops from the eastern to the western hemisphere included citrus fruits, apples, bananas, dates, coffee, figs, melons, carrots, pears, peas, asparagus, and strawberries. Plant introduction remains an important tool of plant breeding and development of new crops and varieties of existing crops. Kinnow mandarin, the mainstay of our citrus industry, all cultivated potato and apple varieties, and many other fruits and vegetable cultivars grown in this country are foreign introductions (Fig. 1.10).

1.5.3 The subcontinent

As with the Romans, in the Indian subcontinent gardening was also a hobby for the rich and the feudal elite. Muslim conquerors from the Middle East and Central Asia influenced the development of horticulture in this region. The Persians remained central figures among the Muslim rulers of India, and Persian was the language of scholarship in India for centuries. Some of the best Persian–style water gardens were created in northern India. Babar directed the layout of the Bagh–i–Vafa under the supervision of Persian architects. In this garden, water channels were created in a symmetrical design, and natural beauty was enhanced by planting orange and pomegranate trees around the water reservoirs. Mughal rulers have left a lasting impression by creating formal gardens along with masterful structures like Hiran Minar, Jahangir's Tomb, and the Taj Mahal. Emperors patronized the art of horticulture through incentives for people with an interest in gardening. Akbar in his time planted a garden of one lakh mango trees at Darbanga which is called Lakh–bagh. Mangoes were also planted in Shalimar Garden. The hobbyists and amateurs of this era did a remarkable job of selecting varieties of mangoes, which have been perpetuated through clonal propagation. Some authors argue that the art of grafting was not very popular until that time. The creative work in the design of gardens, selection of plant and non–plant materials during the time of the Mughals was far superior to developments in Europe during the post–Renaissance period. The work of the Mughals on the design and establishment of small gardens is scientifically recognized.

During the British rule, a major development was the diversion of river water into a system of canals for irrigation. Gardening was encouraged by allowing extra irrigation water for this purpose. Trees and flowers were planted on the roadsides and along the railway tracks and canal banks. In almost all big cities, city parks by the name of Company Bagh were established. The British also initiated experimental agricultural education and research and development in the Indian subcontinent around 1910 when they established the Punjab Agriculture College and Research Institute,

Lyallpur (Faisalabad) and many regional agricultural experiment stations. An agricultural extension service was created to disseminate information on research and technology for agricultural development. Due to the influence of the British Empire, the exchange (introductions) of plant materials both into and out of India occurred frequently.

During the 20th century, the development of agriculture and horticulture has accelerated. The Royal Horticulture Society and American Society for Horticulture Science have influenced the development of horticulture during this century by providing a scientific forum for discussion and dissemination of experimental results.

Production of fruits and vegetables at a subsistence level is centuries old in the subcontinent. However, large-scale commercial production emerged after the introduction of experimental agriculture. After the world wars, especially the economic depression after World War I, circumstances were conducive to change. Urbanization, industrialization, and increasing population are other factors which have promoted the transition from pleasure gardening by the rich and subsistence production by village people on small farms to commercial fruit and vegetable production.

There is a strong association between the development of agricultural information systems and the expansion of the horticultural industries. Table 1.9 lists important events in the development of horticulture in the region, and Table 1.10 presents a chronological record of the area under fruit and vegetable cultivation and production of fruits and vegetables. A major event occurred in 1925 when a Fruit Specialist was appointed. In the next 10–15 years, horticulture became an important discipline of agriculture, with a significant impact on the development of commercial gardening in the country. Important developments took place in the year 1938. However, a serious blow occurred in 1947 with the destruction of many prized gardens and the loss of experienced people, and the pace of development was slow for several years after 1947.

Economic pressures are acute and farmers are very aware of the necessity for increasing production. Demands for more fruit and vegetables have been created by the increasing population and improvement in people's buying power. Significant expansion of the fruit industry took place with the introduction of the Kinnow mandarin in Punjab (Fig. 1.10), cultivation of bananas in Sindh, introduction of new apple varieties in the hilly areas of Punjab, NWFP, and Balochistan, and date palm cultivation in D.I. Khan division of NWFP and lower Balochistan. Similarly, the area under vegetables expanded due to the introduction of Dutch potato varieties and hybrid seeds of many vegetables. The frost-free belt of lower Sindh is exploited to grow most summer vegetables during the winter months. Private–sector nurseries and seed merchants have played an important role by making plants and seed available for sale.

Figure 1.10. The Kinnow mandarin was introduced from the USA in the 1940's.

Fig. 1.11 shows No. 70, a variety of potato developed at the University

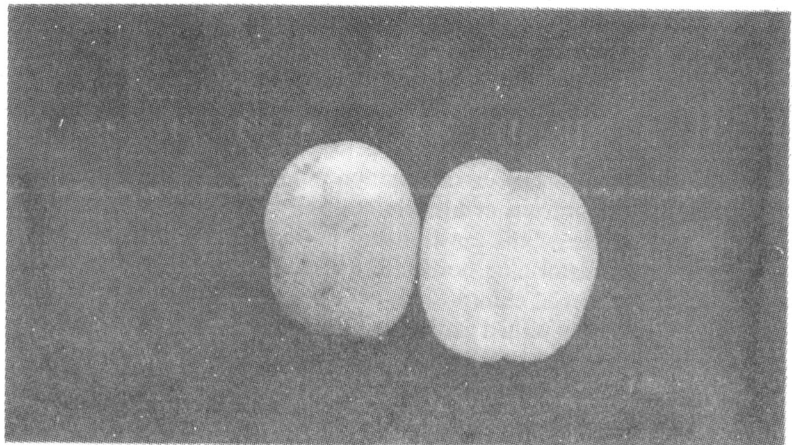

Figure 1.11. Number 70, a white-skinned potato variety developed at the University of Agriculture, Faisalabad.

of Agriculture, Faisalabad. This is the highest–yielding potato available which can be stored on a small scale without cooling. The fact that cold storage is not needed for the seed potatoes makes No. 70 an attractive choice for small farmers of Punjab and Sindh who are not now growing potatoes.

Landscape gardening and floriculture have been patronized since the days of the Mughal emperors. But these subjects were not a part of our education and research system until the start of the B.Sc. (Hons.) programme at West Pakistan Agricultural University (WPAU) in 1962, when all three disciplines of horticulture—fruit culture, vegetable growing, and floriculture—were combined. Degree courses in ornamental horticulture were introduced. Many research topics were investigated and new plant materials brought from overseas. Today floriculture and decorative horticulture are an important wing of horticulture with opportunities for employment and business. Landscape horticulture is playing an important role in the improvement of the environment in urban centres and industrial areas.

Two important historical mistakes need mention here. First was the separation of agricultural research and education in 1962, which diluted the impact of the scientific community on the development of agriculture in this country. This experiment was a failure and has already been reversed in NWFP. The second important step backward was taken by the Punjab government a few years ago when they declared vegetables a non-horticulture discipline, which is against any standard international definition of horticulture. This has not only reduced the number of job opportunities for graduates in horticulture but has also slowed progress in research and teaching on vegetable crops. Because of the lack of opportunities, students and researchers in horticulture have given up work on vegetable crops.

In summary, horticulture has a long history in Indo–Pakistan. The economic importance of fruits and vegetables, developments in research information, incentives like extra irrigation water, the introduction of new crops and varieties of fruits and vegetables, and growing concerns in recent years about the role of the environment in human life have contributed significantly to transforming horticulture from a pleasure or subsistence activity to a commercial enterprise. It is not inappropriate to say that the ancient horticulture of walled gardens has today become a modern industry, with an important role to play in feeding the world's growing population.

1.6 Present status and future scope of horticulture

Of the total cultivated area in this country, only about 6% is under horticultural crops. In the USA, the area under horticultural crops is more than 20% of the total cultivated land. Estimates of area and production of different fruits and vegetables in Pakistan for 1990 are presented in Table 1.11. Of the total country–wide area planted to vegetables, 17% is under potatoes, 18% under melons, 15% under chilies, 12% under onions, and the rest under about 30 different vegetables. A regional distribution map of different fruits is presented in Fig 1.7. The province–wise distribution shown in Table 1.12 indicates the fruit and vegetable contribution of each province. Citrus fruits, which rank in first position (38%), are primarily produced in Punjab (65%). Sindh province leads in mango production, closely followed by Punjab. Bananas and papayas are produced exclusively in Sindh. Grapes are produced mainly in Balochistan. Stone and pome fruits are grown at higher altitudes, especially in NWFP, Balochistan, and the Murree Hills. The Kashmir and FATA areas also have significant horticultural production. There are no statistics available on the area and monetary value of ornamental horticulture.

Except in the cases of potato production in Punjab, mango and banana production in Sindh, and melon growing both in Sindh and Punjab, most growers of horticultural crops are small farmers. There are 215,000 orchards in the country, 96% of which are between 0.5–5 ha in area and cover 66% of the total orchard area. Of the total area, 30% is under 5–20 ha holdings, and only 0.3% of the total orchards are more than 20 ha.

Per capita consumptions of fruits and vegetables are 88 g and 90 g per day, respectively. The World Health Organization (WHO) recommends a minimum of 400 g of fruits and vegetables each day for a healthy human diet. This means a demand of 146 kg per head per year and 17.5 million tonnes for a population of 120 million. Comparing it with our present level of production (7.8 million tonnes), we are currently about 10 million tonnes short of fruits and vegetables just to provide a balanced diet to our own population. This estimate does not include post–harvest losses of fruits and vegetables, which usually run very high (25-50 percent). There are 3.7 million children added to our population every year, which is very close to the Malthusian theory of geometric population growth. Will there be enough food in the next century when we reach a population of 200 million? The government has fixed a modest growth rate target of 6.8 percent for the horticulture sector by the end of this century. Imagine how many fruits and vegetables will be available for our people by that time!

Fruits and vegetables are an important source of foreign exchange. We have been exporting potatoes, mangoes, onions, and other vegetables for a long time. There is now great demand for our Kinnow mandarin in the Middle East, Iran, and Pacific–rim countries. Recently, concentrated kinnow juice has become another exportable item. There has been a steady increase in the quantity and value of horticultural exports (Table 1.13). According to a report of the National Commission on Agriculture, we have a great potential export market for our fruits and vegetables (Table 1.8) in the Gulf alone. In most cases our exports are less than 10% of the potential market.

The demand for horticultural products grows greater with prosperity. This is particularly seen in the urban areas. In addition to their desire for fresh vegetables and fruits to balance their diets, affluent people are spending money on ornamental horticulture.

The existing shortages, the losses between harvest and consumption, the growing population, the need for foreign exchange, and the prosperity in urban and industrial centres point to a very important role for horticulture in the future. Crops like potatoes will become an important supplement to our cereals as a source of carbohydrates. Nutrition consciousness and the desire for a balanced diet can only be met by significant increase in the production of fruits and vegetables. As shown in Table 1.10, the annual growth of the industry has been very high since the 1960's.

Bringing more area under fruits and vegetables would mean reduction in the area under major crops. It is not possible to go on increasing cultivated acreage as has been done in the past. However, there is tremendous scope for increasing the productivity of our existing acreage. For example, the average yield of our citrus plantations is 9 t/ha as compared to 20–30 t/ha in California, Florida, and Brazil. Similarly, our national average yield of potatoes is 10 t/ha against a potential of as much as 50 t/ha. The situation is similar for many other horticultural crops. The overall average of vegetables and fruits is 8.3 and 10 t/ha, respectively. Realization of potential yields through short–term and long–term measures can not only fill the existing gap between supply and demand but also create a considerable surplus. There are, however, several factors which limit the productivity of horticultural crops.

1.7 Factors limiting the growth of horticulture

Chronological data on the area and production of fruits and vegetables (Table 1.10) show that the growth of horticulture has progressed very rapidly. Despite impressive growth, however, we still remain short of the supply needed to meet the minimum required standards of fruit and vegetable consumption. At present, yields are very low and prospects for bringing more area under fruits and vegetables are also not bright. There are several factors and constraints which can be considered both on a short–term and a long–term basis to improve yields and induce more farmers to grow fruits and vegetables. Some of them are discussed in the following paragraphs.

1.7.1 Seed

The cost of seed is a minor component in the total cost of production, with the exception of potatoes, ginger, and a few other crops. Yet the fate of all agricultural investments revolves around the quality and quantity of seed used. Unfortunately, the production, processing, storage, and distribution of seed have been ignored. Most of the vegetable and flower seed in this country either comes from unsalable poor–quality or perished produce, or is imported from abroad. There is very little indigenously produced vegetable seed available. According to a reliable source, we need 2400 tonnes of vegetable seed annually (excluding potatoes), whereas only 83 tonnes of seed is produced in the country. Also, malpractices like fraudulent packaging, lack of guarantees, and adulteration are prevalent in the seed business. These problems affect the returns of growers and sometimes force them to give up vegetable cultivation. We lack a system which can ensure the production and supply of good seed, a badly needed component in the production chain.

The most serious drawback is the failure to breed our own varieties.

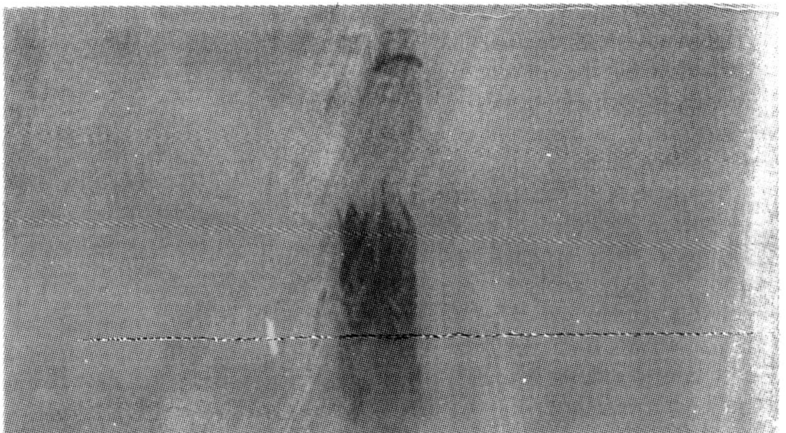

Figure 1.12. Tissue culture, a branch of biotechnology, has become an important tool in horticulture for rapid propagation and improvement of horticultural crops.

This is also a cause of expense in foreign currency to import seed from abroad. The situation is alarming for crops like potatoes, which are totally based on foreign introductions. The majority of commercial potato cultivars in use, e.g. Ultimus, Desiree, Multa, Cardinal, and Diamont, are being imported every year in the form of virus–free seed. The same seed could be produced in this country by tissue culture techniques (Fig. 1.12). We must solve the problem of seed by developing an indigenous technology of breeding, production, processing, and marketing of seeds. Provincial seed corporations, Cargill and Pioneer Seeds of the USA, and a few other private sector companies are in the business. The results will not be significant unless enough research input is ensured to strengthen the seed business.

1.7.2 Nurseries

Most of the plants supplied to the fruit growing industry and for ornamental purposes come from private sector nurseries. There are no laws to regulate the nursery business in Pakistan, and any business-minded person can start a commercial nursery. Most such nurseries are located in big cities. For fruit trees, there is no system for the production and supply of rootstock seed. Most such nurseries lack their own supply of scion bud wood. There is no check on trueness-to-type (pedigree) or freedom from diseases and other health problems. As a result, thousands of hectares of fruit trees have been planted with inferior quality plants. The life span of such orchards is short and the yields are low. The average life of citrus groves in places like California is more than 50 years, whereas it is only 15–20 years in Pakistan; and average yields are three times higher than ours. The ultimate sufferer is the farmer, although the consumer also suffers from reduced supply. The poor returns from such plantations lead the growers to uproot their orchards or intercrop their gardens with other crops and consider the fruit trees as a bonus or side business only.

As a long-term strategy, laws must be enacted and enforced to ensure the supply of quality plant material. In all major citrus producing countries of the world, there are nursery laws. In such countries, certified bud wood is produced in the public sector and supplied to nurserymen. To qualify for business, a nurseryman is required to have a seed orchard to ensure the supply of quality rootstock seed. There is a nursery wing of the agriculture department to enforce laws relating to fair business. Nurseries have to sell plants under warranty.

1.7.3 Capital

Horticulture is a capital-intensive industry. Highly priced inputs and investments in the form of land preparation, purchase of plant materials, fertilizer, irrigation water, pest management, implements, skilled labour, and management costs, plus high risks make farmers hesitate to venture into horticultural enterprises. A fruit grower has to invest for several years before he gets any return. Vegetable growers face marketing uncertainties. In a small-farmer economy like ours, growers are subsistence-oriented, and food for their families and fodder for their livestock are the first considerations. Horticulture, on the other hand, is a profit-oriented industry. One must invest before he expects a profit. The investment potential of our small farms is negligible. To encourage the establishment of more farms and to improve the productivity of existing small farmers in the horticultural business, capital investment must be institutionalized. Most of the successful horticultural entrepreneurs in the vicinity of big cities are actually investors. They offer good examples of return from capital invested in horticulture.

Small farmers in the rural areas can also benefit if the necessary capital and infrastructure in the form of credit, roads, marketing facilities, etc. are provided to them along with technical know–how and capital incentives.

1.7.4 Marketing system

The marketing of horticultural produce has unique characteristics. Most fruits and vegetables are highly perishable, with a short post–harvest life measured in hours or days, and in few cases up to a few months. This forces the producer or the contractor to sell the harvest as early as possible. Seasonal fluctuations in supply are common, with alternating shortages and gluts. Faced with these marketing risks, more than 97 percent of fruit growers sell their standing crop to contractors. Many vegetable growers also engage contractors or share–croppers to market their crops. According to a survey of the marketing of major fruits and vegetables in Sindh and Punjab, the producer's share of the consumer–paid price in many cases is less than 50 percent. Marketing intermediaries take away the rest of the return. There are heavy post–harvest and marketing losses, due to inadequate post–harvest handling and storage, and to malpractices (*chungi*).

There is a need to develop the processing industry to consume the excess during glut periods. An excellent example is the kinnow juice concentrate being produced by the Cargill Citrus Juices Company at Sargodha. Ordinarily, the kinnow is not suitable for juicing. Concentrated kinnow juice is now being exported and blended as an ingredient of mixed drinks. Cargill buys directly from the farmers, eliminating the intermediaries. This has stabilized the prices of kinnow in favour of the farmers. In developed countries, fruit growers sell their produce through cooperatives. Farmers can be encouraged to undertake direct marketing if facilities like good roads, pre–chilling plants, grading and packing houses, cold storage, market intelligence (information on prices and fluctuations), and refrigerated transport are provided. With direct marketing, the benefits are shared by the producer and consumer only.

A public sector organization, Agriculture Marketing and Storage Limited (AMSL), has been in operation for several years with the responsibility of stabilizing the prices of perishable commodities like fruits, vegetables, and livestock products. AMSL has limited its activity to potatoes, onions, and mangoes, that too without much success. The Agricultural Marketing Department and local market committees are responsible for safeguarding the interests of growers and consumers. Unfortunately, the whole system is not functioning properly. The consumer is paying a high price for produce, and the grower is not getting his due share. Market committees are supposed to be run by representatives of the farmers, consumers, and commission agents. In reality, they often belong to the influential people and city elite with political clout and vested interests. As a result, the intermediaries continue to benefit. A fair return to the farmer will be available only if marketing channels are regularized through genuine market committees. Alternatively, the farmers must create their own pressure group by marketing through their own cooperatives.

1.7.5 Management problems

As with any business, management plays an important role in making farming profitable. One can make wise decisions by selecting the right kind of crop according to the soil and climatic suitability and demands, or he can waste his investment by choosing inappropriate crops. Similarly, the right time and method of sowing, proper irrigation, fertilization, pest and disease management, weed control, training and pruning of orchard trees, and intelligent harvesting and marketing decisions can all make significant differences in the profitability of farming. Most of our small farmers lack the skills and information necessary to make the best decisions. Mixed cropping is a rule with the small farmer, making the adoption of different management practices for each crop inefficient, since different crops will have different soil, water, climatic, and management requirements.

The biggest drawback is the lack of research–based information on technology available to the farmers. There is a shortage of irrigation water, and in most cases excess water for orchards is denied. Fertilization has its own problems, which are discussed in a later chapter. Obnoxious weeds have limited the cultivation of melons in Gujranwala district and out–of–season tomatoes in Khushab district. Availability of virus–free seed of potatoes is dependent on import. There is no system for the production and supply of virus–free fruit plants, especially citrus and peaches. Mango malformation is a disease of unknown etiology with identification of its cause nowhere in sight. The banana crop in Sindh has been badly damaged due to diseases, the exact nature of which is still not confirmed. Both nematodes and viruses have been implicated in the serious recent attacks of banana diseases. The production of chilies and guava is also threatened by diseases. There is almost complete lack of a very efficient control system known as integrated pest management (IPM). The lack of know–how among farmers can also be attributed to an inefficient extension service, which needs improvement by creating a separate setup for horticulture.

Another serious problem is the lack of diversity in most horticultural crops. There are usually only a very limited number of cultivars or land races in use, severely limiting growers' choices. For example, Kinnow mandarin is the only popular citrus variety, and it has many drawbacks. Most cucurbits grown in this country are land races. Rootstocks used for fruit trees are also very limited in number and used without proper selection for a particular environment. There is need for new crops to expand our production base. Nuts like pistachio and pecan can be grown in Balochistan. There are appropriate environments for cultivation of olives, strawberries, and kiwi fruit. Entire barani areas in the Pothwar Plateau are suitable for almonds. Tea, a very important import in this country, has a chance in the Mansehra area. Early-maturing grapes like Perlette can be grown in parts of Punjab. There is scope for vegetables like celery, asparagus, and leeks and for expansion of the area under sweet potatoes.

1.7.6 Export production

Export of fruits and vegetables can bring much-needed foreign exchange. The benefits of the higher prices realized by export are passed back to the growers. In the 1970's kinnow mandarin was exported in large quantities to Iran and the Middle East. The higher prices served as a catalyst for the rapid expansion of the area under kinnow. Similar situations can be created for many other fruits and vegetables which are in great demand in foreign markets. For an export-oriented production of horticultural crops, we must design a long-term strategy for export production. To remain in the market, we will have to ensure the constant supply and quality of the produce.

In the past, our exporters have failed to establish their credibility. Their interest has been confined to earning high profit on the first consignment by sending poor-quality, cheap produce, and bad packaging. As a result, orders to such exporters were not repeated. This has not only reduced export possibilities, but also brought a bad name to the country. Agencies like the Export Promotion Bureau must be given the responsibility for quality control. Farmers can be encouraged to specialize in the export production of specific high-quality crops and varieties of fruits, vegetables, and cut flowers by offering them premium prices on a contract basis. Exporters must be provided special rebates, and refrigerated cargo and air-freight facilities.

1.7.7 Public awareness

Our people are not sufficiently aware of the importance of horticultural food and appreciative of its aesthetics. The human diet remains incomplete without the essential vitamins and minerals supplied by fruits and vegetables. An important aspect of vegetable consumption is the loss of vitamins during cooking, about which the general public should be educated. In recent years, the ornamental horticulture industry in Pakistan has expanded exponentially to serve the demand of urban horticulture. However, most houses still lack proper landscape design and environmental aspects. People are aware of good masonry and spend money on concrete, but are not willing to make provision for plants. In developed countries, urban horticulture is a well-recognized field. Increasing awareness and concern about the deteriorating environment is bound to produce an appreciation for the role of plants in the maintenance of a good environment for human health.

TABLES

Table 1.1 Botanical and common names of varieties of some important fruits

Botanical name	English or local name	Family
	EVERGREEN FRUITS	
Citrus sinensis Linn.	Sweet orange	Rutaceae
Citrus paradisi Macf.	Grapefruit	"
Citrus grandis Linn.	Pummelo	"
Citrus reticulata Blanco	Mandarin/tangerine	"
Citrus limon (L.) Burm. f.	Lemon	"
Citrus limetetioides Tan.	Sweet lime	"
Citrus aurantifolia Swingle	Sour lime or *kaghzi nimboo*	"
Citrus limonia Osbeck	*Nasnaran*	"
Citrus Jambhiri Lush.	Rough lemon or *jatti khatti*	"
Citrus Jambhiri Lush.	Rough lemon or *jam-beri*	"
Citrus limon Linn./*C. limonia* Osbeck	Sweet lemon or *mithi*	"
Citrus Jambhiri Lush.	Smooth lemon or *jullundri khatti*	"
Citrus medica Linn.	Citron or *mokari*	"
Citrus karna Raf.	*kharna khatta*	"
Citrus aurantium Linn.	Sour orange or *khatta*	"
Citrus aurantium Linn.	*Gada dehi*	"
Citrus regulosa Jan.	*Atoni*	"
Fortunella margarita Swingle	Oval kumquat	"
Mangifera indica Linn.	Mango	Anacardiaceae
Anacardium occidentale Linn.	Cashew nut	"
Eugenia jambolana Lam. (*Syzygium cumunii* Skeels)	*Jaman*	Myrtaceae
Psidium guajava Linn.	Guava	Myrtaceae
Litchi chinensis Sonn.	Litchi	Sapindaceae
Phoenix dactylifera	Date palm	Palmaceae
Cocos nucifera Linn.	Coconut	"

Botanical name	English or local name	Family
Olea europea Linn.	Olive	Oleaceae
Persea americana Mill	Guatemalan and Indian avocado	Lauraceae
Zizyphus jujuba Lam.	Jujube (Chinese date), *Ber*	Rhamnaceae
Musa sapientum Linn.	Banana	Musaceae
Achras zapota Linn.	*Chiku*	Sapotaceae
Mimusops hexendra	*Khirni*	"
Phyllanthus emblica Linn.	*Amla*	Euphorbtaceae
Averrhoa carambola Linn.	Carambola or *kamrakh*	Oxalidaceae
Eriobotrya japonica Lindi	Loquat	Rosaceae
Ananas comosa Linn.	Pineapple	Bromaliaceae
Ficus carica Linn.	Fig	Moraceae
Morus nigra Linn.	Mulberry	"
Annona reticulata Linn.	Bullock's heart or *Ram Phal, sharifa*	Annonaceae
Annona cherimola Mill	Cherimoya or *Lakshman Phal*	"
Annona muricata Linn.	Soursop or *Mundla*	"
Carissa carandas Linn.	*Karanda*	Apocynaceae
Passiflora edulis Sims	Passion fruit	Passiflorceae
Durio zibethinus Linn.	Durian	Bombacaceae
Carica papaya Linn.	Papaya	Caricaceae
DECIDUOUS FRUITS		
Pyrus malus Linn. (*Malus sylvestris*)	Apple	Rosaceae
Pyrus baccata Linn.	Crab apple	"
Pyrus communis Linn.	Pear (common)	"
Prunus americana Marsh	American plum	"
Prunus domestica Linn.	European plum	"
Prunus amygdalus Stockes	Almond	"
Prunus armeniaca Linn.	Apricot	"
Prunus persica Sieb and Zuce	Common peach	"
Prunus avium Linn.	Sweet cherry	"
Prunus cerasus Linn.	Sour cherry	"
Cydonia oblonga Mill	Quince	"
Rubus spp.	Blackberry, Raspberry, Dewberry	"
Rubus loganobaccus	Loganberry	"
Fragaria ananassa	Strawberry (cultiv.)	"

Botanical name	English or local name	Family
Fragaria virginiana Duchesne	Strawberry (wild)	"
Vaccinium corymbosum Linn.	Blueberry	Ericaceae
Ribes hirtellum Micht	American gooseberry	Sexifragaceae
Vitis vinifera Linn.	European grape	Vitaceae
Vitis labrusca Regel	Eastern grape	"
Castanea sativa	Mill Chestnut	Fagaceae
Corylus avellana Linn.	Hazelnut	Corylaceae
Corylus maxima Mill	Filbert	Corylaceae
Carya illinoensis Kock	Pecan	Juglandaceae
Juglans regia Linn.	Persian walnut	"
Juglans nigra Sarg.	Black walnut	"
Pistacia chinensis Bunge	Chinese pistachio	Anacardiaceae
Pistacia vera Linn.	Pistachio	"
Diospyros Kaki Linn.	Kaki persimmon	Ebenaceae
Diospyros lotus Linn.	*Amlok*	"
Punica granatum	Pomegranate	Punicaceae
Grewia asiatica	*Falsa*	Tiliaceae

Table 1.2 Botanical and common names of important vegetable crops

Botanical name	Common name(s)
MONOCOTYLEDONEAE	
Gramineae (grass family)	
Zea mays var. praecox	Popcorn
Zea mays var. rugosa	Sweet corn
Araceae (arum family)	
Colocasia esculenta	Taro or dasheen
Liliaceae (lily family)	
Asparagus officinalis	Asparagus
Amaryllidaceae (amaryllis family)	
Allium ampeloprasum, Porrum Group	Leek
Allium cepa	Onion
Allium sativum	Garlic
DICOTYLEDONEAE	
Polygonaceae (buckwheat family)	
Rheum rhaponticum	Rhubarb
Chenopodiaceae (goosefoot family)	
Beta vulgaris	Beet
Spinacia oleracea	Spinach
Tetragoniaceae (carpet-weed family)	
Tetragonia tetragonioides	New Zealand spinach

Botanical name	Common name(s)
Cruciferae (mustard family)	
Brassica juncea	Mustard (*sarsọn*)
Brassica napus, Napobrassica Group	Růtabaga
Brassica napus, Pabularia Group	Siberian kale
Brassica oleracea, Acephala Group	Kale, collards
Brassica oleracea, Alboglabra Group	Chinese kale
Brassica oleracea, Botrytis Group	Broccoli, cauliflower
Brassica oleracea, Capitata Group	Cabbage
Brassica oleracea, Germmifera Group	Brussels sprouts
Brassica oleracea, Gongylodes Group	Kohlrabi
Brassica rapa, Rapifera Group	Turnip
Amoracia rusticana	Horseradish
Raphanus sativus	Radish
Leguminosae (pea family)	
Pisum sativum	Garden pea
Pisum sativum var. Arvense	Field pea
Vicia faba	Broad bean
Phaseolus coccineus	Scarlet runner bean
Phaseolus limensis	Lima bean
Phaseolus lunatus	Butter bean
Phaseolus vulgaris	Common or kidney bean
Vigna aconitifolia	Moth bean
Vigna mungo	Urd
Vigna radiata	Mung bean
Glycine max	Soybean
Trigonella foenumgraecum	Fenugreek (*methi*)
Malvaceae (mallow family)	
Abelmoschus esculentus	Okra
Umbelliferae (parsley family)	
Coriandrum sativum	Coriander
Daucus carota var. sativus	Carrot
Petroselinum crispum	Parsley
Apium graveolens var. dulceii	Celery
Convolvulaceae (morning–glory family)	
Ipomoea batatus	Sweet potato
Solanaceae (nightshade family)	
Solanum melongena	Eggplant
Solanum tuberosum	Potato
Lycopersicon esculentum	Tomato
Physalis pruinosa	Husk tomato
Zingiberaceae	
Curouma longa	Turmeric
Zingiber officinale	Ginger
***Capsicum annum* var. annum**	
Cerasiforme Group	Cherry pepper

Botanical name	**Common name(s)**
Conoides Group	Cone pepper
Fasiculatum Group	Red cluster pepper
Grossum Group	Bell pepper
Longum Group	Cayenne, chili pepper
Capsicum frutescens	Tabasco pepper
Cucurbitacea (gourd family)	
Cucurbita maxima	Winter squash
Cucurbita pepo var. Pepo	Field pumpkin, acorn squash
Citrullus lanatus	Watermelon
Cucumis anguria	West Indian gherkin
Cucumis melo, Inodorus Group	Honeydew melon, casaba melon
Cucumis melo, Reticulatus Group	Muskmelon, Persian melon
Cucumis sativus	Cucumber
Benincasa cerifera	Ash gourd (*petha*)
Luffa actangula	*Tori*
Compositae (composite family)	
Cichorium endiva	Endive
Lactuca sativa	Lettuce
Helianthus tuberosus	Jerusalem artichoke

Table 1.3 Production of fruits and vegetables during 1990 (1000 tonnes)

Crop	World	Europe	Latin America	North America	Africa	Asia	Paki-stan
Fruits	341,888	68,045	73,843	24,627	34,608	112,246	3,914
Grapes	59,943	30,875	5,727	5,099	761	8,972	34
Citrus fruits	73,195	8,882	27,597	9,888	2,699	689	1,593
Bananas	45,845	442	18,945	5	5,576	1,506	217
Apples	40,263	13,284	2,789	4,802	408	11,960	2,017
Total nuts	4,461	963	280	890	249	1,908	58
Root crops	597,071	100,092	48,273	21,388	116,997	–	–
Potatoes	269,561	100,004	11,695	20,794	4,743	64,973	670
Vegetables and melons	441,778	71,321	21,937	33,708	18,612	249,099	3,617
Coffee, green	5,964	–	3,700	1	1,204	992	–
Tea	2,522	–	60	–	310	2,015	–
Tobacco	6,634	672	737	795	332	3,847	63

Source: Data from FAO Production Yearbook for 1990.

Table 1.4 Percentage of food nutrients contributed by five categories of horticultural foods (1957–59 averages)

Food energy	Protein	Fat	Carbo- hydrate	Calcium	Phos- phorus	Iron	Vitamin A	Ribo- flavin	Niacin	Ascorbic acid	Thiamin
8.7	7.4	.9	17.1	9.3	11.2	20.5	50.8	18.5	17.5	93.6	18.5

*Total of citrus fruits, other fruits, potatoes and sweet potatoes, dark–green and deep–yellow vegetables and other vegetables including tomatoes.
Source: Halfacre and Barden (1979) from C. W. Basham, Laboratory Activities in Horticulture, Kendall/Hunt, Dubuque, Iowa, 1976.

Table 1.5 Composition of edible portion of selected fruits[1]

Average of all varieties in percent

	Water	Protein	Fat	Ash	Fiber	Sugar as invert	Acid[2] as: malic (m) citric (c)	Fuel value/lb (cal)
Apples	84.1	.3	.4	.29	1.0	11.1	.47 m	290
Apricots	85.4	1.0	0.1	.59	.6	10.4	1.19 m	255
Avocados	65.4	1.7	26.4	1.42	1.8	0.6-		1200
Bananas	74.8	1.2	.2	.84	.6	19.2	.39 m	445
Blueberries	83.4	.6	.6	.28	1.2	9.7	.67 c	310
Cherries (sour)	84.4	1.3	.5	.51		9.5	1.38 m	285
Cherries (sweet)	80.0	1.1	.5	.60	.4	11.6	.68 m	365
Figs	78.0	1.4	.4	.64	1.7	16.2	.17 c	395
Grapefruit	88.8	.5	.2	.42	.3	6.5	1.16 c	175
Grapes (European)	81.6	.8	.4	.46	.5	14.9	.47 m	335

Average of all varieties in percent

	Water	Protein	Fat	Ash	Fiber	Sugar as invert	Acid[2] as: malic (m) citric (c)	Fuel value/lb (cal)
Lemons	89.3	.9	.6	.54	.9	2.2	5.07 c	200
Limes	86.0	.8	.1	.80		.5	5.90 c	240
Mulberries	82.8	1.2	.6	.84	2.0	9.4	.95 m	310
Nectarine	82.9	.5	.1	.50	.4	11.8	1.15 m	305
Oranges	87.2	.9	.2	.47	.6	8.8	.68 c	230
Peaches	86.9	.5	.1	.47	.6	8.78	.64 m	230
Pears	82.7	.7	.4	.39	1.4	8.90	.29 c	315
Persimmons	64.4	.8	.4	.90	1.5	18.90	.19 c	640
Pineapples	85.3	.4	.2	.42	.4	11.90	.72 c	265
Plums	85.7	.7	.2	.51	.5	8.30	1.60 m	255
Strawberries	90.0	.8	.6	.50	1.2	5.27	1.09 c	185
Almonds	4.8	21.0	54.9	2.00	2.0			2940
Pecans	2.7	9.6	70.5	1.90				3330
Dates (cured)	23.0	2.0	3.0	1.00	–	72.0		
Guava	–	1.3	0.5	0.60	4.8	7.5		
Mango	82.0	0.5	–	0.49	–	11.4	0.42 c	

1. Adapted from USDA Circ. 50, Proximate Composition of Fresh Fruits and Food Products.
2. Expressed as malic or citric according to which was considered to predominate.

Table 1.6 Nutritive value of 1 pound of selected vegetable foods, as purchased, with a few other foods for comparison

Kind of product	Refuse (%)	Food energy (cal)	Protein (g)	Fat (g)	Carbohy-drates (g)	Calcium (mg)	Phos-phorus (mg)	Iron (mg)	Vitamin A. (Int'l. units)	Thiamin (mg)	Ribo-flavin (mg)	Niacin (mg)	Ascorbic acid (mg)
Asparagus	25	90	7.5	0.7	13.3	71	211	3.1	3,430	0.54	0.59	3.9	113
Beans, snap	10	172	9.8	0.8	31.5	266	180	4.5	2,560	0.32	0.41	2.5	79
Beets	25	155	5.4	0.3	32.6	92	146	3.4	80	0.11	0.17	1.4	34
Broccoli	39	103	9.1	0.6	15.2	360	211	3.6	9,700	0.26	0.59	2.5	327
Cabbage	27	95	4.6	0.7	17.5	152	103	1.7	270	0.23	0.21	0.9	173
Carrots	12	179	4.8	1.2	37.2	156	148	3.2	48,000	0.27	0.26	2.0	24
Cauliflower	55	63	4.9	0.4	10.0	45	147	2.2	200	0.21	0.22	1.2	141
Celery	37	63	3.7	0.6	10.6	143	114	1.4	0	0.09	0.12	0.9	20
Chard	14	98	5.5	0.8	17.2	410*	140	15.6	10,920	0.22	0.51	0.9	148
Cucumbers	30	46	2.2	0.3	8.6	32	67	1.0	0	0.12	0.28	0.5	27
Kale	36	144	11.3	1.7	21.0	655	180	6.4	21,950	0.35	1.01	(2.4)	335
Lettuce, head	31	57	3.8	0.6	9.1	69	78	1.6	1,710	0.20	0.21	0.5	24
Lettuce, all other	31	57	3.8	0.6	9.1	194	63	3.4	5,060	0.20	0.21	0.5	57
Onions, mature	6	208	6.0	0.9	44.0	137	188	2.1	210	0.15	0.10	0.6	38
Peas, green	55	206	13.7	0.8	36.1	45	249	3.9	1,390	0.72	0.37	4.2	54
Peppers, green	16	112	4.6	0.8	21.7	42	95	1.5	2,410	0.27	0.17	1.4	457
Potatoes	16	325	7.6	0.4	72.8	42	213	2.7	70	0.40	0.15	4.4	64

Kind of product	Refuse (%)	Food energy (cal)	Protein (g)	Fat (g)	Carbohy-drates (g)	Calcium (mg)	Phos-phorus (mg)	Iron (mg)	Vitamin A. (Int'l. units)	Thiamin (mg)	Ribo-flavin (mg)	Niacin (mg)	Ascorbic acid (mg)
Spinach	18	92	8.6	1.1	11.9	301*	205	11.2	35,040	0.44	0.90	2.6	219
Squash, summer	3	83	2.6	0.4	17.2	66	66	1.8	1,140	0.18	0.23	5.0	75
Squash, winter	26	147	5.0	1.0	29.6	64	94	2.0	16,640	0.16	0.26	1.9	28
Sweet potatoes**	14	488	7.0	2.7	108.8	117	191	2.7	30,030	0.37	0.23	2.8	86
Tomatoes	12	91	4.0	1.2	16.0	44	108	2.4	4,380	0.24	0.16	2.5	93
Turnip greens	16	140	11.0	1.5	20.6	987	190	9.1	36,370	0.37	2.15	2.9	518
Turnips	18	136	4.8	0.8	28.0	158	134	2.0	20	0.26	0.24	1.8	113
For comparison:													
Whole–wheat bread	0	1,187	43.1	15.9	217.9	(272)	1,680	11.8	(0)	1.28	0.70	16.1	0
Milk, whole	0	312	15.9	17.7	22.2	536	422	0.3	(720)	0.16	0.78	0.5	6
Eggs	11	636	51.7	46.5	2.8	218	848	10.9	4,590	0.47	1.35	0.3	(0)
Round steak	11	789	78.0	53.0	0.0	44	840	11.7	(0)	0.48	0.61	21.0	0
Apples	12	258	1.2	1.6	59.6	24	40	1.2	360	0.15	0.08	1.0	18
Oranges	28	164	2.9	0.7	36.6	108	75	1.3	(620)	0.25	0.08	0.8	162

*Based on yellow corn.

**If only pale varieties were used, the Vitamin A value would be lower.

Source: Adapted from USDA Miscellaneous Publication, No. 572.

Table 1.7 Average net return per acre for sample grower–marketers (Rs.)

Crop	Cost of seed	Fertilizer	Insecticides/ Pesticides	Packing	Transport	Marketing	Total	Gross receipt	Net return
Apples	–	16	50	1,067	780	977	280	10,214	7,324
Plums	–	–	–	5,357	2,260	4,520	12,137	22,741	10,604
Apricots	–	266	107	3,303	1,119	2,077	6,872	20,640	13,768
Pears	–	–	–	4,049	1,832	2,506	8,387	12,960	4,573
Tomatoes	139	106	111	2,230	697	1,472	4,755	12,710	7,955
Okra (lady fingers)	173	115	58	404	115	519	1,384	7,612	6,228
Peas	259	103	207	362	414	103	1,448	10,963	9,515
Potatoes	1,778	227	64	424	1,482	1,138	5,113	17,955	12,842
Watermelons	79	–	–	151	1,131	528	1,889	7,966	6,077
Muskmelons	147	–	–	221	1,147	618	2,133	20,618	18,485

Source: Integrated Development of Horticulture of NWFP for Export, January 1986.

Table 1.7(b) Net returns per acre and per 40 kg of wheat and rice

Item	Wheat		Rice	
	Sample	Normal	Sample	Normal
Yield per acre (kg)	1,018.80	1,111.09	994.00	909.61
Price per 40 kg (Rs.)	56.30	56.30	80.81	80.81
Gross returns/acre (Rs.)	1,433.96	1,563.86	2,008.13	1,837.64
Net returns/acre (Rs.)	-177.52	-47.62	192.11	21.62
Gross returns/40 kg (Rs.)	56.30	56.30	80.81	80.81
Cost/40 kg (Rs.)	-63.27	58.01	73.08	79.86
Net returns/40 kg (Rs.)	-6.97	-1.71	7.73	0.95

Source: Department of Farm Management, University of Agriculture, Faisalabad.

Table 1.8 Pakistan's production and exports of fruits and vegetables compared to Gulf imports

Products	Pak. prod. 1987–88	Pakistan exports 1984–85	Gulf imports 1992	Pakistan exports 1992	Gulf imports 2000	Pakistan exports 2000
Processed	–	–	624	30	1150	115
Vegetables ($ million)	–	–	631	30	1300	130
Fruit juices ($ million)	–	–	631	30	1300	130
Fresh vegetables (000 MT)	3650	81	1230	147	1840	220
Chilies	97	10	80	20	100	25
Onions	515	66	255	85	275	95
Potatoes	543	–	188	20	240	24
Peas	68	–	7	1	10	2
Others	2427	5	700	21	1215	74
Fresh fruits (000 MT)	3500	59	1529	260	2000	525
Citrus	1373	29	471	160	700	200
Mangoes	692	8	60	30	200	140
Dates	235	12	50	20	70	35
Others	1200	10	948	50	1030	150

Source: Report of the National Commission on Agriculture, 1988.

Table 1.9 Important events in the history of horticulture in Indo–Pakistan

Year	Event
1905	The Royal Commission on Agriculture in India recommends development of fruit industry on scientific lines.
1909	Agriculture College and Research Institute along with network of stations established.
1925	First fruit specialist appointed, and research on vegetables under professor of botany started.
1927	A 25 acre experimental fruit garden established.
1928	Area under fruits in United Punjab 12,800 acres.
1931–35	Regional experimental fruit gardens; establishment of canning hall and juice plant.

Year	Event
1938	Establishment of fruit plant nurseries; initiation of annual fruit show; initiation of a vernacular training class in horticulture (mali class) to produce skilled labour and a training course in fruit preservation, installation of a cold storage for research, horticulture as a major subject offered to B.Sc. (Agri.), establishment of a rootstock experiment station at Sahiwal, a progeny garden at Resalewala, manuring experimental at Atari (India), a new substation for hilly fruits at Palampur (India), which is now an agriculture university.
1940	Post graduate training in horticulture started. Punjab cooperative Fruit Development Board started to function.
1947	Area under fruits in United Punjab 103,983 acres, partition of India, 58,000 acres to the share of East Punjab. Area in Pakistan (West 120,000 acres about 6,000 acres destroyed due to disturbances).
1948	Establishment of agriculture colleges on the pattern of Lyallpur at Peshawar and Tandojam.
1962	Establishment of West Pakistan Agriculture University, Lyallpur, separation of research from education.
1977	Establishment of Agriculture University at Tandojam.
1978	Agricultural College at Bahauddin Zakariya University, Multan
1979	PARC/NARC. Agriculture College at Gomal University.
1981	Upgrading of Peshawar University's College of Agriculture to NWFP Agricultural University at Peshawar.
Present	Punjab: Director of Horticulture, Director of Vegetable Crops, Director of Floriculture, and several regional offices and stations both in research and extension.
	Sindh: Important stations at Tandojam and Mirpur Khas.
	NWFP: Tarnab (Peshawar), Swat, Abbottabad, and D.I. Khan.
	Balochistan: Sariab — developments taking place under FAO Programs.
	Horticulture wings in most government and autonomous bodies like PIA, State Bank, ADA, CDA, LDA, KDA, and many other civic bodies.

Table 1.10 Area and production of major fruits and vegetables in Pakistan, 1960–1990 (area '000' ha, production '000' tonnes)

	1960		1970		1980		1990	
Fruits	**Area**	**Prod.**	**Area**	**Prod.**	**Area**	**Prod.**	**Area**	**Prod.**
Citrus fruits	14.3	88.0	44.5	487.1	94.5	926.2	171.1	1576.3
Mango	20.6	130.0	57.1	543.5	57.2	546.6	82.7	766
Guava	3.2	18.9	14.7	89.5	17.3	123.6	46.2	347.3
Date palm	–	–	21.4	194.5	24.2	194.1	41.8	284.1
Apple	0.8	9.4	5.1	33.7	11.4	107.4	22.4	232.4
Banana	1.2	2.0	10.7	101.1	14.2	130.8	23.5	209.8
All fruits	–	–	–	–	305.8	2532.0	449.8	3881.8

Fruits	1960 Area	1960 Prod.	1970 Area	1970 Prod.	1980 Area	1980 Prod.	1990 Area	1990 Prod.
Vegetables								
Onion	25	80	17.5	186.6	43.2	447.6	58.6	712.9
Garlic	3	8	1.9	24.2	4.9	36.9	5.7	48.3
Chilies	71	26	36.2	56.8	64.0	106.2	71.0	125.5
Coriander	–	–	2.8	11.0	4.9	2.0	7.2	3.3
Turmeric	–	–	0.8	7.4	4.2	31.2	3.3	26.9
Potato	31	99	23.4	241.3	38.0	394.3	80.0	830.9
Tomato	–	–	–	–	11.6	92.1	19.3	193.8
Sugar beet	–	–	12.6	259.9	15.7•	452.5	11.0	342.8
Other vegetables	239	643	126.9	1660.9	123.4	1549.4	206.3	2750.8

Source: 1980 and 1990 data from Agricultural Statistics of Pakistan, 1989–90:86, 102.

Table 1.11 Area, production, yield per hectare, and percentage share of major fruits and vegetables in Pakistan (1989–90)

Fruits and vegetables	Area (1000 ha)	Production (1000 tonnes)	Yield (tonnes/ha)	Percentage share of total fruit and vegetable area
		Fruits		
Citrus fruits	171.1	1576.3	9.2	38.0
Mango	82.7	766.0	9.3	18.4
Guava	46.2	347.3	7.5	10.3
Dates	41.8	284.1	6.8	9.3
Banana	23.5	209.8	8.9	5.2
Apple	22.4	232.4	10.4	5.0
Other	62.1	455.9	–	13.8
All Fruits	449.8	3881.8	–	100.0
		Vegetables		
Onion	58.6	712.9	12.2	12.7
Garlic	5.7	48.3	8.4	1.2
Chilies	71.0	125.5	1.8	15.4
Coriander	7.2	3.3	0.5	1.6
Turmeric	3.3	26.9	8.3	0.7
Potato	80.0	830.9	10.4	17.3
Tomato	19.3	193.7	10.0	4.2
Sugar beets	11.0	342.8	31.3	2.4
Other vegetables	206.3	2750.8	–	44.6
Total	462.4	5035.1	–	100.0

Source: Agricultural Statistics of Pakistan (1989–90).

Table 1.12 Province–wise distribution of area and production of fruit and vegetables in Pakistan (1989–90)

Province	Area (1000 ha)	% of total area	Production (1000 tonnes)	% of total production	Yield (tonnes/ha)
		Fruits			
Punjab	293.5	65.25	2550.8	65.71	8.69
Sindh	83.3	18.52	617.8	15.92	7.41
NWFP	26.4	5.87	293.5	7.56	11.11
Balochistan	46.6	10.36	419.7	10.81	9.00
Pakistan	449.8	100	3881.8	100	8.63
		Vegetables (plus potato and sugar beet)			
Punjab	177.1	59.57	2383.7	60.74	13.46
Sindh	32.2	10.83	191.2	4.87	5.94
NWFP	51.5	17.32	816.3	20.80	15.85
Balochistan	36.5	12.28	533.3	13.59	14.61
Pakistan	297.3	100	3924.5	100	13.20

Source: Agricultural Statistics of Pakistan, 1989–90, p. 45, 84, 86, 102.

Table 1.13 Quantity and value of fruit export from Pakistan

Year	Quantity (1000 tonnes)	Value (million Rs.)
1984–85	66.1	243.2
1985–86	96.4	483.7
1986–87	103.9	542.1
1987–88	104.4	644.4
1988–89	96.0	681.9
1989–90	101.4	785.3

Source: Agricultural Statistics of Pakistan (1989–90), p. 239.

Table 1.14 Marketing losses of produce and producers' share in consumer paid price of major fruits and vegetables

Commodity	Marketing losses (%)	Producers' share (%)	Marketing margin* (%)
Mango	13.74	52.44	47.56
Citrus	20.77	43.48	56.52
Potato		42.00	58.00
Carrot		47.00	53.00
Cauliflower		58.00	42.00
Onion		59.00	41.00
Brinjal		55.00	45.00
Okra		62.00	38.00

* Marketing margins are earned by intermediaries like contractor, commission agent, *pharia*, and retailer.

STUDY QUESTIONS

1. What is horticulture? Discuss both in the light of classical views and the modern outlook.
2. Differentiate between a fruit and vegetable. Name some important fruits and vegetables, and specify their original distribution in Pakistan.
3. List up to ten each of fruits, vegetables, and ornamental plants, sub–categorized as deciduous and evergreens.
4. What is the role of foreign introductions in the history and development of horticulture? Discuss specific examples for Pakistan.
5. Name and discuss some fruits and vegetables which have a totally indigenous origin.
6. Work out cost of production and returns for different horticultural crops according to Table 1.7.
7. What is a balanced diet? Explain the role of fruits and vegetables in our daily diet.
8. How can awareness and appreciation of different branches of horticulture improve our lives?
9. Discuss the significance of post–harvest and marketing losses.

PRACTICAL EXERCISE

Practice identifying and giving both the common and botanical names of important horticultural plants.

REFERENCES

Ashraf, M.; Rana, M. S. K.; Bhatti, M. H. eds. Proceedings of international workshop/training course on vegetable seed production. Islamabad: Pakistan Agricultural Research Council; 1988.

Bailey, L. H. The standard cyclopedia of horticulture. New York: MacMillan; 1939.

Chaudhry, A. M.; Ahmad. B. Cost of producing major crops in Vehari District 1981–82. Faisalabad: Department of Farm Management, University of Agriculture; 1986.

FAO. Production yearbook, 1990. Vol. 44. FAO Statistics Series, 1991, No. 99; Rome: Food and Agriculture Organization of the United Nations; 1991.

Halfacre, R. G.; Barden, J. A. Horticulture. New York: McGraw–Hill; 1979.

Janick, J. Horticultural science. 4th ed. New York: W. H. Freeman; 1986.

McWilliams, M. Food fundamentals. New York: John Wiley and Sons, Inc.; 1966.

Mohyuddin, Q. Marketing of major fruits (citrus and mango) in Punjab. Faisalabad: Department of Agricultural Marketing, University of Agriculture; 1989.

Pak–Swiss Potato Development Project. Production and marketing of potatoes in Gujranwala district. Islamabad: Pakistan Agricultural Research Council; 1991.

Simmonds, N. W. Evolution of crop plants. London: Longman; 1984.

Thompson, H. C.; Kelley, W. C. Vegetable crops. New Delhi: Tata McGraw–Hill; 1978.

University of Agriculture, Faisalabad. Silver Jubilee Celebrations. 25 years of education and research, 1961–1986. Faisalabad: University of Agriculture.

Outline of Chapter 2

2. MORPHOLOGY AND ANATOMY OF FLOWER AND FRUIT

Altaf–ur–Rehman Rao[1]

LEARNING OBJECTIVES

After studying this chapter, the student should be able to:

- Name the parts of a typical flower and discuss their functions
- Discuss the stages of embryo development
- Outline the stages of seed formation
- Explain the processes of fertilization and fruit formation
- Explain how fruits are classified

Morphology is the study of the form and external structure of organisms, both plants and animals. Other related fields of study are:

Anatomy: the area of morphology which deals with the internal structure of root, stem, leaf, ovary, ovule, fruit, and seed.

Ontogeny: study of the life cycle of a plant including developmental stages of plant organs.

Histology: study that deals with the microscopic structure of tissues.

Phylogeny: study of a race or plant type or species in relation to its evolutionary history.

2.1 Flower morphology

2.1.1 Vegetative and protective parts

Angiosperms, flower–bearing plants with enclosed seed, are the largest, most widely spread, and most advanced group of plants. They originated nearly 120 million years ago and came to dominate most parts of the world. During

---

[1]Dr. Rao is Professor, Department of Botany, University of Agriculture, Faisalabad.

this period, along with the plants, the flower itself has undergone gradual evolutionary changes. As a result of this, the shape, size, and structure of the flower has become highly variable. *Wolffia* (duckweed) produces the smallest flower (0.1 mm in diameter) while *Rafflesia*, a root parasite of Java/Malaya, has a flower more than a metre in diameter.

The word 'flower' suggests variable corolla colour patterns, fragrance, scent, beauty, charm, and honey–laden nectaries. However, these are the characteristic features only of showy entomophilous flowers which have to attract insects as their pollinating agents. On the other hand, a large number of anemophilous and hydrophilous flowers, whose pollinators are wind and water, respectively, have neither scent, honey, or coloured corollas, but are green, because they do not need to attract insects.

The flower, a modified vegetative shoot, has become highly specialized over the ages. It still compares closely with an ordinary vegetative shoot in its morphology. Like an ordinary shoot, the flower also arises in the axil of a leaf (bract). Its thalamus is a branch with shortened internode, as a result of which floral leaves get crowded on it. Floral buds may sometimes be replaced by vegetative buds or vice versa. The calyx and corolla sometimes remain green and resemble foliage leaves, as in the green rose. In certain plants, stamens get transformed into petals and vice versa. In some Asteraceae and Rosaceae, the carpels may be modified into leafy structures or stamens, respectively and vice versa. Thus there are inseparable relationships between floral elements and foliage leaves.

Calyx, corolla, androecium, and gynoecium are the four floral whorls of a typical flower. Non–floral parts like nectaries close to the corolla and bracts/bracteoles may be present on the thalamus. The number of floral parts in monocotyledons and dicotyledons is generally a whole number multiple of three or five, respectively. The flower is classified as **complete** when calyx and corolla are present; when either one or both of them are missing, it is **incomplete**. The calyx and corolla together are called the **perianth**. When calyx, corolla, and androecium are inserted below the carpel (gynoecium), the flower is called **hypogynous**, or having a superior ovary. When the floral parts are inserted above the ovary, the latter is called **epigynous**, i.e. having an inferior ovary.

The flower is **bisexual** (hermaphrodite or perfect) when both the reproductive parts, i.e. stamen and carpel, are present. It is unisexual when only one sex is present: **staminate** when only stamens are present and **pistillate** when only the carpel is there, as in cucumbers and gourds. When the stamen is non–functional the flower is called **staminode**; when both sexes of the flower are non–functional it is called **neuter**. When male and female flowers are present on separate branches of the same plant it is known as **monoecious**, e.g. cucurbits, *pinus*; when male and female flowers are borne on separate plants the species is called **dioecious**, e.g. date palm, papaya.

Some specialized forms of corolla characterize certain families.

1. **Cruciform.** The corolla has four clawed petals inserted at right angles to each other, e.g. radishes, *Brassicas*.
2. **Rosaceous.** The corolla has five petals (sometimes more in cultivated roses). The claw of the petal may be small or absent, e.g. rose, apple.
3. **Tubular.** When the united corolla forms a tube–like structure, e.g. disc floret of sunflower.
4. **Rotate or wheel–shaped.** The corolla tube is short, and its limbs are at right angles to the tube below and appear wheel–shaped, e.g. *Cestrum nocturnum* (night jasmine or *raat ki rani*), *Solanum nigrum* (*mako*).
5. **Papilionaceous.** Butterfly–shaped corolla where three sizes and shapes of petals are present in a single flower, e.g. pea family (Papilionaceae).
6. **Bilabiate** or two–lipped. The united corolla appears to be divided into two lips, e.g. *Ocimum basilicum* (*niazboo*).
7. **Personate.** A united corolla, the two lips of which are close to each other towards the mouth of its tube, e.g. Antirrhinum (*kutta phool*).

2.1.2 Reproductive mechanism

In addition to two vegetative and protective parts, the calyx and corolla, the flower has two reproductive parts. They are the **stamens** (male = androecium) and **carpel** or **pistil** (female = gynoecium). Stamens usually consist of a long filament and an anther. The anther has two lobes, each containing two chambers (anther sacs). It produces a large number of pollen grains with a haploid number of chromosomes. The carpel consists of the **stigma**, **style**, and **ovary**, an enlarged basal part having one or more ovules attached to the placentae in its cavity.

Gametogenesis. The stamen consists of one filament, two anther lobes containing two pollen sacs each (four total), and a connective tissue. The pollen sacs may be one–chambered (monothecous), two–chambered (bithecous) or, rarely, four–chambered (tetrathecous). After meiotic division of the pollen mother cells, a large number of pollen grains are produced. On maturity, the anther lobes burst open, forming a pore through which the pollen grains are released.

Pollination. The transfer of the pollen grains to the stigma is known as pollination. Insects, animals, wind, and water are the main agents of pollination. Pollination is of two types. 1) Self–pollination or **autogamy** is transfer of the pollen grains of a flower to the stigma of the same flower or flowers of the same plant. 2) Cross–pollination or **allogamy** is transfer of pollen grains to the stigma of flowers of another plant. In nature, cross–pollination is very important and prevalent. Self–pollination also occurs in nature, but usually in cultivated plants like wheat, rice, and cotton most of the progenitors are cross–fertilized. Many plants have evolved special devices adapting them to cross–pollination. On the other hand, the small flowers of the pea, peanut (groundnut or *moongphali*) and *Oxalis* do not open up at all (**cleistogamy**), thus ensuring self–pollination.

Embryo sac. One or more ovules develop on the placentae inside the ovary. A cell from the nucellus enlarges itself and undergoes meiosis to form four megaspores with a haploid number of chromosomes. Three of these degenerate, and the nucleus of the survivor divides mitotically into two parts, one of which moves to the upper and the other to the lower pole. Each nucleus divides twice more, increasing the number of nuclei to eight, four in each of two groups. One nucleus from each group moves towards the centre, and both fuse to form one nucleus with a diploid number of chromosomes (n+n = 2n). This is called the **definitive nucleus**. The three nuclei at the micropylar end enclose themselves with a thin wall, forming the egg apparatus; while the other three at the chalazar end form the antipodal cells. The egg apparatus is the **ovum** or **oosphere**, the female gamete. The remaining two cells are called **synergids** or helping cells. On fertilization, the egg cell gives rise to the embryo, and the synergids disintegrate. Similarly, the antipodal cells have no function and disintegrate. On fertilization, the definitive or secondary nucleus forms the endospermic nucleus, developing into the endosperm of the seed with a triploid (=3n) number of chromosomes.

Types of ovule. The following four types of ovule are found (Fig. 2.1).

1. **Orthotropous** (straight): The ovule is erect or straight, and its funicle, chalaza, and micropyle lie in one vertical line, e.g. *Polygonum* (*hazardani*), walnut, pepper, and spinach.

2. **Anatropous** (inverted): The ovule bends back along the side of the funicle so that the micropyle comes close to the hilum, micropyle, and chalaza, but not the funicle, which lies in the same straight line; this is the commonest form of ovule.

3. **Amphitropous** (transverse): The ovule is placed transversely, at right angles to its funicle, as in *Chenopodium* (*bathu*).

4. **Campylotropous** (curved): The ovule is transversed and bent like a horseshoe. The micropyle and chalaza are not in a straight line, e.g. peas, beans, and radishes.

Placenta. The placenta is a parenchymatous outgrowth in the inner wall of the ovary, to which the ovules are attached. Placentae may develop along with the marginal suture at the base or apex of a carpel. The arrangement of placentae inside the cavity of the ovary is called **placentation**. The origin

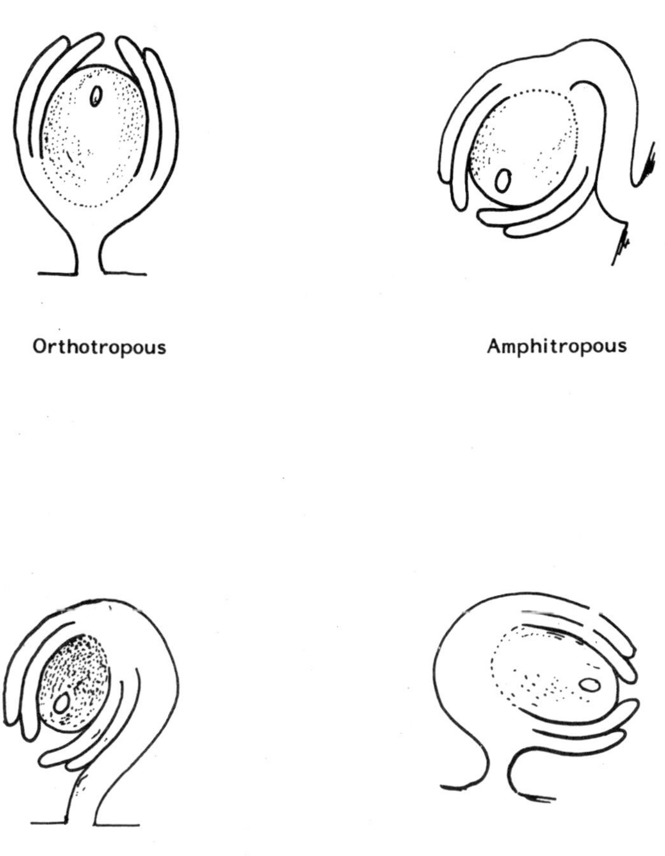

Figure 2.1 Types of ovules

of one or more ovules inside the ovary determines this phenomenon. Accordingly, eight major types of placentation are found (Fig. 2.2).

1. **Marginal**: A simple and common type of placentation. The ovary is single–chambered, and placentae develop along the ventral suture, e.g. peas, gram, pulses, and buttercup.

2. **Axile**: The ovary is multilocular, and the number of loculi corresponds to the number of carpels. Placentae develop from the central axis, hence the name axile. This structure is found in citrus, tomato, and shoeflower. In tomato, false partition of the ovary doubles the number

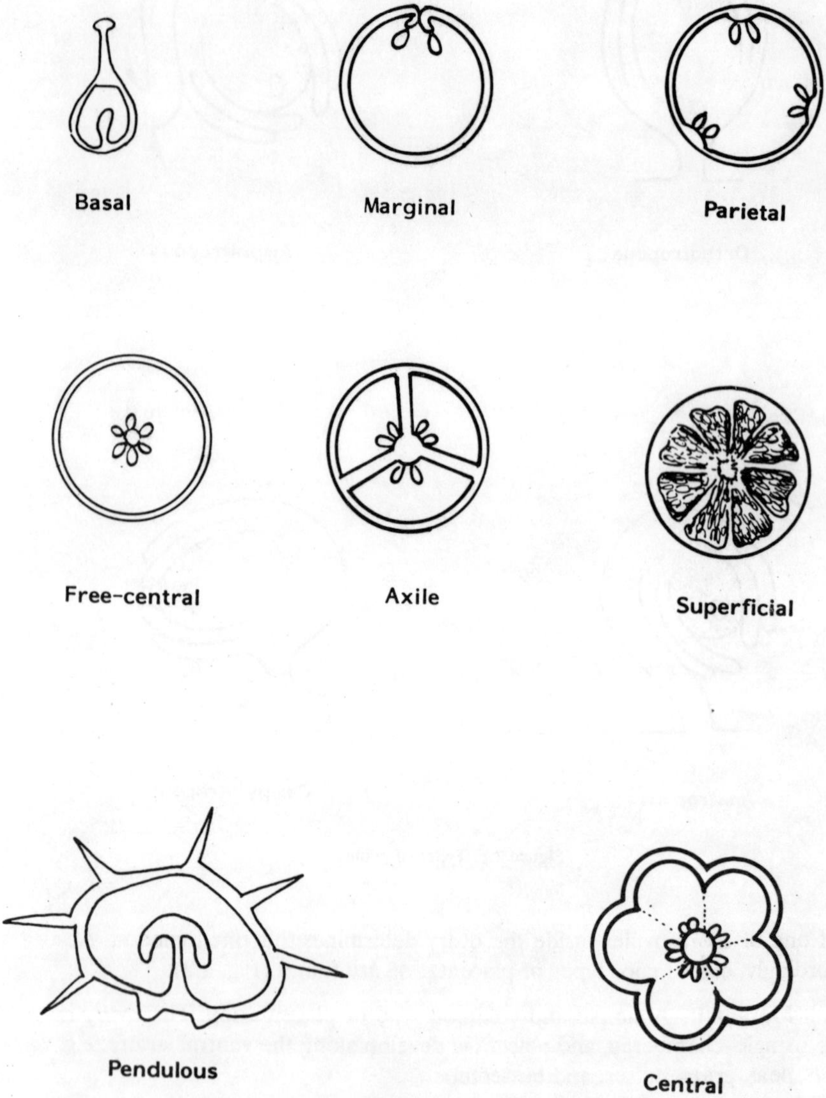

Basal

Marginal

Parietal

Free-central

Axile

Superficial

Figure 2.2 Types of placentation

of loculi to four.

3. **Parietal**: The ovary is single–chambered and placentae develop on its inner wall corresponding to the number of united carpels, e.g. in poppy and *Brassica*. In *Brassica* a unilocular ovary becomes bilocular because of the development of a false septum, called the **replum**.

4. **Central**: The partitioning walls of a young multilocular ovary break down, giving rise to a unilocular ovary. The placentae bearing ovules remain attached to the central axis, as in *Seponaria* and *Stellaria*.

5. **Free central**: Numerous ovules develop on an enlarged central axis bearing ovules all around its surface; the placentae develop from the base of the ovary, e.g. Primula, *Anagallis* (*billi booti*), and *Silene* (*takla*).

6. **Basal**: The ovary is unilocular, and a single ovule develops at its base, e.g. in wheat, rice, and sunflowers.

7. **Superficial**: The ovary is compound and multilocular; ovules develop all around the inner surface of the partitioning walls, e.g. water lily.

8. **Pendulous**: A single ovule is attached to the top of a single–chambered ovary like a pendulum, e.g. coriander and carrot.

2.1.3 Fertilization and formation of seed

Fertilization is the fusion of male and female gametes. In flowering plants, a pollen grain, a product of meiosis with haploid (halved = n) number of chromosomes, is enclosed by two protective layers. The outer rough layer is called the **exosporium** (extine) and inner thin layer the **endosporium** (intine). The pollen grain may start germination before or soon after it is transferred to the stigma of the ovary. On germination, the nucleus of the pollen grain divides into two unequal nuclei. The smaller nucleus further divides into two male gametes; after this, the pollen tube has three nuclei. These nuclei are naked and non–motile. The pollen tube, while germinating, penetrates inside the canal of the style and, through the micropyle, into the ovule located inside the cavity of the ovary. The tip of the pollen tube, after coming in contact with the embryo sac, swells and bursts open, releasing the two male gametes. One of them fuses with the female gamete (oosphere) to form the zygote (oospore), while the other male gamete fuses with the secondary nucleus to form the endosperm. This phenomenon, in which one male gamete fuses with the oosphere and the other with the secondary nucleus, is known as **double fertilization**.

Development of embryo. After fertilization, the zygote (oospore) divides into two cells, and subsequently both divide again and again producing a tissue known as the **proembryo**. The upper terminal cell of the proembryo is called the **embryonal cell**, while the remaining portion forms the **suspensor**. The embryonal cell develops into an embryo by dividing itself at right angles after producing eight cells (octants). Four of them stay close to the suspensor, and the other four move away from it. All of them enlarge and undergo repeated divisions. The four lower cells lying close to the suspensor give rise to the plumule and two cotyledons, while the distant cells give rise to the hypocotyl. The uppermost cell of the suspensor, called the **hypophysis**, develops into the apical growing point of the radicle.

Development of endosperm. The secondary nucleus, a product of three nuclei, with a triploid number of chromosomes (=3n), produces a large number of small nuclei after repeated divisions through a process of free cell formation, giving rise to endosperm or albumen. The nucellus supplies food to the endosperm which then enlarges. After enlargement, the endosperm crushes the nucellus, reducing it to a very thin layer called **perisperm.**

Formation of seed. Endospermic seeds store their food material in the albumen or endosperm, while their cotyledons remain thin, e.g. wheat and maize. In non–endospermic seeds, storage of food material takes place in the cotyledons, which become heavily thickened (e.g. soybean), while the endosperm may be reduced or disappear. The outer integument of the ovule shrinks and forms the seed coat, called the **testa**, whereas the inner integument remains thin and develops into the **tegmen**. The micropyle of the ovule persists, and an outgrowth of the funicle forms the **aril**. A scar left by the funicle is called the **hilum**.

2.2 Fruit development and morphology

2.2.1 Definition of fruit

The definition of 'fruit' presents certain difficulties. In simple botanical terms it is customary to say that a fruit is the product of the matured or ripened ovary (gynoecium). However, both stigma and style may also play a part in fruit formation. There are also many instances where other floral parts—the floral receptacle, calyx, even the axis of inflorescence—may also contribute towards fruit formation. These organs may develop simultaneously with the development of the gynoecium and become an integral part of the ripened fruit. When distinguished botanically, such structures are known as **false** or **spurious fruits.** The function of a fruit is to develop, protect, nourish, and ultimately disperse the seeds contained in it.

The horticultural consumer's definition of fruit is a plant product with aromatic flavour which is either naturally sweet or normally sweetened before eating.

2.2.2 Changes during fruit development

Physical changes. Soon after fertilization, shedding of some floral parts takes place. Among them, the corolla and androecium are the first to fall. In certain cases the calyx may persist with the fruit until its maturity, e.g. in tomato, apple, citrus, and guava. In rare cases, the stigma and style may also become persistent, e.g. clematis. During the development of the embryo, the number of cells increases tremendously. Consequently, this triggers further changes in the size, shape, and physical appearance of the fruit.

The ovary wall develops into the **pericarp** (*peri* 'around', *carp* 'fruit'). While in flower, the ovary wall consists of little-differentiated parenchyma cells, vascular tissues, and outer and inner epidermal layers. During maturation, the pericarp frequently shows an enormous increase in the number of cells; its tissue may either remain relatively homogeneous or differentiate into parenchyma and sclerenchyma. The pericarp of succulent fruits when developed may become differentiated into three parts, more or less distinct morphologically: (1) **epicarp** or **exocarp**, (2) **mesocarp**, and (3) **endocarp**, which are the outer, median, and inner layers of the fruit wall, respectively. However, not all these layers occur in all dry fruits. Sometimes only two layers, i.e. epicarp and endocarp, may be distinguishable. In citrus fruits, the epicarp (**flavedo**) and mesocarp (**albedo**) differentiate into distinct layers, but the endocarp forms separate chambers in which placental hairs store food material and become succulent. These distinctions are useful for the classification of succulent fruits.

Chemical changes. Towards maturity the chemical changes that take place in the fruit involve pericarp colour, carbohydrates, organic acids, nitrogenous compounds, aroma (volatile chemicals), and vitamins. During ripening, in addition to an increase in water content, chlorophyll is destroyed and anthocyanin and carotenoid pigments develop. Chlorophyllases brings about certain changes which induce different colours such as bright-green, olive-green, brown, and variedly coloured, or even colourless. In tomato, on the destruction of chlorophyll, synthesis of carotenoids occurs; a similar change is noticeable in banana. In response to light, anthocyanins and carotenes in apple skin develop red or purple colours. Anthocyanins are generally water soluble and accumulate in the vacuoles of cells.

Carbohydrates present as starches change into different types of sugars, altering the taste and texture of the fruit. This is one fundamental change that makes the edible parts of the fruit sweet. The majority of fruits have fructose, but sucrose is found in dates. Pectin, protopectin, and hemicelluloses break down and weaken the cell walls, as a result of which fruit tissues are softened.

The level of organic acids (e.g. ascorbic acid, malic acid in apple, and citric acid in mango) declines during ripening, and these substances are converted into sugars. Banana and pineapple have lower acid levels than other fruits. Nitrogenous compounds also show conspicuous changes. Amino acid level also decreases as they get incorporated into protein synthesis, however, the level of free amino acids rises at the verge of senescence. The aroma developed by volatile organic compounds tremendously improves the eating quality of fruits.

On maturity, several vitamins develop, prominently ascorbic acid (vitamin C), thiamine (B_1), and riboflavin (B_2). Their peak level, however, declines quantitatively during storage, the level of thiamin and riboflavin changing least.

2.3 Classification and morphology of fruits

2.3.1 Descriptive classification of fruits

Figures 2.3–2.8 diagram the fruit types described in section 2.3.1.

I. **Simple fruit**. Fruit or fruitlet develop from one or more syncarpous carpels of a single flower; other floral part(s) may also be incorporated. It may be dry or succulent.

   A. **Dry fruit**. Gynoecium is composed of one or more united carpels, but pericarp remains dry at maturity. Dry fruits are of two types: (1) dehiscent, and (2) indehiscent.

      1. **Dehiscent fruits**. Pericarp splits open at maturity to release the seeds. Following are the important types (Fig. 2.3).

         a. **Follicle**. Dehisces longitudinally along with the dorsal suture only, many–seeded; fruit develops from only one carpel. e.g. Delphinium (larkspur). Phylogenetically, a follicle is considered the most primitive type of fruit.

         b. **Legume**. (Pod) Generally dehisces longitudinally along with both the sutures (ventral and dorsal) into two halves e.g. pea, bean, gram.

         c. **Capsule.** Develops from two or more united carpels, many–seeded and dehisces in several ways, e.g. cotton, poppy.

         d. **Silique.** A long, many–seeded fruit developing from two united superior carpels with parietal placentation. Splits open longitudinally. Unilocular ovary becomes bilocular due to development of a false septum called replum, e.g. radish, mustard.

         e. **Silicula**. A flattened and much shorter silique, nearly as broad as long, and containing only a few seeds, e.g.

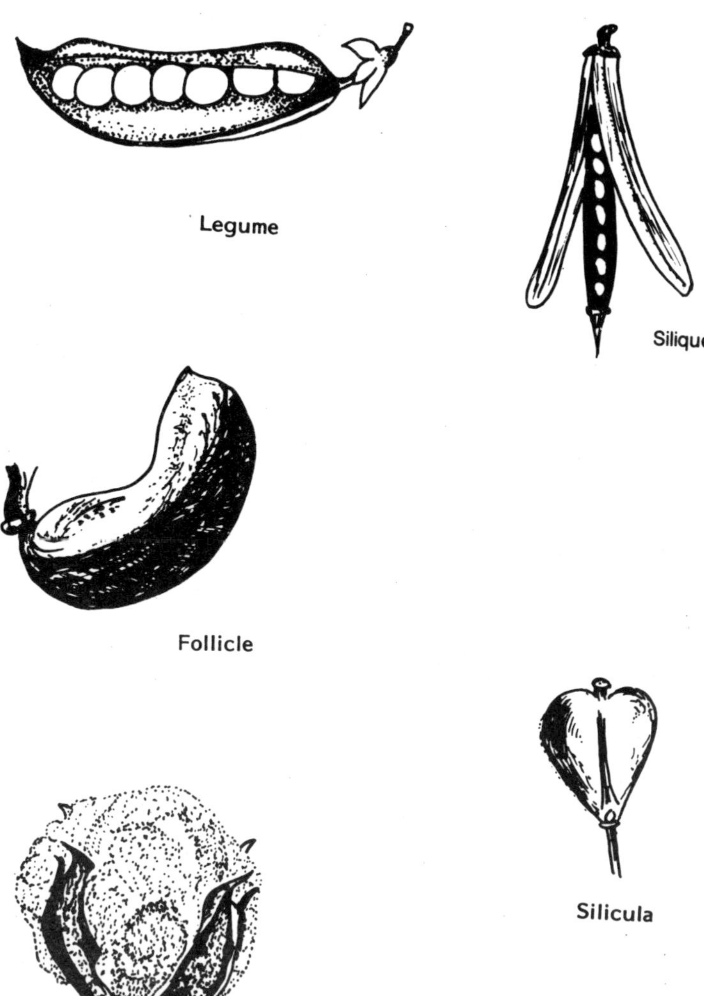

Figure 2.3 Simple dry dehiscent fruits

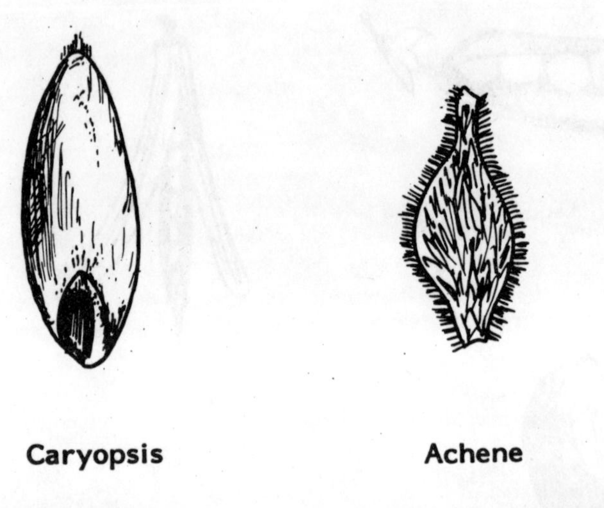

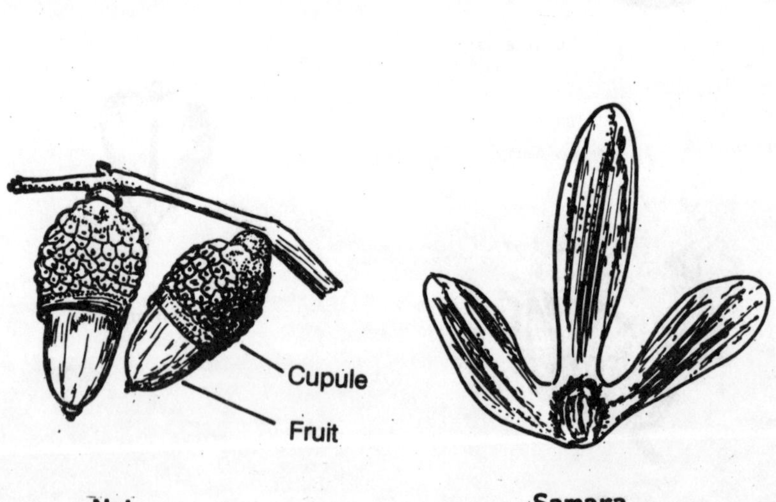

Figure 2.4 Simple dry indehiscent fruits

Coronopus didymus (*jangli halon*).

2. **Indehiscent fruits**. Pericarp does not split open at maturity (Fig. 2.4).
   a. **Achene**. Pericarp is thin, dry, hard, and free from its single seed. Achenes may form an aggregate of fruitlets when developing from an apocarpous ovary, e.g. buttercup, clematis (persistent hairy style).
   b. **Cypsela**. A dry single–seeded fruit developing from a single–chambered bicarpellary inferior ovary surrounded by other floral tissues, e.g. sunflower (Asteraceae).
   c. **Caryopsis**. One–seeded seed–like fruit where testa is fused firmly to thin pericarp, e.g. wheat, maize, and rice.
   d. **Nut**. Structure resembles an achene, but is much larger; dry pericarp becomes thick and hard, e.g. chestnut, oak.
   e. **Samara**. One or two–seeded fruit developing from a superior bi– or tricarpellary ovary with the pericarp forming the wings.
3. **Schizocarpic fruit**. Developing from two or more united carpels each splitting into indehiscent one–seeded parts called mericarps. The mericarp may be attached to the prolonged end known as carpophore. These are of four types (Fig. 2.5):
   a. **Loment**. A legume in which single–seeded parts are constricted or partitioned, e.g. *Acacia* (*kikar*), *Albizzia* (*siris*).
   b. **Cremocarp**. A two–chambered fruit developing from a bicarpellary inferior ovary where each chamber splits apart on ripening, e.g. carrot, coriander.
   c. **Regma**. Trilocular ovary when ripe splits into three single–seeded parts, e.g. castor.
   d. **Carcerullus**. Develops from a poly–carpellary superior ovary which separates into many single–seeded parts. *Althaea rosea* (*gul–e–khaira*).
   e. **Double samara**. Develops from a bicarpellary syncarpous ovary where two winged single–seeded parts separate on maturity, e.g. *Acer* (maple).

B. **Fleshy or succulent fruit**. Develops from a simple or compound superior or inferior uni– or multilocular ovary, with one or more seeds. The pericarp is pulpy or fleshy due to the storage of food material (Fig. 2.6).

C. **Drupe or stone fruit**. Pericarp distinctly three–layered; epicarp thin, forming the skin; mesocarp fleshy, usually the main edible portion; endocarp (pip or stone) is hard and stony enclosing single seed; e.g. mango, sweet cherry, apricot, peach, and almond.

Loment

Cremocarp

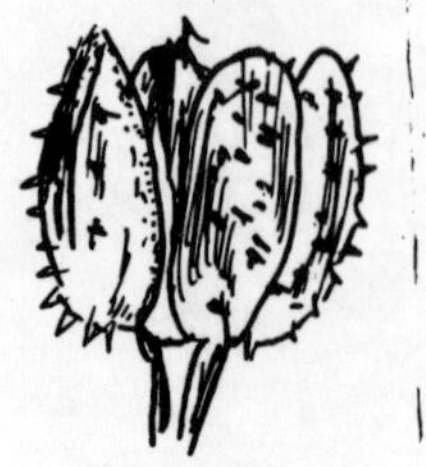

Regma

Carcerullus

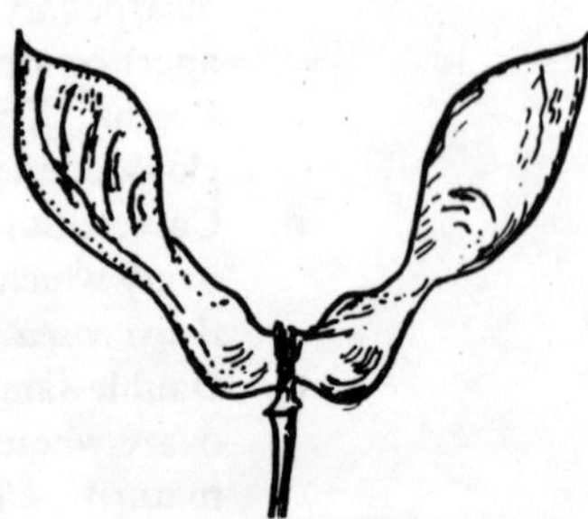

Double Samara

Figure 2.5 Simple schizocarpic fruit

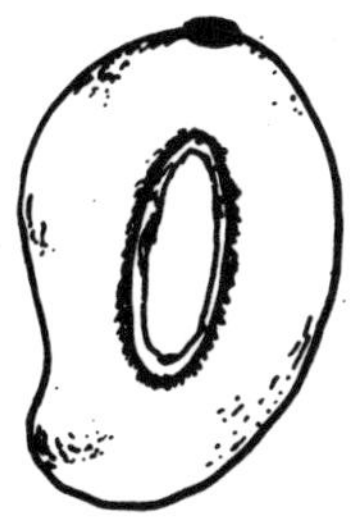

Figure 2.6 Succulent fruit

1. **Pome**. Two or more chambered false fruit developing from a syncarpous inferior gynoecium. Fleshy and edible thalamus surrounds the fruit. Four or five carpels are intimately united to a fleshy receptacle forming the biggest portion of this false fruit. One, two, or more seeds per carpel borne within the papery and leathery endocarp, e.g. apple, pear.
2. **Berry or bacca**. Pericarp is usually thick and juicy, distinguishable as three layers. The outermost skin layer is exocarp containing pigment of the fruit. The relatively thicker middle is mesocarp, which is pulpy and fleshy. The endocarp is soft and papery. This fleshy pericarp may enclose one or many seeds, e.g. date palm, avocado, tomato. Coffee, cucumis, and banana are berries. They are false fruits; they differ from typical false fruits, though, in that extra carpellary parts contribute very little toward the actual fruit. In banana, the skin is composed of tissue through which transverse vascular bundles supply materials to the perianth, stamens, and carpels.
   a. **True (unmodified) berry**. Tomato, papaya, avocado, and date palm exemplify a true and simple berry as defined above.
   b. **Hesperidium**. A modified true berry with a leathery and glandular pericarp; locules contain juicy sacs arising from placental hairs of the endocarp, e.g. orange, mandarin.
   c. **Pepo**. A modified berry developing from an inferior ovary where epicarp alone becomes hard and tough, or brittle, at maturity. Seeds remain attached to the placental wall embedded in the pulp, e.g. cucurbits—cucumber, melon.
   d. **Special berry**. A thick-skinned or leathery berry crowned by the sepals, containing numerous seeds covered with juicy pulp forming the edible part which is derived from the seed coat, e.g. pomegranate.

II. **Aggregate fruit**. An aggregate fruit develops from a single flower with an apocarpous pistil. Being free, each carpel develops into a simple fruit or fruitlet, and their aggregate is known as an etaerio (Fig. 2.7).
   A. **Etaerio of follicles**. Each etaerio consists of two or more follicles, e.g. *calotropis* Ak. and larkspur.
   B. **Etaerio of achenes**. Many achenes have persistent feathery styles occurring in groups, e.g. clematis.
   C. **Etaerio of drupes**. A fleshy thalamus that has the aggregation of several small drupes or drupelets developing from each separate carpel, e.g. raspberry.

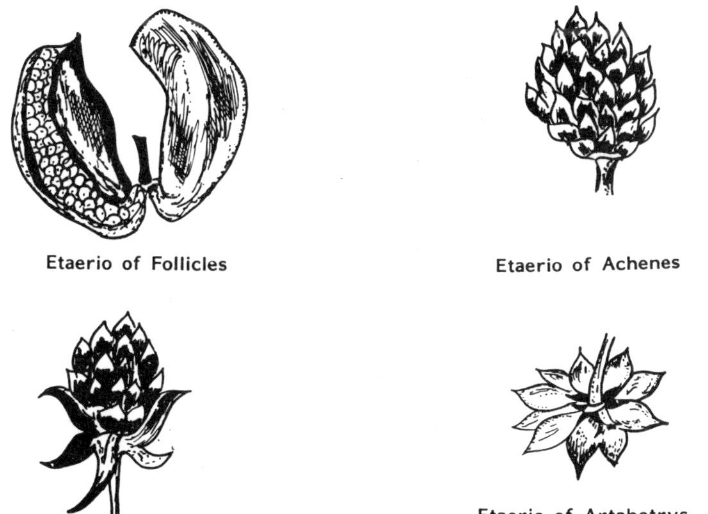

Figure 2.7 Etaerio/aggregate fruits

III. **Multiple or composite fruit**. Develops collectively from numerous flowers where the whole of the inflorescence is involved. Also known as infructescence. It is of two types (Fig. 2.8).

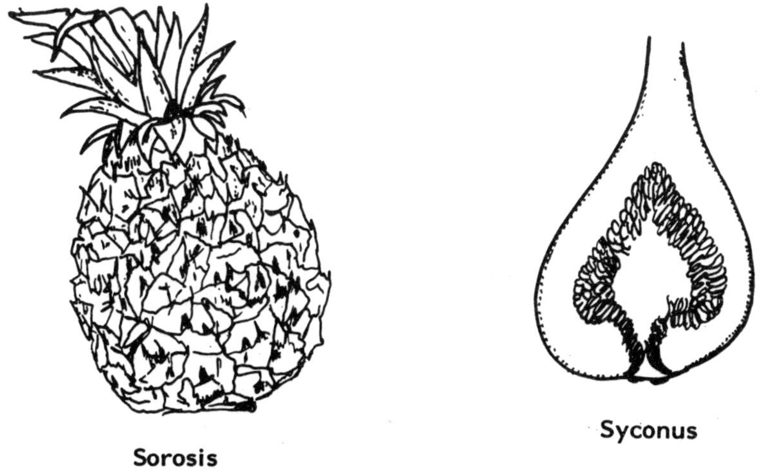

Figure 2.8 Multiple fruit

A. **Sorosis**. Develops from a spike or spadix; small fruits fuse by their succulent calyx, and the mother axis becomes fleshy. Consequently, the whole inflorescence forms one fruit, e.g. pineapple, jackfruit, and mulberry.

B. **Syconus**. Develops from a hollow fleshy and enlarged receptacle enclosing numerous tiny male and female flowers. The receptacle becomes fleshy and edible, e.g. banyan, fig.

Table 2.1 Edible parts of some common fruits: classification and morphology

Fruit	Class	Edible part
Apple	Pome	Fleshy thalamus
Almond	Drupe	Seed (enclosed in endocarp)
Apricot	Drupe	Mesocarp
Banana	Berry	Mesocarp
Cashew nut	Drupe	Peduncle, cotyledons, and oily seed
Coconut	Drupe	Endosperm
Custard apple	Etaerio	Fleshy pericarp of berries
Date	Berry	Epicarp–mesocarp
Fig	Syconus	Fleshy receptacle
Grapefruit, orange, mandarin (berry)	Hesperidium	Juicy sacs or placental hairs of the endocarp
Jackfruit	Sorosis	Bracts, perianth and seeds
Guava	Berry	Thalamus and pericarp
Grapes	Berry	Pericarp and placentae
Litchi	One–seeded nut	Fleshy aril
Loquat	Pome	Fleshy thalamus
Mango	Drupe	Mesocarp
Mulberry	Sorosis	Fleshy sepals
Olive	Drupe	Epicarp and mesocarp
Papaya	Berry	Mesocarp
Papaw	Berry	Mesocarp
Palmyra palm	Drupe (fibrous)	Mesocarp
Pummelo or shaddock	Hesperidium (juicy)	Placental hairs
Pear	Pome	Thalamus
Peach	Drupe	Mesocarp
Plum	Drupe	Mesocarp
Persimmon	Berry	Pericarp
Pineapple	Sorosis	Entire inflorescence, outer portion of receptacle, bracts, and perianth
Quince	Pome	Thalamus

Fruit	Class	Edible part
Walnut	Drupe	Seed
Strawberry	Etaerio of achenes	Fleshy thalamus (receptacle)
Raspberry	Etaerio of drupelets	Pericarp
Sweet cherry	Drupe	Mesocarp
Mangosteen		Aril (*Garcinia mangostana*)
Maize, rice, wheat oats	Caryopsis	Starchy endosperm
Pomegranate	Special berry	Juicy outer coat of seed (aril, outer layer, and testa)

Some fruits eaten as vegetables are artichoke (flower bud), broccoli (inflorescence), and cucumber (mesocarp, endocarp, and placenta).

2.3.2 Flower and fruit morphology of some common species

The flowers and fruits described in this section are diagrammed in Figures 2.9–2.26. The abbreviations used in these figures are listed here.

A: androecium
ar: aril
b: bract
C: corolla
d: disc
e: embryo
en: endocarp
ep: epicarp
f: funiculus
fr: fruit
G: gynoecium
K: calyx
m: mesocarp
P: perianth
pc: pericarp
r: radicle
s: seed
st: staminodes
t: thalamus

***Mangifera indica* (mango)**
FAMILY: Anacardiaceae

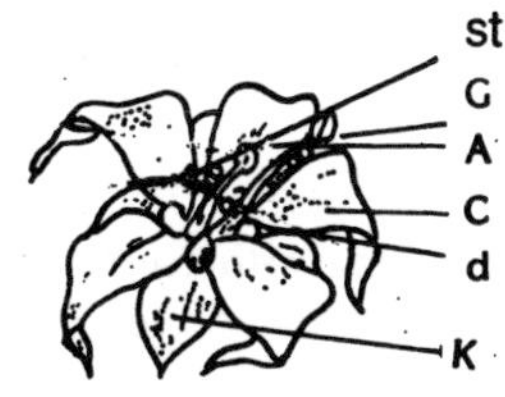

Figure 2.9 Mango flower

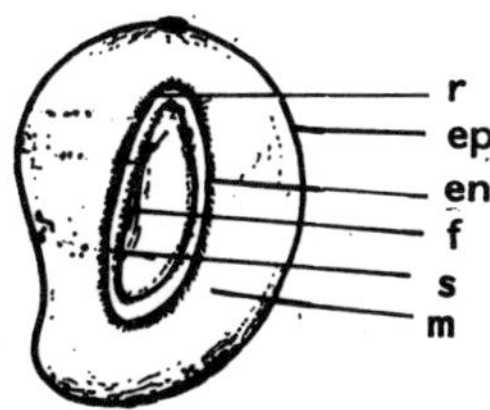

Figure 2.10 V.S. of mango fruit

MORPHOLOGY: Terminal panicles, erect or pendulous raceme arising vigorously with heavy lateral branching. Flowers small and polygamous, generally unisexual or bisexual (Fig. 2.9). Sepals 4–5, deciduous and pubescent. Petals 4–5, disc fleshy and ridged, imbricate, alternating with sepals. Stamens as many as petals (4–5), but only one longer stamen functional. Pistil monocarpellary, ovary unilocular and superior with one pendulous ovule. Fruit an oblong or ovoid drupe (Fig. 2.10).

***Psidium guajava* (amrud)**
FAMILY: Myrtaceae
MORPHOLOGY: Flowers solitary in the leaf axil or cymose, bisexual epigynous or perigynous (Fig. 2.11). Sepals 4–5, persistent. Petals 4–5, white, valvate. Stamens numerous, free with bilobate anthers. Gynoecium 2–5, syncarpous, ovary multilocular with many seeds in each chamber. Ovary may show incomplete union with the receptacle. Fruit is a berry, round to oblong, white to green on the outside.

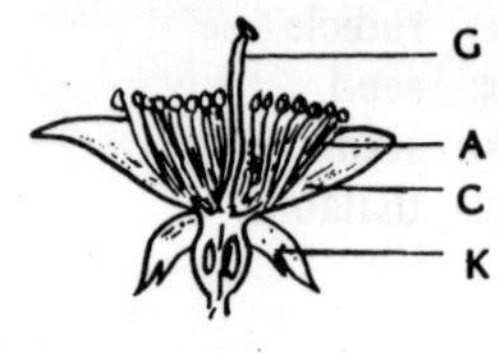

Figure 2.11 V.S. of guava flower

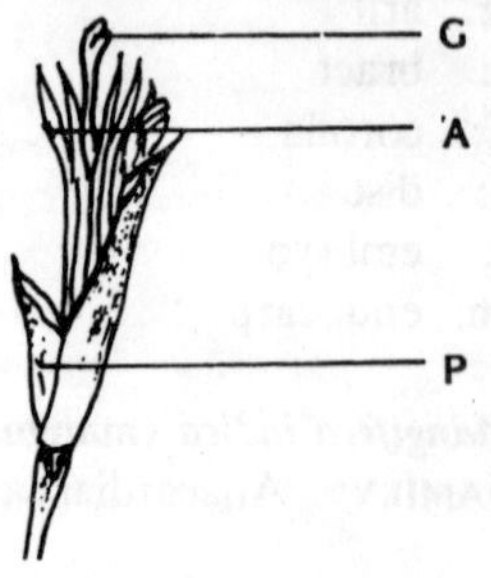

Figure 2.12 Banana floret

***Musa paradisiaca* (banana)**
FAMILY: Musaceae
MORPHOLOGY: Inflorescence terminal, erect or drooping with clustered flowers, spike approximately 1 m long, peduncle thick, bracts ovate, concave, fleshy and dark red, opening in succession. Perianth leaves ovate, acute, fleshy or pinkish (Fig. 2.12). Stamens 5, anthers 2–lobed and linear with one staminode. Pistil tricarpellary syncarpous; ovary inferior and trilocular with axile placentation; stigmas 3. Fruit elongated, oblong berry. Fruits are generally seedless; when seeded the seeds may be embedded in the perisperm.

Citrus aurantium **(orange)**
FAMILY: Rutaceae
MORPHOLOGY: The flower of the orange (*Citrus aurantium*) is an example of a flower which is complete, bisexual, hypogynous, and regular. The calyx consists of 5 united, greenish, and gland–dotted sepals. The corolla is made up of 5 free, gland–dotted white petals. The androecium comprises polyandrous stamens (filaments are united at the base in groups), and free and dorsifixed anthers. The pistil or gynoecium is polycarpellary, syncarpous and superior, situated on a large, nectar–secreting disc. The ovary has a capitate stigma, short style, and about 5 locules, each containing a few ovules (Fig. 2.13). The fruit is a berry (hesperidium) (Fig. 2.14).

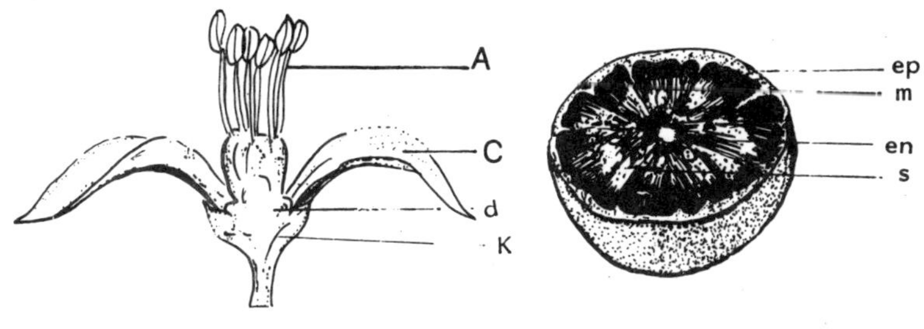

Figure 2.13 V.S. of citrus flower

Figure 2.14 T.S. of citrus fruit

***Malus sylvestris*, syn. *Pyrus malus* (apple)**
FAMILY: Rosaceae
MORPHOLOGY: Inflorescence simple umbel or grouped, cyme (Fig. 2.15). Flower: bisexual and epigynous. Sepals 4–5, united, broad–lobed, campanulate, reflexed and persistent. Petals 5, polypetalous, shortly clawed rosaceous, imbricate. Stamens 20 or more; filaments 2–5, free; stigma truncate; anthers in–curved. Carpels 4–5; ovary polycarpellary, syncarpous, epigynous; styles 4–5, multilocular with two ovules in each loculus at basal placentation. Fruit a pome (Fig. 2.16).

Prunus persica **(peach *aroo*)**
FAMILY: Rosaceae

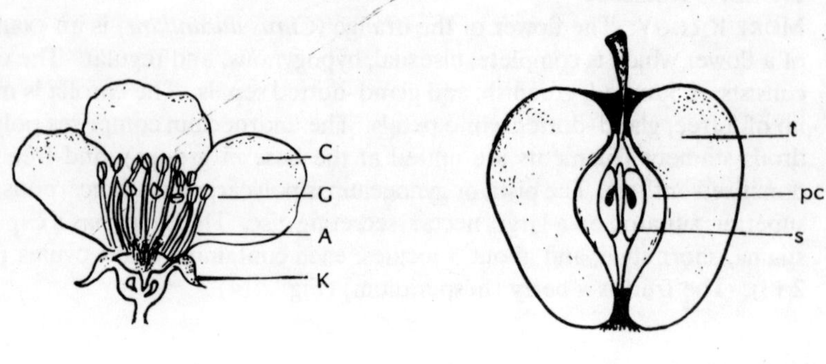

Figure 2.15 V.S. of apple flower

Figure 2.16 V.S. of apple fruit

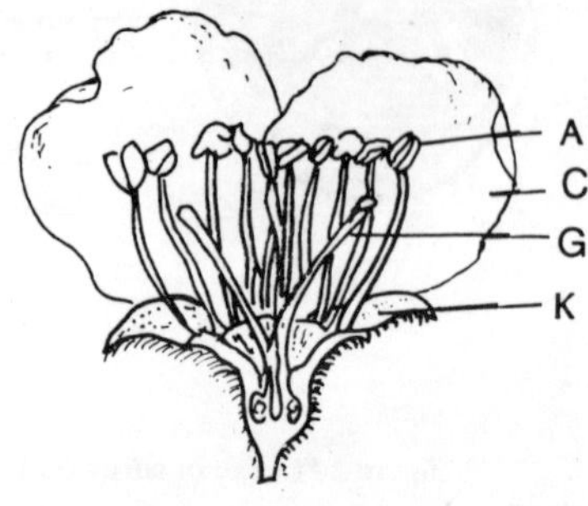

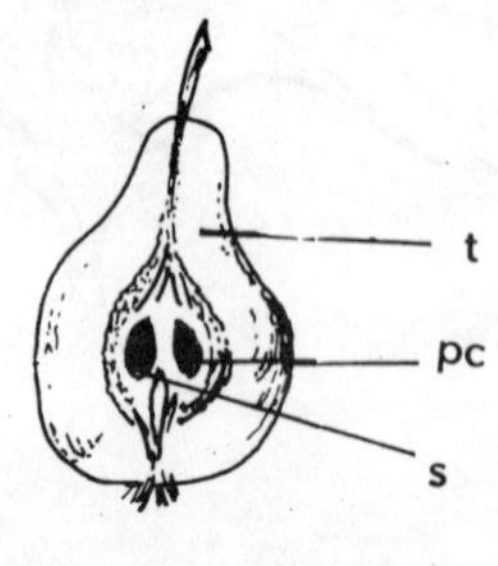

Figure 2.17 V.S. of peach flower

Figure 2.18 V.S. of pear fruit

MORPHOLOGY: Flowers pedicellate, ebracteate, complete, hermaphrodite, regular, perigynous, pink red (Fig. 2.17). Calyx 5, gamosepalous, imbricate. Corolla 5, polypetalous, imbricate. Androecium 15–60, polyandrous, anthers bithecous. Gynoecium polycarpellary, apocarpous ovary superior, unilocular with two pendulous ovules. Fruit a drupe enclosing a hard furrowed one-seeded seed (stone).

Pyrus communis **(pear)**

FAMILY: Rosaceae

MORPHOLOGY: Lower flowers bracteate and upper ones ebracteate, pedicellate, bracteoles on the pedicels, complete, hermaphrodite, actinomorphic, epigynous. Calyx 5, green, gamosepalous, persistent. Corolla 5, white, imbricate. Stamens numerous, polyandrous in three whorls, filaments of variable lengths. Pistils 5, syncarpous; ovary inferior, multilocular, two ovules in each loculus; placentation axile; styles 5, free; stigma terminal and glandular. Fruit a pome (Fig. 2.18).

***Phoenix dactylifera* (date)**

FAMILY: Palmae

MORPHOLOGY: Male and female inflorescences covered by a hard, boat–like membranous bract sheathing the branches. Female inflorescence: large, main stalk flat, glabrous, spikelets numerous. Flowers rounded, green, distant; perianth lobes 6, united at the base. Inner whorl (petal) large. Male inflorescence: small, flat, glabrous. Flowers sessile, white, sweet–scented, much larger than female flowers. Stamens 6 (3+3), anthers erect, filament short, subulate, pistillodes three, minute and scale–like. Fruit a cylindrical berry, seeds stony, acute at the apex, longitudinally grooved from one side (Fig. 2.19).

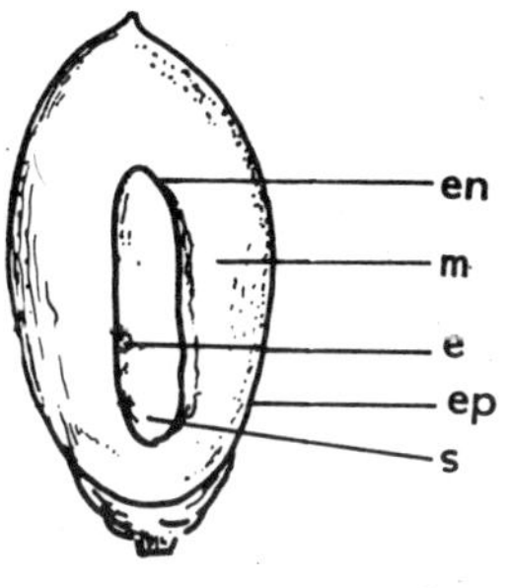

Figure 2.19 V.S. of date fruit

***Diospyros kaki* (Persimmon, 'Japani fruit')**

FAMILY: Ebenaceae

MORPHOLOGY: Inflorescence cauliflorous, dioecious, rarely bisexual or polygamous. Flowers highly variable 3–7–merous. Male cymes 3–flowered; female flowers solitary and large (Fig. 2.20). Sepal segments oval or broadly lanceolate, pubescent. Petals yellow–white, lobes rounded. Stamens numerous 16–24, hairy, staminodes 8–10 in female flower. Carpels 8–10; ovary multilocular; style 4–partite hairy. Fruit a globose, glossy and fleshy berry (Fig. 2.21).

***Punica granatum* (Pomegranate)**

FAMILY: Punicaceae

MORPHOLOGY: Flowers scarlet red or white (Fig. 2.22). Sepals 5, reddish, somewhat succulent, triangular. Petals 5, broadly ovate, wrinkled, alternate with sepal lobes. Stamens, numerous filaments, multiseriate, persistent. Pistil is polycarpellary, syncarpous; ovary subglobose; style thick, reddish; stigma simple, slightly bilobate; ovules many. Fruit a berry, globose pale–red to scarlet, partitioned by thin leathery yellow septa. Seeds red to pink, angular; testa thick, fleshy, and juicy (Fig. 2.23).

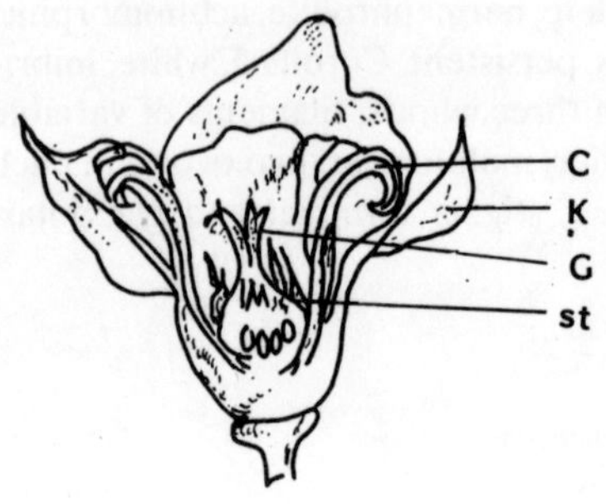

Figure 2.20 V.S. of persimmon flower

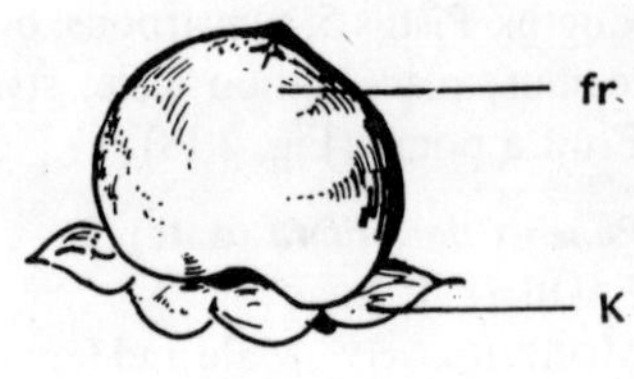

Figure 2.21 Persimmon fruit

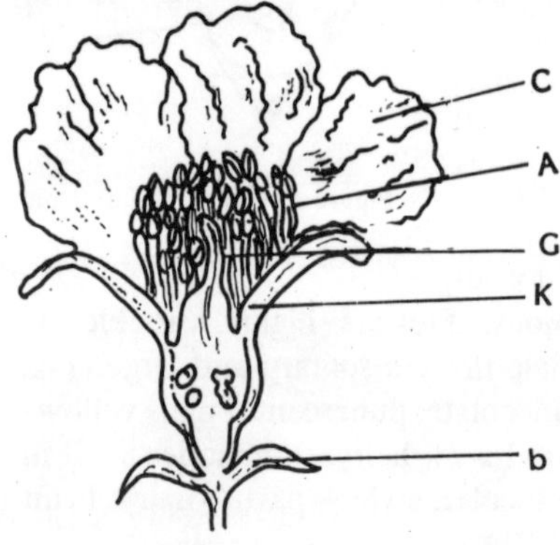

Figure 2.22 V.S. of pomegranate flower

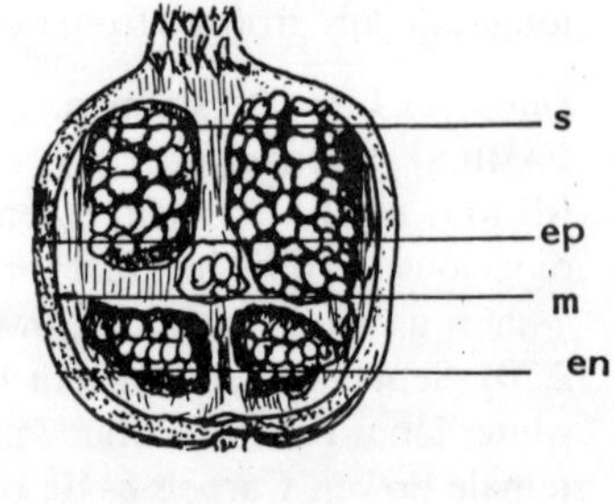

Figure 2.23 V.S. of pomegranate fruit

***Litchi chinensis* (litchi)**

FAMILY: Sapindaceae

MORPHOLOGY: Flowers small, unisexual, actinomorphic. Sepals 3–5 unequal. Petal 3–5 or absent, equal or unequal often with scales. Stamens 5–10 with few staminodes. Pistils 3 or more, syncarpous, trilocular; 1–2 ascending

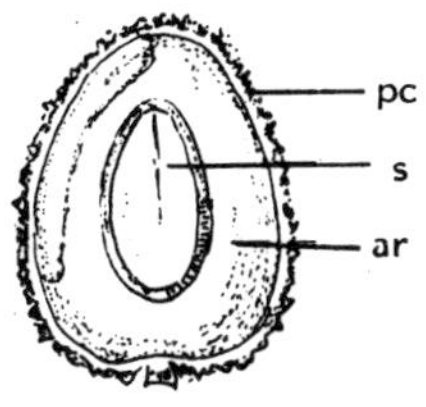

Figure 2.24 V.S. of litchi fruit

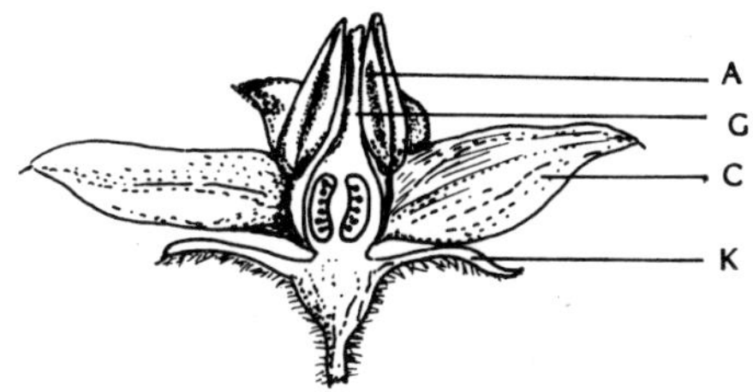

Figure 2.25 V.S. of potato flower

ovules in each loculus; style terminal. Fruit a 1–seeded nut with fleshy edible aril and leathery pericarp (Fig. 2.24).

***Solanum tuberosum* (potato)**

FAMILY: Solanaceae

MORPHOLOGY: Flowers ebracteate, pedicellate, complete, hermaphrodite, actinomorphic, hypogynous (Fig. 2.25). Calyx 5, gamosepalous, persistent. Corolla 5, gamopetalous, rotate. Androecium 5, polyandrous, epipetalous, alternating with petals; anthers coherent, dehiscing by apical pores. Gynoecium bicarpellary syncarpous, ovary obliquely placed, superior; several ovules in each loculus; placentation axile. Fruit a berry with many seeds.

***Lycopersicon esculentum* (tomato)**

FAMILY: Solanaceae

INFLORESCENCE: A few–flowered cyme. Calyx 5–partite, size enlarging in fruit, triangular, acuminate. Corolla 5, rotate, yellow, triangular–acute, minutely ciliate. Androecium 5, attached near the corolla throat; anthers elongated, adnate, dehiscing longitudinally. Filaments shorter than the anthers. Gynoecium is bicarpellary, syncarpous with superior ovary, bilocular, placentation axile, style single with two–lobed stigma. Fruit a berry, globose and depressed with many seeds.

***Brassica rapa* (turnip, *shalgam*)**

FAMILY: Cruciferae

MORPHOLOGY: 30–40 flowered corymbose raceme; flowers dull to golden yellow. Calyx 4, sub–equal, outer oblong obtuse. Corolla 4, alternating with sepals, free, cruciform. Androecium 6, tetradynamous, outer pair short and inner 4 stamens long, filament linear. Nectar glands often green. Ovary cylindrical. Fruit a silique, linear, beaked, septum (replum) membranous and veinless. Seeds 8–12 in each locule.

***Abelmoschus esculentus*, Syn. *Hibiscus esculentus* (okra, 'lady's finger')**

FAMILY: Malvaceae

MORPHOLOGY: Flowers axillary solitary, pedicellate. Epicalyx segments 7–12, linear to lanceolate, caducous. Sepals 5, adnate. Petals 5 spathaceous, caducous, yellow or yellowish–white with purple spot at the base. Androecium 5, or staminal column antheriferous all over the style. Gynoecium 5, stigma 5, ovary pubescent, 5–loculed. Capsule cylindrical, 5–angled, acuminate, densely hairy, usually hispid. Seeds ovoid to reniform, glabrous, black to dark brown coloured.

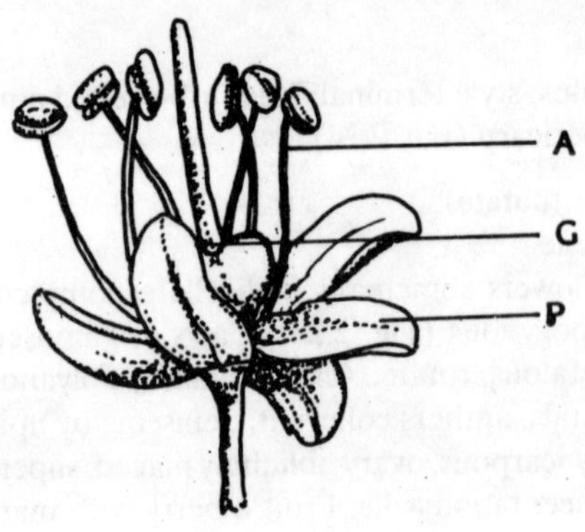

Figure 2.26 Onion flower

***Allium cepa* (onion)**
FAMILY: Alliaceae
MORPHOLOGY: Cymose head (umbel). Umbels spherical, densely flowered. Flowers stellate, pedicellate, bracteate, regular, complete, bisexual, hypogynous, white, perianth 6, 3 in each whorl, gamophyllous (Fig. 2.26). Androecium 6 (3 + 3) in two whorls, polyandrous; anthers long, 2–celled, dorsifixed. Gynoecium tricarpellary, syncarpous. Ovary trilocular, placentation axile with 2 ovules in each loculus, style short, stigma minute. Fruit a membranous capsule. Seeds small, endospermic.

STUDY QUESTIONS

1. What is morphology? Describe other fields of study related to it.
2. Define flower. Name the smallest and largest flowers among the angiosperms.

3. Specify the distinguishing features of floral and vegetative shoots.
4. Name the parts of a typical flower and briefly discuss their functions.
5. Give an account of the different modes of arrangement of floral leaves on the thalamus. What are the principal forms of corolla?
6. Write a note on gametogenesis.
7. Describe various types of ovules giving examples in each case.
8. What is placentation? Describe its various forms.
9. Describe the process of double fertilization in seed plants.
10. Name the parts of a pollen grain. How does a pollen grain germinate?
11. How does an embryo develop after fertilization?
12. List the developmental stages of an embryo.
13. Describe the stages of seed and fruit formation.
14. Distinguish between dry and succulent fruits.
15. What are the differences among simple, aggregate, and multiple fruits.
16. Classify the edible portions of common fruits.
17. Name some fruits eaten as vegetables, and classify their edible parts.
18. Describe the fruits of strawberry, cashew nut, walnut, pineapple, mulberry, fig, mango, citrus, date palm, guava, and pomegranate, giving the fruit class, and the part of the flower that develops into the fruit.
19. Describe the physical changes that take place during fruit development.
20. What chemical changes occur during fruit development?

REFERENCES

Ahmad, S. S. A textbook of intermediate botany. Lahore: Sheikh Sons Publishers, 42 Urdu Bazar; 1985.

Dutta, A. C. Class book of botany. London: Oxford University Press; 1959.

Lawrence, G. H. M. An introduction to plant taxonomy. New York: Macmillan; 1955.

Maheshwari, P. An introduction to the embryology of angiosperms. New York: McGraw–Hill; 1950.

Nasir, E.; Ali, S. I. Flora of West Pakistan. Karachi: Department of Botany, University of Karachi; 1972.

Nawaz, M. A text book of general botany, Vol. II. Angiosperms (I). 3, Kabir Street, Urdu Bazar, Lahore: Talimi Markaz Ltd.; 1980.

Rendle, A. B. The classification of flowering plants. Vol. I. Gymnosperms and monocotyledons. 2nd ed. Cambridge: Cambridge University Press; 1959/1971.

Rendle, A. B. The classification of flowering plants. Vol. 2. Dicotyledons. Cambridge: Cambridge University Press; 1959/1971.

Outline of Chapter 3

3. PLANT METABOLISM

Noor Badshah[1]

LEARNING OBJECTIVES

Photosynthesis and respiration are among the most important biological phenomena on earth. Photosynthesis is the process through which green plants fix CO_2 in the presence of light into useful organic compounds which are available for their growth, development, and maintenance. Respiration is the oxidative degradation of organic compounds which yields usable energy. Oxygen is essential for respiration since it is the terminal electron acceptor. During respiration, glucose is oxidized to form pyruvate (glycolysis), which through a series of reactions, is oxidized to CO_2 and water within the mitochondria.

After studying this chapter a student should be able to:

- Summarize the biochemical processes involved in photosynthesis and respiration
- Outline the steps in the metabolism of carbohydrates
- Discuss the photochemical aspects of photosynthesis and respiration of plants
- Explain the biological significance of photosynthesis and carbohydrate metabolism

3.1 Photosynthesis

3.1.1 Definition of photosynthesis

Photosynthesis is the process which converts light energy into useful chemical energy. It includes a series of reactions in which electromagnetic energy is converted into chemical free energy that can be used for biosynthesis (Kamen 1963).

Radiant energy is trapped by pigments of green plants and utilized to reduce atmospheric CO_2 to sugars, which in turn can be oxidized, releasing energy for growth, development, and maintenance of living organisms (Ting 1982). The overall reaction of photosynthesis is represented as follows:

---

[1]Associate Professor of Horticulture, NWFP Agricultural University, Peshawar, Pakistan.

$$6CO_2 + 12H_2O \xrightarrow[\text{chlorophyll}]{\text{light}} C_6H_{12}O_6 + 6O_2 + 6H_2O$$

The oxygen produced in photosynthesis and the electrons from hydrogen which are used to reduce CO_2 to sugar ($C_6CH_{12}O_6$) both come from water. Thus, 12 molecules of water, each with an oxygen atom, participate to yield six diatomic O_2 molecules. It is estimated that about 1.55×10^{11} tons of dry matter are produced annually by photosynthetic plants.

3.1.2 History of study of photosynthesis

Although organized agriculture existed in ancient times, the process of photosynthesis was not recognized until the 16th century. Early Greek farmers thought that plant material came from the soil, and the practice of adding plant and animal debris to soil was common. The first important plant physiologist was Stephen Hales, who in 1727 studied the effects of air and light on plant growth. He suggested that plants derived some useful substance from the air and that light was necessary for growth.

In 1771, Joseph Priestley implicated O_2 when he found that green plants could renew air made stale by the breathing of animals. However, he was unaware of the role of CO_2 or the importance of light. Antoine Lavoisier's studies of combustion showed that O_2 is taken up and CO_2 evolved. On the basis of the discoveries of Lavoisier, a Dutch scientist, Jan Ingen–Housz suggested that CO_2 in the air was the source of carbon for organic matter. Then in 1779, he recognized that light was necessary for the purification of air. In 1782, Jean Senebier demonstrated that CO_2 produced by animals and plants in darkness stimulated production of purified air (O_2) by plants in the light. N. T. de Saussure (in 1804) demonstrated the simultaneous uptake of CO_2 and evolution of O_2 by plants in the light. He found that plants gained in dry weight during photosynthesis. He also made the first quantitative measurement of photosynthesis and noted that approximately equal volumes of CO_2 and O_2 were exchanged during photosynthesis.

The development of the Law of Conservation of Energy by Robert Mayer in 1884 showed that the sun was the source of energy for plants and animals, and the energy converted to chemical energy by green plants was light energy. In the 20th century, Blackman discovered that photosynthesis consists of two reactions: a fast light–dependent reaction, and a slower process that occurs in the dark. In 1937, Robin Hill demonstrated that in the presence of light, water, and an artificial electron acceptor (potassium ferricyanide) chloroplasts evolve O_2. This reaction is called the Hill Reaction in honor of his discovery. The process is schematized as follows.

$$2H_2O \longrightarrow O_2 + 4H^+ + 4e^-$$

3.1.3 The primary processes of photosynthesis

The primary reactions of photosynthesis occur during a period of 10^{-15}–10^{-9} seconds. A photon is absorbed by the chlorophyll pigment, resulting in an excited pigment molecule. The excitation usually lasts for about 10^{-9}–10^{-5} seconds.

$$\text{CHL} + \text{light} \longrightarrow \text{CHL}^*$$

Energy in excited electrons of various pigments is transferred to an energy–collecting pigment, a reaction center.

The next reaction involves the conservation and stabilization of the energy of the excited chlorophyll molecule.

$$\text{CHL}^* + \text{X} \longrightarrow \text{CHL} + \text{X}^*$$

X^* is a compound which can reduce NADP.

$$\text{NADP} + \text{X}^* \longrightarrow \text{X} + \text{NADPH}$$

The whole series of reactions can be summarized as follows:

$$\text{NADP} + H_2O + \text{light } (\textit{chloroplast}) \rightarrow \text{NADPH} + H^+ + 1/2\ O_2$$

The reducing agent (NADPH) formed during the reaction is used to reduce CO_2 to glucose. All the above reactions take electrons (H^+) from water. The overall reaction is given as under:

$$6CO_2 + 12\text{NADPH} + 12H^+ \longrightarrow C_6H_{12}O_6 + 12\text{NADP} + 6H_2O$$

The organic compounds in the above reactions are converted into cellular constituents, i.e. cellulose, hemicellulose, starches, membrane, and other cytoplasmic substances.

Photosystem I, Photosystem II, and the Z–Scheme. The two light–gathering photo centers (at 680 and 700 nm) are integral parts of two systems that are referred to as photosystem I (PS–I) and photosystem II (PS–II). The absorption maximum for chlorophyll *a* in PS–I is about 700 nm and for PS–II about 680 nm.

Both systems have chlorophyll *a* and accessory pigments, including chlorophyll *b*, carotenoids, and some other pigments. Chlorophyll *a* absorbs farther into the red end of the spectrum than chlorophyll *b* (Fig. 3.1).

PS–I is more sensitive to far–red light than to red light and absorbs light at 700 nm as well as shorter wavelengths. The light–absorption center at 700 nm is called P700 (Fig. 3.2). It reduces NADP to NADPH and hence forms a strong reductant (Ting 1982). The light gathering center of PS–I (P700) has a special type of chlorophyll molecule, chlorophyll *a*, which has an absorption maximum at 700 nm. The center has about 200 chlorophyll molecules of which one (or two) is of the P700 type. PS–II also has 200

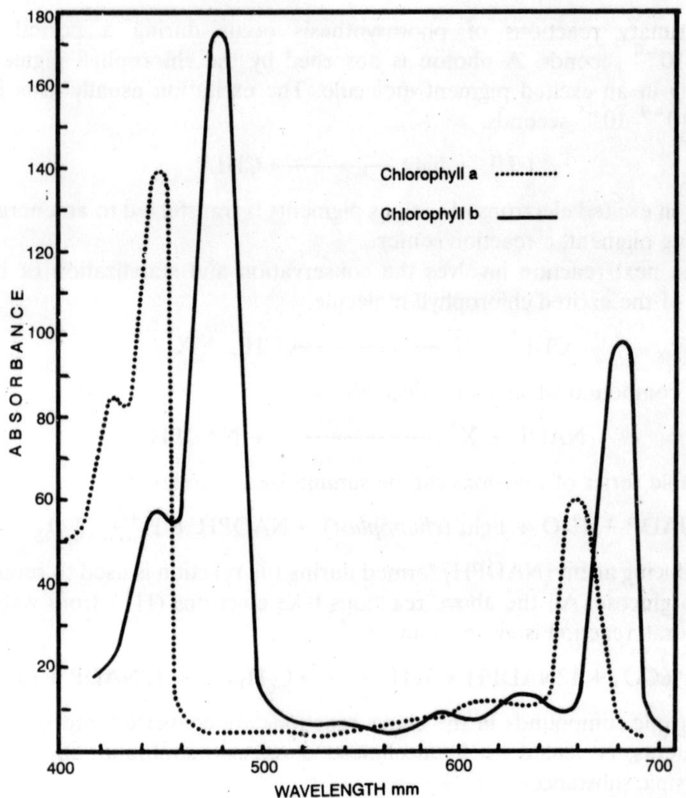

Figure 3.1 Absorption spectra for chlorophyll *a* and chlorophyll *b*. Redrawn from Zschiele and Comar (1941).

chlorophyll molecules. Transfer of energy in PS–II is by molecular collision. The strong reductant produced in PS–I reduces NADP to NADPH. The electron is transferred from water at PS–II.

PS–II has the light–gathering center that functions in the Hill reaction. The center contains chlorophyll *a* that absorbs light at 680 nm (P680). The center also has chlorophyll *b*, but chlorophyll *a* is the primary trap while chlorophyll *b* works as an accessory pigment. Here water is oxidized to form molecular oxygen as electrons are removed. PS–II forms a very strong oxidant and is more sensitive to red light than to far–red light (Fig 3.2).

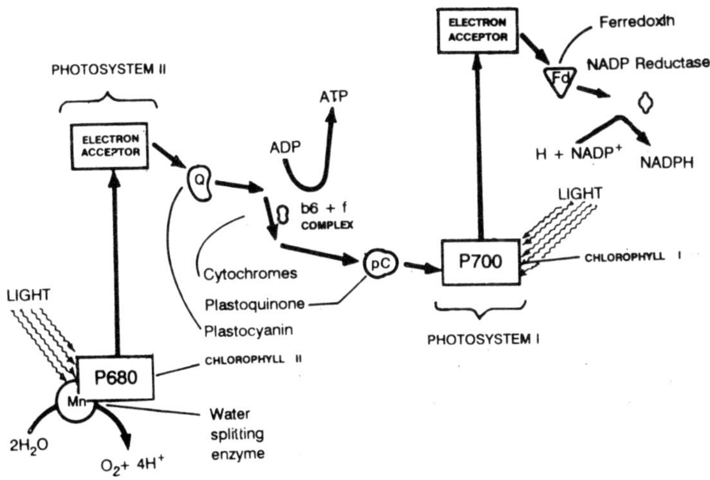

Figure 3.2 The Z-scheme for the photochemistry of photosynthesis, showing the two photosystems acting together to produce ATP and reduce NADP to NADPH. Drawing by Habib-ul-Rahman.

Electron–transport system in photosystems I and II. The two photosystems are coupled by a series of electron–transport components similar to those of the electron–transport system of mitochondria (Anderson 1982). Within the electron–transport system of the mitochondria there are flavoprotein, a quinone, and cytochromes of the a, b, and c types similar to those of chloroplast membranes. Also present are copper–containing proteins, plastocyanin, and nonheme iron protein such as ferredoxin. Ferredoxin or another iron–sulfur protein is believed to be the primary electron acceptor of PS–I.

Plastocyanin serves as an electron carrier and is loosely bound to the inside of thylakoid membranes (Salisbury and Ross 1985). When its copper becomes reduced from Cu^{++} to Cu^{+} by PS–I, it can move along the membranes carrying an electron to PS–II where it is reoxidized to Cu^{++} form. The PS–II center also has about 50 carotenoid molecules, about four plastoquinones, two cytochrome b, and six manganese. In addition to the known components, the center has an unidentified electron donor called Z (strong oxidant) and an unidentified electron acceptor called Q (the weak reductant). Q may be plastoquinone. Finally, PS–II contains one or more proteins, called the manganese protein. According to Ting (1982), chlorides also have an essential role in PS–II.

The electron–transport system that couples photosystems I and II can be diagrammed as below. There is a redox potential decrease from PS–II to

PS–I of about 0.43 volts, sufficient to produce ATP. This drop in potential as electrons flow from PS–II to PS–I brings about photophosphorylation.

H_2O → (P680) → Plastoquinone → Cytochrome f +/– Plastocyanin → (P700) → NADP

3.1.4 The photochemistry of photosynthesis

One of the most important features of the photochemistry of photosynthesis is the ultimate trapping of usable energy in the form of NADPH.

The overall process of photosynthesis when summarized is as follows:

i. CHL + Light + NADP + ADP + Pi ⟶ ATP + NADPH + CHL

ii. CO_2 + H_2O + NADPH + ATP ⟶ CH_2O + NADP + ADP + Pi

In the first reaction light is trapped, and in the presence of ADP and NADP the rich component ATP is produced. In the second reaction carbohydrate is produced. Incorporation of carbon into organic acids and sugars is discussed under the separate section on carbon metabolism.

The Emerson enhancement effect. The rate of photosynthesis decreases rapidly at wavelengths over 700 nm and increases at shorter wavelengths. However, if short wavelength light is provided at the same time as the longer red wavelengths, photosynthesis increases faster than could be expected when either wavelength is provided alone. This enhancement is named after Robert Emerson (1950's), who noticed that the quantum requirement for photosynthesis remained reasonably low below 680 nm but increased greatly above 680 nm. The fact that light of less than 680 nm and greater than 680 nm when given together result in synergism led to the suggestion that photosynthesis must include at least two light systems working together: one sensitive to red light, and the other to far–red light (Ting 1982).

Photophosphorylation. Photophosphorylation involves a series of electron–transport coupling factors (proteins) which transfer electrons along an electrochemical gradient, yielding ATP in the presence of light. In the presence of light, chloroplást membranes pump protons to the inside rather than to the outside. In this reaction the pH of the medium tends to increase and becomes alkaline as is the case with mitochondria during respiration. The pH gradient across the membranes of the chloroplast provides a powerful form of chemical potential energy responsible for driving photophosphorylation.

3.1.5 Chloroplasts

Photosynthetic cells have 15–20 or even more chloroplasts. **Chloroplasts** are plant organelles surrounded by two bilayered membranes. They arise from tiny **proplastids**, which divide actively as the embryo develops. When the chloroplasts are exposed to light, the enclosed membranes of the basic unit structure, the **thylakoids**, are stacked into arrays collectively called **grana**. When thylakoids extend from stack to stack to form intergranal lamella, they are called **frets**. The material surrounding the thylakoids and within which the grana and fretwork are imbedded is called **stroma**. It mostly consists of soluble protein, lipid drops, starch grains, amino acids, sugars, RNA, and DNA. Chloroplasts also have the capacity for protein synthesis.

3.1.6 Photosynthetic pigments

Photosynthetic pigments are some of the most important chemicals of the photosynthetic apparatus. A **pigment** is a molecule that absorbs light; the light absorbing portion of a pigment is called the **chromophore**. The pigments involved in photosynthesis are chlorophyll and other accessory pigments such as carotenoids. Other pigments such as flavoproteins, plastocyanin, the cytochromes, ferredoxin, and the quinones, which are also found in chloroplasts, are probably not directly involved in light gathering but rather in the electron–transport system. The pigments involved in the photochemical reactions are discussed below.

Chlorophyll. Chlorophyll is the green pigment present in thylakoid membranes. It consists largely of two kinds of green chlorophylls—*a* and *b*. The chlorophyll *a* molecule is the primary absorbing pigment of green plants, although chlorophyll *b* frequently occurs with chlorophyll *a*. The absorption spectra of chlorophyll *a* and *b* are shown in Fig. 3.1.

Chlorophylls are magnesium porphyrins. Because of the presence of the phytol chain ($C_{20}H_{39}OH$), they are hydrophobic molecules. As stated above with regard to the primary processes of photosynthesis, the primary light traps of photosynthesis have absorption spectra in the red (PS–I at about 700 nm and PS–II at about 680 nm). Because of the nature of the protein complexed with chlorophyll *a* molecules within the photosynthetic unit, chlorophyll *a* molecules may have absorption spectra shifted toward the blue, that is, have maxima at shorter wavelengths. These accessory chlorophylls absorb radiant energy and transfer it by excitational energy transfer to the chlorophylls that absorb at longer wavelengths. Ultimately, the energy is transferred to the primary trap, either P680 of PS–II or P700 of PS–I.

Carotenoids. Carotenoids are yellow to orange pigments which contain linear tetrapyroles and are found in the photosynthetic unit. They exist in the chloroplast envelope, giving it a yellow color. There are two kinds of carotenoids: the pure hydrocarbon carotenes and the oxygen–containing xanthophylls. Certain carotenoids (especially violaxanthin, a xanthophyll) also exist in the chloroplast envelope. Because of their association with chlorophyll in chloroplasts, they are assumed to be accessory pigments in the process of gathering light and transferring energy to chlorophyll. They are also important in protecting against photo–oxidation, as has been observed in mutants that lack carotenoids, where there is severe light–induced injury.

3.2 Respiration

Respiration is the controlled oxidation process by which all organisms ultimately obtain the energy stored in organic compounds. The energy is conserved in the form of ATP or other compounds which are used in growth, reproduction, and maintenance. Respiration is an active process of all living organisms absorbing O_2 and releasing CO_2. It is the oxidative degradation of organic compounds (carbohydrates, fats, organic acids, and proteins) to yield usable energy. Malic acid and citric acid are the predominant organic acids found in fruits and vegetables. They are metabolized later on to form other products during respiration. Oxygen is the terminal electron acceptor for the oxidation process. Glucose is oxidized to pyruvate in a series of reactions called glycolysis. Pyruvate is decarboxylated to form acetate, which is oxidized to CO_2, water, and energy within the mitochondria. The common respiration of glucose can be written as follows.

$$C_6H_{12}O_6 + 6O_2 \longrightarrow 6CO_2 + 6H_2O + \text{energy}$$

The complete combustion of glucose to CO_2 and water yields about 2900 kilojoules mole^{-1} of hexose. Much of the energy released during respiration is heat, which may stimulate metabolism and benefit growth of certain species, but usually is transferred to the atmosphere or soil with little consequence to the plant. Most importantly, energy is trapped in ATP and used later for many essential biochemical processes of growth, reproduction, and maintenance. As the breakdown of glucose proceeds, carbon–skeleton intermediates are provided for a large number of other essential plant products, such as amino acids for proteins, nucleotides for nucleic acids, and carbon precursors for porphyrin pigments (chlorophyll and cytochromes), fats, sterols, carotenoids, anthocyanins, and certain other aromatic compounds (Salisbury and Ross 1985).

3.2.1 The respiratory quotient

The ratio of CO_2 released from the living cells to O_2 uptake during respiration is called the **respiratory quotient** or RQ. It is an indication of the substrate being oxidized. The RQ's for different organic compounds obtained from their complete oxidation are given below.

1. Carbohydrates (hexose)

$$C_6H_{12}O_6 + 6O_2 \longrightarrow 6H_2O + 6CO_2$$

$$RQ \text{ of } CO_2/O_2 = 6/6 = 1.0$$

2. Fatty acid (oleic acid)

$$C_{18}H_{34}O_2 + 25.5O_2 \longrightarrow 18CO_2 + 17H_2O$$

The RQ for this reaction is 18/25.5 = 0.71

3. Organic acid

$$2C_4H_6O_4 + 7O_2 \longrightarrow 8CO_2 + 6H_2O \quad \text{(succinate)}$$

$$RQ = 8/7 = 1.14$$

$$C_6H_8O_7 + 4.5O_2 \longrightarrow 6CO_2 + 4H_2O \quad \text{(citrate)}$$

$$RQ = 6/4.5 = 1.33$$

The RQ is often very close to unity, which is an indication of carbohydrate respiration.

3.2.2 Reserve carbohydrates

Starch is the main reserve carbohydrate in tubers, seeds, and leaves of most plant species. Starch is stored as water-insoluble granules (grains) that consist of amylopectin (branched molecules) and amylose (unbranched molecules). Sucrose and other non-reducing sugars are converted after translocation into starches in different organs of plants. These starches are also the principal respiratory substrates for plant organs at certain stages of development. The potato tuber is an example where during sprouting starch is converted into hexose sugars which are then oxidized releasing energy for sprout growth. Parenchyma cells in stems, roots, and tubers commonly store starch: in perennial species, the starch stored during the growing season is kept as reserve during the winter months and then used in new growth the following spring. Starch is also present in the endosperm or cotyledon storage tissues of many seeds; most of this reserve starch is used in respiration during the germination of seeds. During respiration, glucose molecules derived from starch are oxidized to CO_2 and H_2O. The remaining glucose is converted into sucrose molecules and moved from the endosperm (scutellum) into the growing root and shoot, where some are totally oxidized and others are converted into cell-wall materials, proteins, and other substances needed for growth of the seedling.

3.2.3 The metabolism of starch

Starch is a glucose polymer, and on breakdown it forms glucose molecules. The metabolism of starch can be separated into hydrolysis and biosynthesis, which are catalyzed by different enzymes. Most steps in the degradation of starch to glucose can be catalyzed by three different enzymes, although still others are needed to complete the process. These three enzymes are alpha–amylase (α–amylase), beta–amylase (ß–amylase), and starch phosphorylase.

The amylases occur in reserve and storage tissues and function largely in starch degradation for sugar mobilization during growth. Phosphorylases are more closely associated with respiratory processes; they hydrolyze starch to form the highly active phosphorylated sugars that act as substrates for respiration. The breakdown of starch is not by incorporating water into the products as amylases do, but incorporating phosphate as shown below.

$$\text{Starch} + H_2PO_4^- \rightleftharpoons \text{Glucose–1–P}$$

The amylases hydrolyze starch to maltose.

$$\text{Starch} + nH_2O \longrightarrow \text{n Maltose}$$

Maltose is then further hydrolyzed to glucose units by the action of maltase.

$$\text{Maltose} + H_2O \longrightarrow \text{2 Glucose}$$

Sucrose, which is the main carbohydrate of transport to various receiving (sink) cells, must also be degraded to glucose and fructose before respiratory breakdown can continue. The major reaction of sucrose degradation is irreversible hydrolysis to free glucose and fructose by invertases.

$$\text{Sucrose} + H_2O \longrightarrow \text{Glucose} + \text{Fructose}$$

The ADP–glucose substrate can be formed from sucrose by two means, either directly by catalysis with sucrose synthetase or by glucose inversion by the enzyme invertase to form glucose and fructose followed by phosphorylation and ADP–glucose synthesis.

$$\text{Glucose} \xrightarrow{\text{ATP}} \text{Glucose–6–P}$$

$$\text{Glucose–6–P} \xrightarrow{\text{Hexokinase}} \text{Glucose–1–P}$$

$$\text{Glucose–1–P+ATP} \longrightarrow \text{ADP–glucose} + P_iP_i$$

The conversion of glucose, glucose–1–P, glucose–6–P or fructose to pyruvic acid during respiration is called **glycolysis**. Glycolysis is the first of three closely related phases of respiration and is followed by the Krebs cycle. The complete reactions of glycolysis are discussed in the subsequent section.

Glycolysis. Glycolysis is the formation of hexose sugars from reserve carbohydrates. The glycolytic process occurring within the cytosol, the liquid phase of the cell, begins with phosphorylated glucose, glucose–6–P. Stored starch and sucrose are hydrolyzed to glucose, which is the immediate precursor for respiration. Starches found in the amyloplasts of storage organs after translocation of sucrose or other non–reducing sugars are the principal substrates for respiration. The oxidation of glucose to form pyruvate (glycolysis) is illustrated in Fig. 3.3.

In the first series of reactions, glucose is phosphorylated to glucose–6–P, which is cleaved into two 3–carbon sugars: glyceraldehyde–3–P and dihydroxyacetone–P. All reactions are catalyzed by specific enzymes. In the subsequent steps, the 3–carbon sugars are converted into pyruvate. The overall reaction sequence (glucose–pyruvate) for the production of pyruvate from glucose is as follows.

$$\text{Glucose} + 2\text{ADP} + 2\text{H}_2\text{PO}_4 + 2\text{NAD}^+ \longrightarrow 2\text{pyruvate} + 2\text{ATP} + 2\text{NADH} + 2\text{H}^+ + 2\text{H}_2\text{O}$$

In the above reaction, 2 net moles of NADH and 2 moles of ATP are produced. When sugars are oxidized by respiratory processes, they are activated by phosphorylation. Phosphorylation of glucose to form glucose–6–P is catalyzed by the enzyme hexokinase (Fig. 3.3). The conversion of glucose–6–P into fructose–6–P requires the enzyme phosphoglucoisomerase, while phosphofructokinase is needed for the next reaction (fructose–6–P —— fructose–1,6 bisphosphate). Phosphofructokinase is an important regulatory enzyme of respiration.

In the subsequent reaction, fructose–1,6–bisphosphate is cleared by the enzyme aldolase into two triose–P compounds. The two compounds glyceraldehyde–3–P and dihydroxyacetone–P are in equilibrium through the enzyme phosphotriose isomerase, with about 97% dihydroxyacetone–P (Fig. 3.3). Oxidation of glyceraldehyde–3–P to 1,3–bisphosphate is catalyzed by the enzyme glyceraldehyde–3–P dehydrogenase. During the reaction NAD is reduced to NADH.

The next step in glycolysis is the phosphorylation of ADP. The 1,3–bisphosphoglycerate provides a substrate for the phosphorylation, while enzyme phosphoglycerate kinase catalyzes the reaction to produce 3–phosphoglycerate and ATP. The 3–phosphoglycerate is converted to 2–phosphoglycerate, catalyzed by phosphoglyceromutase.

Next, 2–phosphoglycerate is changed into phosphoenolpyruvate by dehydration. Phosphoenolpyruvate has a high standard energy change which yields ATP on phosphate hydrolysis. Pyruvate kinase is the enzyme which catalyzes the reaction. The complete reaction sequence of glycolysis is given in Fig. 3.3.

Functions of glycolysis. Glycolysis has several functions related to many basic life processes. These functions are as follows:

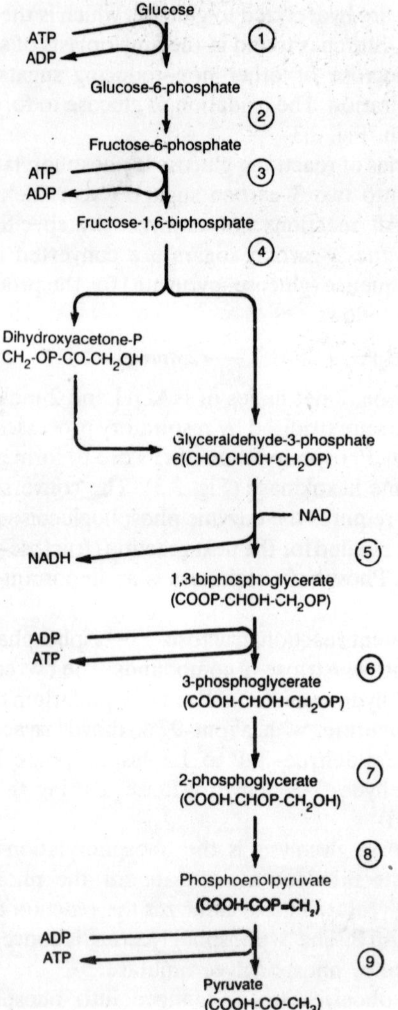

Figure 3.3 Glycolysis, the anaerobic oxidation of glucose to form pyruvate. Diagram by Habib-ul-Rahman.

1. Yielding of NADH: NADH molecules drive various anabolic and reductive processes. NADH is formed at only one step in glycolysis, during the oxidation of 3–phosphoglyceraldehyde to 1,3–bisphosphoglycerate.

2. ATP Formation: There is a total production of four ATP per hexose molecule, and a net production of two ATP per hexose. ATP is an energy–rich source necessary to drive several reactions.

3. Glycolysis provides several intermediate compounds from its pathway which are used to synthesize several other constituents for plant growth.

3.2.4 Fermentation

Oxygen is the terminal electron acceptor in aerobic respiration, but in fermentation oxidized organic compounds function as terminal electron acceptors. When O_2 is limited, NADH and pyruvate begin to accumulate. Anaerobic organisms or aerobic organisms under anaerobic conditions accumulate either lactate or ethanol. The process is called **fermentation**. There are two separate reactions involved in fermentation. In the first, there is decarboxylation of pyruvate to form acetaldehyde, then rapid reduction of acetaldehyde by NADH to form ethanol. In the second, pyruvate is reduced to lactate. The reaction is catalyzed by alcohol dehydrogenase, which uses NADH to reduce pyruvate to lactate. Alternatively, pyruvate can be reduced directly to lactate by lactate dehydrogenase. The products of fermentation are ethanol and lactate (Salisbury and Ross 1985). The reactions are schematized as follows.

$$\text{Pyruvate} \longrightarrow \text{Acetaldehyde} \longrightarrow \text{Ethanol}$$

$$\text{Pyruvate} \longrightarrow \text{Lactate}$$

3.2.5 The pentose phosphate pathway (PPP)

In many cellular biosyntheses, such as biosynthesis of fatty acids, NADPH is the reducing agent and not NADH. This alternate route for the oxidation of glucose to triose, pentose, ribulose, erythrose, and xylulose sugars is important for biosynthesis of other organic compounds and NADPH for cellular biosynthesis. Thus the pentose phosphate pathway (PPP) may also play a role in the production of reducing power for biosynthesis. Several compounds of the PPP are also members of the Calvin cycle, in which sugar phosphates are synthesized in the chloroplasts. Reactions of the PPP are outlined in Fig. 3.4.

The first reaction (Fig. 3.4) takes glucose–6–P as a substrate, forming phosphogluconate. Phosphogluconate is further oxidized to form ribulose 5–P. Oxidation of glucose–6–P also yields CO_2 and NADPH. Subsequently,

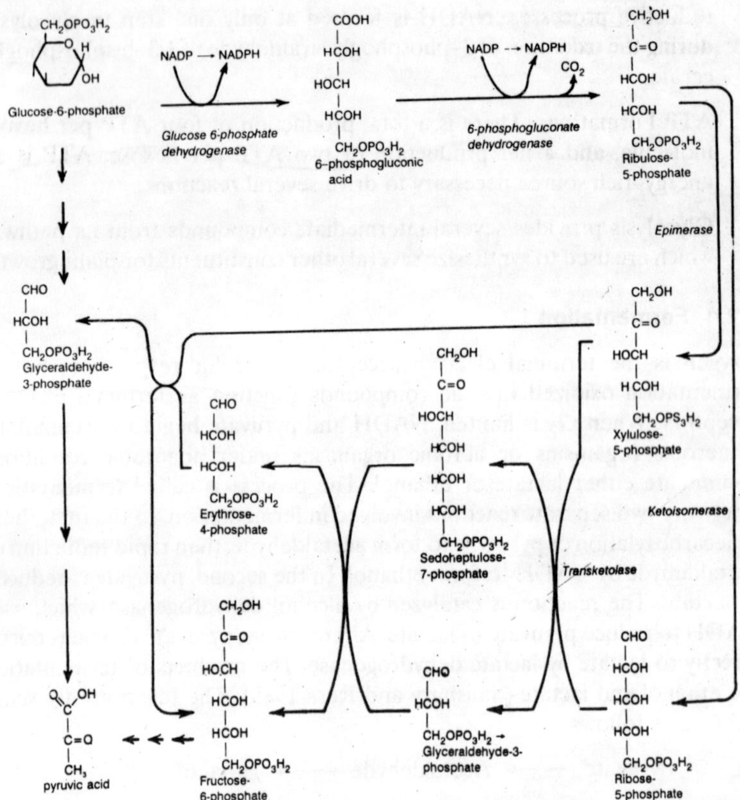

Figure 3.4 The pentose phosphate pathway. Diagram by Habib-ul-Rahman.

ribulose–5–P is isomerized by phosphoriboisomerase to ribose–5–P and then to xylulose–5–P. Next, two xylulose–5–P donate C_2 fragments to two ribose–5–P acceptors to form two moles of glyceraldehyde–3–P and two moles of sedoheptulose–7–P. Glyceraldehyde–3–P is also an intermediate of glycolysis.

Next the two sedoheptulose–7–P react with the two glyceraldehyde–3–P to form two moles of erythrose–4–P and two moles of fructose–6–P. As a result, the PPP can be considered an alternate route to compounds subsequently degraded by glycolysis. The two moles of erythrose–4–P then react with two additional xylulose–5–P formed from ribulose–5–P and two moles of fructose–6–P. The production of ribose–5–P is important, as it is the precursor of the ribose and deoxyribose units in nucleotides, including those in RNA and DNA.

Although the PPP could conserve energy in the form of ATP, an additional important metabolic role would be the production of NADPH for cellular biosynthesis. NADPH is the reducing agent which works in the biosynthesis of fatty acids. Clearly, the PPP is just as essential to plants as glycolysis and the Krebs cycle.

3.2.6 Mitochondria

It is helpful to understand some of the properties of mitochondria before we discuss how pyruvate and NADH are oxidized by mitochondria. Mitochondria, like chloroplasts, are double–membrane–bounded organelles found in plant cells (Fig. 3.5). The drawing in Figure 3.5 is a conceptual model of a mitochondrion showing the arrangement of the double membrane to form the outer–membrane and the inner–membrane cristae. The outer membrane, which is slightly thicker than the inner membrane, has pores. The inner membrane is frequently highly folded to form sheet–like or tube–like extensions called **cristae** (singular=*crista*). Cristae contain most of the enzymes that are associated with the electron transport system and oxidative phosphorylation. The protein–rich matrix within the mitochondria is the site where Krebs cycle reactions occur. Thus the mitochondrion is the subcellular

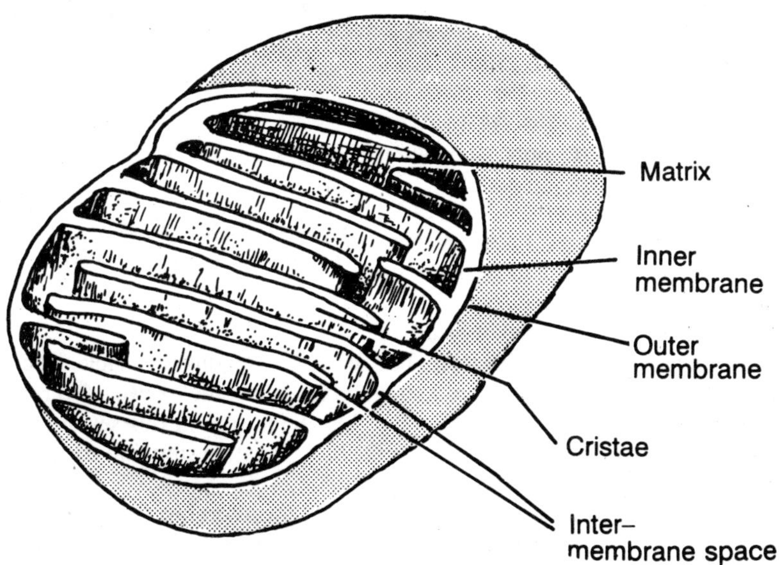

Figure 3.5 Structure of a plant-cell mitochondrion.

organelle where the aerobic respiratory process occurs that oxidizes acetate, generating the energy for ATP production.

3.2.7 The tricarboxylic acid (TCA) cycle—Krebs cycle

The complex type of enzymes (pyruvate dehydrogenase complex) catalyzes the decarboxylation of pyruvate to form acetate. It is acetate that is oxidized to CO_2 and water within the mitochondria. The decarboxylation of pyruvate to form acetyl CoA releases energy which is conserved in NADH. The oxidation of acetate to CO_2 and water through the reactions of the TCA cycle was first described by Hans Krebs in 1937. He called this series of reactions the citric acid cycle, because citric acid is an important intermediate. The TCA cycle is also called the Krebs cycle, after Hans Krebs.

The TCA cycle yields reduced NAD and FAD. These two compounds, in the form of NADH and $FADH_2$, ultimately transfer electrons to oxygen through the electron transport chain, conserving energy in the form of NADH or $FADH_2$. The total energy produced from one mole of glucose equals 36 ATP (Figs. 3.6 and 3.7).

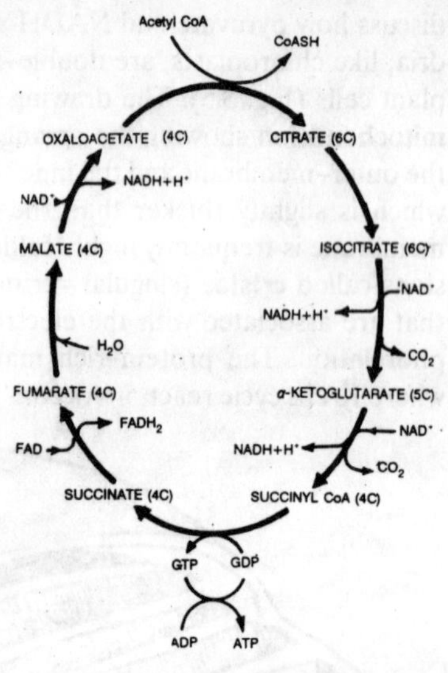

Figure 3.6 Tricarboxylic acid (TCA) cycle [Krebs cycle], which oxidizes acetate to CO_2 and water.

In the first step of the TCA cycle, active acetate (acetyl CoA) is condensed with oxalacetate to form citrate through catalysis by the enzyme citrate synthetase. Citrate is metabolized to isocitrate by the enzyme aconitase. The reaction is readily reversible. Isocitrate is then decarboxylated to α–ketoglutarate in a complex two–step reaction that generates reduced NAD (Fig. 3.6). Each step in this set of reactions requires specific enzymes. This is the first oxidation step that conserves energy in the form of NADH.

Next, α–ketoglutarate is decarboxylated to form succinyl CoA and CO_2. Succinyl CoA is converted to succinate; succinate is oxidized to fumarate by the enzyme succinate dehydrogenase; fumarate to malate by fumarase; and next malate is oxidized to oxalacetate (OAA) by the enzyme malate dehydrogenase. Oxalacetate again enters into the cycle shown in Fig. 3.6. Oxalacetate, the acceptor of acetate, initially formed citrate. The whole series of

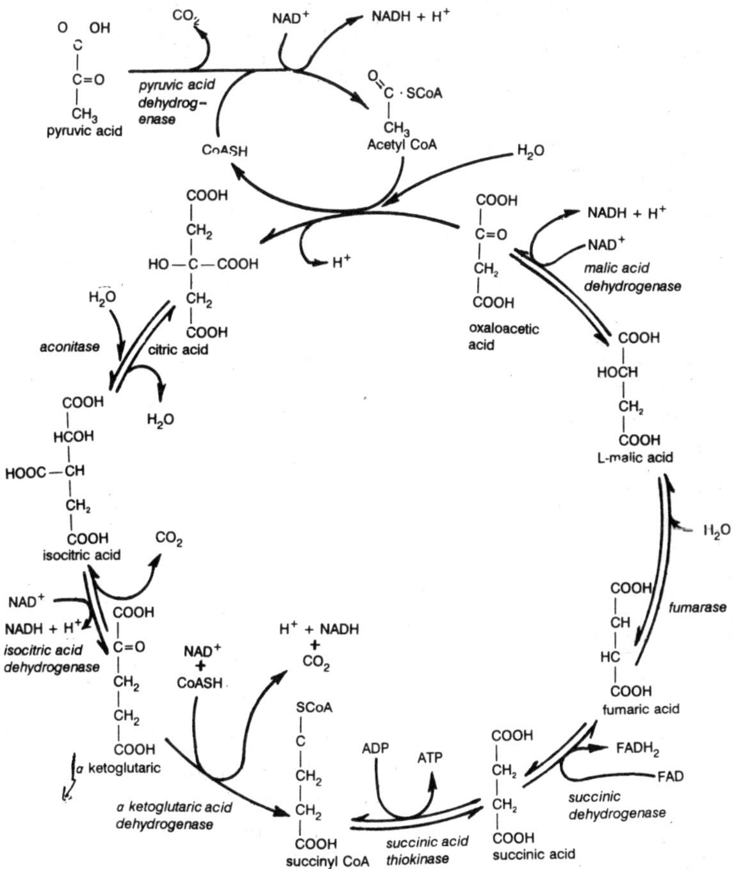

Figure 3.7 Tricarboxylic acid (TCA) or Krebs cycle, showing action of enzymes and co-enzymes. Adapted and re-drawn by Habib-ul-Rahman following Salisbury and Ross (1985).

reactions (four oxidation–reduction reactions) generates three moles of NADH, one mole of $FADH_2$, one mole of ATP, two moles of CO_2, and two moles of water (Figs. 3.6, 3.7). The release of CO_2 in the Krebs cycle accounts for the product CO_2, but no O_2 is absorbed during any Krebs cycle reaction.

Primary functions of TCA (Krebs) cycle. The important primary functions of the TCA cycle are:

1. Reduction of NAD^+ and FAD to NADH and $FADH_2$, which are subsequently oxidized to yield ATP.
2. Formation of carbon skeletons that can be used to synthesize certain amino acids which in turn are converted into large molecules (Fig. 3.7).

Thus α–ketoglutarate is converted into glutamic acid and other amino acids in the Krebs cycle. Other amino acids such as glutamine and aspartic acid are subsequently formed. Amino acids are the basic building blocks for protein synthesis. Thus the Krebs cycle is essential for growth processes, because it replenishes organic acids converted into large molecules and allows the cycle to continue (Salisbury and Ross 1985).

Factors affecting respiration. Several environmental factors influence the efficiency of respiration.

1. SUBSTRATE. Plants respire at the cost of substrate availability. Plants with more carbohydrate reserve will respire more than plants with less reserve, depending upon other environmental factors. Monosaccharides are the primary substrates for respiration.

2. OXYGEN. Oxygen is important for the electron transport system and the TCA cycle during respiration; without oxygen they cannot function. Furthermore, products of fermentation, anaerobic respiration which occurs in the absence of oxygen, accumulate and may become toxic.

3. TEMPERATURE. Respiration increases with increase in temperature. Q_{10} is a term used to denote the rate of respiration. This is the rate of increase in respiration with every increase of 10° in temperature.

4. AGE AND SPECIES. The respiration rates of different species of plants vary differently with age. There is also large variation in rates of respiration among different plant tissues or organs. Root tips and other regions having meristematic tissues have higher rates of respiration than other organs. Respiration remains high during the period of most rapid vegetative growth.

3.3 Carbon metabolism

Fixation of CO_2 by green plants and its incorporation into organic compounds was explained by Calvin in the 1950's. Calvin and his colleagues demonstrated how CO_2 is reduced to form sugar and other organic compounds. They showed that ribulose bisphosphate (RuBP) was the acceptor molecule for CO_2 that produced phosphoglycerate (PGA), and that after a series of reactions the ribulose bisphosphate was again regenerated and net phosphoglycerate was produced. The enzyme which catalyzes the reaction is ribulose biphosphate carboxylase.

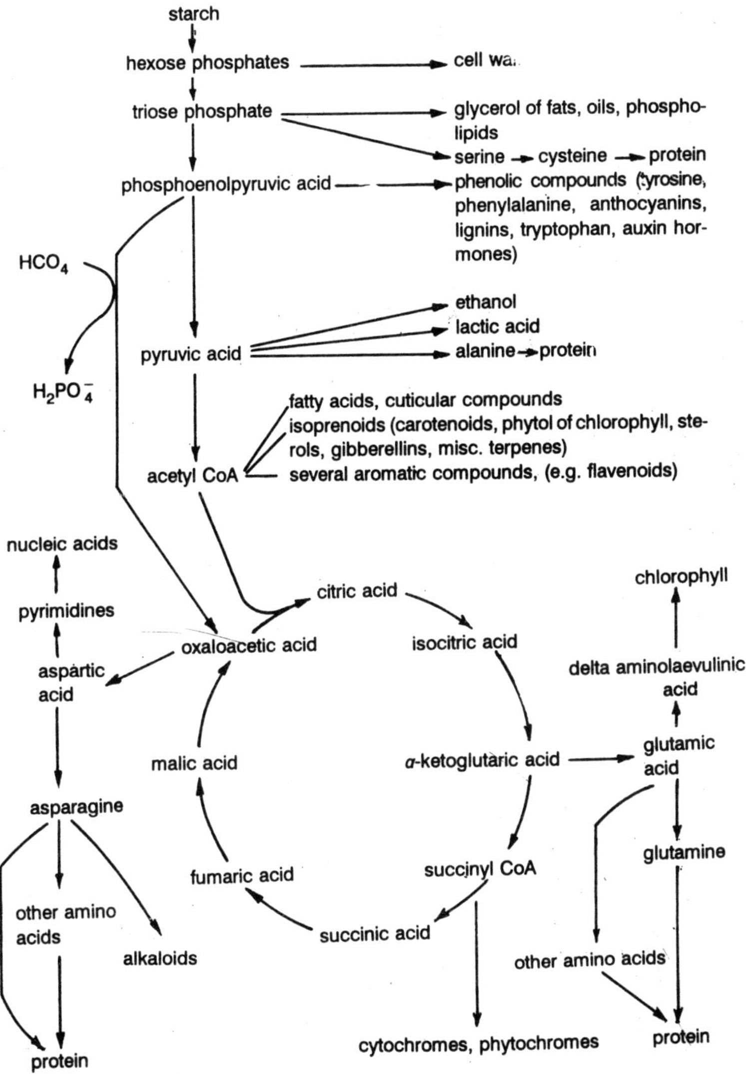

Figure 3.8 Glycolysis and TCA cycle schematized together to show their role in formation of essential compounds. Adapted by Habib-ul-Rahman from Salisbury and Ross (1985).

3.3.1 The first product of CO_2 fixation

Ribulose bisphosphate, which is a five–carbon compound, serves as an acceptor for CO_2, which yields two molecules of phosphoglycerate (a three–carbon compound) as illustrated below.

$$\begin{array}{l} CH_2OPO_3^{=} \\ | \\ C{=}O + CO_2 \\ | \\ CHOH \\ | \\ CH_2OPO_3^{=} \end{array} \longrightarrow \begin{array}{l} CO_2^{-} \\ | \\ CHOH + \\ | \\ CH_2OPO_3 \end{array} \quad \begin{array}{l} CO_2^{-} \\ | \\ CHOH \\ | \\ CH_2OPO_3 \end{array}$$

Ribulose–1,5–bisphosphate Phosphoglycerate

Next, phosphoglycerate is reduced to glyceraldehyde phosphate, the first sugar of photosynthesis. Glyceraldehyde–3–P dehydrogenase is the enzyme that catalyzes the reduction of the first product, phosphoglycerate, to the first sugar of photosynthesis, glyceraldehyde phosphate (Ting 1982). The reaction follows.

$$\begin{array}{l} CO_2^{-} \\ | \\ CHOH + ATP \\ | \\ CH_2OPO_3 \end{array} \longrightarrow \begin{array}{l} O \\ \| \\ C{-}O{-}PO_3^{=} \\ | \\ CHOH + ADP \\ | \\ CH_2OPO_3^{=} \end{array}$$

The basic reactions have been discussed under glycolysis, where glyceraldehyde phosphate is oxidized to phosphoglycerate. Ribulose bisphosphate is regenerated through a series of reactions for further CO_2 fixation. The important enzymes involved in all these reaction are RuBP carboxylase, glyceraldehyde–3–phosphate dehydrogenase, and RuBP kinase. The enzyme ribulose–5–P kinase catalyzes the biosynthesis of ribulose bisphosphate from ribulose–5–P and ATP, regenerating the five–carbon acceptor for photosynthetic carboxylations.

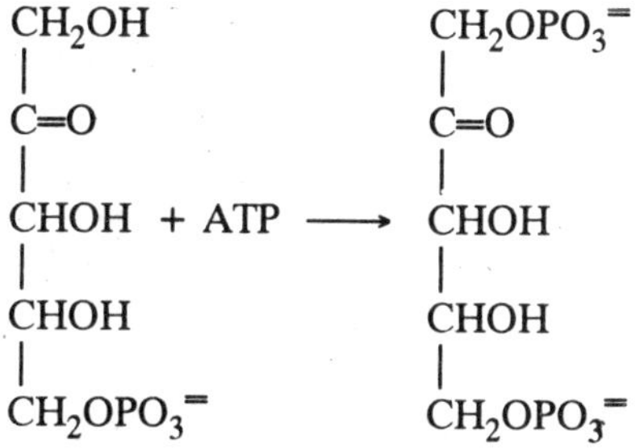

$$
\begin{array}{lll}
CH_2OH & & CH_2OPO_3^{=} \\
| & & | \\
C{=}O & & C{=}O \\
| & & | \\
CHOH \; + \; ATP & \longrightarrow & CHOH \\
| & & | \\
CHOH & & CHOH \\
| & & | \\
CH_2OPO_3^{=} & & CH_2OPO_3^{=} \\
\text{Ribulose–5–}PO_4 & & \text{Ribulose–1,5–P}
\end{array}
$$

This step is the second phosphorylation of the carbon assimilation reactions. Overall, the primary reactions of photosynthetic carbon metabolism are:

1. generation of ribulose bisphosphate, CO_2 acceptor,
2. carboxylation reactions, and
3. reduction of phosphoglycerate to glyceraldehyde phosphate (Ting 1982).

These reactions are summarized as follows:

1. Ribulose–5–P + ATP → Ribulose Bis–P + ADP
2. Ribulose Bis–P + CO_2 → 2 Phosphoglycerate
3. 2 Phosphoglycerate + 2 ATP + 2 NADPH →
   2 Glyceraldehyde–P + 2 NADP + 2 Pi

Three moles of ATP are required for the uptake of each mole of CO_2. Two moles of NADPH are also needed. Overall, the production of one net glyceraldehyde phosphate requires three moles of CO_2, nine moles of ATP, and six moles of NADPH (Ting 1982; Salisbury and Ross 1985). These reactions were discovered to work in a cyclical pathway that uses 3–PGA to form other sugar phosphates and that also converts some of its carbon back to RuBP. The recycling of glyceraldehyde–P back to ribulose through a series of reactions is known as the Calvin–Benson cycle or Calvin cycle. Calvin was awarded a Nobel Prize in 1961 for his work. The Calvin cycle occurs in the stroma of chloroplasts and has three main parts: carboxylation, reduction, and regeneration, as explained above.

3.3.2 The Calvin cycle and the reductive pentose cycle

Calvin cycle, reductive pentose cycle, or **photosynthetic carbon-reduction cycle** are terms used to refer to all the reactions involved in CO_2 fixation and sugar-compound formation. After the reduction of PGA, several sugar phosphates containing four, five, six and seven carbon atoms are formed. These include tetrose (4 carbon) phosphate, erythrose–4–phosphate, pentose phosphate, ribose–5–phosphate, ribulose–5–phosphate, xylulose–5–phosphate, fructose–6–phosphate, fructose–1, 6–bisphate, glucose–6–phosphate, and sedoheptulose–7–phosphate. The whole series of reactions has been discussed in the preceding section. The reduction cycle is diagrammed in Fig. 3.9.

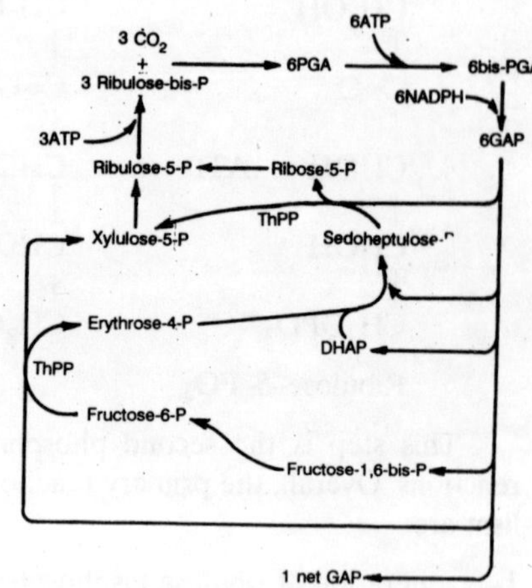

Figure 3.9 The reductive pentose cycle, as outlined by Melvin Calvin and his associates.

STUDY EXERCISES

1. Explain the initial reactions involved in photosynthesis.
2. What are the important factors influencing photosynthesis?
3. How does photosynthesis differ from respiration in regard to sugar metabolism?
4. Identify the organic compounds that fix CO_2 and predict the first formed sugars.
5. Explain the relationship of respiration to photosynthesis.
6. Which organic acids are formed during respiration? Explain the process of their formation.
7. What is the overall significance of respiration?
8. How are different amino acids formed during the metabolic processes?

REFERENCES

Anderson, J. M. Distribution of the cytochromes of spinach chloroplast between the appressed membranes of grana stacks and stroma–exposed thylakoid regions. FEBS Letters 138:62–66; 1982.

Arnon, D. I. Photosynthetic activity of isolated chloroplasts. Physiol. Rev. 47:317–358; 1967.

Avron, M.; Neumann, J. Photophosphorylation in chloroplasts. Ann. Rev. Plant Physiol. 19:137–166; 1968.

Barber, J. ed. Primary processes of photosynthesis. Amsterdam: Elsevier; 1977.

Bassham, J. A. Kinetic studies of the photosynthetic carbon reduction cycle. Ann. Rev. Plant Physiol. 15:101–120; 1964.

Bearden, A. J.; Malkin, R. Primary photochemical reactions in chloroplast photosynthesis. Quart. Rev. Biophys. 7:131–177; 1975.

Beevers, H. Respiratory metabolism in plants. Evanston, Illinois: Row, Peterson; 1961.

Bjorkman, O. Comparative studies on photosynthesis in higher plants. In: Giese, A. C. ed. Photophysiology. New York: Academic Press; 1973.

Black, C. C. Photosynthetic carbon fixation in relation to net CO_2 uptake. Ann. Rev. Plant Physiol. 24:253–286; 1973.

Black, C. C.; Burris, R. H. eds. CO_2 metabolism and plant productivity. Baltimore: University Park Press; 1976.

Calvin, M.; Bassham, J. A. The photosynthetis of carbon compounds. New York: W.A. Benjamin; 1962.

Chang, J. H. Corn yield in relation to photoperiod, night temperature, and solar radiation. Agric. Met. 24:253–262; 1981.

Davies, D. D. The central role of phosphoenolpyruvate in plant metabolism. Ann. Rev. Plant Physiol. 30:131–158; 1979.

Downton, W. J. S. The occurrence of C_4–photosynthesis among plants. Photosynthe. 9:98–105; 1975.

Dunn, G. A model for starch breakdown in higher plants. Phytochem. 13:1341–1346; 1974.

Erectinska, M.; Wilson, D. V. Topical review. Regulation of cellular energy metabolism. J. Membrane Biol. 70:1–14; 1982.

Evans, J. R. Nitrogen and photosynthesis in the flag leaf of wheat (*Triticum aestivum L.*). Pl. Physiol. 72:297–302; 1983.

Fork, D. C.; Amesz, J. Action spectra and energy transfer in photosynthesis. Ann. Rev. Plant Physiol. 20:305–328; 1969.

Gaastra, P. Photosynthesis of crop plants as influenced by light, carbon dioxide, temperature and stomatal diffusion resistance. Meded. Landbhogesch. Wagenigen 59:1–69; 1959.

Gaffron, H. Energy storage: photosynthesis. In: Steward, F. C. ed. Plant physiology: a treatise. Vol. IB. Photosynthesis and chemosynthesis. New York: Academic Press; 1960.

Gifford, R. M.; Evans, L. T. Photosynthesis, carbon partitioning, and yield. Ann. Rev. Plant Physiol. 32:485–509; 1981.

Givan, C. V.; Harwood, J. L. Biosynthesis of small molecules in chloroplast of higher plants. Biol. R. 51:365–406; 1976.

Govindgee; Govindgee, R. The primary events of photosynthesis. Scient. Am. 231:64–82; 1974.

Gregory, R.P.F. Biochemistry of photosynthesis. 2nd ed. London: Wiley; 1977.

Halmer, P.; Bewley, J. D. Control by external and internal factors over the mobilization of reserve carbohydrates in higher plants. In: Loewus, F. A.; Tanner, W. eds. Plant Carbohydrates I: Intracellular Carbohydrates. Encyclopedia of Plant Physiology, New Series, Vol. 13A. Berlin: Springer–Verlag; 1982.

Hatch, M. D.; Osmond, C. B.; Slatyer, R. O. eds. Photosynthesis and photorespiration. New York: Wiley–Interscience; 1971.

Hatch, M. D.; Slack, C. R. Photosynthetic CO_2–fixation pathways. Ann. Rev. Plant Physiol. 21:141–162; 1970.

Inada, K. Action spectra for photosynthesis in higher plants. Plant Cel. P. 17:355–365;1976.

Kahl, G. Metabolism in plant storage tissue slices. Bot. Rev. 40:263–314; 1974.

Kamen, M. D. Primary processes in photosynthesis. New York: Academic Press; 1963.

Kelly, G. J.; Latzko, E.; Gibbs, M. Regulatory aspects of photosynthetic carbon metabolism. Ann. Rev. Plant Physiol. 27:181–205; 1976.

Kirk, J. T. O. Chloroplast structure and function. Ann. Rev. Biochem. 40:161–196; 1971.

Kluge, M.; Ting, I. P. Crassulacean acid metabolism: analysis of an ecological adaptation. New York: Springer–Verlag; 1978.

Kok, B. Photosynthesis: the path of energy. In: Bonner, J.F,; Varner, J. E. eds. Plant biochemistry. 3rd ed. New York: Academic Press; 1976.

Kortschak, H. P.; Hart, C. E.; Burr, G. O. Carbon dioxide fixation in sugar cane leaves. Plant Physiol. 40:209–213; 1965.

Leopold, A. C.; Kriedemann, P. E. Plant growth and development. 2nd ed. New York: McGraw–Hill; 1975.

Lieberman, M.; Baker, J. E. Respiratory electron transport. Ann. Rev. Plant Physiol. 16:343–382; 1965.

Long, S. P. C_4 photosynthesis at low temperatures. (Commissioned review). Plant, Cell and Environment 6:345–363; 1983.

Marcelle, R. Environmental and biological control of photosynthesis. The Hague, Netherlands: W. Junk Pub.; 1975.

Mooney, H. A. The carbon balance of plants. Ann. Rev. Ecol. 33:72–86; 1972.

Preiss, J. Regulation of the biosynthesis and degradation of starch. Ann. Rev. Plant Physiol. 33:431–454; 1982.

Preiss, J. Starch, sucrose biosynthesis and partition of carbon in plants are regulated by orthophosphate and triose–phosphates. Trends Biochem. Sci. 9:24–27; 1984.

Rabinowitch, E. Photosynthesis, Vols. I, IIA, IIB. New York: Academic Press; 1948–1954.

Radmer, R.; Kok, B. Energy capture in photosynthesis: photosystem II. Ann. Rev. Biochem. 44:409–433; 1975.

Salisbury, F.B.; Ross, C.W. Plant physiology. 3rd ed. Belmont, CA: Wadsworth; 1985.

Szarek, S.; Ting, I. P. The occurrence of crassulacean acid metabolism in plants. Photosynthe. 11:330–342; 1977.

Ting, P. I. Plant physiology. Menlo Park, CA: Addison–Wesley; 1982.

Trebst, A.; Avron, M. eds. Photosynthesis I: Photosynthetic electron transport and phosphorylation. Encyclopedia of plant physiology, New Series Vol. 5. New York: Springer–Verlag; 1977.

Zschiele, F. P.; Comar, C. L. Influence of preparative procedure on the purity of chlorophyll components as shown by absorption spectra. Bot. Gaz. 102:463–481; 1941.

Outline of Chapter 4

4. PHASES OF PLANT GROWTH

Abdul Fatah Baloch[1]

LEARNING OBJECTIVES

After studying this chapter, the student should be able to:

- Name the different ontogenetic stages of plant life
- Understand the process of gametogenesis
- Explain the process of fertilization and development of the plant embryo
- Recount the growth cycle of a fruit and assess its importance to horticulture
- Evaluate the natural and synthetic growth inhibitors and their mode of action on growth
- Understand the kinds of pollination and their importance in horticulture
- Assess growth behavior of plants and factors affecting them
- Differentiate between annual, biennial, and perennial plants
- Differentiate between the vegetative and reproductive phases of plants
- Discuss the concepts of maturity, ripening, and senescence

4.1 Vegetative growth and development

Growth is an obvious phenomenon of life. In spite of this it is not easy to give a specific definition of it. After sowing, a kernel of wheat becomes a plant within a few weeks, and the seed of the gum yielding tree babul (*Acacia nilotica*) becomes a big tree after a few years. In both cases growth has taken place and there is increase in size. Increase in size, however, does not always imply growth. You must have observed that after rain the wooden doors and windows of the house absorb water and increase in size. It then becomes difficult to close them, but when the weather becomes dry the

[1]Associate Professor of Horticulture, Sindh Agricultural University, Tandojam, Sindh.

water absorbed by the wood evaporates and the doors again become normal in size. There was increase in size, but it was not permanent.

Growth is the product of metabolic processes that take place in the plant. When conditions are favourable, anabolic (constructive) processes are dominant over catabolic (destructive) ones, with the result that there is an overall increase in size. The growth of living plants is actually the permanent deposition of material which is utilized by the plant to construct new tissues. When conditions are unfavourable, and the rate of catabolic processes is greater than that of anabolic ones, then growth will cease and the plant will ultimately die.

In short, growth is a vital process that brings about permanent and irreversible increase in size and weight. Some plants exhibit growth which is rapid but restricted to certain parts of the plant only, e.g. the tendrils and some fruits of some members of the Cucurbitaceae family and certain fruits such as okra or lady's finger (*Abelmoschus esculentus*), a member of the Malvaceae family.

4.1.1 Factors affecting growth

Adequate water is essential to maintain the turgidity of growing cells, which is basic to growth. Protoplasm is able to work properly only when it is saturated with water. Water is absorbed by the plant roots from the soil and supplied to the leaves through the xylem for the process of photosynthesis. A small amount is utilized by the plants for different processes including growth, and the rest is lost during transpiration.

Oxygen is indispensable for the process of respiration of all living plant tissues, particularly the growing parts. Respiration is a process of oxidation that releases energy which is utilized for the many activities carried out by living tissues. Growth takes place only when the tissues of growing parts are supplied with nutrients such as carbohydrates, proteins, vitamins, hormones, mineral salts, and micronutrients.

During the initial stages of growth, light is not very essential. Actually, the process of growth takes place more rapidly at night or during darkness than in light. In fact, in some cases light has an inhibiting effect on growth, for example the germination of seeds. Plants require a certain rate of light intensity and wavelength to remain healthy and develop normal sturdy stems and green leaves. The process of photosynthesis takes place during the light, and continuous absence of light is harmful to plants. Plants grown in darkness or in dim light become weak, with yellow or pale green leaves, tall lanky stems, and a sickly appearance; they rarely produce flowers and fruits. Plants showing the above symptoms are said to be **etiolated**.

For optimum growth, plants require soil suited to their individual characteristics. The texture of the soil is one of its most essential properties.

Loam consists of about 40% silt, 40% sand, and 20% clay. The capacity of soil to retain and distribute water depends upon the arrangement of various soil particles. Well-aerated soil is suitable for plant growth because the roots of the plant get sufficient oxygen for respiration. Soils having a low rate of gaseous diffusion hamper the respiration of roots, which has an adverse affect on plant growth. Loam is considered to be the ideal for the growth of beneficial microbes and plants. It has the capacity to retain enough water, and permits free movement of water without preventing aeration.

4.1.2 The seed

The seed is one of the most important products of the plant; it maintains the continuity of the species. The seed performs the essential functions of:

- REPRODUCTION. Normally, most flowering plants reproduce through seeds, except those that are vegetatively propagated.
- CONTAINER OF EMBRYO. The seed is a container in which the embryo develops and reaches maturity.
- PROTECTION OF EMBRYO. The embryo remains within the seed and is protected from excess cold, rain, heat, and other natural calamities.
- STORAGE OF FOOD. The food material for the growing embryo is stored in the cotyledons or in the endosperm, especially in monocotyledonous plants.

Seeds are the most essential source of food for human beings, animals, and birds. The members of the family Gramineae (Poaceae) provide cereals such as wheat, rice, maize or corn, sorghum, oats, and barley. The second most important family in this regard is the Leguminosae (Fabaceae). Its members provide protein-containing pulse seeds such as gram, beans, lentils, peas, pigeon peas, mung, and mash. Industrial oil is obtained mostly from castor and linseed. Cotton seed provides oil as well as livestock feed.

The seed of almond (*Prunus dulcis*) is ovate-shaped and is enveloped by a brown-coloured seed coat and hard endocarp. The seed coat consists of two layers fused together. The brown, comparatively thick, outer layer is called the **testa**, and the thin, whitish, inner one the **tegmen**. On removing the seed coat from the soaked seed, the white fleshy embryo is exposed. It is made up of two fleshy cotyledons and the shoot–root axis, to which the two cotyledons are attached. The part of the axis directed towards the pointed end is the **radicle** (young root), and the portion lying within the two cotyledons is the **plumule** (young shoot) (Fig. 4.1).

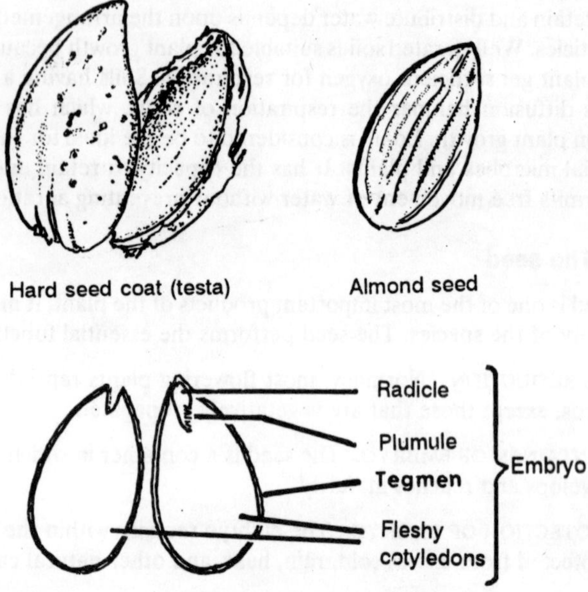

Figure 4.1 Parts of almond seed

4.1.3 Germination

When a seed is provided with moisture, proper temperature, and air (oxygen), the dormant embryo (young plant) starts growing. This process is called **germination.** There are several types of germination, but here we are considering only **epigeal and hypogeal germination.**

Epigeal germination. The embryo of the sown seed, for instance a melon (*Cucumis melo*), absorbs moisture and swells. Pressure develops from inside, due to which the seed coat ruptures and the radicle (young root) first emerges out of the seed coat through the **micropyle** and goes down into the soil. It becomes the primary root and soon gives out lateral roots and fixes the young seedling in the soil. The portion of the axis below the cotyledons is known as the **hypocotyl**; this grows relatively fast and pushes the cotyledons *above* the surface of the soil, hence the term *epigeal* germination. Sometimes the ruptured seed coat is also carried along with the cotyledons. When the cotyledons are exposed to sunlight or artificial light, they become green and through photosynthesis manufacture food material for the young seedling. Meanwhile, the plumule elongates and gives out lateral leaves. This type of germination is observed in muskmelons, gourds, luffa, beans, and

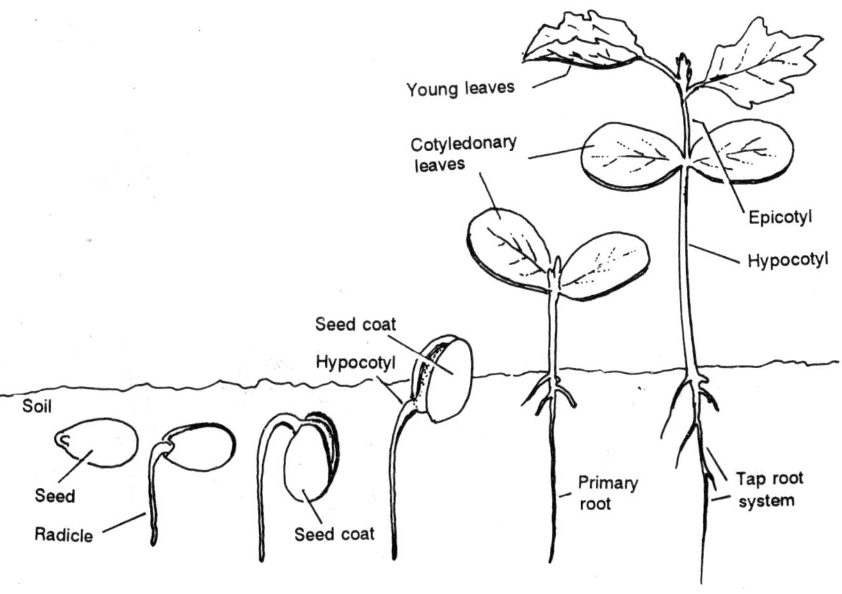

Figure 4.2 Epigeal germination

cucumbers (Fig. 4.2).

Hypogeal germination. A fairly fresh mango seed (*Mangifera indica*), presoaked over night, is sown in the soil. The huge embryo imbibes moisture, and the cells of the embryo become turgid. The pressure exerted by this turgidity splits the hard endocarp, which opens at the pointed end. First the radicle comes out and goes towards the force of gravity in the soil. It establishes itself in the soil by sending out small hairs and lateral roots to absorb water and dissolved mineral salts. The **epicotyl**, the part of the axis situated above the cotyledons, grows rapidly and carries the plumule above the surface of the soil. The large cotyledons which contain the reserve food material do not come above the surface of the soil; they remain *underground* (hence the term *hypogeal* germination), providing nutrition to the growing tissue, and ultimately disintegrate. The **plumule** (young shoot) first gives out brownish leaves which turn green when they are exposed to sunlight and start manufacturing food material like ordinary green leaves. In this way the young seedling is established. Hypogeal germination is also found in litchi, groundnut, peas, gram, and jackfruit (Fig. 4.3).

To observe the stages of germination of seeds, they can be covered with

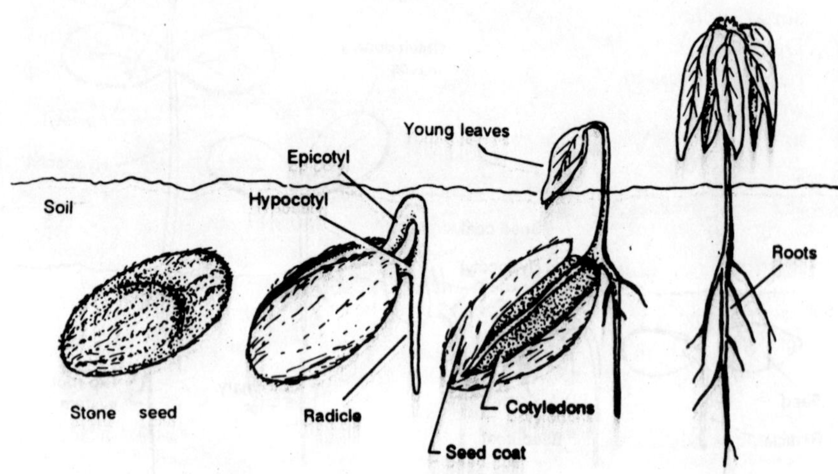

Figure 4.3 Hypogeal germination

a 2–3 cm thick layer of moist sand, sawdust, or loamy soil. The thickness of the layer depends upon the size of the seeds; small seeds need a thinner layer than large ones. Presoaking seeds overnight helps germination.

4.1.4 Vegetative growth patterns

Herbaceous plants are small plants with soft stems. Generally, they are classified according to their life span as: (1) annuals, (2) biennials, and (3) perennials.

Annuals are those plants which germinate, attain their vegetative growth in one season or survive for a few months or a year, giving out flowers and bearing fruits and seeds within that period. Almost all seasonal flowers, cereals, and most vegetable and edible oil seed plants are annuals.

Biennials are also herbaceous plants; they complete their vegetative growth in the first year. In temperate climates they produce flowers, fruits, and seeds in the second year, but in tropical climates they complete their life cycle in one year like annuals. Carrot, radish, turnip, cabbage, and beet are common biennials of Pakistan.

Perennials survive for many years. They can be classified into two categories: (1) herbaceous perennials, and (2) woody perennials. Under favourable conditions, **herbaceous perennials** give out aerial shoots from dormant underground stems like rhizomes, corms, bulbs, and tubers. The aerial shoots, like annual plants, bear flowers and fruits containing seeds. During unfavourable seasons, the aerial parts die but underground parts persist and remain alive. When the season again becomes favourable, they will repeat the same pattern of growth. Ginger, turmeric, arrowroot, crocus, and saffron tulip are examples of such perennials.

Woody perennials may be shrubs or trees. Shrubs are woody, smaller than trees, and bushy in habit. They give out several equally strong branches without a definite single trunk. They survive for many years. Usually they are grown as hedges, climbers, and ornamental plants, e.g. rose, shoeflower, night jasmine, crotons, duranta, and lantana.

Perennial trees are bigger than shrubs and most have a single trunk which bears lateral branches in various patterns. Perennial species, therefore, adopt different shapes. Some grow tall and become cone shaped like casuarina, pinus, cedar, and eucalyptus. Others form a canopy–like crown, like mango, banyan, and peepal.

4.2 Reproductive phase

4.2.1 Floral development

The reproductive phase usually starts when the plant reaches maturity. Most of the apical meristems on the branches cease to initiate leaves and start producing floral parts in accordance with the traits of the species. The formation of flowers is usually the final event during the activity of the apical meristem. In annual plants, the beginning of the reproductive phase is an indication of the completion of the life cycle. In perennial plants, flowering is repeated each season. Flowers may occur at the apex of the main central shoot, at the lateral branches, or on both in the form of solitary blooms or an inflorescence. Flowering does not continue indefinitely, however; the plant gradually, season by season, flowers, produces fruit, and ultimately reaches senescence. Senescence varies in different plants and depends upon environmental factors and the longevity of particular species.

Initiation of flower buds in temperate–region species. The beginning of flower buds in temperate–region fruit species such as apples, peaches, pears, cherries, or plums starts between late spring and late summer during the year preceding the bloom. Flowering starts after the new shoots have reached a certain diameter and a good number of their leaves have matured. The fruit yield depends on the number of buds modified from vegetative to reproductive forms. This state depends on the health of the plant and the nutrition provided it. Plants that are pruned, supplied with nitrogen fertilizers, and heavily watered will generally produce long, fleshy vegetative shoots. Buds appearing on the shoots do not form flower buds for the following year. On the other hand, weak thin shoots, especially on older trees that have low vigour, produce few flower buds, particularly when the number of leaves is reduced by insects or diseases. Serious and lingering drought during the optimum period of bud initiation in the beginning of summer can produce water shortage in the trees, which hampers flower bud formation. A small water deficiency is not harmful; rather, it enhances flower initiation, decreases shoot elongation, and causes carbohydrate storage. It is quite clear that the care fruit plants receive greatly affects their yield.

Once the flower parts are fully formed in the buds of deciduous fruit trees, the buds enter a physiological rest period. During that period they will not open even if the plants are provided with favourable temperature, moisture, and light conditions. The same is true for vegetative buds on the same tree which do not differentiate into flower buds.

The initiation of this rest period varies with species and depends upon the health of the plant. The beginning of the rest period ranges from midsummer to late autumn. The physiological rest condition probably begins due to the storage of natural growth retardants like abscisic acid and the loss of growth–stimulating agents like gibberellin. Chilling of vegetative and flower buds by freezing temperatures in winter is required for the reversion of the rest effect. When the chilling period is prolonged, bud formation ceases. The advent of warm spring temperatures, and sufficient soil moisture stimulate the vegetative and flower buds, which grow with vigour.

Initiation of flower buds in subtropical species. In some subtropical fruits like citrus species and olives, flower bud induction and development are quite different.

The olive tree (*Olea europaea*) survives in parts of the world where there is a lingering growing season and minimum winter temperatures higher than −9°C. Olive varieties generally thrive well where there are chilling temperatures in winter, and produce a good number of flowers and fruits. Some freezing days during late winter directly influence the vegetative growing points so that they are modified into flower buds. Without freezing temperatures, almost all the buds remain vegetative. Because of this plant habit, olive trees are not grown in warm parts of the world.

In the citrus species grown in subtropical environments, flower bud initiation starts in mid–winter and clear evidence of flower parts appears about one month later. In regions where there is a cool summer and moderate winter, some varieties of citrus bear flowers throughout the year, but the most prolific blooming is in spring. In tropical regions close to the equator, citrus flower buds initiate and bloom throughout the year. This applies to citrus fruit such as oranges, lemons, and pomelos. For these fruits, freezing temperature during winter is not required to break bud dormancy.

Initiation of flower buds in some tropical fruit species. The beginning of flowering in mango (*Mangifera indica*) appears to depend on environmental factors that retard growth, like a lengthy dry or cool period. Hormones produced in the leaves are transported to the shoot to initiate flowering in the buds. This begins in late winter or early spring. The initiation of flowers in pineapple plants (*Ananas comosa*) can be brought about by applying Ethephon to the growing points, which liberates ethylene. In this case, induction of flowering is stimulated by the hormone. The initiation of flowering in coffee trees (*Coffea arabica*) begins with the appearance of the flower primordia. The flowers bloom almost one month later. The period of flowering depends upon the environmental conditions prevailing at that time. After a long dry period, flowering begins within one month after the start of the rains. With sufficient soil moisture and favourable growing temperatures, flowering and fruiting take place intermittently throughout the year.

Flower morphology. The flower is the reproductive phase of the plant which is responsible for the production of fruit and seeds. Parts of the flower generally include **calyx** (sepals), **corolla** (petals), **androecium** (stamens), and **gynoecium** or pistil (carpels). The morphology of several flowers is presented in Chapter 2.

Gametogenesis. Gametogenesis is the formation of male and female gametes in the flower. The process of gametogenesis varies significantly among major classes of plants. It comprises two processes: (a) **microsporogenesis**, and (b) **megasporogenesis**.

In angiosperms, **microsporogenesis** takes place in the anthers of flowers. In the beginning some central cells of the pollen chamber or pollen sac are activated; these cells are called **microsporocytes** or **pollen grain mother cells** and are diploid (2n). The nucleus of each microsporocyte enlarges, and meiosis or reduction division takes place. Two nuclei are formed, each of which has half the number of chromosomes of the original cell [haploid (n)]. Immediately after this the two newly–formed nuclei divide mitotically and form four nuclei, but this time the number of chromosomes is not reduced; each has a haploid (n) number of chromosomes. A cleavage of the cytoplasm develops separating the four nuclei into four clear compartments or pollen cells. The original wall of the pollen mother cell degenerates. Each pollen cell gets nourishment from surrounding **tapetum** cells and develops a thick outer wall called the **exine**, and thin inner wall known as the **intine**. Meanwhile, the nucleus of each pollen grain again divides mitotically and two nuclei are formed. One of these is called the **vegetative nucleus** or **tube nucleus**, and other is known as the **generative nucleus**. In this form, the mature pollen grains are released from the anther when dehiscence takes place. The formation of pollen grains or microspores in the above manner is called microsporogenesis (Fig. 4.4).

The type of development which takes place in the female part of the ovule (ovary) to give rise to female reproductive cells within the ovule is known as **megasporogenesis**. Inside the ovule at the early phase of the development of the nucleus, one of the cells of the nucellus becomes active and larger than the other cells. It is called the **megasporocyte** or **embryo sac mother cell**, and is diploid (2n). Meiosis takes place and it divides into two cells each containing a haploid number of chromosomes (n). The two newly–formed cells now divide mitotically, i.e. their chromosome number remains haploid (n). In this way a vertical row of four haploid cells called **megaspores** is formed. After this, the upper three disintegrate, but the lowest one survives and becomes the embryo sac. Its nucleus divides three times mitotically, and eight haploid nuclei are formed. Three of these migrate towards the chalazal end and are known as **antipodal nuclei** or cells. Another three go towards the micropylar end. Of these three, the central one becomes the **ovum** or egg or female gamete, and the two lateral ones are called **synergids**. The remaining two nuclei fuse in the middle of the embryo sac and form the polar nuclei, fuse, and form a diploid nucleus (2n) (See Fig. 4.5).

4.2.2 Pollination

The transfer of pollen grains from the anther to the stigma by a pollinating agent is called **pollination**. There are two main mechanisms of pollination: (a) **autogamy or self–pollination**, and (b) **allogamy or cross–pollination**.

Autogamy or self–pollination. Pollination occurring within a hermaphrodite or bisexual or perfect flower, or between two such flowers situated on the same plant of the same variety or cultivar, is known as **self–pollination**.

Allogamy or cross–pollination. The transference of pollen grains by pollinators such as bees, butterflies, moths, wind, water, squirrels, birds, bats and animals, or even humans from the anther of one flower to the stigma of another flower situated on a different plant of the same species, or on cultivars of related species is called **cross–pollination**. Two parents are involved in this process, and two different haploid (n) sets of chromosomes are brought together. In this process the combination of two sets of parental traits or characters is possible. The production of hybrid offspring is thus possible and probable. In comparison to inbreds, hybrids are healthier and higher yielding; therefore horticulturists use this method to get new varieties or cultivars of fruits, flowers, and vegetables.

Insect–pollinated plants generally have showy, coloured flowers, scent, or nectar, e.g., queen of the night, jasmine, rangoon creeper, poinsettia, larkspur, salvia, snapdragon, and garden nasturtium. Wind–pollinated plants usually do not have showy flowers and do not possess scent. The anthers

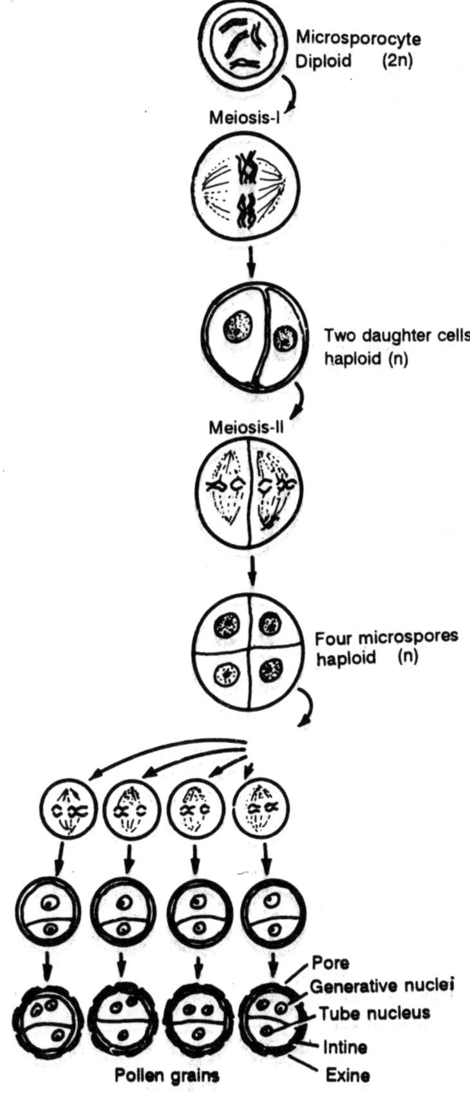

Figure 4.4 Gametogenesis in angiosperms — microsporogenesis

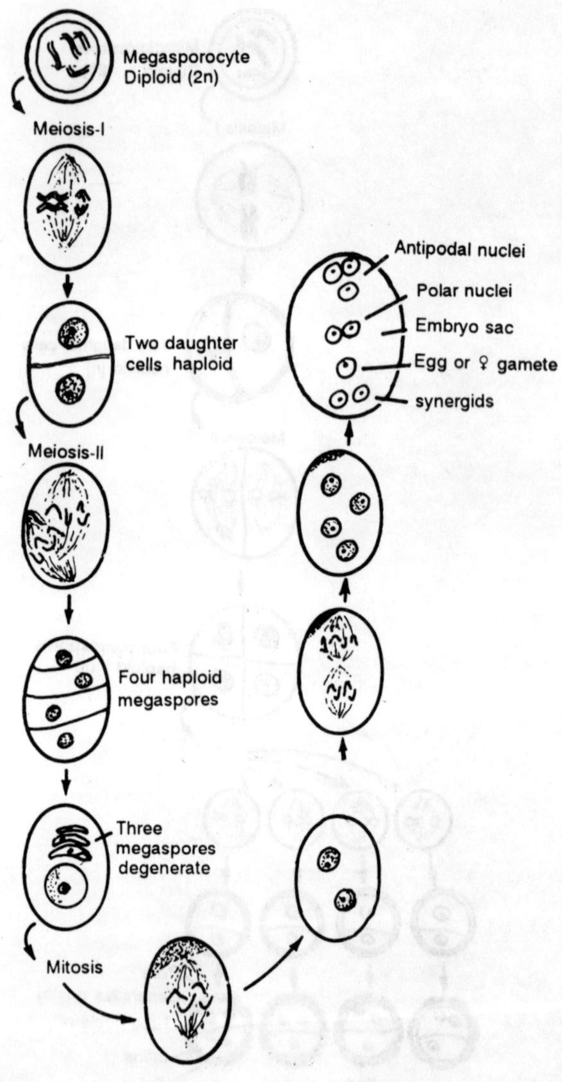

Figure 4.5 Gametogenesis in angiosperms — megasporogenesis

produce pollen grains in great quantity. These pollen grains are light in weight and dry so that they are carried easily by the wind to great distances. Papaya, maize or corn, date palm, and coconut are wind–pollinated species. Water–pollinated plants generally have large flowers and leaves which float on the surface of the water. The female stigma receives floating pollen grains, as in the water lily and giant water lily, while animal and bird–pollinated plants usually have abundant nectar. Bats, squirrels, and birds visit flowers for nectar, and pollen grains stick to their bodies. When they go from one flower to another and from one plant to another, pollen grains are carried from one plant to another and cross–pollination takes place. The silk cotton tree, coral tree, and rose apple are pollinated by animals.

Advantages and disadvantages of self and cross–pollination. Self–pollination has the advantage that it takes place in bisexual flowers, where both male and female parts mature at the same time and there is great possibility of fertilization and fruit formation. There are a number of suggested advantages of cross–pollination. (i) Hybrids possess greater adaptability to the environment. (ii) They have better germinating ability. (iii) Offspring produced are more capable of surviving. (iv) Variation is increased. Two disadvantages are: (i) In self–pollinating species the vigour of the offspring may decrease generation by generation. (ii) There is a possibility that pollination may fail to occur.

4.2.3 Fertilization and development

Fertilization. Fertilization is the union of male and female gametes (sperm and ovum). After the transfer of pollen grains to the receptive stigma, they are invigorated, and the inner wall (intine) of each starts growing through one of the germ spores and forms a pollen tube. The receptive stigma secretes a viscous sugary substance that accelerates the growth rate of the pollen tube inside the style. It carries the tube nucleus and the generative nucleus. The vegetative or tube nucleus disintegrates, and the generative nucleus divides mitotically into two male gametes or sperms. In some cases the generative nucleus divides even before pollination. The pollen tube generally grows and goes towards the micropylar end of the ovule, carrying the two male gametes through the micropyle into the embryo sac. The tip of the pollen tube dissolves, setting free male gametes inside the embryo sac. One of them fuses with the egg or ovum (n); this is known as fertilization. The fertilized ovum becomes diploid (2n) and eventually develops into the embryo (young plant) of the second generation. The other male gamete unites with the polar nuclei (2n) and becomes triploid (3n). The union of the second male gamete with the polar nuclei is known as **double fertilization.** These three nuclei collectively develop into the endosperm, the food for the young embryo. Meanwhile, the synergids and antipodal nuclei degenerate. The actual function of these is not yet fully understood. The whole ovule ultimately develops into the seed of the second generation and the ovary into fruit. In some cases, other parts of the flower, especially the receptacle, also grow along with the ovary and form the fleshy edible part of the fruit.

Sometimes the fruit is formed without fertilization; this process is known as **parthenocarpy** and such a fruit is called **parthenocarpic**. Such fruits do not contain seeds, like some banana cultivars, grapes, mulberries, and pineapple.

Development of the embryo. In dicotyledonous plants, when the processes of fertilization and double fertilization are over, the fertilized ovum is covered by a cellulose wall and becomes a zygote. The zygote remains inactive for several days inside the endosperm. Essential changes such as the synthesis, multiplication, and organization of ribosomes; decrease in the size of the vacuoles; formation of the endoplasmic reticulum; and other changes in cell organelles appear to make the zygote ready for intensive metabolic activity. It then divides mitotically into two upper and lower cells. The lower cell divides in one plane forming a row of cells called the **suspensor**. The suspensor becomes larger, penetrates deeper into the endosperm, and obtains food material for the developing young embryo. During the subsequent processes of development, the suspensor becomes disorganized. The upper cell is known as the **embryonal cell**. It increases in size and divides mitotically into three planes forming eight cells. Four of these cells divide and form the posterior octant, and the other four cells opposite the micropylar end form the anterior octant. Each octant then divides, by a wall parallel to its curved surface, forming a superficial layer of cells and a central mass of cells that forms an embryonal mass. The superficial cells divide radially only and form a single layer of root apex and stem apex. The embryonal mass tissue divides further, forming various tissues, each of which differentiates into a specialized part of the embryo, such as the plumule, two cotyledons, the hypocotyl, and the radicle (Fig. 4.6).

Further cell divisions of the stem apex give rise to a globular mass of cells filled with cytoplasm. Its surface cells form the **protoderm**, which is the precursor of the epidermis. The central core of cells becomes elongated and forms the **precambium**, which is the beginning of the vascular system. The remaining cells in the centre of the globule divide actively and give rise to the ground–tissue system. The differentiation of tissues and organ formation begin with the appearance of the cotyledons. The initial embryo becomes almost heart–shaped. At the central position of the two cotyledons, a cluster of cells becomes organized and gives rise to the apical meristem. Another cluster of cells at the opposite side of the embryonic axis forms the root meristem. These two meristems are situated in a position inside the endosperm that permits their continuous growth. Vitamins, auxins, and cytokinins needed for the normal development of the young embryo are provided by the endosperm.

The apical meristem of the shoot gives rise to new organs and tissues like shoot apex and young leaves. The initial leaves are close to each other because the internodes are short. The leaves are situated at regular spaces surrounding the stem. In due course of time, the shoot is exposed. Under a microscope, it looks like a dome–shaped mound or a flattened cone, or in angiosperms depressed in shape like a saucer. When a longitudinal section is observed under a microscope, this region appears very organized. One or more layers of cells cover the surface; and in the axils of the leaves the primordia and small pockets of meristematic tissues are visible. These develop into small shoot apices and eventually into axillary buds.

The arrangement of cells is more regular in the root apex than in the shoot apex, with clearly defined layers which correspond in number and arrangement to the regions of the mature root. The apices of root and shoot possess sufficient food material for growth. The root apex controls its own activity and has the potential of forming the mature root.

The modification of vegetative primordia into reproductive primordia is brought about by complex biochemical changes which are not fully understood. This mysterious change alters the mode of differentiation from vegetative bud and stem tissues to the tissues that form the reproductive organs such as gynoecium (pistil) and androecium (stamens) and accessory parts of the flower like sepals and petals. When a meristem is being altered from the vegetative to the reproductive form, microscopic modifications in its structure become visible. This structural change is permanent and flower formation continues until the flower is fully mature. At this juncture the flower is ready for the development of the fruit. There are two main patterns of endosperm development: (i) nuclear, and (ii) cellular.

I. NUCLEAR PATTERN. In this pattern, a triploid (3n) endosperm nucleus divides a number of times without corresponding cell–wall formation. As a result, several hundred nuclei are generally formed, and a large vacuole is developed in the embryo sac. Cytoplasm along with nuclei are pushed towards the border of the embryo sac and the nuclei are almost uniformly distributed. Cell–wall formation is centripetal, and in this way a mass of cells is formed, each of which has a nucleus and a cell wall, and is filled with food substances. This accumulated mass of cells is known as the **endosperm**. It grows and fills up the embryo sac.

II. CELLULAR PATTERN. In this pattern, a triploid (3n) endosperm nucleus divides several times forming numerous nuclei; and around each nucleus a cell wall is formed. The wall may be longitudinal, transverse, or oblique, depending upon the plant species. In this way, numerous cells are formed, each having a nucleus and food materials. The tissue of the seed is called

endosperm. In non–endospermic seeds such as garden peas, French beans, gram, sunflower, mango, orange, and guava, the cotyledons store up the endosperm and become thick and succulent, while in endospermic seeds such as papaya, palm, lilies, custard apple, four o'clock plant, wheat, maize, rice, and barley, there is a separate chamber surrounded by the seed coat and fruit wall (pericarp). The accumulated food, whether it is in the cotyledons or in the endosperm, is utilized by the embryo during germination.

The two integuments of the ovule change into the two parts of the seed coat: the outer one becomes the **testa** and the inner one the **tegmen**. In some plants such as the four o'clock (*Mirabilis jalapa*) and coffee, the nucleus also survives and develops into a nourishment–providing tissue like endosperm which is known as **periderm**. The coffee bean consists of periderm tissue. The ovule finally becomes the seed and the ovary the fruit.

Fruit growth. The treatment of fruit growth here is mostly restricted to succulent fleshy fruits because of their horticultural and commercial importance. Generally, the growth of fleshy fruits adopts two patterns. In one set of fruits like tomatoes, apples, pears, and lemons, cell division and cell enlargement are rapid in the beginning and gradually slow down as the fruit approaches maturity. The other group of fruits, e.g. mangoes, plums, Indian plums, pears, and apricots, exhibit continuous cell division at the initial stage, slow down during stone (endocarp) formation, and again resume continuous cell division and enlargement until the fruit matures. In some cases like apples and pears, some accessory parts like the receptacle also grow along with the fruit. These form the succulent edible part of the fruit and contain more intercellular spaces than other parts of the fruit.

The period of growth varies in different types of fruits, generally ranging from four to five months. Fruit development is regulated by numerous internal and external factors. The most important among them are pollination effect, fertilization effect, and development of the embryo and seed. Growth regulating substances such as hormones, auxins, gibberellins, and cytokinins greatly influence fruit size and weight. Their action on growth is mostly physiological; it varies in different types of fruits and is very complex.

Pollination is one of the most essential steps in fruit development. It performs two separate functions: one is the procurement of male gametes or sperms necessary for fertilization; the second is provision of an enzyme that alters auxin precursors available in the stigma into auxins. The growth of the pollen tube inside the style is affected by the availability of proper osmotic concentration and is also accelerated by the presence of certain inorganic chemicals like manganese sulphate, boron, and calcium. The presence of sugary substances secreted by the female organs enhances the growth of the pollen tube. When the pollen tube does not grow rapidly inside the style, abscission of the entire flower and ultimately the failure of fruit development are likely. This is generally overcome by the application

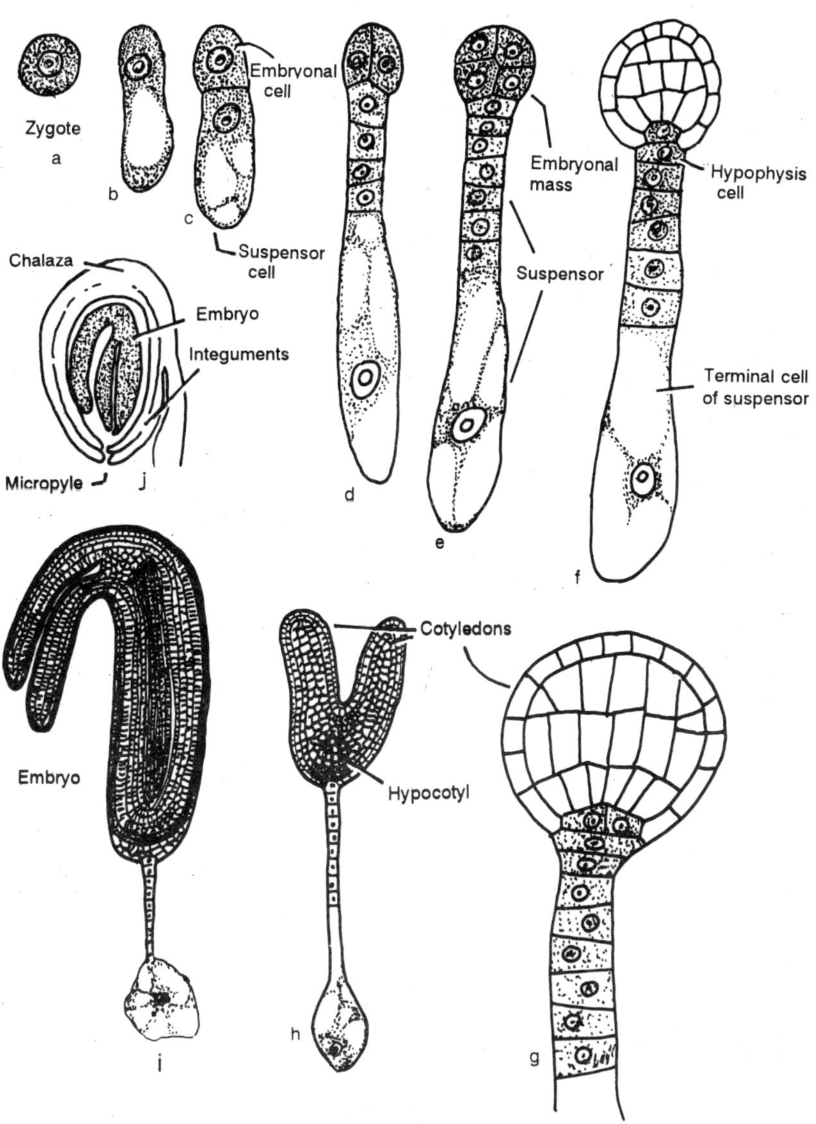

Figure 4.6 Development of typical dicotyledonous embryo

of α–naphthalene acetamide, one of the set of auxins.

There is an interesting phenomenon known as **metaxenia**. The term refers to the direct effect of pollen grains on the female tissues of the ovary of a related cultivar of the same species. This inference is based on the observation that the hormone content of the ovary increases after pollination. In some fruits and seed coats the skin colour is observed to be affected by the source of pollen; this happens with fruits of the Cucurbitaceae family, for example. In oranges, the colour and flavour of fruit are influenced by the pollen parent. In dates, the size and the time of ripening are influenced by the kind of pollen (Markhand et al. 1991). Metaxenia differs from **xenia**, the direct influence of pollen grains on endosperm tissue colour, observed in maize, for example.

The development of the fruit is also affected by a stimulus from the developing seeds within the fruit. Uneven distribution of seeds causes irregularly shaped fruits. In fruits such as tomatoes and strawberries, a direct relation exists between number, weight, and size of seeds. With strawberries, for example, if all the achenes except one are removed, the receptacle tissue that is immediately below that particular achene enlarges. When more than one achene are present, only the portion of the receptacle tissue which has achenes will develop. The growth–inducing effect of the seeds can be restored in the seedless part of the receptacle by the application of auxins. With tomatoes, when the liquid extract obtained by squeezing pollen grains or immature seeds is applied to the stigmas of unpollinated pistils, it accelerates fruit growth. This effect of pollen or seed extract can be duplicated by applying a paste containing an auxin. The stimulating effect of seeds or pollen on the growth of other fruits such as peppers, eggplant, and figs is also altered by the application of auxins. In addition, fruit setting is stimulated in grapes, apricots, apples, and pears by the use of gibberellin. Similarly, cytokinins enhance fruit setting in grapes.

Auxins are usually used with apples, grapes, and pears; they are especially effective on the stone fruits like peaches, apricots, olives, plums, and cherries. The extent of action is related to the concentration of chemical applied, the stage of development of the fruits, the variety or species, and the temperature and humidity prevailing at the time of application.

The adequate availability of nutrients and moisture at appropriate stages of development greatly influences the number and size of fruit. For example, it has been calculated by horticulturists that on a healthy mature apple tree, about 40 leaves provide sufficient nutrients for the normal growth of one apple. When this ratio is abnormally altered, the quality and size of fruits are affected. If the number of fruits is high their size will be reduced. Therefore, fruit growers generally artificially reduce the number of fruits in order to maintain size and quality. This reduction of crop load is called fruit thinning. Crop bearing may also be decreased with chemicals.

However, the mode of action of chemicals such as naphthaleneacetic acid (NAA) and its derivatives is not clearly understood. Probably it retards the process of fertilization, or causes inhibition of embryo development or abortion of the embryo after fertilization. The result is natural fruit thinning.

4.2.4 Maturity, ripening, and senescence

Maturity. Maturity and ripening are two separate processes. Maturity of fruit is the final stage of development that must take place when the fruit is still attached to the mother plant. The enlargement of cells and storage of carbohydrates is taking place. Aroma and taste–producing substances are being accumulated, and acidity is decreasing. Generally, the fleshy edible portion is not yet softened. Ripening of fruit involves certain biochemical changes which take place in the fruit after full maturation. The softening of the edible portion is near completion. Production of the fruit's characteristic aroma and sweet juice increases. Ripening may take place before or after harvesting.

For example, a popular grafted mango variety Sindhri, grown in Sindh, is usually harvested before it is ripe on the tree. Generally, the fruit is considered to be mature and ready for harvesting when one or two naturally ripe, yellowish mangoes drop from the tree. The harvested mangoes are mature, but are still green, sour, hard, and not ready to be eaten. The harvested green fruit is kept in wooden crates or porous new earthen pots between layers of wheat or rice straw, hay, grass, old newspapers, etc. These crates or earthen pots are stored in closed storage rooms for ripening, which is usually complete in a week. The ripened fruit possesses an attractive yellow colour and aroma, and becomes sweet, juicy, and tasty; at this point it is brought to market and fetches a good price. Generally, Sindhri mangoes are transported to big cities like Karachi, Lahore, Hyderabad, and Islamabad, and exported to the Middle Eastern countries packed in wooden crates in the above manner. Ungrafted varieties of mango are harvested for making pickles before maturity when the endocarp (stone) becomes hard.

Ripening. Physiological changes indicate the termination of the mature stage and the initiation of senescence of fruit. The period in which a number of specific biochemical changes—a significant and abrupt increase in natural respiratory rate, and autocatalytic production of ethylene—take place before senescence is called the **climacteric**. After the climacteric, the rate of respiration again decreases and reaches a level equal to or less than the rate prevailing before the climacteric. For example, in some varieties of apple, pear, and banana, the pigment changes from green to yellow. The ripening of pears synchronizes with the end of the climacteric. With apples and bananas, the stage of ripening at which the fruit are best to eat is reached immediately after the climacteric.

There are different views regarding climacteric increase in rate of respiration. One idea is that the acidity of the tissue cytoplasm decreases to a critical stage that increases the permeability of the cell membrane. Thus fructose already stored in the vacuoles possibly enters into the cytoplasm and becomes available for the enhanced rate of respiration. A second view is that adenosine diphosphate (ADP) and adenosine triphosphate (ATP) play a role in energy production during respiration. A portion of energy is kept for the chemical change of ADP to ATP. It is supposed that during fruit maturation, cells enlarge quickly and require more energy for protein synthesis; as the fruit matures ADP is readily made available for the increased rate of respiration.

Temperature has a significant effect on fruit maturation and the climacteric. For example, the rate of respiration of the apples at 22°C is five times more than that of those at 2°C; and the time required to reach climacteric is 25 times more for the apples kept at 2°C.

Fruit ripening is a DNA–regulated activity similar to other plant development processes. There is an increase of RNA and enzymes such as hydrolase, synthetase, and oxidase in climacteric fruits. Ethylene plays an active part in fruit ripening and accelerates the process; ripening can be retarded by the removal of ethylene from the tissues of preclimacteric fruits. However, the mode of action of ethylene on fruit ripening is not fully understood.

Senescence. Senescence is a physiological aging activity in which plant tissues degenerate and ultimately die. The activity is one of the most mysterious plant growth processes, and takes place in almost all organs of the plant. It can be considered under two categories: partial senescence and complete senescence.

Partial senescence. Partial senescence involves the degeneration and death of aerial plant parts like leaves, branches, flowers, and fruits, but not underground plant parts—e.g. the deterioration of cotyledons after germination observed in almost all plants. With deciduous plants in unfavourable seasons, especially in winter, the leaves fall. In the commercially important woody perennial shisham or *tali* (*Dalbergia sissoo*), the leaves become senescent, die, and fall off in winter. The rest of the plant survives, gives out new leaves in spring, and bears flowers and fruits in the ensuing summer. Shoots of herbaceous perennial plants die during unfavourable seasons, but the underground parts survive and remain viable in the soil. This happens with the corms of taro (*Colocasia esculenta*), gladiolus, crocus (*Colchicum autumnale*), tubers of potato (*Solanum tuberosum*), and Jerusalem artichoke (*Helianthus tuberosus*); the bulbs of onion (*Allium cepa*), garlic (*Allium sativum*), and lilies; and the rhizomes of turmeric, ginger, and arrowroot.

Complete senescence. This is the aging process in which all parts of the plant except the seeds ultimately die. It is observed in almost in all cereals, pulses, vegetables, and ornamental seasonal flowers. In other words, annual or seasonal plants complete their life cycle after maturity of their fruits or seeds. In perennial shrubs and trees, a gradual aging process takes place. Their older organs degenerate and ultimately die, and new shoots arise, mature, and bear flowers and fruits. In this manner plants survive for many years. In addition, some perennial woody plants are invigorated by severe pruning and by provision of fertilizers to stimulate the growth of adventitious buds around the pruned branches. For example, in *ber* (*Zizyphus mauritiana*) after the fruiting season is over in late spring, the old branches are pruned in early summer. New adventitious buds arise around the pruned branches in summer, mature in late summer to autumn, and bear flowers in the form of inflorescence in the axils of new leaves on current shoots in the ensuing winter. Therefore, in this plant regular pruning is necessary to induce healthy growth and obtain maximum fruit yield.

Considerable study has been done on the senescence of different parts of plants, especially on abscission of leaves. It is thought that during senescence photosynthesis, starch, chlorophyll, DNA, RNA, proteins, gibberellins and auxins decrease. There are different hypotheses regarding the cause of the senescence of annual or seasonal plants. The most important relates to the quantities of nutrients produced by the plants during various activities. A considerable amount of nutrition is consumed during flowering, fruiting, and seed formation. Therefore, after these processes take place the plants deteriorate and finally die due to lack of sufficient nourishing substances.

4.3 Modifying plant growth

4.3.1 Photoperiodism

The recurring cycle of daylight and dark periods of regularly changing duration is crucial for normal growth and reproduction of plants. The number of daylight hours is referred to as the **photoperiod**. The response of plants to the relative length of daylight or darkness is known as **photoperiodism**. The effect of daylength on growth and reproduction of plants was studied by Garner and Allard in America. They have classified plants mainly into three categories, according to the effect of photoperiod: (i) **short–day** or **long–night** plants, e.g.: cosmos, soybean, chrysanthemum, and poinsettia; (ii) **long–day** or **short–night** plants, e.g. spinach, althea, winter wheat, and oats; (iii) **day–neutral** plants, e.g. cucumber, kidney bean, pea, and tomato. The dividing line between daylengths favourable to vegetative growth and those inducing flowering and seed formation is called the **critical light period** for a specific plant.

The period of daylength from sunrise to sunset differs throughout the world depending on the time of year and geographical zones, except on the spring and fall equinoxes, when days and nights are of equal length throughout the world. Generally, 12 to 14 hours of continuous sunlight or artificial illumination is the critical light period for a particular plant. One hour or even a half–hour difference may cause major physiological changes in the plant.

The phenomenon of photoperiodism is important in controlling the flowering of a good number of horticultural and agricultural plants. Shortening of the period of daylight by artificially shading the plant with coarse black cloth or black plastic, or lengthening the period of daylight by electric illumination can induce early flowering. For example, aster and chrysanthemum, short–day ornamental plants, are forced into flowering by commercial floriculturists at will by covering them in the morning and evening in summer with coarse black cloth to shorten the long photoperiod and provide flowers commercially throughout the year. Long–day plants such as spinach and sugar beet are brought into flowering by additional artificial illumination in winter. Photoperiodism is an important consideration in raising plants in seasons when the requisite daylength does not occur naturally.

4.3.2 Plant hormones

Plant hormones are natural substances produced by plant tissues in small quantities, especially at the growing points, and transported to other regions where they are required to regulate the activities of the plant. **Plant growth regulators** are synthetic products, which when applied to plants produce reactions almost identical to those caused by natural hormones. Therefore in general both plant hormones and plant growth regulators are classified together in five categories: **auxins**, **gibberellins**, **ethylenes**, **cytokinins**, and **abscisic acids**.

Auxins are formed in abundance in growing regions such as terminal and lateral buds, elongating internodes, and the young embryo in its developing stage inside the seed. They are translocated through the vascular bundles to other parts of the plant, especially from apical to basal regions. The accumulation of auxins in terminal buds inhibits the growth of lateral buds. To avoid this, apical tips are pruned, thereby inducing a number of lateral buds below the cut portion of the stem or branches. In this way hedges and edges are thickened. There are numerous natural and synthetic auxins: indoleacetic acid (IAA), indolebutyric acid (IBA), naphthaleneacetic acid (NAA), and 2,4–dichlorophenoxyacetic acid (2,4–D). Synthetic auxins are utilized for several purposes in horticulture and agriculture:

I. BEGINNING ADVENTITIOUS ROOTS: Stem cuttings of various fruit species, for example mulberry, pomegranate, fig, grape, and olive, are treated with IBA and NAA for early initiation of adventitious roots.

II. KILLING WEEDS: 2,4–D is used for annihilation of broadleaf weeds such as bindweed (*Convolvulus arvensis*), lamb's quarters (*Chenopodium album*), wild radish (*Raphanus sativus*), and prickly poppy (*Argemone mexicana*) in cereal crops.

III. MICROPROPAGATION OF PLANT SPECIES: Auxin or 2,4–D in small quantity is used in plant tissue culture under aseptic conditions in glass vessels containing explants (minute pieces of tissue) for accelerated beginning of roots and shoots in them. The explants are then separated for developing into full size plants. In this way thousands of plants are produced in a short time in the laboratory. Nowadays plum (*Prunus domestica*), apple (*Pyrus malus*), rose species, and many other commercially important plants are propagated true-to-parents by this method.

IV. SETTING OF FRUITS: In tomato (*Lycopersicon esculentum*), spraying of auxin 4–chlorophenoxyacetic acid (4–CPA) in adequate quantity before flowering enhances fruit setting.

V. CHECKING PRE-HARVEST FRUIT DROP: In oranges, pears, and apples, fruit often drops to the ground before it can be picked. To control pre-harvest drop, the synthetic auxin NAA is sprayed in dilute form.

Gibberellins, a group of natural plant hormones, are obtained from the fungus *Gibberella* and many higher plants. These are present in the apices of shoots and leaf primordia of plants, and in embryos and cotyledons of immature seeds and fruit tissues. A good quantity of gibberellin is formed in roots. Gibberellins are easily transported both upward and downward throughout the plant. Gibberellins are manufactured commercially in crystalline form as acids or potassium salts. They perform numerous regulatory functions such as stimulation of growth enhancing cell division, cell elongation, beginning of flowering, sex determination, fruit setting, fruit growth, ripening of some fruits, senescence of leaves, and breaking the dormancy period of seeds and buds. Gibberellins are said to be the most effective and essential set of plant hormones, despite the fact that little is known about their physiological mechanism of action. Some of essential uses of synthetic gibberellins in horticulture and agriculture are:

I. ENHANCING SEED GERMINATION AND SEEDLING GROWTH: Soaking seeds of numerous plants such as pea (*Pisum sativum*), French bean (*Phaseolus vulgaris*), barley (*Hordeum vulgare*), grape (*Vitis vinifera*), and apple (*Pyrus malus*) overnight in different concentrations of gibberellic acid has proven very effective in stimulating germination and seedling growth.

II. ELIMINATING COLD TREATMENT: Plants like radish (*Raphanus sativus*) and turnip (*Brassica rapa*) do not flower and bear fruit in temperate climates due to cold. A few applications of gibberellic acid in low concentration on the leaves stimulates flower bud initiation, and the plants bear flowers and fruits as they do in tropical regions.

III. INITIATING MALE FLOWERS IN CUCUMBER (*Cucumis sativus* L.): For hybridization purposes, the leaves of different cultivars of cucumbers are sprayed once with a low concentration of gibberellic acid. In this way the number of male flowers is increased, and anthesis starts in them when the female flowers are ready for pollination. Thus hybridization becomes possible for the production of hybrid seed.

Ethylene, in comparison to other hormones, is a minute molecule (C_2H_4). Like carbon dioxide (CO_2), ethylene in the gaseous state spreads easily through the plant, and affects it even in small quantities. Because of its tiny molecules, it moves rapidly through the plant in water solution. It is produced in meristematic tissues of plants, germinating seeds, withering flowers, maturing and ripened fruits, and in injured plant tissues. A synthetic commercial product, Ethephon, produces a good amount of ethylene automatically when applied to plants. There are a number of views about the mode of action of ethylene. One of them is that it regulates certain kinds of DNA transcription of RNA. In this way RNA–guided protein synthesis ultimately alters the enzyme forms. Ethylene and the ethylene–releasing Ethephon have numerous horticultural uses such as:

I. INDUCING FRUIT MATURITY: Ethylene in gaseous form is forced into storage houses for ripening bananas, watermelons, muskmelons, and tomatoes. Ethephon is sprayed before harvesting on apples, figs, cherries, coffee beans, and pineapple to induce uniform ripening of the fruit. Ethylene–releasing chemicals are also sprayed on mature but green fruit to bring it to its natural ripened colour to increase its commercial value.

II. INITIATING FLOWERS: Ethylene gas obtained from Ethephon is utilized for initiation of flowers in a number of ornamental flowers, pineapple and banana.

III. ALTERING THE SEX OF FLOWERS: Members of the Cucurbitaceae family bear more male flowers than female ones on the same plant. Spraying Ethephon on cucumber and pumpkin vines during the vegetative stage increases the number of fertile female flowers which develop into fruits, thus increases the yield.

IV. CHANGING THE GREEN COLOUR OF CITRUS FRUITS: When citrus fruits reach maturity, their rind is still green. To bring them to their fully–ripened colour they are treated with ethylene or dipped in Ethephon solution before packing and marketing. During transit, their green colour changes du to modification of chlorophyll, yellow carotene (orange–red), and xanthoph (yellow) pigments.

V. ABSCISSION–INDUCING EFFECT: Some fruit like walnuts, cashew nuts, and sour cherries are harvested by shaking the trees. These fruits are tightly attached and do not drop easily. Spraying Ethephon a week before harvesting induces abscission in the stalks of the fruit. In this way a good number of fruits drop or can be pulled away easily by hand. With some commercially important flowers like tulips, carnations, and geraniums, a small quantity of ethylene may sometimes produce harmful effects. To overcome this they are treated with silver thiosulphate.

Cytokinins. These are a set of plant hormones which support plant–cell division and take an active part in numerous physiological activities of plant growth and development, namely dormancy, cell elongation, tissue differentiation, various aspects of flower formation and fruit development, and inhibition of leaf deterioration processes. There are a number of synthetic as well as natural cytokinins. One of the prominent natural cytokinins is zeatin, which is obtained from maize seed and coconut milk. Some common synthetic cytokinins are kinetin and benzyladenine. Cytokinins are found in various plant tissues in the free hormonal state or as one of the component forms of transfer ribonucleic acid (tRNA). A good quantity of cytokinins is present in embryos, germinating seeds, in early stages of developing fruits, and in meristematic tissues. Cytokinins formed in root tissues are transported towards the shoot. The mode of physiological action is not fully understood. Probably they indirectly enhance enzyme activity and thereby increase the DNA produced by the plant tissues.

Some uses of cytokinins in horticulture are:

I. TO STIMULATE SHOOT DEVELOPMENT: A small quantity of cytokinin is put in culture media under aseptic conditions to stimulate shoot development.

II. DELAYING SENESCENCE: When cytokinins are applied to green tissues of plants, their senescence is delayed.

III. ACCELERATING BUD GROWTH: Growth of axillary buds in roses and chrysanthemum is accelerated, probably by hampering the activity of natural bud inhibiting agents.

Abscisic acid (ABA). This is a natural plant hormone, which instead of promoting growth retards it. It is widespread in plant tissues; cotton fruits, leaves of the sycamore tree, and especially the fruits of rose (hips) are rich in abscisic acid. It is also found in the leaves of plants which are under water stress. It is obtained by alcoholic extraction from the above plant tissues. It moves rapidly within the plant and takes an active part in leaf and fruit abscission, in the initial stages of the resting period of vegetative and reproductive buds of perennial shrubs and trees, and the induction of dormancy in seeds. Abscisic acid is believed to act against the growth–enhancing actions of gibberellins and auxins and bring about change in nucleic acid and the protein–synthesizing system. It is used for experimental purposes, where only a small quantity is required. The commercial use of ABA in horticulture is restricted because it is difficult and costly to synthesize.

Plants store a good number of other growth–retarding or inhibiting chemicals which do not play any definite part in the metabolic process, despite the fact that they take an active part in growth retardation and have toxic effects on insects and animals.

There are numerous commercial synthetic growth inhibitors.

i. Daminozide negatively affects growth and activates flowering of herbaceous and woody ornamentals, and enhances their colour and fruit size. Dwarf plants of chrysanthemum and azaleas so produced have an attractive shape.

ii. Chlormequat is active in inhibiting height increase in some ornamental plants like red poinsettias, azaleas, geraniums, and bougainvillea. Applied as a drench to the soil or as a spray to the foliage in autumn, it spurs flower–bud initiation.

iii. Cycocel (CCC) is used generally to prevent lodging of wheat plants when applied between the tillering and rapid shoot growth stages. Internode elongation is retarded, and in this way sturdy short plants are obtained which are resistant to lodging.

iv. Ancymodol. This growth inhibitor affects the height of potted plants like lilies, poinsettias, chrysanthemums, and tulips. The compact plants thus obtained look attractive and showy.

v. Paclobutrazal. This synthetic growth retardant inhibits vigorous shoot growth in herbaceous and woody plants thereby dwarfing the plants and making them showy in appearance. Various synthetic products with different trade names are produced.

Researchers have worked on the action of plant growth regulators individually and in combination with other hormones and on growth promoting substances on various plant parts throughout the world. Studies on the effect of growth regulators (Aron and Planofix), micronutrients (Greenzit), and chemical fertilizers (NPK) were carried out to observe both the individual effects of plant hormones and their effect in combination with chemical fertilizers on growth and fruit yield in Kinnow mandarin (*Citrus reticulata blanco*). Maximum growth was obtained when the three types of substances were applied together; Planofix and Greenzit increased fruit yield both when they were applied together, and each in combination with NPK (Gilani et al. 1989).

Research was also carried out on malformation in mango inflorescence. The percentage of female flowers was increased on malformed bunches of inflorescence by 51–60 percent with foliar sprays of GA3 (one of the gibberellin group) at 400 ppm, and NAA at 50–100 ppm. With both chemicals, treatments were given three times: the first in August, the second during the last week of January, and the third at the end of February (Akhtar 1973). Fruit drop in mangoes was attributed to the formation of an abscission layer and other factors e.g. rain, humidity, mango flower mildew, mango hopper attack, and anthracnose (Malik and Raza 1985). Investigations were carried out to determine the possibility of reducing mango fruit drop with Planofix spray at concentrations of 0.5 ml, 1.0 ml, 1.5 ml, and 2.0 ml per gallon of sterilized water 15, 30, and 45 days after fruit set. A 1.0 ml dose 30 days after fruit set was found to be the most effective (Jagirdar and Choudhry 1967).

In peas no effect of CCC and IAA was observed on the time required for germination. Plant height responded positively to IAA concentrations and negatively to CCC. Maximum height was produced by a dose of 200 ppm of IAA. Stem thickness was increased by higher concentrations of CCC and decreased by higher concentrations of IAA. Dry weight decreased at lower CCC concentrations, but increased with increasing concentration of IAA (Muhammad et al. 1988).

4.3.3 Temperature

Plant tissues require proper temperature to carry out various activities. At low temperatures almost all functions performed by the tissues slow down. The higher the temperature, the more the activity; of course, within a certain range of temperature all processes are performed normally by the living cell. The optimum temperature varies in different plant species of different regions and also differs at various stages of growth. On the whole, the average temperature for maximal plant growth is 28–30°C, while the activities of plant tissues cease at 4°C maximum. Temperatures of 45–50°C coagulate and kill the living substance of plant cells, however, growth continues in most plants at 10°C.

The influence of temperature on sexual reproduction of plants is known as **vernalization**. Without a certain cold temperature and exposure to an adequate photoperiod, the induction of flowering does not take place in some annual plants, e.g. winter wheat. It is beneficially utilized by the Russians, who obtain a crop of wheat in about two months in Siberia, where the severe period of freezing temperature lasts for almost ten months. They sow winter wheat before the first snow. The plants establish themselves under the snow. As the snow begins to melt, they continue the growth that was interrupted by the long period of cold. In some biennials such as carrot (*Daucus carota*), exposure to cold temperatures is replaced nowadays by the use of gibberellin.

One of the traits of plants of temperate regions is their marked periodicity (periods of rest or dormancy alternating with periods of active growth). It is particularly prominent in woody deciduous plants which lose their leaves in autumn and remain dormant until the following spring. Deciduous fruit trees like apples, peaches, and cherries require lingering freezing rest periods, are therefore grown only in temperate climates. Some plants like walnuts, grapes, and almonds also require low temperatures for varying lengths of time. Certain plants require freezing temperatures to end the rest period, while in others a similar effect is produced by temperatures which are low but above freezing. Generally, tubers, bulbs, corms, and rhizomes require a short rest period.

Sometimes it becomes desirable to reduce dormancy or break the rest period in ornamental or food crops. The procedures used to do this are warm or cold storage, or treatment with different chemicals. For example, dormant potato tubers will sprout within a week to a fortnight after harvest ing if they are treated with ethylene. Plant physiologists take advantage of this fact by growing seed potatoes in greenhouses under controlled environments for early detection of certain potato diseases present in some tubers. Ethylene is also used to break the rest period in gladiolus corms, which are normally dormant for six to seven months; by this method gladiolus flowers are induced to bloom during winter.

4.3.4 Fertilizers

At first, the fruit growers of Sindh were under the mistaken impression that bananas could not be grown profitably there. In the late 1960's, attempts were made by research workers at Mirpurkhas, Sindh to explore the possibility of growing bananas on a commercial basis in Sindh. Several variables: spacing of suckers in rows 6′ apart at plant-to-plant distances of 4′, 6′, and 8′; irrigation intervals of 6, 10, and 14 days; and levels of nitrogen at 50, 75, and 100 pounds as ammonium sulphate per acre were tried. After completing their juvenile period, suckers planted 8′ apart produced the largest number of young suckers. At maturity, the grade of the bunches and size of the fingers obtained from main plants irrigated at intervals of 6 and 10 days were superior. High doses of nitrogen in the form of ammonium sulphate improved the bunch grade, accelerated maturity, and increased plant growth (height and basal circumference). Plants irrigated at six-day intervals with a single nitrogen dose of 100 pounds matured about three months (83 days) earlier than the plants given other treatments (Jagirdar et al. 1963).

The effect of three levels: 0.0 (control), 12.5, and 25.0 g per plant of P_2O_5, and K_2O, with a uniform application of 12.5 g N, significantly increased the dry fruit yield of chilies per plant. The application of 12.5 each of P_2O_5 and K_2O along with a basic dose of 12.5 g per plant produced a good yield of red chilies (Baloch et al 1989).

Studies were carried out to evaluate the effect of potassium and magnesium on the vegetative growth and yield of tomatoes. The experiment was split in seven K and Mg combined treatments: (1) K 0 Mg 0 (control); (2) K 200 Mg 50; (3) K 400 Mg 50; (4) K 200 Mg 100; (5) K 400 Mg 100; (6) K 200 Mg 150; (7) K 400 Mg 150. The combined doses did not display any significant effect at the early juvenile stage of plant growth. However, after a period of 50, 71, and 92 days of transplanting, plant height, number of leaves, and branches were remarkably increased. Only stem diameter was higher under K 400, Mg 150 ppm combined treatment. At maturity, K 200, Mg 100 ppm under combined application produced higher yield (Baloch and Hargitai 1986).

Results of four different row and plant spacings 25×10, 20×10, 20×8, and 15×8 cm (plant populations of 40, 50, 62, and 83 plants/m^2) on the growth and development of onions suggest that a plant population ranging from 50 to 62 plants/m^2 appears to be optimum for getting the maximum quality and quantity yield of bulbs. The results further suggest that a population of 62 plants/m^2 is the maximum plant density which yields good economic returns. A density of 83 plants/m^2 reduced the size of bulbs and yield per unit (Mangrio and Baloch 1985).

QUESTIONS

1. Define growth and explain the factors affecting growth.
2. Describe the parts of the almond seed and discuss the role of each part.
3. Discuss in detail the functions of the seed.
4. What is germination? Explain hypogeal germination, illustrating your discussion with an appropriate diagram.
5. Explain the difference between epigeal and hypogeal germination with the help of suitable diagrams.
6. Define annual, biennial, and perennial plants. Cite suitable examples of each.
7. Explain the stages of the reproductive phase of plants.
8. Describe the parts of the citrus flower; illustrate your description by drawing and labelling a diagram.
9. Describe the process of gametogenesis; illustrate with an appropriate diagram.

10. What is pollination? Discuss the importance of cross–pollination in horticulture.

11. (a) Differentiate between cross–pollination and self–pollination. Give examples of plants pollinated in each of these ways. (b) Discuss the advantages and disadvantages of self and cross–pollination for the horticulturist.

12. (a) Name the various pollinating agents. (b) How is pollination of apple carried out?

13. Describe the process of fertilization and explain its importance for the plant.

14. (a) Describe the various parts of the ovule. (b) Draw and label a diagram to illustrate your description.

15. Describe the growth cycle of a fruit, citing suitable examples.

16. Discuss in detail the development of the plant embryo.

17. Explain photoperiodism and its importance for horticulturists.

18. Discuss plant hormones, being sure to deal with the following points: (a) what they are; (b) different types; and (c) their effects on plants.

19. What is meant by partial senescence and complete senescence? Differentiate between them by citing suitable examples.

20. List the functions of synthetic auxins in horticultural plants. Cite an appropriate example of each.

21. Describe the uses of ethylene in horticulture, giving suitable examples.

22. In fruit, how does maturity differ from ripening? Explain by citing appropriate examples.

23. What are natural and synthetic growth inhibitors? Explain their mode of action on growth.

24. Explain vernalization and discuss its relation to rest period and dormancy.

25. How are natural and synthetic gibberellins obtained? Describe their uses in horticulture.

26. What is the effect of temperature on fruit maturity?

27. What do you understand by the term climacteric? Explain different views regarding it.

28. Differentiate between xenia and metaxenia by citing appropriate examples.

REFERENCES

Akhtar, M. I. Effect of growth regulating substances on malformation in mango inflorescence. Lyallpur: University of Agriculture; 1973. M.Sc. thesis.

Alam, S. M. Influence of saline irrigation water on the growth and mineral composition of cauliflower (*Brassica oleracea*) Var. Chin Ka Moti. Pakistan J. Agric., Agri. Engg., Vet. Sc. 6(1–2):94–96; 1990.

Baloch, A. F.; Baloch, M. A.; Deho, N. A. Effect of phosphorus and potassium on the yield of red chilies (*Capsicum frutescens* L.) Pakistan J. Agric., Agri., Engg., Vet. Sc. 5(1-2):74–76; 1989.

Baloch, A. F.; Hargitai, L. The effect of potassium and magnesium on the growth and yield of tomato. Pakistan J. Agric., Agri. Engg., Vet. Sc. 2(1–2):7–10; 1986.

Carl, L. A.; Kriedemann, P. E. Plant growth and development. 3rd ed. New Delhi: Tata McGraw–Hill, Ltd.; 1985.

Esau, K. Anatomy of seed plants. 2nd ed. New York: John Wiley and Sons; 1977.

Gilani, A. H.; Ali, Y.; Tariq, M. A.; Muhammad, F. Studies on the effect of growth regulators and chemical fertilizers on the growth and yield in Kinnow mandarin (*Citrus reticulata* Blanco). Sarhad J. of Agric. 5(1):47–51; 1989.

Hartmann, H. T.; Kofranek, A. M.; Rubatzky, V. E.; Flocker, W. J. Plant science: growth, development and utilization of cultivated plants. 3rd ed. Englewood Cliffs, New Jersey: Prentice–Hall; 1989.

Jagirdar, S. A. P.; Bhutto, M. A.; Shaikh, A. M. Effect of spacing, interval of irrigation, and fertilizer application on Basrai banana (*Musa cavendishii* Lampert). W. Pakistan J. Agr. Res. 1(2):5–19; 1963.

Jagirdar, S. A. P.; Choudhry, M. R. Fruit drop control in mangoes by hormonal spray. W. Pakistan J. Agr. Res. 5(4):33–42; 1967.

Malik M. N.; Raza, M. Effect of different doses of NPK and NAA, and time of deblossoming on the intensity of malformation of mango inflorescence. J. Agric. Res. 23(2):97–104; 1985.

Mangrio, H. K.; Baloch, A. F. Plant population studies on onion (*Allium cepa* L.). Pakistan J. Agric., Agri., Engg., Vet. Sc. 1(2–4):81–84; 1985.

Markhand, G. S.; Larik, A. S.; Hafiz, H. M. I.; Maher, A. R. Effect of metaxenia on the size, weight and ripening of Aseel date fruit. Pak phyton 3:165–170; 1991.

Muhammad, T., Mohar, T. A.; Ishfaq, M. Effect of cycocel and indole acetic acid on the growth of peas (*Pisum sativum* L.) J. Agric. Res. 26(4):295–299; 1988.

Shaikh, A. A.; Baloch, M. H.; Baig, M. S.; Dehraj, I.B. Effect of different fertilizer doses and sources on growth and production of banana. Sindh J. of Agric. Res. 6(1):40–46; 1986–87.

Outline of Chapter 5

5. CROP IMPROVEMENT

Musahib-ud-Din Khan[1]

Habib-ul-Rahman Mian

LEARNING OBJECTIVES

After studying this chapter the student should be able to:

- Explain the relationship between the mode of a plant's reproduction and the plant breeding methods appropriate for it
- Describe the mode of reproduction of important horticultural plants such as maize, citrus, potato, and mango
- Discuss the sources of both natural and induced variation
- Discuss various causal hypotheses explaining heterosis
- Outline the steps of the backcross method of breeding
- Test the viability of pollen by two different methods
- Select two parent plants for vegetable or fruit cultivars and cross them

5.1 Introduction

The primary purpose of plant breeding is to obtain or develop varieties or hybrids that are efficient in their use of plant nutrients, that give the greatest return of high-quality products per unit area as judged on the basis of cost and ease of production, and that are adapted to the needs of the grower and consumer. It is also desirable that cultivated varieties be able to withstand adverse conditions of cold or drought, and show resistance to disease and insect pests.

The progress of the science of plant breeding has been closely associated with the development of other basic plant sciences. Thus sciences like genetics, cytogenetics, cytology, taxonomy, physiology, morphology, anatomy,

---

[1]Musahib-ud-Din Khan is Director General (Rtd.), Ayub Agriculture Research Institute, Faisalabad. Dr. Habib-ul-Rahman Mian is Professor of Plant Breeding at the NWFP Agricultural University, Peshawar.

and biochemistry have close relationships with plant breeding. Since diseases and insect pests cause losses in the production of economic crops, knowledge of plant pathology and entomology is also essential. In an effort to breed better varieties, the plant breeder employs the science of biometry to analyse data that reveal differences among the developed and old varieties.

The most fascinating property of living organisms is reproduction, a sequence of events through which the continuity as well as the multiplication of cells and individuals of every species is maintained. Plants are peculiar in that they reproduce by many different methods. For this reason, the method of perpetuating a particular character, modified by a plant breeder, is primarily determined by the mode of reproduction of that plant. It is essential, therefore, that a plant breeder be fully conversant with the mode of reproduction of the plant species in his breeding programme. Plants can reproduce by (a) asexual, and (b) sexual methods.

5.2 Asexual reproduction

It is not always convenient to propagate crop species from seed. In such cases vegetative methods of propagation are used. This method is sometimes adopted because certain plants, like banana, ginger, or chrysanthemum, do not regularly produce seeds for genetic or physiological reasons. In many cases vegetative propagation is employed to ensure the genetic purity of the line. Asexual methods of reproduction may be used if a plant is a hybrid with meiotic irregularities or is sterile because of other genetic factors.

Different plant parts are used in vegetative propagation for various species. Some of these are listed below.

Stem cuttings: ornamental plants (e.g. roses, bougainvillaea, pelargonium); hedge plants (e.g. euphorbia, duranta, clerodendron); fruit plants (e.g. sweet lime, pomegranate, fig, *falsa*)

Tubers: potatoes, yams

Corms: *colocasia*

Bulbs: onions, garlic

Rhizome: ginger, turmeric

Suckers and offshoots: banana, aloes, agave, date palm, pineapple, and chrysanthemum

Grafts and buds: several fruit and flower species, e.g. mango, citrus, apple, pear, plum, apricot, roses

Layering: flowering plants like rose, jasmine, magnolia

Other methods of vegetative propagation involve the use of root suckers, tuberous roots (dahlia), bulbils (dioscorea), and polyembryony (citrus).

Cloning. A plant population that is propagated vegetatively from a single original stock is called a **clone**. A clone may be more precisely defined as a population of identical genotypic composition produced by mitosis traceable to a single ancestral zygote. In the case of fruit trees, if a mutation in a vegetative bud occurs and it proves superior, clonal propagation is sought and a new variety can be obtained. Washington Navel orange and Cayenne pineapple were developed in this manner.

5.3 Sexual reproduction

The most common mode of reproduction in plants is by sexual means. This involves the formation of seed through the union of two gametes—the sperm (male) and the egg (female). Pollen grains carry the sperm, and the ovules carry the egg. Pollination, the physical transference of pollen from the anthers to the stigma, is essential for fertilization and seed formation. Flowers are the organs of sexual reproduction, and there is a direct relationship between the floral biology of a plant and the technique to be used for hybridization work. Study of the structure and arrangement of various parts of the flower and their maturation times is thus of great importance in plant breeding. The more common types of flowers are described in Chapter 2, section 2.2.

Relationships between the maturation times of various parts of the flower are characterized as follows. In **dichogamous** flowers, the stigmas and anthers of the same flowers mature at different times (*dacus*, *delphinium*, *lactuca*), and in **homogamous** flowers, the androecium and gynoecium mature simultaneously. In **protogynous** flowers, the stigmas mature ahead of the anthers (colchicum, magnolia, and mango).

5.3.1 Pollination

According to the mode of pollination, horticultural crops can be grouped into several categories: (a) naturally self–pollinated, (b) naturally cross–pollinated, (c) dioecious, and (d) self–sterile plants.

1. Naturally self–pollinated plants. Some plants of this type are soybeans, peanuts, tobacco, tomatoes, peppers, eggplant, peas, asters, and marigolds. With such plants, self–pollination is the rule, and a genetic character once fixed (having become homozygous) will continue to breed true without any artificial effort to protect from cross–pollination. However, a low proportion of natural cross–pollination may occur in an established self–pollinated population. Such plants are called **autogamous**.

2. Naturally cross–pollinated group. Prominent crop plants in this group are maize, sugar beets, cucurbits, various fruit, and most root vegetables. These plants have anemophilous or entomophilous habits; enormous quantities of pollen are produced, which in most cases ensures 100 percent cross–pollination. Such plants are **allogamous**.

3. Dioecious plants. Date palm, papaya, spinach, asparagus, and willow belong to this group. In these species the male and female flowers are on separate plants. In such plants cross–pollination is the only possibility.

4. Self–sterile plants. Because of genetic or physiological conditions, sometimes fertilization cannot occur by the flower's own pollen, as in some orchids, tea, and apple. In **herkogamous** flowers, the relative positions of stigma and anthers do not allow self–pollination to take place, because of differences in the lengths of style or filaments. This condition is also known as **heterostyly**.

Flowers may show different adaptations with respect to means of pollination. Pollen (which contains male gametes), can be transferred from one flower to another by wind, water, and insects, which are attracted to flowers by their colour and scent. Plants which are pollinated by insects are **entomophilous**, and those in which pollination is effected by wind are called **anemophilous**. Flowers offer nectar and pollen as a protein source to insects. Honeybees and bumblebees are the most important pollinators of cultivated plants, although other insects, especially flies, also play a significant role in pollination of many crops and are the major pollinators of mango and Umbelliferae such as carrots and parsnips.

Perhaps the best–known association of insect and floral pollination is between the *Blastophaga* wasp and the fig. Wild figs bear three kinds of closed saclike inflorescences: male, female, and a specialised sterile type known as the **caprifig,** which functions solely to nurture the larvae of the wasp. Most cultivated figs are not of the caprifig type, and when fertilization is necessary for fruit development as in the variety Smyrna (used for drying), it is customary to tie some caprifigs in the plantations to attract wasps to effect pollination. This is, however, unnecessary for cultivars which produce edible fruits without fertilization.

Wind is a major agent of pollination in some Chenopodiaceae including beets and spinach; and in many catkin–bearing trees such as hazel–filbert, papaw (*Asimina triloba*), and mulberry, pollination is also effected by insects.

Hand pollination and emasculation. For controlled crossing it is often necessary to hand pollinate flowers. The seed (female) parent is emasculated by carefully removing the anthers or entire stamens with forceps. This must be done before the anthers have dehisced, either before the flower buds have opened or shortly afterwards. The forceps must be sterilized by dipping in 75% alcohol for 10–20 minutes and allowed to dry when different flowers are to be emasculated. In some families (e.g. Leguminosae, Rutaceae, and

Malvaceae), the filaments are joined to form a tube around the style, and therefore one must emasculate the flower carefully without damaging the female parts. In lettuce the flowers are very small but they are protandrous, and a useful technique is to wash them off with a jet of water shortly after they have opened. Pollen is often applied to the stigma with a fine sable–hair brush, or better, the dehisced anther is rubbed directly on the stigma.

After pollination the flowers must be protected by bagging. Bags of muslin cloth or Kraft paper are most useful. These can be tied to a wooden support for weak–stemmed plants. With some plants that have large flowers, such as *Lilium*, a small piece of aluminum foil wrapped around the stigma after pollination prevents contamination from foreign pollen. Crosses should be labelled with tags or on the bag, indicating date and parentage, and using permanent ink or pencil that will not fade or wash off.

Male sterility. Some species are male sterile; that is, the pollen is not functional. This condition in many cases is controlled by cytoplasmic factors, e.g. onion and leek (*Allium*), beet (*Beta)*, sweet pepper (*Capsicum)*, carrot, (*Daucus)*, sunflower (*Helianthus*), and tomato (*Lycopersicon)*. In other plants, genetic male sterility is found (e.g. cabbage, fennel, lettuce, and field beans). Sometimes more than one gene controls the condition, as in carrots. The above mentioned species are normally hermaphrodite, but non–functional anthers or pollen are produced. Male sterility is useful to the breeder in controlling crossing and production of hybrid cultivars. Cytoplasmic male sterility is the most useful, because it is possible to maintain male–sterile lines easily by pollinating seed plants with fertile pollen from sister lines which give all male–sterile plants.

Tests for pollen viability. (i) Pollen should be examined under a microscope in a drop of normal saline (0.85% aqueous sodium chloride solution). (ii) Staining with acetocarmine can distinguish viable pollen from immature grains. (iii) The best test for pollen viability is by germinating it in an artificial culture medium at 20–25°C. For dates and mangoes, 10–20% aqueous sucrose solution can be used. For germination of mandarin pollen, a 5–10% solution of glucose is the best. Adding 10 ppm borax to the culture medium enhances pollen germination.

5.3.2 Fertilization

Embryo sac and egg cell. The embryo sac is located within the ovule in a tissue of thin–walled cells, the **nucellus**. It is commonly oblong in shape and contains three groups of cells. Close to the micropyle end of the ovule is the egg, accompanied by two specialised cells called **synergids**. The synergids usually die shortly after fertilization. More or less in the middle lie two polar nuclei, which after fusion with one of the generative nuclei grow into the **endosperm**. At the other end of the embryo sac there are generally three

antipodal cells. The exact function of the synergids and the antipodal cells is not known. The fusion of one generative nucleus with the egg, that produces the zygote, and the other with the two already fused polar nuclei is known as **double fertilization**.

Parthenocarpy. Some plants can experience normal fruit set and development without any fertilization or seed formation. Several types of parthenocarpy occur, some requiring pollination and others not. Parthenocarpy may be VEGETATIVE or STIMULATIVE. In vegetative parthenocarpy, the fruit develops without any pollination as in tomatoes, peppers, pumpkins, and cucumbers, and the consistently seedless citrus, banana, and pineapple. In stimulative parthenocarpy, pollination occurs but fruit development is stimulated without the pollen tube ever reaching the ovule and effecting fertilization. A prominent example is parthenogenetic species of Poa. Fruits of triploid plants which are genetically sterile also fit into this group. Seedlessness can also occur as a result of the abortion of the embryo before the fruit reaches maturity. Embryo abortion is common in some cherries, peaches, and grapes.

Apomixis. Apomixis is a reproductive process in which seeds or propagules are produced from an unfertilized egg, or from nucellar or embryo-sac cells other than the egg. It is of many types.

1. In a few cases the seed develops directly from the nucellus tissue, no embryo sac is formed, and several processes are completely bypassed. This occurs frequently in some species of *Rubus*, *Allium*, and *Opuntia*, and is a barrier to hybridization since the new seedlings produced are genetically identical with the mother. In certain circumstances apomixis has been used to propagate apple rootstocks which are difficult to propagate vegetatively.

2. A normal embryo sac is formed and the egg cell develops into an embryo without fertilization. The embryo and eventually the seedling then have a haploid number of chromosomes of maternal origin only. In most cases pollination and sometimes fertilization of the polar nuclei takes place but the egg cell remains unfertilized. The developing endosperm which grows from the polar nuclei stimulates the unfertilized egg to grow.

3. The most common type of apomictic seed production is when an embryo sac is formed from a diploid cell without reduction division. In many cases pollination, which provides a stimulus, is necessary before seed is produced. This type of apomixis is referred to as **pseudogamy** or false fertilization.

4. **Polyembryony** refers to the situation when more than one seedling sprout from one seed. It is found in mangoes and citrus. One embryo in polyembryonic seeds develops from the zygote and the remaining ones are apomictic. The multiple seedlings from polyembryonic seeds will, therefore, include one seedling of sexual origin and the rest of vegetative (non-sexual) origin.

Because of this, it is difficult to locate the sexual variant among the progeny when doing hybridization of mango and citrus.

5.3.3 Incompatibility

In the normal course of events, pollination is achieved by transferring viable pollen to the stigma of a flower with a viable embryo sac. However, the processes leading to fertilization and seed development may be blocked if pollen germination or pollen tube growth down the stylar tissue fails. The blockage occurring between pollen germination and the release of sperm in the embryo sac is generally referred to as **incompatibility**. Many plant species have gene-controlled natural incompatibility which prevents or deters self-fertilization or fertilization between siblings, consequently preventing inbreeding.

Incompatibility reactions may be due to various causes. (i) The stigma may inhibit the effectiveness of pollen by suppressing its germination. (ii) Pollen germination may take place, but pollen tube growth through the tissue of the style be inhibited at some stage along its length. Failure may even occur when the tube reaches the micropyle but does not penetrate it. Inhibition of pollen tube growth in the style may be due to the formation of inhibitors which diffuse out from the ovary. In citrus flowers, for example, incompatible pollen germinates normally on the stigma and grows through the style but fails to penetrate the ovule. (iii) Genetic incompatibility is controlled by multiple alleles of which two are present in the diploid stylar tissue and one in the haploid pollen. If the latter is identical with either of the stylar alleles, incompatibility results.

5.3.4 Fruit set and development

Developmental changes which mark the transition of the flower into a young fruit begin soon after pollination, and are called **fruit set**. Flowers that fail to set fruit wither and fall off. The term *fruit* is difficult to define accurately, but may be described very broadly as the enlarged or altered wall of the ovary containing one or more seeds.

Fruit and seed development are often interdependent, so that failure of either one affects both. This can be very important to the plant breeder wishing to achieve a certain cross. In apples and pears, for example, the developing seeds produce hormones which stimulate fruit growth. If hormone production falls to a low level because very few seeds have been set, it may be insufficient to maintain fruit growth; in such cases an abscission layer is formed, the fruit is shed, and the seed lost. This situation has sometimes been overcome by applying auxins like gibberellic acid (GA) or naphthalene acetic acid (NAA) to the ovary by spray or its sodium salt in lanoline paste or dissolved in water in the early stages of development. Cytokinins have also been used for the same purpose.

A single mango inflorescence may contain as many as 5000 flowers, and an average set of five fruits to the cluster would provide a heavy crop. However, on the average, the actual percentage is much lower. The maximum number of flowers noted in any inflorescence of Langra mango was 2453, but the average was 885. Out of these, 0.38 fruits per inflorescence were obtained; that is, about three inflorescences yielded one fruit. In the case of Dusehri, the maximum number of flowers per inflorescence was 498, and the average was 324 flowers. Only 1.2 ripe fruits were obtained per inflorescence (Khan 1943).

5.4 The genetic basis of variability in crop species

Variation in chromosome number. Each species has a characteristic number of chromosomes. Most fruit, vegetable, and ornamental plants are diploid, containing two sets of homologous chromosomes—one haploid set donated by the male and the other by the female parent. However, variation in the number of sets of chromosomes or fraction thereof is commonly encountered in nature and in some cultivated horticultural species (e.g. potato, strawberry, banana). It is estimated that one–third of the angiosperms have variations of chromosome numbers different from the standard diploid set.

5.4.1 Euploids

This term applies to organisms with chromosome numbers that are multiples of some basic or genomic number designated as x. The haploid chromosome number of a species is indicated by the letter n. The hexaploid sweet potato is, therefore, $6x = 90$, with a haploid number (n) of 45. The genomic number (x) in this species is thus 15. The diploid varieties of sweet potato have $2x = 2n = 30$ chromosomes in the nuclei of their somatic cells.

a. Monoploid. One set of chromosomes (n) is characteristically found in the nuclei of some lower organisms such as fungi. In higher organisms, monoploids are usually smaller and less vigorous than the normal diploids. Monoploid plants are known but are usually sterile. They can be developed by breeding or other techniques which have been used to produce homozygous lines. At the University of Agriculture, Faisalabad (UAF), haploid plants have been produced by pollen culture of tobacco (Khan 1990). Homozygous plants have been developed in potato and asparagus by the production of doubled haploids.

b. Polyploid. This term is applied to any organism with more than 2x chromosomes, or more than 2n chromosome sets of the diploid species.

Ploidy levels higher than tetraploid are not commonly encountered in natural populations, but some of our most important crops are higher polyploids. Some strawberries, for example, are octaploid (8x). Some triploids as well as tetraploids exhibit a more robust phenotype than their diploid counterparts, often having larger leaves, flowers, and fruits (gigantism). Many commercial fruits and ornamentals are polyploids.

Rao (1962), working on the cytology of citrus, found a 72–chromosome polyploid in grapefruit, a 68–chromosome polyploid in Nasnarn (*C. aurantifolia*), and a 45–chromosome polyploid in Baramasi (*C. microcarpa*). Nasnarn and Baramasi are citrus species not commonly used for commercial cultivation. The genomic number of chromosomes in *Citrus* is 9.

c. Triploid. Three sets of chromosomes (3x) can originate from the union of a monoploid gamete (n) with a diploid gamete (2n). During meiosis the extra chromosome set of the triploid is distributed erratically and produces unbalanced gametes. Because of the sterility which characterises triploids, they are not commonly found in natural populations. However 10 percent of apple varieties (Baldwin and Gravenstein) were found to be triploid (17x3=51 chromosomes). At least 12 varieties of pears, some Japanese cherries, and tiger lily are also known to be triploid.

The seedless triploid watermelons artificially produced by Kihara in Japan have 22+11=33 chromosomes, and the *bajra* × Napier grass hybrid produced at the Ayub Agricultural Research Institute, Faisalabad is also triploid with 21 chromosomes. Of these, 14 chromosomes are derived from Napier grass and seven from *bajra*. Triploid hyacinths are known which have 3x=24 chromosomes. The gametes from these plants contain a variety of chromosome numbers ranging from 8 to 16, most of which are imbalanced, thus leading to sterility.

d. Tetraploid. Four sets of chromosomes (4x) can arise in a cell by somatic doubling or by the union of unreduced diploid (2n) gametes. In nature, doubling takes place spontaneously, but it can be induced artificially by treatment with colchicine. Two general types of tetraploids occur, each exhibiting different genetic and cytogenetic behaviour.

Chromosome doubling in somatic cells of a diploid species with chromosome sets each designated as, for example, AA, would lead to an **autotetraploid** AAAA. Here the four chromosome sets are homologous since they come from the same source. If, on the other hand, a hybrid AB produced from the cross of species A with genomes AA × species B with genomes BB undergoes chromosome doubling, an **allotetraploid** AABB is produced. In this case, the basic chromosome sets (genomes) are essentially derived from unrelated sources (Fig. 5.1).

In autotetraploid plants, each chromosome is present in four replications. During meiosis at prophase I, the chromosomes show great variation in pairing pattern, ranging from a complete failure of synapsis to regular

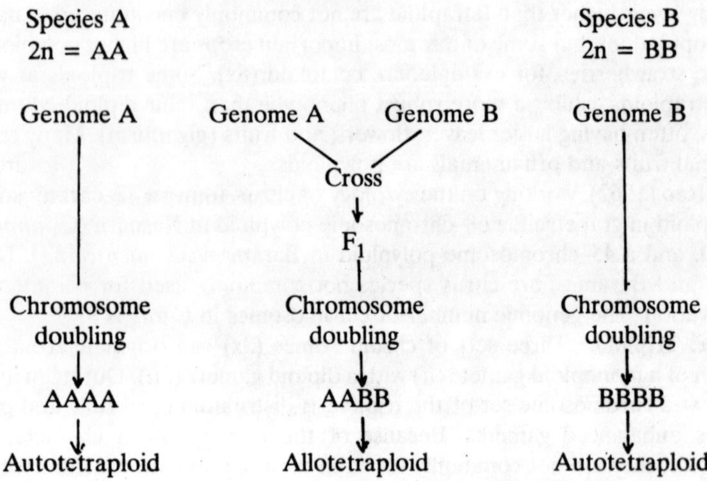

Figure 5.1 Autotetraploidy and allotetraploidy

bivalents of homologues, or association in groups of three (trivalents) or four (tetravalents). This causes irregular distribution at anaphase I, leading to a high degree of pollen sterility. Despite these abnormalities, autotetraploids are not uncommon in vegetable or fruit crops. Some prominent examples are tetraploid types of plums, tomato, apple, peanut, and coffee.

Meiotic irregularities are less common in allopolyploids because the homologous chromosomes are present only in pairs which synapse as bivalents similar to those of the diploid species.

5.4.2 Aneuploidy

Variations in chromosome number may occur which involve not whole sets of chromosomes but only parts of a genome or basic set. Thus the change involves the addition or removal of one or two chromosomes from the diploid number.

a. Monosomics. Diploid organisms which have one chromosome missing from any homologous pair are called monosomics, and have the genome formula 2n−1. Monosomics can form two kinds of gametes: (n) and (n−1). In plants, the (n−1) gametes seldom function; particularly the pollen fails to function, thus a high degree of male sterility is observed.

b. Trisomics. Trisomics are diploids which have one extra chromosome, so that triplicate members of that chromosome and a trivalent structure may be formed during meiotic prophase I. Trisomy can produce different phenotypes depending on the particular chromosome of the complement which is represented in triplicate. Trisomic plants produce two types of gametes, (n+1) and (n).

c. Nullisomics. An organism which has lost both members of a homologous chromosome pair is nullisomic (2n−2). The result is lethal to the carrier if it is a diploid. Some *polyploids*, on the other hand, can lose two homologues of a set and still survive. For example, several nullisomics of hexaploid wheat (6x=42) exhibit reduced vigour and fertility but can survive to maturity because of the high genetic redundancy in polyploids.

5.4.3 Variation in the arrangement of chromosome segments

a. Translocations. Chromosomes occasionally undergo spontaneous breakage or can be induced to rupture by exposure to high doses of ionising radiation. The broken segments may rejoin with non–homologous chromosomes. Several types of translocations are recognized. Reciprocal translocations are probably the most frequent and best understood. They occur when breaks in two non–homologous chromosomes produce an exchange of chromosome segments between them. During meiosis at the pachytene stage, reciprocal translocation forms a cross–shaped configuration which affects pairing of all homologous segments. Translocations have profound genetic and cytological effects. Genetically, they disturb linkage and create new gene orderings; cytologically, they produce meiotic irregularities, giving rise to imbalanced gametes or producing gene deficiencies and duplications (Fig. 5.2).

b. Inversions. Inversion takes place when a section within a chromosome breaks and rejoins in the same location but in reversed sequence. The size of the inverted segment, whether it involves the centromere or not, and whether it is present in both members of a homologous chromosome pair will determine the behaviour during meiosis and thereafter. If the plant is homozygous for inversion, chromosomes will not have any difficulty in pairing. In a plant heterozygous for inversion, one chromosome will have a segment with inverted gene order, and its homologue will be in the normal order. During meiosis the chromosomes attempt to achieve maximum pairing between homologous regions in the two chromosomes, which is usually accomplished by the formation of a loop in one of the chromosomes.

5.4.4 Variation in the number of chromosome segments

a. Deletions (deficiencies). A portion of the chromosome is broken and lost (Fig. 5.2). Deletions can be useful when some deleterious gene is lost from the chromosome. If the lost segment contains dominant alleles, the recessive phenotype is expressed—a situation known as **pseudo–dominance.** A deletion heterozygote may be detected cytologically during meiotic prophase

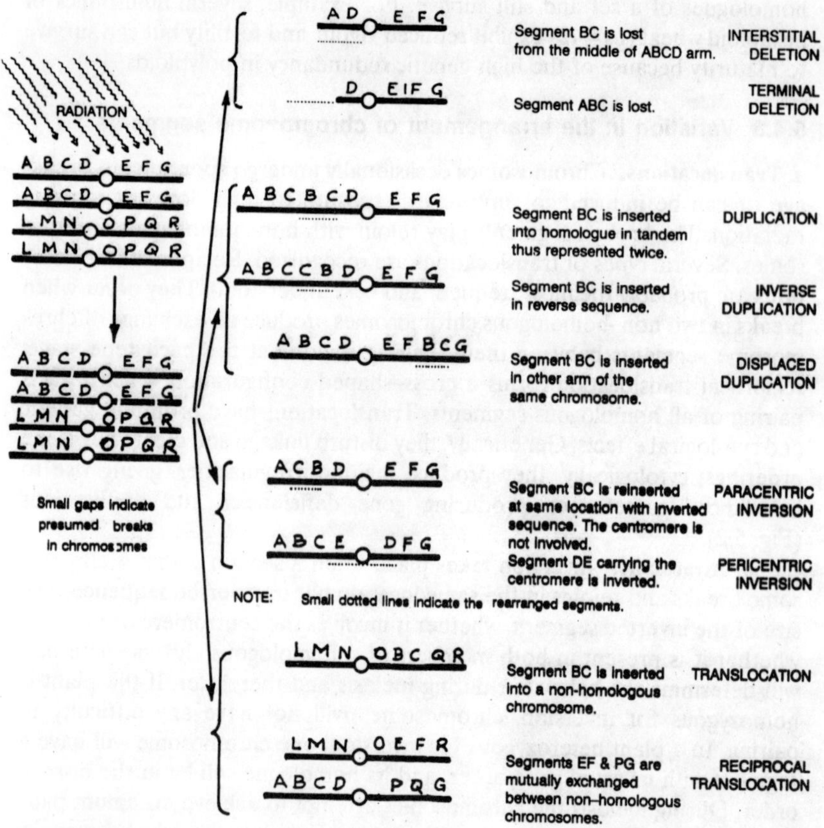

Figure 5.2 Chromosomal aberrations

when pairing causes the normal chromosome segment to bulge away from the region in which the deletion has occurred. A **terminal deletion** occurs at one of the ends of a chromosome, and an **intercalary deletion** somewhere in the middle. Terminal deletions usually do not survive and are lost during meiosis or within a few cycles of mitosis. All deletions capable of being maintained are, therefore, of the intercalary type. Large deletions cause drastic pollen sterility; such deletions may, however, be transmitted to the progeny through the egg.

b. Duplications (additions). A duplication is a doubling of a chromosome segment, which may be located in a single homologous pair or may be transposed to a non–homologous chromosome (Fig. 5.2). Duplications are not as deleterious to the organism as deletions, and are therefore better tolerated. Genetic redundancy, of which this is one type, may protect the organism from the effects of a deleterious recessive gene or from an otherwise lethal deletion. During meiotic pairing, a duplication heterozygote of the chromosome bearing the duplicated segment forms a loop similar to that formed in deletion.

5.4.5 Artificial induction of genetic variation

Chromosomal alterations or other small heritable changes are known as **mutations**. Mutations provide a valuable source of variation in plant material from which the breeder can make selections. They can arise naturally, but this happens very infrequently. It is possible to induce these changes with uV or X–rays, gamma radiation, thermal shock, or certain chemicals. Some of these procedures are described here.

1. Radiation. Natural mutations are usually rare, and it is advantageous for the plant breeder to increase the rate of mutation to obtain more variation. The most effective way to achieve this is to subject plant material to ionising radiation such as X–rays or gamma rays. These radiations can change the genes and chromosomes at random and at a high rate. Radiation treatment does not, of course, affect every cell of the tissue in the same way, so that only parts of the plant may be mutated, and a partially mutated tissue, or a sectorial chimera is then produced. The results of induced mutation in plant breeding have fallen short of the early enthusiastic claims, and to date only a few cultivars have been produced in this way.

Of special interest to horticulturists are examples like induced fruit–colour changes in apples, compact mutants in apple and pear, upright habit in black currant, and flower–colour changes in several ornamentals. The technique of X–ray induction of flower colour mutants is well proven and effective for several vegetatively propagated florist's crops. The effect of irradiation is generally to cause a mutation of the gene or genes controlling the particular character.

2. Colchicine treatment. The use of colchicine has permitted the production of the vast number of polyploids that have been developed in the past 20–30 years. Colchicine is an alkaloid obtained from the crocus plant (*Colchicum autumnale*). Colchicine selectively arrests spindle formation during mitosis, with the result that the duplicated chromosomes do not pass to the anaphase and telophase stages. They thus stay within the same nucleus, which now contains twice the original chromosome number.

It has been applied in various media: solution in water or glycerine, as lanolin paste, or in agar preparations. Relatively more success is achieved when it is applied to meristematic tissue, germinating seeds, and actively growing buds. For seed and seedling treatments, the concentration of colchicine aqueous solution can easily be varied. The most effective concentration range is from 0.1–0.4% with an exposure time from 30 minutes to three hours. After treatment, seeds are thoroughly washed. For growing buds, 0.2–0.5% solution is used and the treatment is given intermittently for two to six days.

5.4.6 Desired traits incorporated by hybridization

Varieties with desirable characters are selected and crossed. How they behave in breeding F_1, F_2, and other generations depends upon the inherent genetic makeup of each variety and the way it is expressed in hybrid combinations. In hybrid recombination, reciprocal exchange of genes by crossing over can produce novel characteristics, and by selection one may get a new hybrid variety that combines the good characters from each parental variety. Varieties can be crossed easily and fertile hybrids produced, but interspecific or intergeneric crosses are difficult and occasionally impossible.

A cross between *Citrus* and *Poncirus trifoliata* resulted in citranges. Poncirus has typical trifoliate leaves; the plants are thorny and deciduous; the fruits are small and fuzzy with many seeds and inedible. The hybrid, although of no commercial value, expresses many of the good characters of citrus. Crosses between *Fortunella* (kumquat) and *Citrus* resulted in the 'limequat'. Similarly, the famous variety Kinnow is the result of a cross between King Tangor and Willow Leaf Mandarin.

5.5 Plant breeding methods

5.5.1 Acclimatization

Centres of diversity or origin of crop plants. Plant varieties originating in a particular part of the world have not remained confined to that zone, but have been spread to other areas by several agencies, primarily man. The work of Vavilov, a Russian plant explorer, during the early part of this

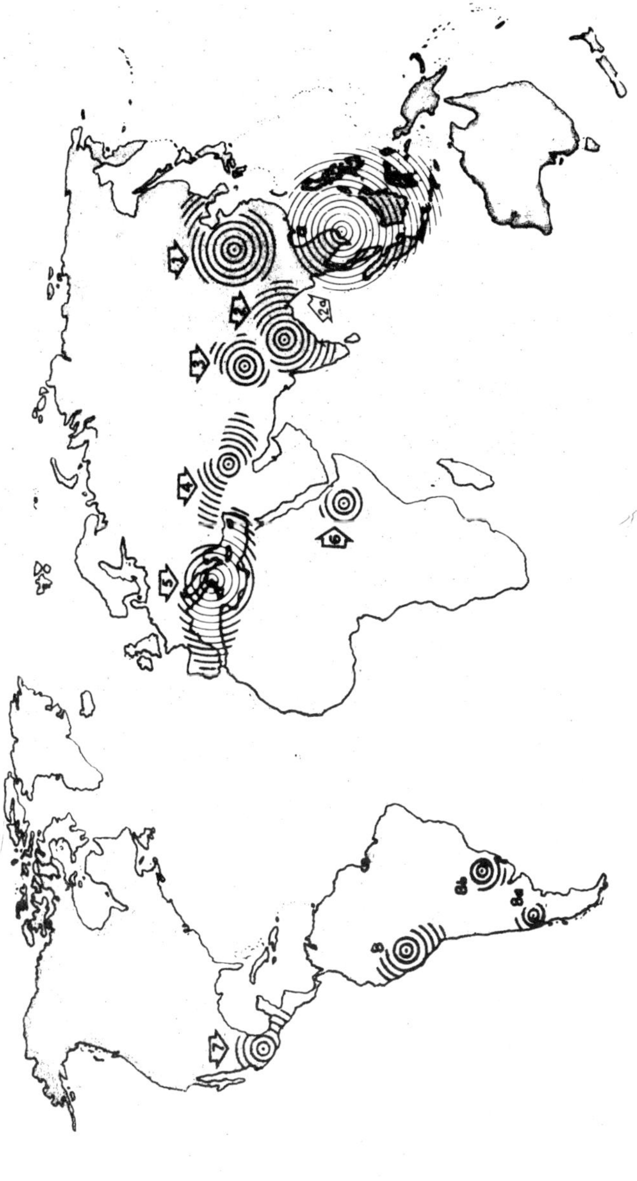

Figure 5.3 Centres of origin of some important crop plants

century showed that certain areas of the world are centres of much of the diversity in crop plants (1951:20-45). Vavilov identified the following eight centres of origin (Fig. 5.3).

1. CHINESE. The mountainous and sub–mountainous area of central China is one of the richest dispersion centres. *Pyrus* (pear) and *Prunus* species (peach, apricot), walnut, litchi, and loquat are prominent examples.

2. INDIAN. Burma and Assam – mango, jaman, tangerine, lemon, sweet orange, brinjal.

2a. INDO–MALAYAN. Islands of Java, Sumatra, Borneo, Philippines, and Indo–China. Jackfruit, grapefruit, banana.

3. CENTRAL ASIATIC. The region consisting of Pakistan (Punjab, NWFP, Kashmir), Afghanistan, Tajikistan, Uzbekistan, and western Tien–Shan. Onion, garlic, spinach, carrot, cantaloupe, radish, pistachio.

4. NEAR EASTERN. Iran, heights of Turkmenistan, Transcaucasia. Many species of cucurbits, turnip, cabbage, lettuce, walnut, almond, grapes, cherry, and pomegranate originated here.

5. MEDITERRANEAN. Barley, wild wheat, beans, chick peas.

6. ABYSSINIAN. Watermelon.

7. SOUTH MEXICAN AND CENTRAL AMERICAN. Maize, sweet potato, *Ananas*, papaya, guava, avocado.

8. SOUTH AMERICAN (Peru–Ecuador–Bolivia). Peanut, pepper, squash, potato, pineapple.

These centres are separated by deserts, mountain ranges, or large bodies of water. Of the crop plants that Vavilov associated with these centres, over 500 species belong to the old world, and approximately 100 to the new world. In the old world, southern Asia is the source of over 400 crop plants. Some plants originate from yet other centres: the date palm, of unknown origin; sunflower from the Great Plains area of North America; the macadamia nut from Australia; and several forage species of Eurasian origin.

The intentional transfer of an individual plant or a plant population from one ecological zone to another to establish it as an economic proposition is called **acclimatization**. This has been an easy approach to the improvement of crop plants. Sometimes a good ready–made plant with superior traits is found in an imported mixture. Sometimes imported plants adapt readily to the new environment and can be released directly for general cultivation after preliminary testing. Almost any species or variety is worth importing if it has not been imported before, but its commercial success will depend entirely on whether its genotype is adaptable to the new circumstances. Thus, high quality and good yield in one country do not necessarily guarantee the same in a new location. For this reason it is advisable to import a great many diversified varieties to enhance the chances of finding a suitable one. If the imported varieties do not prove useful by themselves, they can, after suitable evaluation, be utilized as a source of superior germ plasm in subsequent plant improvement projects.

A large number of fruit, vegetable, and ornamental plant species have been introduced from exotic sources and acclimatized in Pakistan. Some of these are: 148 varieties of grapes introduced from Europe and America, 35 varieties of dates introduced from the Middle East, 80 varieties each of bananas and mangoes, and 85 varieties of apples. A large number of mulberry varieties, and pomegranates from Syria and Afghanistan were also imported to find varieties suited to the agro-climatic conditions of Pakistan, but did not succeed, as the climate was not suitable for them. Several elite varieties of citrus, mangoes, dates, and bananas now form an integral part of the horticulture industry in Pakistan. World-famous varieties of apple like Golden and Red Delicious imported from Lebanon did well in the hilly tracts of the country. The Skyspur variety of apples has recently been added to the approved list.

Hundreds of varieties of peas, potatoes, tomatoes, radishes, and cauliflower have been introduced in the country for acclimatization. The famous Patronus, Desire, Multa, Cardinal, and Wilja varieties of potatoes; Roma, Red Top, T-10, and Moneymaker tomatoes; and Shao-Mi radishes have brought huge economic returns to the farming and business communities (Hussain 1983). A number of varieties of roses such as Cleopatra, Iceberg, and Queen Elizabeth; chrysanthemums like Anemone (large-flowered and reflexed); as well as varieties of carnations and marigolds have done extremely well throughout the country.

5.5.2 Selection

An essential step in picking up or isolating a trait which suits one's objectives from a large plant population is selection. Whenever we want to breed by way of selection, the most important step is to have sufficient genetic variability in the plant population. Selection acts only on genotypical differences and not on variations due to environment, which are of no value in breeding. If two plants have the same genotype and differ only because of environmental conditions, it makes no difference whether the larger or the smaller plant is used for seed because the offspring will have no significant difference. A plant with a small variation of high genetic value will be a better choice than a plant which apparently looks much different but possesses less useful genes.

Before a selection is made, exhaustive observations on the amount of natural variability in a particular population is necessary. Unimproved crops are generally a mixture of plants that are good, bad, or indifferent from the economic point of view. The method of selection to be adopted will vary with the plant's natural mode of pollination.

1. Breeding of self-pollinated crops (pure-line selection). Pure-line selection consists of repeated selection for several years of individual plants possessing the desired characters, and then raising the progeny of each plant separately. This is used in self-pollinated crops like peas and beans. The essential steps of this method are as follows. (i) A large number of individual plants are selected from a field having a genetically mixed population, keeping in mind the specific objective, e.g., disease resistance, quality, or yield. The seeds of individual plants are harvested and kept separately. (ii) The seeds of selected individual plants are sown in separate rows and the comparative purity of the offspring so obtained is studied. From a large population the most desirable plants are again selected. (iii) A similar selection of the few outstanding individuals is again made in the subsequent generation. (iv) In the third, fourth, and fifth generations the selected plants are tested in larger plots using the existing or standard varieties as control. (v) If the selected plants prove promising, their seed is multiplied, large-scale trials are made, and after thorough testing the variety is released for general cultivation.

This method is very simple and easy, but it can only be used in those populations which carry a large number of useful heritable characters. A pure line (a homozygous, true-breeding variety), because of its narrow genetic base, may not be suitable to a wide range of soil and climatic zones. It must also be clearly understood that by this method we do not create a new variety but only select a variety from existing ones. Also, no further improvement is possible in a pure line, unless genetic variability is introduced into it by artificial means.

2. Breeding of cross-pollinated crops (mass selection). In cross-pollinated crops the method applied is mass selection, although breeders make modifications to suit their specific objective. In this case, either a large number of desirable plants are selected and saved for seed for subsequent crops, or alternatively, undesirable plants are removed before they produce flowers. A number of the best plants are picked, based on phenotype, pedigree, or performance. In contrast to pure-line selection, which essentially consists of the progeny of a single selected plant, in mass selection, seed from a majority of plants is collected and bulked together. This seed is then sown together for further testing. This method is also useful because of its safety and the rapidity with which initial improvement can be made in a crop variety carrying an abundance of genetic variability.

3. Clonal selection. In some crops, sexual reproduction is not used. Potatoes, asparagus, chicory, many varieties of apples, citrus, mangoes, and ornamental plants like begonias and delphiniums are propagated asexually by budding, grafting, tubers, or cuttings. These techniques are also known as clonal propagation.

A **clone** is a population propagated asexually from a single individual. It is a biotype that has genetically identical members, which may be either homozygous or heterozygous. Clonal propagation usually leads to the perpetuation of a genetically uniform progeny. It is, however, known that in clones heritable variations do sometimes occur, although very rarely. These changes may be picked up and propagated separately. This method closely resembles the pure-line method, and a clonal variety apparently resembles a pure line.

Vegetatively propagated cultivars have their special problems. Many of them are especially susceptible to disease, and vegetative propagation often carries over disease to the new plants, whereas seed propagation usually screens out pathogens. Virus diseases especially are often difficult to eliminate from clones, and many well-known clonal cultivars have disappeared because of a buildup of deleterious viruses or mycoplasmas. Nowadays it is possible, in many cases, to eliminate viruses from clones by new techniques, such as meristem culture or heat shock or a combination of both.

5.5.3 Hybridization

Hybridization is the sexual mating of plants containing desirable characters. The object is to combine in a single variety the useful characteristics of two or more lines, varieties, or species. The first and most important step in hybridization is the wise selection of parents for a cross, which is made on the basis of the traits to be combined and their mode of inheritance. Usually crosses are made either to introduce desirable characteristics into an already established variety, or to replace that variety by another one which is expected to be evolved by mating thoroughly studied individuals, i.e. plants which have proven their worth through several generations. It is thus important that when selecting for parents to be crossed, the genetic behaviour of the characteristics in question be clearly understood.

Hybrids may be interspecific, intergeneric, or intervarietal. Interspecific and intergeneric hybridization present numerous difficulties. They are generally characterized by genetic abnormalities, of which sterility is the most common. It is due either to the failure of chromosomes to form normal pairings at meiosis or to genetic imbalance. Gene segregation in such hybrids is very characteristic, and every individual plant in F_2 is likely to differ. This situation may continue for several subsequent generations as well. Moreover, it is hard to achieve all the possible combinations. In some cases, especially when the chromosome numbers of the species are different, special techniques have to be employed for successful crossing.

Techniques. Crossing work may be carried out in the field or in the greenhouse. Since it is essential to provide optimum temperature, moisture, and plant nutrient supplies, many workers have suggested crossing in the greenhouse. Not only can environmental factors be easily controlled in the greenhouse, but it also offers better protection. Sometimes in interspecific hybridization the flowering times of the two parent species differ, so that it becomes necessary to sow them on different dates in order to synchronize the flowering periods of the two parents. Crossing should be done after making a thorough study of the anthesis and pollination behavior of the flowers of the individual species.

First an unopened flower is selected on a healthy branch of the desired female plant. This flower is carefully emasculated and a suitable paper or cloth bag is securely tied over it to prevent foreign pollen from falling on the stigma. When the stigma becomes receptive (which should be determined from previous observation), the bag is gently removed and pollen collected from the anthers of the male plant is deposited on the stigma. This should be done with a fine, soft hair brush. The bag is then re-tied and information concerning parentage, date, and other relevant points is written on it with a lead pencil or water-resistant marker.

Handling the hybrid materials. After the crosses are successfully secured, the next step is to maintain the progeny carefully. All possible efforts should be made to make the F_1 plants produce the maximum possible amount of seed for raising the subsequent F_2 generation. In F_2 there will be segregation of characters, and selection has to be initiated right from then. In order to make it more effective, selection should be done in the smallest possible units of characters. In order to regain stability of the material and obtain new true-breeding types, it is necessary to inbreed the selected lines again for five or six consecutive generations to fix the characters and regain the desired level of homozygosity. However, if the plants are propagated vegetatively or by apomictic means, this inbreeding will not be necessary. In this event, the desirable phenotypes selected in F_2 can be multiplied vegetatively and carried for further evaluation.

Although several attempts have been made in Pakistan to produce horticultural varieties by hybridization, they have met with scant success. Crossing work of different citrus varieties and species is in progress both at the UAF and Ayub Agricultural Research Institute (AARI) in Faisalabad. Hybridization work on mangoes is in progress at the Mango Research Station, Shujabad, and is expected to yield good results. Workers at AARI have succeeded in producing a hybrid pea, Samarina Zard, from a cross between Meteor and FC–3954. This hybrid is an early, short-duration, high-yielding variety. It is on the approved list of the Department of Agriculture and recommended for cultivation as a general-purpose variety (Hussain 1990). In India, a new variety of mango, Malika, has recently been produced by hybridizing Dusehri and Neelum, combining the good qualities of both the parents (Nijjar 1972).

Inbreeding depression. When plants are inbred, after several generations of selfing or by sib-crossing (brother–sister mating) the progeny become progressively more homozygous, but frequently they suffer inbreeding depression. The inbred generations are less vigorous, and from a horticultural point of view, less productive than the original parent material. There may be more than one explanation for this condition, but primarily it is related to the accumulation of deleterious or even semi-lethal genes in homozygous condition. These do not have the same effect in the original parents because there they are in the heterozygous condition. The degree of inbreeding depression varies with species. Peas and French beans, for example, show very little loss of vigour on inbreeding and cucurbits almost none. At the other extreme, Brussels sprouts rapidly lose vigour after two or three generations of inbreeding. The loss of general vigour may also be accompanied by impaired fertility, resulting in poor fruit and seed set.

Heterosis. When the parents involved in a cross are genetically remotely related, the progeny may express **heterosis**, or hybrid vigour, a fascinating biological phenomenon. The seedlings of the F_1 generation produced by crossing inbred lines may be more vigorous than the parents. Heterosis is generally assumed to result from bringing together a maximal number of growth factors. The phenomenon may be due to the concentration of a large number of dominant genes which are favourable for growth and development, some contributed by one parent and some by the other. Another explanation is that in a heterozygous condition the alleles at any specific locus develop a divergent function. Thus in a heterozygous condition they interact to produce a better function than in the homozygous condition. The large number of genes and their linkage in the hybrid virtually prevent the perpetuation of the same degree of hybrid vigour in subsequent generations. Most quantitative traits such as height, weight, and size are multigenic, i.e. each is controlled by a large number of genes. These genes are likely to be located on different chromosomes; therefore no line is expected to have all the possible favourable genes. Probably most varieties have a moderate proportion of favourable genes, and if the majority of these genes have some degree of dominance, even though it is small, F_1 is the only generation with the maximum number of such genes. This is why maximum heterosis is expressed in F_1, and in subsequent generations the genes begin to segregate, which rapidly leads to a decline in heterosis as well. It is necessary, therefore, to repeat the parental cross each time to maintain maximum vigour.

In any study of hybrid vigour it is important to realize that besides the major characters inherited in simple Mendelian fashion, there are minor genes determining quantitative characters. The problem of hybrid vigour is complex, and consequently in most plants it is not open to a simple explanation. Actually there may be positive or negative heterosis, so that the synonymous use of *hybrid vigour* and *heterosis* is not always appropriate. Heterosis is the developmental stimulation resulting from the union of different gametes, while hybrid vigour is one manifestation of heterosis.

The maximum effects of hybrid vigour are always evident in the generation immediately following a cross, i.e. F_1. It is expressed in different ways in different plants and therefore has to be measured differently. In tomatoes and eggplant, heterosis is expressed not so much by an increase in size of the fruit as by an increase in their number. With mangoes, a cross between Neelum and Dusehri resulted in Malika, which was superior to both the parents in average weight, pulp percentage, total soluble solids (TSS), and keeping quality (Singh et al. 1977). Studies in the Vegetable Section of the AARI on onions, radishes, and brinjals have given interesting results. With radishes, the hybrid from the cross of Desi x English Long Red gave a root length up to 8 cm, while each of the parents had roots only 5 cm long. Similarly, crossing purple–red and white varieties of onions produced a hybrid with an average bulb weight of 50 g as against 20 g of purple–red and 12 g of the white variety. In brinjal, there was an average increase of 9 percent in the fruit weight of the hybrid (Khan 1960).

Commercial utilization of heterosis. Since maximum heterosis appears in F_1, it is necessary in plants raised from seed to repeat the parental cross each time if maximum vigour is to be maintained. Material benefit in increasing the yield of crops can only be derived from heterosis if the hybrid seeds are produced in large quantities. Special techniques and methods have been developed to overcome the difficulties of economical hybrid seed production. Heterosis has been utilized commercially in (1) cross–pollinated crops: sugar beets, radishes, sunflowers, onions, cabbage, spinach, Brussels sprouts, cucumbers, and snapdragons; and (2) self–pollinated crops: tomatoes and brinjals.

5.5.4 Backcross method

The backcross technique of breeding consists primarily of making successive backcrosses of the progeny with a parental variety which has to be improved. This method has proved indispensable for introducing one or several simply–inherited characteristics from foreign or wild sources to an improved and adapted variety which may already be good in many other characteristics. For instance, the potato variety Ultimus has a high yield and possesses good cooking and storing qualities, but is susceptible to frost and chill. A few wild species, *S. demissum*, *S. acaule*, and *S. andigenum*, have been imported from South America and have been sown in the Murree Hills. In the potato breeding programme of the Vegetable Section, the backcross method is being used. Briefly, the steps of the procedure are: (i) crossing of *S. tuberosum* and *S. andigenum*; (ii) crossing of F_1 obtained from this cross back to the recurrent parent, *S. tuberosum;* (iii) selecting from this progeny desirable individuals manifesting frost resistance (Khan 1960).

5.6 Biotechnology, genetic engineering, and crop improvement

Biotechnology techniques, often called *genetic engineering*, include *in vitro* cell culture, cell selection, protoplast fusion, and recombinant DNA. A shortcut to utilizing variability is to select a population of cells rather than a population of plants. For the technique to be successful, it must be possible to regenerate plants from individual cells or groups of cells obtained from suitable plant organs, e.g. leaf, meristem, or root. A large population of cells, called a **callus**, can be grown in a very small space like a petri dish or a test tube. The callus can also be induced to develop roots and shoots and eventually a complete plant.

The problem is the difficulty of selecting variants at the cellular level that will be expressed as desirable characteristics in the full–grown plant. The culture system itself appears to generate variants with a higher frequency than would be expected as a result of spontaneous gène mutation. Such variants are called **somaclonal variants**. They may be due to gene and chromosomal changes, or to non–genetic adaptive changes; the precise cause is still unknown. Somaclonal variation induced by cell and tissue culture appears to be a new source of variation available to the plant breeder.

5.6.1 Protoplast fusion

Plant hybridization has been shown to be feasible with protoplasts. Protoplasts can be isolated from cells by digesting the cell walls with appropriate enzymes. A small but significant percentage of protoplasts placed under suitable conditions can be induced to fuse. Thus if the protoplasts of two genetic backgrounds are mixed, some of the fused protoplasts will be nuclear hybrid, some will have mixed cytoplasm (**cybrids**), and some may be both. Then the hybrid protoplasts must be isolated from protoplasts of the same genotype. One method is to use a medium in which only hybrid products will grow. Mature plant hybrids have been produced from cellular hybridization in tobacco and petunias. Refinement of this process may make it possible to achieve wide crosses that are impossible with traditional sexual methods of hybridization.

5.6.2 Anther culture

Pollen can be cultured to produce haploid cells or even haploid plants. Haploid cell lines of plants are extremely valuable for selecting useful recessive alleles and mutation breeding, because when dominance is eliminated, all mutated recessive genes will be immediately expressed. Further doubling such haploids produces 'instant' homozygous diploids, a process that may require many years of inbreeding with normal sexual breeding.

5.6.3 Recombinant DNA

The basis of recombinant DNA technology is the ability to isolate DNA fragments containing one or more genes and replicate them in a different host, such as bacteria. Enzymes called **restriction endonucleases** permit DNA to be cleaved at desired points and into fragments of desired sizes that will hybridize, i.e. attach to other complementary DNA fragments. The vector used is a circular bacterial DNA called a **plasmid**, which in certain bacterial species is present as episomes (in addition to the normal chromosome). The plasmids are modified to accept foreign DNA, which is incorporated by genetic recombination and then labelled with biochemical markers to allow them to be identified after replication in the bacteria. The replication and subsequent selection of these bacteria containing the recombinant DNA is called **gene cloning** (See Fig. 5.4).

Examples of successful gene cloning include: (i) The human insulin gene transferred to bacteria by genetic engineering produces a human insulin called **humulin**. (ii) Herbicide resistance genes have been successfully transferred to certain plant species by genetic engineering.

KEY STEPS FOR RECOMBINANT DNA FOR CROP IMPROVEMENT

1. **Precise identification and isolation of the gene of interest.** The DNA of the cell's genome is cleaved by using appropriate restriction enzymes; in this way a library of DNA fragments of various sizes is produced. These fragments are compared by separation with electrophoresis techniques, and then their relative size is determined. From these fragments the gene of interest can be picked up by special techniques of mRNA–DNA hybridization.

2. **Producing recombinant DNA.** The isolated gene is inserted into the chromosome of a vector (an appropriate virus or plasmid). This vector carries and transfers the gene into other, host cells. The gene will now be replicated along with the plasmid or viral genome inside the host cell.

3. **Gene cloning.** Occasionally the vector carrying the gene integrates into the host cell genome. Such cells are identified and maintained separately. Each such cell replicates to produce a large colony or clone. Further screening of clonal libraries can be done if necessary to obtain pure geneti-

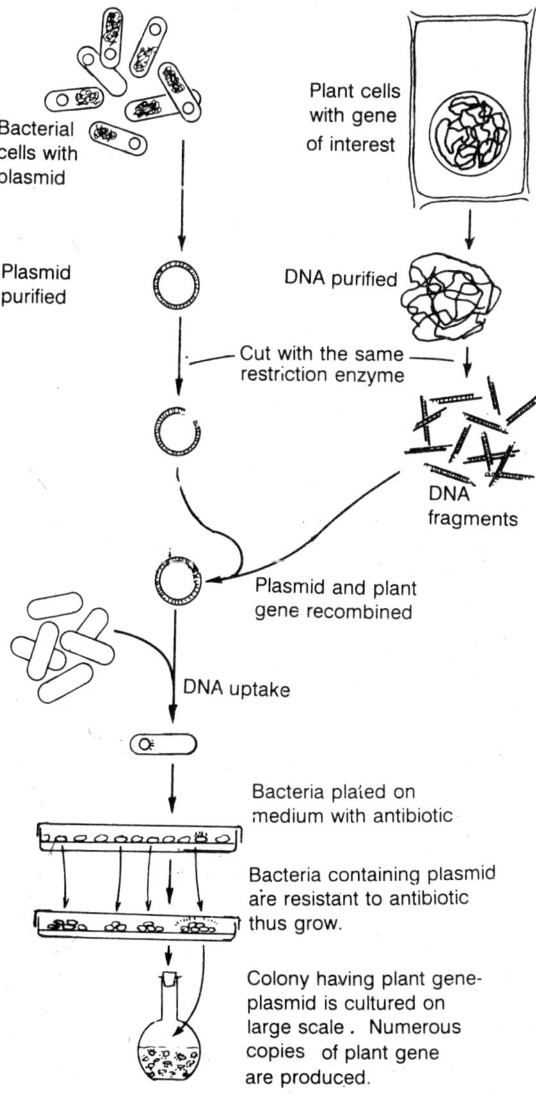

Figure 5.4 Gene cloning

cally–engineered cell lines that contain the gene of interest.

4. Plant regeneration. Regeneration of the transformed cells into complete plants which express the new gene introduced through the above steps.

5.7 Breeding of individual fruits, vegetables, and ornamentals

5.7.1 Apple

The apple (*Pyrus malus*) belongs to the family Rosaceae, sub–family Pomoideae, with chromosome number 2n=34. The Pomoideae may be allopolyploids derived from a double hybrid between two remote ancestors, one with the basic number 8 and the other 9, giving rise to a 34–chromosome species. Triploids, 2n=51, are also occasionally found among the cultivars normally grown in orchards. These have arisen naturally from the fertilization of unreduced gametes. Ten percent of the commonly grown cultivars are triploids, e.g. the Baldwin, Gravenstein, Blenheim, and Mutsu varieties. They are vigorous and bear large fruit, but produce only a small amount of poorly formed pollen, so a diploid is necessary to pollinate the triploid. Aneuploid plants are also found, but they are weak and slow–growing, and seldom develop into trees.

Floral biology. Flowers are usually borne on spurs, in clusters each containing five petals, five sepals, and 20 stamens. The pistil is divided into five styles. The inferior ovary has five carpels each bound together by the hollow floral receptacle so that only the styles are free. Each locule contains two ovules; if all the seeds develop there are 10 of them. The fruit resembles a berry (pome), and on maturity the fruit–receptacle becomes fleshy and succulent.

Breeding objectives. Apart from consumer preferences and requirements of the processing industry, the primary objectives in apple breeding are high yield, an annual crop of good quality fruits, disease and pest–resistance, and firm juicy fruit with pleasant flavour. Large (but not excessively so) fruit commands premium prices.

PROCEDURES IN CROSSING

Emasculation. All apple flowers are hermaphrodite (contain both male and female sexual parts). Flowers that are to be used in breeding as female have to be emasculated. This consists of the removal of stamens before the anthers dehisce. To facilitate the procedure, the calyx and corolla are also removed. For this purpose, two flowers per cluster are sufficient. Fruit from unemasculated flowers can be identified, as fruit from emasculated flowers has no calyx. Experiments have shown that insects do not visit flowers when the stamens and petals have been removed, and there is thus no need to protect them from insects. However, fruit set is improved if the emasculated flowers are protected with suitable paper bags, since the desiccation of wounded floral parts is lessened. Such protection is particularly desirable in windy and exposed situations.

Pollination. Most apple varieties are naturally cross–pollinated, with insects and wind playing a key role in pollination. For artificial crossing, pollen from the flowers of one tree is transferred to the stigma of another tree. Flowers are chosen at the balloon stage just before the anthers dehisce. The anthers are removed and collected in a clean, dry petri plate which is placed in a warm and sunny position until they dehisce and liberate pollen. If necessary, pollen should be stored in a desiccator with $CaCl_2$ and kept cool in a refrigerator. If it is to be stored for a long period, it should be put in a deep freeze unit at −15°C. If proper conditions are maintained, dry pollen will remain viable for up to a year. Dry pollen can be sent to different places hermetically packed in cellophane. The quality and viability of pollen can be tested by following the procedures described in section 5.3.1.

Pollination is achieved by lightly brushing the pollen onto the stigma of an emasculated flower when it becomes receptive. To prevent contamination, the brush is cleaned by washing in 70% alcohol and drying before using for another cross. Pollination can be done on small trees planted in pots or suitable plastic containers, preferably kept in a cool greenhouse. The whole tree can be pollinated with one parent, or a number of crosses may be made on different parts of the tree. If this is done, each different cross should be labelled carefully, indicating parentage, date, and other relevant information.

Problems of sterility and incompatibility are sometimes encountered in crossing. Sterility can be due to inviable pollen or genetic reasons. Incompatibility, the failure of functional pollen to grow down the style, occurs widely in apples. Both self–incompatibility and cross–incompatibility occur. All apple cultivars are self–incompatible to some extent; some are completely so. If they are cross–pollinated, they will set more fruit with more seeds. Selfing of the varieties Cox's Orange Pippin, Golden Russet, and Northern Spy on a large scale set only 1.5 percent fruit, but when these varieties were crossed more than 7 percent fruit set was observed.

Stratification of seed. Stratification is the storing of seeds in a cold humid environment for some time. This allows an after–ripening process during which changes take place in the embryo, and is commonly practised to improve germination of hybrid seeds. The temperature range for stratification is from 0–10°C, the optimum being 3–5°C. The time required is 6–14 weeks, depending on the temperature. Seeds to be stratified should be extracted immediately after fruit harvest. A 5 mm thick plastic foam sheet is placed in a tray, and seeds are placed on it in a single layer. Then the seeds are covered with a second plastic sheet. Water is added to just moisten the lower plastic foam. This tray is then put in a polyethylene bag to prevent drying and placed in a refrigerator. After six weeks the seeds should be examined and used for propagation.

5.7.2 Citrus

The citrus-producing regions occupy a belt spreading approximately 35 degrees north and south of the equator. There are about 50 major citrus producing countries in the world, and citrus fruits are grown on over 1.6 million hectares. Oranges account for 75% of this total, and lemons and grapefruit 10%.

The genus *Citrus* belongs to the family Rutaceae and sub-family Aurantiodeae. Other related genera are *Poncirus*, *Fortunella*, *Eremocitrus,* and *Microcitrus*. Citrus species are shrubs or small to medium size evergreen trees. Most species have thorns, and the leaves are unifoliate. Flowers appear in clusters in the leaf axil, and produce nectar and a strong perfume. They are self-pollinated or cross-pollinated by bees.

Floral biology. In subtropical zones with cool winter temperatures, citrus species flower once a year during early spring. In the tropics and in coastal areas, flowering may occur several times a year. The ovary of citrus is composed of 6–14 carpels joined to each other and to a central axis (Fig. 5.5). The flowers of Rutaceae are regular with superior (hypogynous) multilocular ovaries. A ringlike or lobed disk surrounds the ovary around which 4 or 5 velvety petals develop. There are 4 or 5 free sepals, and 8–10 (or even more) free stamens inserted in a disk. The 6–14 carpels are usually raised by a gynophore. The seeds may or may not contain endosperm; this character varies from species to species. In some species polyembryonic seeds also develop. The stigma is receptive from one to a few days before anthesis and for several days afterwards. Modified epidermal cells on the stigma secrete viscous fluid that aids in retention and germination of the pollen.

Sex of flowers. The proportion of staminate flowers is highly variable, depending on the cultivar and growth conditions. Low temperature during flower formation is reported to increase the percentage of aborted pistils. Leafless inflorescences have a higher percentage of staminate flowers than leafy inflorescences. A large percentage of perfect flowers also may fail to set fruit. Heavy loss of set fruit also occurs, much of it shortly after initial set. With lemons, only 7% of the buds develop mature fruit. In sweet oranges, the percentage varies from a low of 0.2% to over 5% depending on the cultivar. While studying the erratic behaviour of fruit formation in sweet lime, Ibrahim (1988) found 30,134 flowers on a tree, out of which 86% were male and 14% female; 3.89% fruit was set.

Anthers and pollen. Normal anthers are bright yellow at maturity. Self-pollination may occur due to nearness of the anthers to the stigma; however,

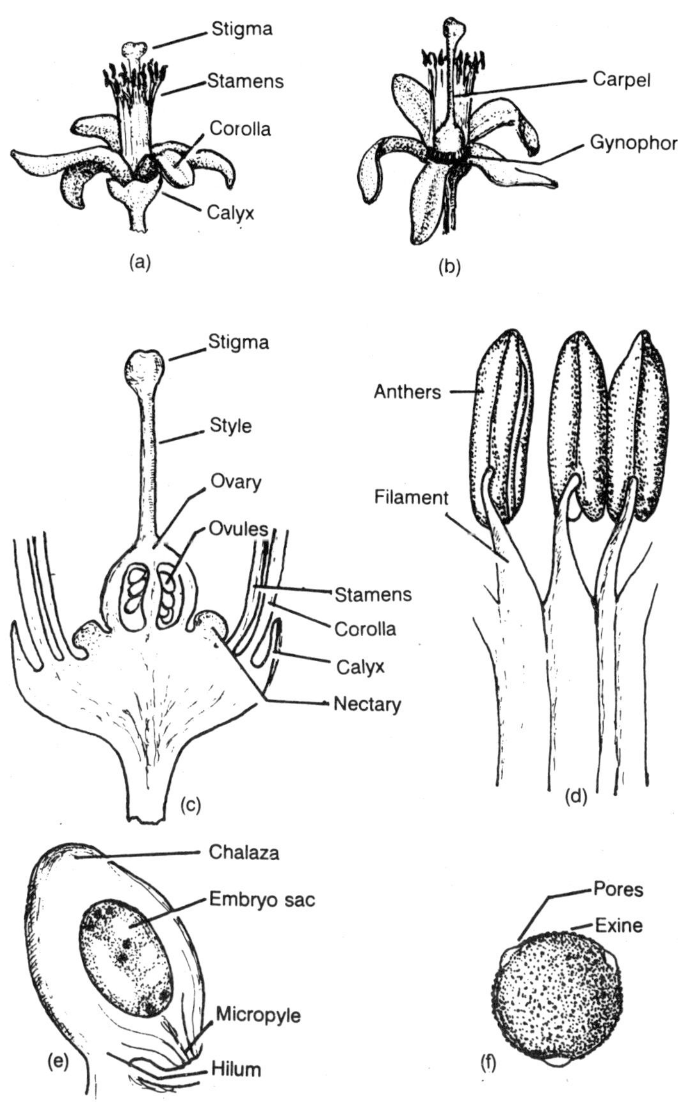

Figure 5.5 Citrus flower. (a) intact flower, (b) portion of polyadelphous stamens removed to show attachment of ovary, (c) section of ovary axile placentation, (d) detail of anthers, (e) ovule, (f) pollen grains.

cross–pollination is more common because the conspicuous corolla with its strong perfume and abundance of pollen and nectar attracts insects. Honeybees are the major pollinators of citrus. Fertilization occurs eight days after pollination. Washington Navel oranges produce no viable pollen and Satsuma mandarin very little. Marsh grapefruit produces 5–15%, and lemon and oranges 25–50% well–developed pollen. Mandarins and grapefruit produce a much higher percentage of functional pollen.

Breeding objectives. Most citrus orchards are raised from budded stock. Bud wood is obtained from mature branches of the current year's growth of virus–free plants. This bud wood is called the **scion**. Some of the most desirable characters in a scion are: vigour; longevity of tree; regularity of crop; reduced tree size; cold hardiness; fruit of medium size and shape without stem–end necks or blossom–end nipples; deep–orange or red–orange rind colour for oranges and mandarins; rind easy–to–peel, smooth, and thin; ratio of TSS and TA high for oranges and mandarins.

The rootstock should induce scion longevity with high yields and good fruit quality. It should lead to reduction of tree size without affecting yield or scion health; should be tolerant of salinity, cold, phytophthora, and tristeza; should produce many seeds with high polyembryony. The selection of suitable rootstock is extremely important since stock and scion must be compatible to produce long–lived trees, and the two plants should have the same growth rate so that the trunk diameter is similar above and below the graft. For these reasons it is very essential to do thorough testing before a cultivar is selected for a certain region. Some rootstocks have been found widely successful; sour orange (*C. aurantium*) is favoured because of its deep root system, drought tolerance, fruit quality, and other characteristics.

BREEDING TECHNIQUES

Pollen collection. Controlled pollination in citrus is relatively easy to achieve. Pollen is collected from chosen flowers of the male parent. An alternative method is to collect the whole flowers before they open. The anthers are allowed to dehisce, which usually occurs within 12–24 hours at room temperature. Then pollen is collected and applied to the stigma.

Emasculation and pollination. Dehiscence in intact flowers usually occurs soon after the petals separate. In grapefruit it often occurs when the bud is still closed. The first–appearing and larger flowers of a cluster are selected, emasculated, and covered with insect–proof bags. Several flowers may be emasculated on each selected twig; the other blossoms but not the leaves are removed. The petals are gently separated with fine forceps which should be sterilized in 70% alcohol when handling other flowers. After emasculation, the flower is covered with a flat–bottomed paper bag and securely tied with twine. Cotton pads may be wrapped around the twig at the point where the bag is tied. Plastic or airtight bags are not suitable because they cause suffocation and failure of fertilization.

Pollination may be carried out immediately after emasculation, provided that the stigma is receptive and the flowers do not show any signs of senescence. Receptivity of the stigma is indicated by a sticky secretion. Paper bags should be promptly put back after pollination. Later, the developing fruit can be protected with open-mesh cloth bags or perforated plastic bags.

Seed harvesting. Cut the fruit through the rind around the equator about halfway into the fleshy interior, then twist the two halves and squeeze out the seeds. Seeds should be immersed for 10 minutes in well agitated warm water. Remove surface moisture with absorbent cloth or paper. Treat with ARASAN or a comparable fungicide. The surface of the seeds should be dried in the shade in an airy place. Treated seed can be stored in polythene bags at 0–10°C for up to eight months. *Poncirus trifoliata* seeds may benefit from chilling. They have been stored for over two years with little reduction in viability. Most seeds give highest germination if planted shortly after extraction. Citrus seeds are very sensitive to drying, and germination may be reduced by drying them even for a few hours. However, after harvesting seeds should not be kept for a long time in their own juice or water.

Viability tests for seeds. One of two methods can be used. (i) Remove seed coat without injuring the interior, and germinate in a petri dish on wet filter paper. It is best to place the petri dishes in an incubator at a constant temperature (25°C). (ii) Seeds are cut in half longitudinally and placed with the surface down in a 1% aqueous solution of 2,3,5–triphenyl tetrazolium chloride (TTC). Viable seeds should turn bright pink within 24 hours.

Germination. Germination requires two to four weeks at a temperature range of 20–25°C. It can be hastened by removing the seed coat or cutting the chalaza end of the outer coat. It is recommended to sow seed at a depth of 1.5–2 cm. Immature or small embryos can be grown in a tissue-culture medium under aseptic conditions.

Distinguishing zygotic and nucellar seedlings. (i) When the pollen parent is well-identified and has morphological characteristics that clearly distinguish it from the seed parent, zygotic seedlings can be identified with some degree of accuracy. For example, if hybrids are produced using trifoliate orange as the male parent, the progeny can be distinguished by their trifoliate leaves. This is because the seedlings emerge from zygotic embryos that were produced by fertilization. Such plants will exhibit a blend of both the pollen and seed-bearing plants. If, however, parents are genetically similar, it becomes nearly impossible to separate the sexual and apomictic seedlings arising from any one polyembryonic seed. (ii) Culture of partially developed embryos may also provide a means of identifying the zygotic embryos in seeds produced on highly polyembryonic cultivars.

Polyembryony. Polyembryonic seeds are common in citrus and are a great problem in breeding citrus varieties. Such seeds contain one sexual and the remainder asexual (apomictic) embryos. The multiple embryos are usually crowded together at the micropylar end; often some of them are small with poorly developed cotyledons. In polyembryonic seeds, the seed surface may be corrugated because of unequal development of the embryos.

Some citrus varieties like grapefruit and citron produce seeds which are of zygotic origin, i.e. produced after fertilization. Some mandarins are also monoembryonic but they are nucellar, and develop by apomictic processes. Plants arising from such seeds exhibit only the maternal characters. Lemon and lime yield a small percentage of zygotic seedlings along with the apomictic ones. The pollen parent can influence the proportion of zygotic seedlings, the number of asexual embryos varying considerably between species and even varieties. Rough lemon when self-pollinated can produce almost exclusively nucellar seedlings; in contrast, when pollen of *Poncirus* was used, 46% of the seedlings were zygotic. Seedling counts on various types of citrus have shown an average number of seedlings from Musambi of 4.0 and a maximum of 8; from Kinnow the average was 4.67 and the maximum 11; from Jatti Khatti the average was 2.86 and the maximum 6.

Polyploidy. Diploidy is the general rule in citrus and related genera with a chromosome number of 2n=18. The genera *Fortunella*, *Poncirus*, *Microcitrus*, and *Eremocitrus* are all diploid. Hong Kong wild kumquat *Fortunella hindsii* was the first form to be identified as tetraploid. Triploids have been identified in zygotic hybrids of diploid parents. In addition, some aneuploids and a few individuals with a higher ploidy level have also been reported. Tetraploids also arise spontaneously. In the course of breeding and varietal studies of oranges, lemon, and grapefruit, about 1–6% of spontaneous tetraploids have been found as nucellar seedlings.

The triploids show considerably more average vigour in plant growth than the tetraploids and have thick rounded leaves, but in most other characters they are intermediate between (2n) and (4n) parents. The rate of sterility is very high; however, seeds are occasionally formed, which usually grow into unproductive plants. Nevertheless, a few of them may set fruit well. This kind of behaviour is a serious problem in developing citrus polyploids. Common varieties of Kinnow are diploid, but tetraploid stocks are also available and are being maintained for use in breeding. The large number of seeds in Kinnow is an undesirable character, therefore the Horticulture Department at the UAF is making efforts to produce a triploid Kinnow which is either seedless or with a low seed count (Malik 1991).

Mutations. Spontaneous somatic mutations occur frequently in citrus, as limb sports and sectors on fruits. They are also detected occasionally in nucellar seedlings or their budded progeny. These changes involve both mutation and a chimeral system and are usually unfavourable, but useful mutations do occasionally emerge. The famous varieties Washington Navel, Marsh grapefruit, Shamouti orange (originated in Palestine by bud mutation), Salustaniana orange of Spain, and Marrs of Texas are somatic variants. Thompson grapefruit was discovered in 1913 in Florida as a limb sport from the White Marsh. The Thompson grapefruit is pink, and from it the Ruby or Red Blush variety has originated by somatic mutation.

Some breeding achievements. In 1893 two tangelos, Sampson and Thornton, were developed by crossing the Dancy tangerine and grapefruit. Sampson is recommended as a rootstock for vigour, tolerance to gummosis, and a high degree of nucellar embryony. A cross of sweet orange and trifoliate produced Rusk and Mortan and other citranges. Grapefruit crossed with mandarins produced Orlando and Minneola, which are being grown successfully. Crosses between Orlando and Clementine resulted in a new variety, Fair Child. Fremont is the product of a cross between Clementine and Ponkan, and Fortune is a cross between Clementine and Dancy. These crosses are under trial in Pakistan. In 1914 Frost made a cross between King and Willow Leaf mandarins which resulted in Kinnow, the most popular and extensively cultivated mandarin in Pakistan (Frost 1935). In Japan and Russia, research efforts are being directed to develop cold–tolerant varieties, for which wide hybridization is being adopted. Several inter–specific and inter–generic crosses have been made and designated as:

Tangelo	=	Tangerine	x	Grapefruit
Tangor	=	Tangerine	x	Orange
Orangelo	=	Orange	x	Grapefruit
Citrange	=	*Poncirus*	x	Sweet orange
Citradia	=	*Poncirus*	x	Sour orange
Citrange–quat	=	Citrange	x	Kumquat

In each cross the parents have a distant genetic relationship, and it is thus expected that useful variants can be found among the hybrid progeny.

5.7.3 Mango

The mango is native to southern Asia or the Malay Archipelago; the Indo–Burma region is its centre of origin. Phytogeographical studies have shown that the largest number of species of both these groups, separately as well as taken together, occurs in the Malay Peninsula. The greatest diversity and wealth of forms in the genus—about 20 out of the 41 species—are concentrated in this area. The genus spread to India and Ceylon in the west, and to the Philippines and New Guinea in the east through the Sunda Archipelago and Borneo. The most interesting feature of distribution is that out of the 41 species only seven occur in Eastern Malaysia.

Cytological investigations of the species found in the Indian subcontinent show that all of them have the chromosome number n=20. However, in a variety known as Villa Kolamban, n=40. Mango is an allopolyploid and highly heterozygous. Morphological studies have shown that the genus *Mangifera* consists of two groups of species based on the character of the flower disc. The first, which includes 34 species, has flowers with well-developed, swollen discs; whereas the second, consisting of seven species, has absolute or pedicellate discs. Further delimitation of the species within these groups has been made on the basis of leaf morphology and floral and inflorescence characteristics. The available evidence suggests that the second group of species diverged earlier than the first, either from a common ancestor or from some of the present species, possibly *M. duperreana* or *M. lagenifera*, which are the most primitive species within this genus.

Floral biology. Mango exhibits five growth flushes; the maximum inflorescences are borne by shoots which grow in the March–April flush. The inflorescences are polygamous, bearing hermaphrodite and staminate or pistillate flowers on the same panicle. The flowers are sessile. The calyx is usually five-partite, and the corolla consists of five petals. The androecium consists of one fertile and four sterile stamens. Perfect flowers are easily distinguished by the presence of pistils. The ovary is small, greenish, located on a disc, and consists of a single locule with a short style and a minute protogynous stigma. It is highly cross-pollinated by insects. Ali and Khan (1974) noted two functional stamens in Samar Bahisht and Alfonso. They observed less than 3% purely female flowers in Langra, Neelum, and Alfonso, but as high as 7.3% female flowers in Sobi Di Ting. In an earlier study it was shown that Langra has 67% bisexual and 33% staminate flowers, while Dusehri has 78% and 22% bisexual and staminate flowers, respectively.

Emasculation is a difficult task, as the flowers are small. A new technique has been developed to facilitate crossing by growing the female and male parents side by side and covering them with plastic cages. Crossing is accomplished by ordinary house flies, which are raised separately. As no nectar is present, honeybees are not attracted and play practically no role in pollination. Self-incompatibility has been found in Dusehri, Langra, Chaunsa, and Bombay Green.

Breeding objectives. Existing varieties are good in quality and production, but suffer from many problems which require immediate attention and intensive work: bearing of fruits in alternate years, malformation of inflorescences, premature fruit drop, and susceptibility to mildew and anthracnose. Dusehri, Samar Bahisht, Langra, and Malda bear notoriously irregularly, while Neelum and Tota Pari bear regularly. Bearing in bunches is a dominant character of Dusehri.

The objectives of breeding are, therefore, to evolve varieties which are regular bearers and are resistant to the various maladies mentioned. Dwarf tree habit, long cropping season, immunity to fruit drop, and long shelf–life of fruit are also desirable characteristics. Several crosses of the elite varieties have been started in the Horticultural Research Station, Shujabad, Pakistan to evolve varieties better than the existing ones.

Juvenility. Mango trees grown from seeds have a long vegetative or juvenile phase. Plants may bear fruit only after 9-10 years, while grafted trees bear after 5–6 years. To shorten the juvenile period of a seedling, grafting wood can be taken from it and grafted on a mature, full–bearing tree. This is being tried experimentally at the UAF, with some indications of success. Other techniques such as chilling the seed, defoliation of experimental seedlings, and application of cytokinins to seedlings can be used.

5.7.4 Potato

Only South American species of the tuber–producing genus of *Solanum* have been used in breeding potatoes. Most cultivars are tetraploids (2n=48) assigned to *Solanum tuberosum*. A related species, *S. andigena*, is also tetraploid and is considered to be the primitive type. It requires short days for tuber production, while *S. tuberosum* requires long days. A few other species have contributed to the improvement of the modern potato, notably the hexaploid Mexican *S. demissum* (2n=72), the tetraploid Mexican *S. stolonidenum* (2n=48), and the tetraploid South American *S. acaule* (2n=48).

The main objectives in potato breeding have been increased yield and resistance to diseases and pests such as blight, wart disease, scab, and viruses. Potato cultivars differ markedly in various plant characters. Plants show a range from prostrate to erect growth habit. The size, form, and number of leaves show extensive variations. The stem may be green or coloured. Tubers have different sizes, shapes, and colours; their flesh shows white, yellow, pink, red, or blue colouration. These and many other morphological characters are used as selection markers. Tuber quality is profoundly important, and recently selection has been made for characteristics required by specialised potato processing industries. Potato chip production, for example, requires high dry matter and low sugar content. Most potato breeding involves hybridization within the *S. tuberosum* group.

The crop has already been remarkably improved by breeders in many parts of the world; now, therefore, very large numbers of seedlings have to be raised and tested to find any which appear to have advantages over existing cultivars. Although wild potato species flower profusely, many tetraploid cultivars bloom sparsely, and some, like King Edward and the Dutch cultivar Bintji, do not flower in the field. Furthermore, flower buds tend to fall off prematurely. Not all cultivars produce berries (fruit), therefore potatoes are normally grown from tuber cuttings. However, true potato seed is now being produced commercially, which may prove very useful in extending the growing areas of potato.

Problems. Most cultivars freely cross and are self-fertile, although selfing leads to weak progeny. Nevertheless, some diploid types and wild species have strong incompatibility and cannot be selfed. Even when pollination has been effected, the flowers may drop off prematurely. Dry potato pollen can be kept viable for about a month in a refrigerator and for considerably longer in a deep freezer. Pollen sterility occurs widely, and is a major drawback in hybridization; consequently, many desirable crosses cannot be achieved. However, most pollen-sterile cultivars have a high ovule-fertility, and these can thus be used as female parent. In all circumstances, stringent measures must be taken to prevent virus infection of the breeding material during the actual breeding process and the early stages of testing.

The main source of blight resistance, an extreme hypersensitivity reaction, is *S. demissum*. Crosses of this species with tetraploid cultivars give mainly pentaploid seedlings, and these have to be backcrossed to *S. tuberosum* to regain tetraploids. The situation is, however, complicated by the occurrence of different races of the fungal pathogen *Phytophthora infestatans*, and control of the disease by other means is generally incomplete. Unfortunately, good resistance is associated with late maturation.

5.7.5 Brassica crops

Cabbage, Brussels sprouts, cauliflower, broccoli, kohl rabi, and most kales are all forms of *Brassica oleracea* and all rapidly intercross among themselves. Summer cauliflower, and to a lesser extent early round-headed summer cabbage, is self-fertile and does not suffer excessively from inbreeding, but with plants of nearly all other forms selfed progeny exhibit considerable inbreeding depression, or self-incompatibility is found. To effect selfing it is usually necessary to pollinate by hand before anthesis at the bud stage to overcome the natural incompatibility. *Brassica oleracea* will hybridize with difficulty with turnip and radish, but even if a cross is successful the hybrid progeny are mainly sterile unless chromosome doubling is induced. The first of these combinations is similar to the turnip, and the second, called *Raphnobrassica*, is of some academic interest but has not yielded any useful horticultural cultivars.

Mass selection and pedigree breeding are still used for early cauliflower and summer cabbage. Mature cauliflower plants selected for breeding often produce seed with difficulty because relatively few flowers develop and survive. This has been overcome either by vegetatively propagating the plants from cuttings of curd portions, each with a small section of leaf, or by grafting curd portions on seedling cauliflower plants. Most forms of *B. oleracea* can be maintained as perennials by vegetative propagation from stem, leaf–bud, and sometimes root cuttings when it is desirable to keep a certain genotype as a clone for breeding purposes. These are grown at temperatures greater than 15°C to maintain the plant in the vegetative phase. Hygienic conditions must be maintained to prevent rotting, and precautions should be taken to prevent infection by cauliflower mosaic, turnip mosaic, and broccoli necrotic yellow viruses, all of which are transmitted by aphids or mechanical injury.

Most cultivated Brussels sprouts are selections from F_1 hybrids, and this technique is now being used for all other forms of *Brassica oleracea* including the recently evolved summer cauliflower. Breeding objectives for all horticultural *brassicas* include increased yield, uniformity in size and maturity, and absence of premature bolting. Some objectives for specific crops are:

Brussels sprouts
- firm, medium–size sprouts with few rough outer leaves
- not subject to internal browning (a physiological disorder of complex inheritance)

Cabbage
- firm heads with short internal stems which do not split when overmature
- late forms not sensitive to frost damage or rotting of the outer leaves

Cauliflower
- white curds without bracts in the curd
- not subject to riciness (a kind of bitterness) or lack of buttoning (premature development of very small, poor quality curds).

5.7.6 Ornamentals

Chrysanthemum. The cultivated chrysanthemum can be traced to 500 BC. Earliest forms were probably all yellow–flowered types derived from *Chrysanthemum indicum* (tetraploid 2n=36). Hybridization with *C. morifolium* (hexaploid 2n=54) introduced other colours and flower forms. Cultivars have a fairly high rate of natural mutation, and in this way many new forms have originated. Chromosome numbers of cultivars are often irregular and range from 51–68. Radiation has been used successfully in the last decade to speed up mutation and thus to produce new cultivars.

Dahlia. Dahlia cultivars are said to have been mainly derived from *Dahlia variabilis* which is an octaploid (2n=64). Most species are tetraploids (2n=32). *D. pinnata* with semi–double flowers, and *D. coccinea* have probably also been utilized in breeding of present day dahlia cultivars, and a rogue plant in a batch imported from Mexico in 1872 that was called *D. popenovii* gave rise to cactus–like cultivars. Another species, *D. merckii*, which has an extra two chromosomes (2n=34), has also been used in breeding. Dahlia flowers are usually strongly self–incompatible and do not need to be emasculated before hybridization. Four major genes for flower colour are known. In recent years new mutant forms have been obtained by irradiation.

Rose. Although there are several indigenous European rose species, modern rose cultivars are largely derived from Asia. *Rosa gallica*, said to be the red rose of Lancaster, was one of the first cultivated species and crossed with *R. phoenicea*, gave the Damask rose. The musk rose *R. moschata* was also used in early hybrids; later roses, notably the cabbage rose, in which all stamens are petaloid, and the moss rose, owe some of their character to the native dog rose *(R. canina)*. Later the China rose *(R. chinensis)* hybridized into the group gave a longer flowering period, and the Hybrid Tea rose and Hybrid Perpetual were born. The Pernettiana roses were hybrids with the above group, and the Austrian briar *(R. foetida)* and these provided the main gene pool for the modern *Floribunda* types.

Most modern rose cultivars are tetraploids or higher polyploids and apomixis is not uncommon within the genus. Although the seed often requires stratification, the raising of seedlings to the flowering stage usually takes about 18 months.

PRACTICAL EXERCISES

1. Test the viability of the pollen grains of mango, citrus, peas, and rose by at least two methods.
2. Dissect the ovaries of peas and show the placentation, indicating the type of ovule.
3. Macerate the pollen of any available cucurbit and try to find any of the stages in meiosis. Sketch where necessary.
4. Learn the techniques of pollen germination. Prepare the media in the tissue culture laboratory.
5. Germinate the seeds of rough lemon, and grapefruit in pots in a greenhouse and make weekly observations. Count the number of seedlings. Can you distinguish nucellar from zygotic seedlings?
6. Dissect the seeds of rough lemon, ponkan (Nagpuri Sangtra), and trifoliate orange. Separate and count the various types of embryos.
7. Select two parents of vegetable and fruit cultivars and practice crossing.
8. Study the distinguishing characteristics of the most important varieties of mango, citrus, potato, and marigold.

9. Prepare a list of the important rose varieties grown at your university and at the Ayub Agricultural Research Institute.

QUESTIONS

1. Define plant breeding and its relation to other sciences.
2. Draw a map showing the centers of origin of modern cultivated crops.
3. Compare and contrast the flowers of citrus, mango, and peas showing the various essential and non–essential parts.
4. Describe the modes of reproduction of crop plants. Explain the significance of the mode of reproduction for plant breeding.
5. Describe briefly the different types of chimeras. Name some important fruits where chimeras are generally found.
6. Distinguish between self–pollinated and cross–pollinated crops. Give examples of each.
7. Show by a diagram the differences in the leaves and fruits of diploid and polyploid citrus cultivars.
8. Define gene. What is the most recent concept of the gene?
9. What is a chimera? Explain its role in the evolution of new varieties, giving illustrative examples.
10. Write a detailed note on sources of variation both natural and induced.
11. Distinguish between autopolyploidy and allopolyploidy.
12. Define a mutagen. Give examples of artificially produced mutations by X–rays, gamma rays, or chemicals.
13. Define male sterility. Has it been commercially utilized? If so, in which crops?
14. What is a normal fruit set? Explain with examples.
15. Discuss the methods of plant breeding. Which method has been mostly applied in Pakistan and what successes have been achieved?
16. Write a brief note on clonal selection as applied to asexual crops. Discuss its merits and demerits.
17. What precautions are necessary in crossing two selected parent varieties? Discuss with special reference to citrus fruits.

18. Define heterosis, explaining its mechanism in detail. Can it be utilized commercially?

19. Discuss the backcross method of breeding with reference to potatoes.

20. Discuss the various problems involved in the improvement of crop varieties of potatoes, mangoes, citrus, and chrysanthemums.

21. Define the following: (a) Parthenocarpy; (b) Incompatibility; (c) Polyembryony; (e) Apomixis; (f) Genetic engineering.

REFERENCES

Ali, N.; Khan, M. D. The mango in Punjab. Lahore, Punjab: Government Printing Press; 1974.

Briggs, F. N.; Knowles, P. F. Introduction to plant breeding. New York: Reinhold; 1967.

Dutta, A. C. A class book of botany. 12th ed. Lahore, Karachi: Oxford University Press; 1965.

Frost, H. Four new citrus varieties: The Kara, Kinnow, and Wilking mandarins and Trovita orange. Univ. Calif. Agri. Exp. Stn. Bull. 597. Berkeley, CA: Agricultural Experiment Station; 1935.

Hussain, A. Vegetables of Punjab. Lahore: Punjab Department of Agriculture; 1983.

Hussain, A. A new pea hybrid. Annual Report, Vegetable Research Institute. Faisalabad: Ayub Agri. Res. Inst; 1990.

Ibrahim, M. C. Studies of the physiological aspects of erratic bearing in sweet lime (*Citrus limettioides*). Faisalabad: University of Agriculture; 1988. Ph.D. dissertation.

Janick, J.; Moore, J. N. Advances in fruit breeding. West Lafayette, Indiana: Purdue University Press; 1975.

Juliano, J. B. Embryos of Carboa mango (*Mangifera indica L.*) Department of Agri. (Botany), Philippines. Punjab Fruit J. 23 (82, 83); 1969.

Khan, A. R. Hybrid vigour in onion, radish and brinjal. Fifty years of agricultural education and research, Vol. I. Department of Agriculture, Lyallpur, West Pakistan; 1960.

Khan, D. A. Germination trials on mandarin pollen. Punjab Fruit J. 26, 27:90–99; 1962–64.

Khan, I. A. Production of haploids from anther culture of tobacco. Annual report. Faisalabad: Dept. of Hort., University of Agriculture; 1990.

Khan, M. D. Bud formation and differentiation in mango. Lahore: University of Panjab; 1943. M.Sc. thesis.

Khan, M. D. Kinnow and Feutrell's Early: two promising varieties of citrus. Ann. Rep. of the Fruit Section. Lyallpur (Faisalabad): Punjab Agriculture College and Research Institute; 1943.

Khan, M. D.; Habib–ul–Rahman. Elements of genetics and plant breeding. Lyallpur (Faisalabad): Punjab Agriculture College Press; 1960.

Khan, M. D.; Habib–ul–Rahman. Genome relationship and chromosome behavior in allotriploid hybrid of *Pennisetum typhoides* and *P. purpurium*. W. Pakistan J. Agr. Res. 1(3); 1962.

Leopold, A. C.; Kriedemann, P. E. Plant growth and development. 2nd ed. New York: McGraw–Hill; 1975.

Malik, M. N. Production of triploid Kinnow through tetraploid and diploid hybridization. Ann. Res. Rep. Faisalabad: University of Agriculture; 1991.

Nijjar, G. S. Breeding in India. 2nd ed. New Delhi: Oxford and IBM; 1972.

North, C. Plant breeding and genetics in horticulture. New York, London: Macmillan; 1979.

Raghava Kurup, C. G. A handbook of Indian Council of Agri. Res., New Delhi, Calcutta: Sarswati Press Limited; 1976.

Rao, A. R. Some cyto–taxonomic studies in the family *Rutaceae*. Lyallpur (Faisalabad): Agriculture College; 1962. M.Sc. Thesis.

Singh, R.N.; Mujumdar, P.K.; Sharma, D. K. Malika – A new mango variety. New Delhi: Division of Hort. and Food Technology, IARI; 1977.

Sturrock, T. T. Mango breeding in Florida. Boca Raton, Florida: Florida Atlantic University; 1977.

Vavilov, N. I. The origin, variation, immunity and breeding of cultivated plants. (translated by K. Starr Chester). New York: The Ronald Press Company; 1951.

Outline of Chapter 6

6. PLANT PROPAGATION

Saeed Ahmed[1]

LEARNING OBJECTIVES

After studying this chapter, the student should be able to explain how to:

- Establish a nursery of fruit plants
- Establish progeny plantations stocked with improved clonal varieties
- Propagate plants by both sexual and asexual methods
- Reproduce ornamental plant species, annuals, perennials, shrubs, and flowering and shade trees
- Produce seeds of important vegetable crops
- Set up a plant tissue culture laboratory
- Reproduce plant species *in vitro*

6.1 Prerequisites for successful plant propagation

Since numerous crop species are involved, plant propagation is a very vast field. This chapter is limited to the most common methods of propagating the most important horticultural crops. Plant propagation is both an art and science. A successful plant propagator must understand genetics, growth characteristics, adaptability to local climate, soil, biotic conditions, and principles and limitations of each mode of perpetuation of plant species. To attain a high percentage of success with different methods of reproduction, grafters and budders should be well practiced in the different methods and techniques of plant propagation.

In Pakistan, the average yields of horticultural crops are estimated to be only one–third to one–fifth of proven potentials. There is great room for improvement in the quality of produce as well. This low orchard efficiency, resulting in below optimum yields and quality, is largely attributed to poor quality seed material and non-pedigreed nursery plants being multiplied and distributed to raise vegetables, fruit crops, and floricultural plant species.

---

[1]Dr. Saeed Ahmed is a member of the Pakistan National Rose Society.

Establishment of nurseries. It is imperative that the importance of good seed material and nursery plants in the production of horticultural crops is recognized. To eliminate the multiplication and distribution of inferior cultivars and strains, planting material must be raised under controlled conditions using improved methods and techniques.

Establishment of seed–production facilities and a well–equipped nursery is the first step in the reproduction and multiplication of horticultural crops. The following points must be considered and taken care of in the development of the infrastructure (Garner and Chaudhri 1976).

- Nurseries should be located in important production areas.
- Soil should be deep, fertile, well–drained, and free from any soil–borne pathogens. Adequate water supply is a basic necessity.
- Climatic conditions must be considered in relation to the crops involved.
- The site should be well connected by different means of communication and must be easily accessible.
- An adequate number of field personnel, budders, and grafters should be available to handle different operations.
- Fertilizers, fungicides, pesticides, growth regulators, grafting waxes, budding knives, secateurs, pruning shears, plastic, tools, and other related equipment and maintenance facilities should be readily available.
- According to need, the nursery should have lath houses, glass and greenhouses, seed germinators, and mist propagation facilities.
- The nursery should have progeny of selected trees to furnish seeds to raise seedlings and rootstock material.
- The nursery should also have its own resources to provide bud wood and other propagating material for vegetative propagation.

6.2 Modes of reproduction

Plant species perpetuate in two ways, sexually or asexually. Reproduction through seed is sexual, except in some apomictic species. Seeds of some species, including mango, give rise to more than one seedling per seed. In such cases, however, only one seedling is zygotic; the rest are produced from nucellus cells and are called **nucellar seedlings**. This phenomenon is called **polyembryony**. Nucellar seedlings can be utilized to raise uniform plants.

6.2.1 Reproduction through seed

Raising horticultural plants from seed has several drawbacks. Ordinarily when cross–pollinated plants produce seeds, the seedlings do not reproduce the characters of their parents. Fruit tree seedlings take a long time to come into bearing and have a tendency to grow tall, which adds to maintenance and harvesting costs. However, propagation through seed also has some merits. Seedlings are generally hardy, bear heavy crops, and are long–lived. Most of the fruit varieties in commercial production are chance seedlings having desirable characteristics of yield, quality of fruit, or disease resistance, which have been propagated vegetatively and established as clonal varieties. Reproduction through seeds is also essential in the breeding of hybrids and new varieties and in raising seedlings to be used as rootstock in vegetative propagation. In some species, seeds are the only economical means of reproduction.

For the propagation of forest species, vegetables, and annual and biennial flowers, seeds are the most convenient, easiest, and cheapest method of reproduction. The method is usually employed for the following purposes.

- Establishment of seed gardens: Seed gardens are established for the selection of promising chance seedlings with desirable yield and fruit characteristics to be perpetuated as clonal varieties.
- Evolution of promising new varieties through controlled cross-breeding.
- Crop production: Most vegetable crops, annual and biennial flowers, and some fruit crops such as papaya, coconut, and *falsa* can be produced through seed conveniently and cheaply. Some citrus species and a few mango varieties are polyembryonic. The zygotic seedlings or variants are generally weak and retarded in growth and are rogued out to maintain a uniform crop of nucellar seedlings identical in all respects to the seed parent. Since most of the viruses in citrus trees are not transmitted through seed, the phenomenon of polyembryony is being used in some countries to screen out virus diseases and to establish commercial plantations using nucellar seedlings.
- Raising rootstock: In vegetative propagation, particularly in budding and grafting operations, seedlings of congenial rootstock species are always needed (Singh and Krishnamurthi 1967).

Preparation for sowing. There are several factors which must be considered before actually sowing the seed.

I. COLLECTION OF SEED: Collection of seed from reliable sources minimizes the chances of seedling variation. The fruits selected for seed extraction should possess good fruit and keeping–quality characteristics. In most cases they should be fully mature and taken from healthy, vigorous plants with a high degree of disease resistance.

II. STORAGE OF SEED: Seeds of most of the evergreen fruit species lose viability soon after extraction, and should be sown without prolonged storage. In case it becomes necessary to store the seeds, they should be carefully washed, surface dried, mixed with an equal part of ground charcoal, and packed in suitably sized wooden or metallic containers, which should be stored in a dry, cool place with temperature between 3.0°C and 13°C (Singh and Krishnamurthi 1967).

III. PRE-GERMINATION SEED TREATMENT: In some deciduous fruit species, like apple, pear, peach, plum, and cherry, the seeds need a certain period of rest after extraction before they will germinate. Such seeds must be stored with alternate layers of moist sand, at a controlled cold temperature and suitable moisture conditions before sowing. This process is called **stratification**; it allows the embryos in the seeds to complete their development.

Seeds with hard shells like *ber* (jujube) and guava need to have the seed coat cracked or softened by soaking in water for several hours or days. Seeds of mango, citrus, papaya, *falsa*, jaman, and loquat do not need any treatment before sowing (Garner and Chaudhri 1976).

Sowing. The seed is either sown in pots, trays, or plastic bags; or in beds in nursery rows. With species which do not tolerate transplanting well, seeds are sometimes sown in specially prepared holes, at the recommended distance from plant to plant and row to row for the species and variety which is to be used as scion. The seedlings thus raised are budded or grafted *in situ*.

When a small number of seedlings is required, they are raised in earthen pots, wooden trays, plastic bags, or in advanced countries 'jiffy pots', discs of compressed peat encased in netting, which swell to several times their original size on absorption of water. The containers are filled with a potting mixture of equal parts of topsoil, sand, and well–rotted leaf mold or FYM.

In nurseries, seeds are sown in well–prepared beds, where the soil has been thoroughly worked, pulverized, and mixed with well–rotted FYM or leaf mold. The distance allowed varies with crop species. Seeds of most species of fruit can be sown in lines 10 cm apart. A sowing depth two to three times the diameter of the seed is usually satisfactory. The seeds are covered with sand or leaf mold, and watered after sowing.

6.2.2 Methods of asexual propagation

Asexual or **vegetative reproduction** uses a part of the plant for multiplication. Plant parts used are stems, leaves, buds, roots, bulbs, corms, rhizomes, suckers, and tubers. Asexually propagated progeny are identical to the parent plant in all respects. Vegetatively propagated plants remain relatively low–headed and bear earlier than sexually reproduced ones, with uniformity in fruit quality. These traits result in savings in cultural operations and make harvesting of the crop and its marketing easier.

The use of congenial rootstock allows wider adaptability of a variety to soil, climatic, and biotic conditions. The influence of rootstock on scion variety can also be utilized to regulate tree size, yield, quality of produce, and time of ripening. Sometimes newer varieties become popular, and growers want to replace their old orchards. In such situations, trees of older stands can easily be top-worked using bud wood or graft wood of superior varieties or strains. This process saves time and expenditure in the establishment of plantations of new varieties to meet the requirements of changing demand trends.

There are only a few disadvantages to using vegetative propagation techniques. The budded and grafted plants are generally less vigorous and short–lived than seedlings. Also, new varieties cannot be evolved through asexual propagation.

There are many methods of vegetative propagation. Some are quite simple, while others are complicated and laborious, requiring considerable knowledge and skill. The methods most prevalent in Pakistan are discussed in this chapter (See also Childers 1969 and Hartman and Kester 1968).

6.2.3 Layering

Layering means rooting of shoots, stem, or branches while they are still attached to the parent plant. Some trees and bushes, like blackberry, layer themselves naturally. Strawberry produces runners, another form of natural layering. The advantage in layering is that the young plant continues receiving nourishment from the parent plant. Layering can be carried out in the spring or late summer months. Different methods of layering are practiced in the propagation of fruit and perennial flowering shrubs (Garner and Chaudhri 1976).

Air layering. This method involves shoot tips which are buried 3–5 cm in the soil to encourage the development of fibrous roots, after which the root tip portions are separated. The technique is usually employed for blackberry and other bushes.

Air layering or marcottage. This method is locally known as *ghooti*. It is common for vegetative propagation of guavas and litchi in areas where rainfall is abundant and humidity very high. It is done by making a round incision with a knife, a ring of very small diameter, or an ordinary cut in a mature and healthy vigorously growing branch just below a node. When an ordinary cut is made, a very small stone is pushed slightly into the cut to prevent the two sides from reuniting. The cut is then covered with sphagnum moss or wrapped with a mixture of sawdust and mud. A pot is hung just above the layering so that water drips on it and keeps it moist. The need for providing dripping water is now being eliminated by using sphagnum moss and polythene as wrapping material. The plastic film allows the exchange of gases but does not allow moisture to escape. Thus it helps the retention of moisture by the moss without interfering in respiration. Rubber strips are used to tie the ends of the polythene enclosing the moss. This operation is done successfully preferably in spring or in the rainy season when there is a lot of humidity in the air.

Since the films are transparent, the initiation and development of roots within the wrap can be seen without disturbing the layering. Sufficient roots develop within three to four months, and the layers are carefully detached from the parent plant and the wraps removed. The layers are then placed in pots containing well decayed leaf mold or soil mixed with well–rotted FYM. The pots are kept in shade under humid conditions. The soil in the pots is not allowed to dry out.

Ground layering. In ground layering, instead of wrapping, the rigged or notched portion is buried in the ground or in soil prepared in pots. The shoot is kept in position by tying it with a peg or by putting a weight on the surface of the soil. After two months, a half cut is made in the parent branch just below the layering. If no wilting is noticed, it indicates that the new layer has developed roots. The layering is separated and set out as an independent plant (Singh and Krishnamurthi 1967).

Serpentine layering. With some species which give out long and flexible branches, the shoot to be layered is covered with soil at several places to encourage rooting at more than one point. The method is thus known as serpentine layering.

Mound or stool layering. This technique involves cutting down the parent plant to ground level. The stubs or stools are earthed up to encourage the growth of shoots. The emerging shoots are earthed up gradually until one–third to one–half of their length is covered with soil. Growth of shoots in complete darkness results in blanching (etiolation). This etiolation induces development of roots at the basal ends. These are detached and transplanted in the nursery. The best results in layering follow banking of the shoots with earth as early as possible and before the tissue becomes woody. This method is employed generally in the production of material to be used as clonal rootstocks in mango and guava, and dwarfing rootstocks of apple and pear.

6.2.4 Division

Division involves separating vegetative parts like rhizomes, offsets, crown, runners, or suckers from the parent plant and establishing them as independent plants. The method has wide application in the propagation of flowering shrubs. Among fruit crops, strawberries are propagated by runners, date palms by suckers, pineapple by offsets and crown, and bananas by suckers or rhizomes.

6.2.5 Cuttings

A cutting is a piece of a plant: a shoot, stem, leaf, bud, root, or merely a tiny piece of meristem. Though incomplete in themselves, cuttings have the capacity to produce roots and shoots, essential to the development of plants identical to the parent which furnished the cutting. Cuttings should only be taken from strong and healthy plants. Grapes, sweet limes, lemons, limes, roses, and other perennial flowering shrubs are commonly propagated through cuttings.

Root cuttings. To prepare root cuttings, the roots are cut into pieces 10–25 cm long, which are planted horizontally in well–prepared beds or in pots. Many perennial shrubs, guava, breadfruit, apple, pear, cherry, and persimmon can be reproduced from root cuttings. However, the method is seldom used in Pakistan.

Stem cuttings. Cuttings of either soft wood or hard wood are used to propagate a number of fruits and flowering shrubs.

SOFT–WOOD CUTTINGS: Soft–wood cuttings, as the name suggests, are taken when the stem is still soft and succulent in spring or early summer, and generally consist of terminal portions of the shoots. Sometimes middle and basal portions are also used, provided they are soft and sappy. Soft–wood cuttings are mostly used to propagate flowering shrubs and guava. Cuttings of guava from terminal shoots root easily in misty conditions.

HARD–WOOD CUTTINGS: Hard–wood cuttings are taken from the previous season's mature growth, when the stem has shed leaves and is hard and lignified. Hard–wood cuttings are used in the propagation of sweet lime, lime, lemon, grapes, pomegranate, fig, and plum. Cuttings of most plant species which are prepared from mature wood are of pencil thickness varying from 18–25 cm in length. The upper cut, about 5 cm from a node, is made at a right angle, to reduce the size of the wound. A basal slanting cut close to the node increases the surface for nutrient absorption. The cuttings are planted 15 cm from each other in rows 30 cm apart. Any closer planting induces damping off. Generally two–thirds of hard–wood cuttings and one-half of soft-wood cuttings should be inserted in the soil in a slanting position to prevent rain water from entering through the cut surface. To expel air, the soil around each cutting must be pressed and firmed up. The beds or pots are kept moist by daily irrigation. Root growth is favoured by a soil temperature of 18–21°C.

In evergreen species, cuttings are prepared and planted in early spring or the rainy season. However, in deciduous species, the cuttings are made during the dormant season and are stored to be planted during early spring.

Plant growth regulators. Cuttings of some plant species do not root readily. Plant hormones, such as indoleacetic acid, indolebutyric acid, naphthaleneacetic acid, and naphthaleneacetamide have successfully been used to stimulate rooting in cuttings. These chemicals are used as solutions, powders, or in lanolin. Aqueous solution of about .01% or 50–1000 parts per million (ppm) is generally used, depending upon the plant species. The basal ends of the cuttings are either dipped in the solution, or powder is applied to the rooting ends (Singh and Krishnamurthi 1967).

6.2.6 Grafting

Unlike with layering, division, and cuttings, a grafted plant is composed of two parts. The basal part which provides the root system for anchorage and absorption of moisture and nutrients is called the **rootstock**. The **scion** provides top and fruit–bearing surfaces and synthesizes food which is transported to other parts of the plant. Scion and rootstock should be compatible and closely related, belonging to the same genus.

Grafting is a process by which a piece of scion is attached to a rootstock in such a way that the cambiums of both scion and rootstock come in firm contact, so that the new secondary tissue resulting from cambial cell division in the scion and rootstock is closely knitted. The stock usually influences the size and vigour of the tree, and the scion the yield and quality of fruit it will bear. Grafting allows utilization of varying climatic, soil, and biotic conditions with advantage. When a single bud is used as scion and inserted into the rootstock, the method is called **budding**, whereas when a piece of stem or branch carrying one or more than one bud is used the process is known as **grafting**. Several important methods of grafting used to vegetatively propagate fruit and ornamental trees are discussed below.

The first step in the budding and grafting operation is to raise seedlings of compatible rootstock species, or to produce clonal rootstocks to avoid variation otherwise encountered in seedling populations. If seedlings are used, only strong, vigorous, and healthy ones should be selected to be used as rootstock (Childers 1969; Gourley and Howlett 1949; Hartman and Kester 1968; Kains and McQuesten 1952).

The selection and care of scion wood is of the utmost importance. It must be collected from selected progeny trees which are true–to–type and have desirable fruit and yield characteristics. Twigs that have shown growth during the previous season usually furnish the best scion wood. The buds should be well–developed, plump, mature, and in dormant condition.

If the bud wood is not to be utilized immediately, it can be stored for a limited time. The sticks of bud wood are tied in small lots and properly labelled for identification. Both the ends are sealed in melted wax to prevent excessive evaporation. The bundles are then put in sphagnum moss, wrapped in polythene bags, packed in a suitable box, and stored in a cool place.

Only the most compatible scion and rootstock combinations should be used in grafting and budding. Incompatible combinations result in poor bud take, weak graft or bud union, overgrowth of stock as compared to scion or vice versa, low disease resistance, excessive leaf drop, early decline, low yields of poor quality fruit, and short life span.

6.2.7 Budding

Budding is relatively easy to do and is extensively used in the vegetative propagation of vast numbers of species of fruit and flowering trees, particularly evergreens. Four different methods of budding are generally identified.

Shield budding or 'T–budding'. This is by far the most common method of vegetative propagation of citrus, jujube, apple, loquat, roses, and many other ornamental trees and shrubs.

The rootstock seedling and bud wood are pre-conditioned and prepared when the sap is moving freely. For convenience in handling, before the buds are actually removed, the leaf blades are cut off, leaving the petioles intact. A narrow shield of bark 3–4 cm in length, bearing a single bud is removed with a budding knife. The shield will have a thin layer of wood, which is removed from thornless buds but is kept intact with thorny buds, as with some of the citrus species. A vertical cut 3–4 cm in length is made just through the bark on the rootstock. At the top end of this cut a horizontal cut about 1.25–1.5 cm long is made, so that the cut resembles the letter 'T'. In areas with heavy rainfall, sometimes an inverted 'T' cut is used to avoid the entrance of excessive moisture. Care should be taken not to injure the wood underneath. The height of budding on stock varies with stock and scion being used, climatic conditions, and the prevalence of viruses and diseases. Using the plastic or bone spur of the knife, the bud is inserted under the bark, which has been loosened from the wood. To hold the bud firmly in position and to exclude air and moisture, the bud union is carefully wrapped using suitable plastic strips, keeping the bud exposed.

In most cases, the bud union is complete in about four weeks. When the buds begin to swell and sprout, the top of the budded stock is cut close to the bud union to encourage the rapid growth of budlings. After the sprouts have grown about 10 cm, the tying material is removed and the budded plants are trained to their proper size and shape. They are carefully tended in the nursery until they are disposed of for transplanting (Chandler 1958; Ginai 1968; Yusuf 1989).

Ring budding. In this method, a ring of bark 1.5–2 cm in length, containing a well–developed bud is loosened on the scion shoot and is gently pulled out from the thinner end of the shoot. In pulling off the ring, a handkerchief is used to avoid possible injury to the bud. The top of the stock seedling is removed and the bark is peeled off downward to a point where the ring of the bark bearing the bud fits tightly. No tying is necessary in ring budding. Uniformity in the thickness of the rootstock and scion shoots is desirable for higher percentage of success. This method can be employed in the propagation of *ber*, mulberry, peach, and plum.

Chip budding. A single bud with a large piece of wood is placed on a corresponding cut made on the stock and tied firmly. This method is rarely used.

Flapped–patch budding. A rectangular or square flap of bark on the stock is loosened on three sides. The corresponding bud shield is inserted underneath the loosened flap of bark. The flap is then placed over the bud shield and wrapped. It takes about two months for the buds to sprout, after which the wrapping material is removed.

6.2.8 Grafting methods

Inarching or approach grafting. Unlike other grafting operations, in inarching the scion is attached to the stock while it is still attached to the parent plant. In Pakistan, the method has been extensively used in the vegetative propagation of mango. The method involves potting healthy one–year–old seedlings in earthen pots, which are usually 12 in deep and 8 in wide at the top. The earth bali along with root system is kept intact and placed in the pot. The unfilled space in the pot is then filled with a mixture of well–rotted manure and canal silt. The soil is firmly pressed around the earth ball and the transplanted seedlings are then watered. Potting of seedlings is usually done one month ahead of actual inarching operations. This allows sufficient time for the seedlings to set before grafting.

Inarching can either be done in the spring months or in the rainy season (July–August). July and August are preferred over the spring season because after inarching operations in the spring months the wind storms in the early summer reduce the percentage of success. High wind velocity knocks over the pots and loosens the graft union.

The potted seedlings are brought near the parent plants for inarching, hence the name *grafting by approach* has also been given to this method of grafting. If shoots of desirable size are available at ground level, potted seedlings are put underneath the parent scion trees and inarching operations are completed. However, if shoots are only available at a certain height, the pots are raised and provided support.

Every care must be exercised in the selection of rootstock and scion shoots. The size of the rootstock and of the scion shoot should be the same. A slanting cut about 5 cm long and 2 cm deep is made on the stock seedling at a height of 15–20 cm. A similar cut is made on the scion shoot and the corresponding cuts are then brought together and tied with plastic film. To prevent drying up, the union point is covered with moist sphagnum moss and wrapped with polythene film.

While tying, care is taken to bring the corresponding cambium surfaces into contact. The union takes place within three to four months. When the graft is still attached to the parent plant, the pot is watered twice a day during the summer months. Removal of the graft from the parent plant is a gradual process. Prior to the final cutting away of the graft from the parent plant, an incision is made on the scion some 5 cm below the graft union. The period between the incision and removal of the graft should be seven to ten days. Steps in the process are shown in Figure 6.1.

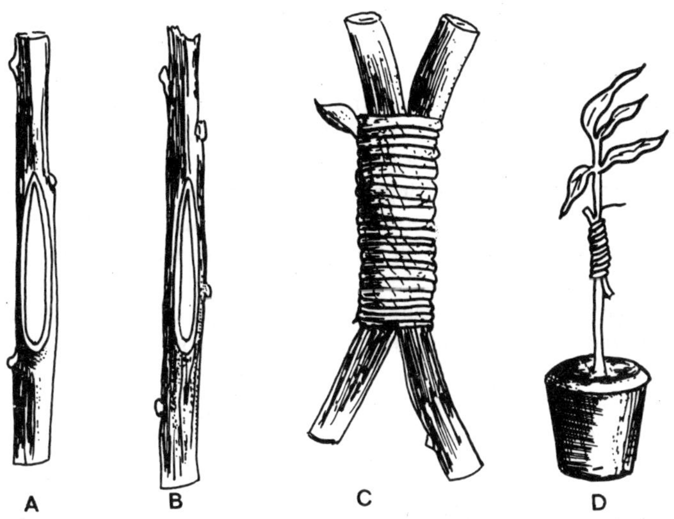

Figure 6.1 Inarching. a) Stock; b) Scion shoot, sliced to expose the cambium in an equal pattern; c) Exposed portions of stock and scion, brought together and tied; d) Stock in pot and sprouted scion.

After the graft has been removed, it is placed in the shade of trees. The top of the stock is also gradually removed after they have been stored in the shade for about two weeks. They are then kept for about four to six months in the nursery before they are planted (Ahmed 1960; Hartman and Kester 1968; Singh and Krishnamurthi 1967).

Wedge or cleft grafting. This method is generally used in the propagation of deciduous fruits and top–working older trees of inferior varieties. Branches of the rootstock 1.25–5 cm thick or even larger are cut at the top, and a vertical split 5–8 cm deep is made in the centre of the stub with a heavy chisel or knife. The split is kept apart until scions containing at least three buds are inserted. The basal end of the scion is cut into a long sloping wedge about 5 cm long. The side of the wedge which is to face outward is made slightly wider than the inside edge. One scion is inserted in each side of the split. After adjustment of the scions, the tool used to keep the split open is removed to allow the full pressure of the split stock to bear on the scions at the point where the cambium of stock and scion are in contact. No tying is usually needed, however, all the wounded surfaces are thoroughly waxed over. Cleft grafting is most successful when done in early spring when the buds are beginning to swell but before active growth has started (Childers 1969; Hartman and Kester 1968).

Splice grafting. A slanting cut about 10 cm long is made at the basal end of the scion, which is placed on a corresponding cut made on the rootstock seedling at a height of 30–45 cm from ground level, and then tied together. The cambial areas of scion and stock should be in close contact; to achieve this both should be of the same thickness. The top of the plant down to the graft union is covered with a polythene bag, loosely tied at the basal end to prevent excessive transpiration. The union takes place within three weeks to three months, depending upon the plant species involved and the age and condition of scion and rootstock. Splice grafting has been employed in grafting mango in the Philippines and Brazil, and for some varieties of apple.

Tongue grafting. The scion and rootstock are cut exactly in the same way as in splice grafting. To provide strength and exposure of greater cambial area for union, tongue–shaped slits are then made in both scion and stock. The secondary slits or cuts are half the length of the original slanting cuts. The scions and stocks are brought together, firmly interlocking. The graft wounds are waxed over, with or without tying.

Bark grafting. Unlike in the cleft graft, the stub of the stock is not cut horizontally. Instead, several scions with their basal ends trimmed as wedges are inserted between the bark and the wood. The operation is done during spring when the bark can be loosened more easily.

Bridge grafting. Bridge grafting is practiced to save valuable trees with diseased or damaged trunks. To prepare the tree trunk for bridge grafting, the damaged part is trimmed back to healthy tissue by removing dead or torn bark from the affected parts. Bridge grafting is best performed during early spring when the bark can be lifted easily. Scions of sufficient length to bridge over the damaged area are taken. One long slanting cut is made at each end of the scion, both cuts on the same side. A second cut is made on the back side. The bark on both upper and lower sides of the wounded area is loosened as in bark grafting, and the scions are then inserted between the bark and wood of the stock and are nailed down (Figure 6.2). All cut surfaces are carefully coated with wax, and any sprouting on scion surface should be removed. Complete healing takes place in a few years time. (Childers

1969; Garner and Chaudhri 1976; Hartman and Kester 1968).

Veneer grafting. Veneer grafting and side veneer have been successfully employed in vegetative propagation of mango and guava. The methods being followed in Pakistan are described here. Healthy, two to three–year–old vigorous stocks are selected. Scion shoots with terminal buds are used as graft wood. The scion wood is prepared by clipping off the leaves leaving petiole stubs .5 cm long intact. The pre–conditioned shoots of the scion are then girdled at a place 1.5–20 cm below the apex. The grafting wood is ready in about 15 days in spring and 12 days in autumn, when the buds begin to swell. Graft wood 10 cm in length is detached from the scion trees and grafted on the stock seedlings at a point having almost the same girth as that of the scion. The grafting operation is done by making a tangential cut 5 cm long through the bark of the stock, making a notch at the base. A similar cut is made on the graft wood with a wedge at the base. The cut surfaces of the stock and scion are united. The scion wood is tied with waxed cotton tape. It is then wrapped in a polythene sheet and both ends of the plastic are tied. After the union has taken place and new growth appears, the wraps are loosened. After the first flush of the scion, the cotton tape is untied. The stock is then cut back partially. The heading–back of the stock is completed after the growth of the second flush of the grafted scion (Ahmed 1960).

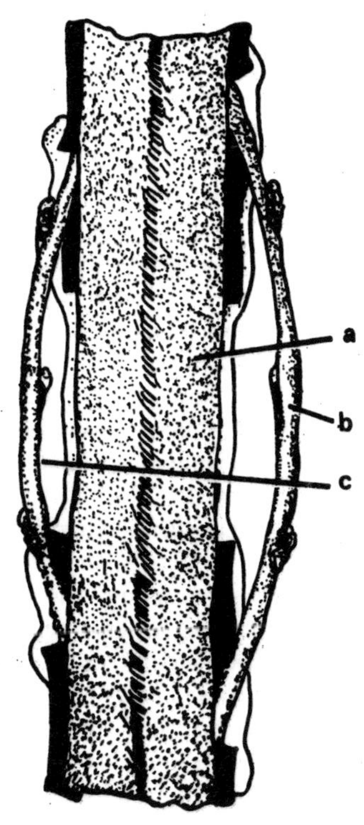

Figure 6.2 Bridge grafting. Longitudinal section showing details: a) wounded portion of the trunk; b-c) scion twigs with trimmed ends inserted under the bark of trunk on both sides of wounded portion.

Side grafting. Vigorous stock shoots are selected. Fresh and mature scion sticks about 12–13 cm long are collected from a desired variety and their leaves are clipped off. A two to three–year–old stock tree planted in orchard rows is selected. At the base of the scion on one side a longitudinal cut of 5–8 cm is made, cutting away half of the girth at the base and thus making a wedge. An oblique cut is made in the stock at an angle of 20-30 degrees and about 2.5 cm deep, and the scion is inserted in the cut portion and gently pushed down until it is fixed in the bark. The union is tied with waxed cotton tape and wrapped with polythene. After the graft sprouts, the polythene is removed. The other precautions are the same as in budding.

This method can be performed in both spring (March-April) and autumn (August-September), but the maximum success rate of 90 percent is achieved during spring (Ahmed 1960; Khan and Ahmed 1972).

T-grafting. In the technique of T-grafting, the scion wood is prepared as for veneer grafting. The shape of the cut on the stock is, however, different from that made for veneer grafting. A 'T'-shaped cut is made in the stock bark, into which the scion is inserted and firmly wrapped with waxed tape. The graft is covered with polythene (Khan and Ahmed 1972).

Top working. Top working of large, inferior mango seedlings to a more desirable variety is possible, though the practice is not very common. The method practiced is to dehorn the tree branches or main limbs to within 25 cm or so of the main trunk of the tree. This is done in early spring. The shoots which grow from the stump become fit for budding in the following summer or autumn months. Since in the plains there is danger of sunburn injury, the naked trunks are protected by whitewashing. In order to provide shade until the new top is established, sometimes a few of the weaker branches are left and not dehorned. Thus the complete changeover of the top may extend over more than one season. The remaining branches can be pruned back in the same way and shield–budded in the subsequent season. Young plants in the field with a trunk diameter from 2.5–20 cm can easily be veneer–grafted on the main branches at a height of 25–50 cm from ground level. If the trunk is too large or difficult to work, the lower main limbs can be veneer–grafted (Ahmed 1960).

6.2.9 Factors influencing success in vegetative propagation

- Graft union takes place readily between varieties and species. It becomes unlikely when plants belonging to species in different genera are involved, and impossible in the case of different families.
- Only compatible stock/scion combinations should be used.
- Use seed and grafting material from well–tested, true–to–type parent trees with a high degree of disease resistance, and desirable yield and

fruit characteristics. For better contact between cambial surfaces to essure grafting surface, rootstock and scion should be of the same vigour and thickness, and of suitable age.

- Grafted seedlings in pots or nursery rows should be carefully tended. They should also be protected from desiccating winds, direct sunlight, rain, and frost injury.
- The material used as bandages and for wrapping graft union should be removed once the bud sprouts have made some growth (Childers 1969; Garner and Chaudhri 1976; Singh and Krishnamurthi 1967).

6.3 Plant tissue culture

Plant tissue culture commonly refers to *in vitro* and aseptic cultivation of any plant part (cells or tissues) on a nutrient medium in an artificial environment. Research efforts in plant-cell and tissue culture have increased dramatically worldwide in recent years, including efforts in the developing nations. Plant propagation using shoot-tip culture is the most widely used area in plant-cell and tissue culture and is currently being used to propagate elite genotypes of several plant species in a number of countries (Garner and Chaudhri 1976; Hartman and Kester 1968; Yusuf 1989).

Clonal propagation. True-to-type plants can be raised by shoot-tip/meristem culture. The terminal 2–3 mm portion of plants (meristem) is almost free from viruses, for the reason that cell divisions in such parts are very active. Virus-infected cells, on the other hand, divide comparatively slowly and lag behind. On nutritive media under controlled environmental conditions, such meristems can be made to grow into complete healthy shoots.

6.3.1 Commercial considerations

Some important commercial considerations in tissue culture are, briefly:

Capital investment. The initial cost of a tissue culture laboratory is higher than the cost of using conventional methods.

Expertise. Well-trained personnel are required for this technique. They should be conversant with standard microbiological techniques and have sufficient botanical knowledge of the material and of its *in vitro* behaviour.

Range of applicability. Although in theory it would seem that any plant can be cultured by supplying the appropriate metabolites in suitable *in vitro* conditions, so far only a small proportion of plants can be cultured well enough to be effectively propagated. The time and expense involved would probably be worthwhile for the major crop plants, but not necessarily for the minor crops.

6.3.2 Technical problems

Phenolics and other inhibiting factors. A number of plants produce excessive amounts of phenolic substances whose oxidation products not only darken both the tissues and the medium but often inhibit growth. Where the problem is confined to the reaction of the initial explant it may be prevented by treatment with ascorbic or citric acid.

Vitrification. An occasional problem with cultured plants is the development of swollen, distorted leaves which become irreversibly translucent and eventually necrotic, a condition that may lead to the death of the shoot cluster. This has been referred to as **watersoaking** (Khan et al. 1987; Murashige 1977) and is more recently described as **vitrification** (loc. cit.). The phenomenon occurs in shoots cultured on media containing cytokinin and can generally be prevented by making sure that the concentrations are not higher than necessary.

Bacterial contamination. A persistent difficulty in tissue culture laboratories is the occurrence of contamination by slowly-growing and often macroscopically invisible microorganisms, particularly bacteria.

Quality of plants propagated by *in vitro* techniques. It has been observed that when proper culture media and conditions are not provided to the explants certain changes may occur in plants raised through tissue culture. These changes are due to epigenetic effects and genetic differences.

Supply of electricity. A continuous supply of electricity is very essential for tissue culture techniques. Since there is frequent load-shedding in this country, an alternate power supply through generators is required for tissue culture laboratories.

6.3.3 Techniques

There are three main steps or stages in the tissue culture multiplication technique.

1. Establishment stage. The explant or tissue fragment is sterilized in a disinfectant and placed in a carefully selected and formulated nutrient medium for a period ranging from four to eight weeks. The purpose is to confirm sterility and to establish growth of the explant.

2. Multiplication stage. Plant material resulting from the establishment stage is divided and each part replanted in separate culture tubes. As multiplication occurs, the fragments are again divided and replanted in separate fresh containers. This procedure may be repeated several times to rapidly propagate shoots. Since each transplanted shoot has the ability to produce a complete new plant, it will produce new shoots of its own, each of which also has the potential to produce shoots and roots.

3. Pre–transplant stage. Before transplantation, the prepared plants are generally hardened by deleting the hormones from the medium and by increasing the light intensity. They are then transferred to soil in the greenhouse. About two to four weeks are usually required to condition or harden a plant before potting it. Special greenhouse procedures are required for about four to eight weeks until a complete transition from laboratory to greenhouse is completed.

6.3.4 Preparation for tissue culture

In preparation for the above stages, the following procedures and precautions are followed.

1. Selection of tissue. For maximum success in using the tissue culture method, the explant must be chosen systematically. The success of this method is directly related to the organ source—its age, the season in which explant is taken, size of the explant, and overall quality of the plant from which the explants are obtained. Shoot tips have been satisfactory for many plants, whereas sections of root, leaf, inflorescence and other organs may be better for other plants. Tissue obtained from young plants tends to reproduce better than that from older source plants. Tissue taken from the growing point has the power of rapid cell division and is generally free from viruses and virus–like disorders. The explants should be free from visible attack of pests and diseases.

2. Sterilization. The primary objectives of this step of *in vitro* propagation are the establishment of a microbe–free explant culture, and the initiation of new growth. The outer surfaces of plants growing under natural conditions are normally infested with spores and other microbial cells. Mature inner tissue may harbour fungi and bacteria without showing visible symptoms. Plants grown in the greenhouse or growth room are somewhat free of these maladies. Before culturing the explants, it is essential to disinfect the material and then wash it four to five times with autoclaved water to remove the residual toxic effect of the chemical. Some disinfectants commonly used for this purpose, and their concentrations, are listed below.

Disinfectant	Concentration	Dipping time
Sodium hypochlorite	0.5–5%	5–20 min
Ethyl alcohol	75–80%	Several sec
Benzalkonium chloride	0.01–0.1%	Several min
Hydrogen peroxide	3%	15–30 min
Mercuric chloride	0.01–0.1%	10–20 min

Each plant behaves differently with these disinfectants.

Since all plants have a waxy outer surface, it is usually necessary to add a small amount of detergent to the disinfectant to allow better penetration. Tween is a detergent being widely used for this purpose (Khan et al. 1987; Murashige 1977; Qureshi et al. 1991; Street 1977).

Inoculation area. For this purpose a special room is prepared which should be free from dust and microorganisms; it should have ultraviolet lights and be air–conditioned to maintain the desired room temperature. It should be equipped with a ventilation unit with laminar flow and positive pressure. A laminar flow cabinet is a special piece of equipment with a 0.3 UM HEPA filter of 99.97–99.99 percent efficiency. It should be fitted in such a way that the air is forced into the cabinet through a dust filter and is then directed over the working area at a uniform rate. Before the cabinet is used, it should be thoroughly cleaned and sterilized with 70–80% ethyl alcohol.

Culture facility. Plant tissue to be cultured on any nutritive medium should be incubated under proper conditions of well–controlled temperature and light for photoperiod and humidity control and air circulation. All these factors are directly related to the success of this technique.

Media for plant tissue culture. It is very important that tap water never be used for dissolving the chemicals because it contains many impurities such as cations, anions, mineral salts, microorganisms, and gases. Double distilled water should always be used for this purpose.

Four different media—MS, B5, White, and Heller—are used in the propagation of horticultural plants by this technique. Of these, the medium named after Murashige and Skoog (ms) is more common (Khan et al. 1987; Murashige 1977; Qureshi et al. 1991; Street 1977).

Culturing of tissues. After samples of explants are collected from healthy young plants at the juvenile stage, and before proceeding further, the worker should disinfect his hands and tools with 70-80% ethyl alcohol to avoid any contamination. The samples are disinfected with a proper disinfectant along with a few drops of detergent under a laminar flow cabinet, and then rinsed thoroughly with autoclaved water to remove the residual toxic effect of the disinfectant. The terminal and lateral buds are removed from the sample with a scalpel and are placed in autoclaved test tubes or jars containing solidified or liquid nutritive medium. Then these test tubes/jars are covered with autoclaved polythene sheeting and closed with rubber bands.

The tubes/jars are labelled with a marker, indicating the treatment, variety, date, and any other essential information. The trays of culture tubes/jars should be placed in the culture room for growth.

6.3.5 Progress of tissue culture in Pakistan

Hardly ten years ago the term 'plant tissue culture' was totally new to most universities and research organizations in Pakistan. There were, however, sporadic reports from the Department of Genetics, University of Karachi; Central Cotton Research Institute, Multan; and Department of Botany, University of Peshawar. From 1980 onwards there was a quantum leap in the number of tissue culture laboratories in various parts of Pakistan, and now a stage has been reached where the private sector is willing to invest in this new technology. Plant tissue culture has emerged as a rapidly developing science because of its proven success in agricultural improvements: elimination of diseases, evaluation of new varieties in a reduced span of time, and a shortened breeding cycle.

At present at least 18 tissue culture or genetic engineering biotechnology laboratories have been established. The tissue culture laboratories at the AARI at Faisalabad and NARC, Islamabad have reported substantial success in *in vitro* propagation of important fruit trees and ornamental plants, including guava, apple, grape, Kinnow mandarin, date palm, chrysanthemum, and bougainvillea. The Potato Research Station at Abbottabad utilizes *in vitro* propagation to increase new potato cultivars.

6.4 Propagation of vegetables

Most vegetables are reproduced from seed. The importance of good seed in truck-farming cannot be overemphasised. Good seed and planting material are produced under controlled conditions. Seed should be pure, free from dust, weed seeds, and diseases; and it should have a high degree of viability. Since most vegetable crops are cross-pollinated, varieties and species are segregated in the fields to avoid unwanted cross–pollination.

In the plains, many vegetables do not set seed because of harsh climatic conditions. Areas at higher elevations in the mountains offer a congenial environment for raising vegetables to produce quality seed.

From the point of view of reproduction, vegetables can be put into five general categories.

1. Vegetatively propagated: Many vegetables like potato, sweet potato, garlic, turmeric, ginger, and cocoyam do not seed under normal conditions. They are propagated vegetatively. For high yields and good quality produce, it is of utmost importance that planting material in these crops be free of viruses.

2. Self–pollinated: Vegetables such as tomato, chili, brinjal, lettuce, and peas are largely self–pollinated. Complete flowers are borne on the

same plants and self-pollination takes place before they open. As a precaution, different varieties meant to produce seed must be planted 40–50 m apart.

3. Partially self-pollinated: Vegetables with complete flowers in which a good nectar supply is easily accessible attract insects which effect cross-pollination. Crops like carrot, celery, Brussels sprout, okra, and onion are partially self-pollinated. For pure seed production, different varieties should be isolated and planted at least 500 m apart.
4. Cross-pollinated: Vegetables like muskmelon, watermelon, cucurbits (*kadu*, *ghia tori*, pumpkins, bitter gourd, *tinda*) bear male and female flowers on the same plant, a condition conducive to cross-pollination by insects. To maintain purity, varieties and species are isolated, and for seed production are planted at least 500 m apart.
5. Self-sterile: Cabbage, cauliflower, radish, and turnip bear complete flowers, but suffer from self-sterility because of incompatibility of the female and male parts. They are freely cross-pollinated by wind and insects. For seed production, varieties are planted allowing at least 500 m distance between stands of different varieties.

The most important factor to be considered in the production of seed of cross-pollinated species on a field scale is isolation from nearby sources of pollen of the same species and variety.

6.4.1 Onion

Onion is largely cross-pollinated, but some self-pollination does occur. For seed production, isolation of 400–600 m is needed. In Pakistan, onion bulbs are planted in October, at a distance of 10 cm in rows 60 cm apart. Two tons of bulbs are needed to plant one hectare. The field is irrigated immediately after planting, and afterwards at intervals of two weeks. Weeds are rogued out. The seed is ready for harvest when the first formed seed heads begin to shatter. The heads are cut and laid on canvas to dry. Much seed falls from the capsules during drying, and the remainder is removed by rubbing the heads over a fine sieve. The seed deteriorates under warm, humid conditions. Dried seed is stored in airtight containers or in cold storage (Hussain and Said 1967; Kernick 1961; Ware and McCollum 1975).

6.4.2 Cauliflower

Cauliflower is mainly cross-pollinated. For seed production, varieties must be isolated from each other and any other species of *Brassica* or *Oleraceae*. Poor or diseased plants are rogued from the field meant for seed production. Flowering branches emerge from the compact heads to set seed. The heads are usually cut before the mature pods dehisce. They are cut about 45 cm from the ground and piled on canvas to save the shattered seed. The seed is dried to 10 percent moisture content for storage (Kernick 1961).

6.4.3 Cucumber

Cucumber is partially self and partially cross-pollinated; isolation is needed for seed production. Fruits meant for seed are allowed to ripen fully and are harvested when they take on a pale-yellow or golden hue with the seeds free from the flesh. The seeds are separated from the pulp by fermentation with water or are freed mechanically from the gelatinous sac. Sun-dried seed retains its viability for several years when it is stored in a cool dry place.

6.4.4 Okra ('lady's fingers')

Okra is partially self and partially cross-pollinated. For seed production, isolation is needed. Agronomic practices for a seed crop are essentially the same as for a market crop. The pods are allowed to dry on the plant, harvested periodically to avoid shattering of seed, and then sun dried and threshed. The seed is dried to 8 percent moisture and stored in a dry cool place. Under normal storage conditions, seed remains viable for two years or so (Kernick 1961).

6.4.5 Tomato

Tomato is normally self-pollinated. For pure seed production, different varieties should be isolated. Seeds should be extracted from fruit showing definite signs of ripening. Seeds are extracted by pulping the fruit by hand and removing the mucilage from the seed by fermentation. In about two days, the heavy seed sinks to the bottom leaving clear juice above. The juice is strained off and the seed washed thoroughly and sun dried rapidly on canvas. Under normal conditions of storage in temperate regions, seed retains its viability for almost four years (Kernick 1961).

6.4.6 Brinjal

Brinjal is partially self and partially cross-pollinated. For pure seed production, isolation distances of 500, 200, and 100 m are allowed for nucleus, foundation, and certified seed production, respectively. Off-types and undesirable plants at different stages of plant growth should be rogued out. For seed production, fruit should be harvested when fully ripe and turning yellow. The fruits are cut, crushed, and pulped; and the seeds are separated, washed, and sun dried for storage (Kernick 1961; Ware and McCollum 1975).

6.4.7 Spinach

Spinach is cross-pollinated by wind. The monoecious condition is best for seed production, since the need for male plants is eliminated, but this is rare. For most varieties a dioecious condition with a 1:1 ratio between male and female plants prevails. For seed production, different varieties require isolation, and a certain percentage of male plants needs to be retained for effective pollination. For seed production, the plants should be allowed to grow and set seed after three to four cuttings. The seed ripens unevenly; harvesting is best done when most seed is firm, before shattering. Plants are gathered in bundles in the field and allowed to dry. Sun-dried seed stored in a cool dry place remains viable for two to three years. Spinach can also be propagated by cuttings (Hussain and Said 1967; Kernick 1961).

6.4.8 Peas

In Pakistan, peas are grown throughout most of the year. Peas are mainly self-pollinated and reproduction is through seed. Isolation is not necessary for seed production. The plants are usually cut when the first pods are fully ripe, and allowed to dry in the field. The mature seeds are immediately removed from the pods by threshing to avoid shattering loss. Sun-dried seed is stored in a dry place at a low temperature (Hussain and Said 1967; Kernick 1961; Ware and McCollum 1975).

6.4.9 Cucurbits

Cucurbits are a large group of vegetables like gourds, pumpkins, cucumbers, watermelon, and muskmelon. In Pakistan, *karela* (bitter gourd), *kadu*, *tinda*, and *kali-tori* are typical vegetables falling in this group. Most of them are grown during the summer months; however, some are now being grown under plastic, thus extending the period of their availability. The agronomic requirements of most of these crops are very similar.

Crops are raised from seed. They are cross-pollinated by insects, chiefly honeybees. The plants are monoecious; both male and female flowers are produced separately on the same plant. High temperatures and long days keep the plants in the male phase, and short days with low temperatures induce the plants to produce female flowers. For seed production, isolation is required. Seed is ready for harvest when the fruits are fully ripe, turning brown or red. At this stage the seeds readily separate and only immature seeds still adhere to the pulp. The fruits are cut and the seeds and pulp scooped out and dried in the sun. The seed is separated and winnowed to remove light matter and sun dried in trays. The seed stores well for three to four years in a cool, dry place (Hussain and Said 1967; Kernick 1961).

6.4.10 Potato

Potatoes are an important crop for local consumption and export. In Pakistan, we were at first entirely dependent on imported potato seed (tubers). However, more recently fairly dependable technology has been developed locally to produce disease-free seed. Seed is used in the lay sense here; it is the tuber which regularly serves as the propagating unit. Plants grown from true potato seed are only recently becoming available. Three crops are sown in a year and enough seed is available to meet requirements to a large extent. However, we are still not self-sufficient in this respect.

In the plains, the spring crop is sown during the month of January with seed either from a sub-mountainous crop or imported. The autumn crop is sown from the middle of September to the middle of October. Seed requirements for this crop are met from the spring crop or seed kept in cold storage from the previous autumn crop. The sub-mountainous crop is raised during April–May. Tubers are dormant for about three months after maturity and normally will not sprout unless treated to break dormancy. To reduce sowing costs, a medium-sized tuber is cut into four to six pieces, each piece bearing one or more eyes. The cut pieces can be dipped in potassium gibberelate (1–2 ppm) for improved results.

Studies have been done on the production and multiplication of disease-free basic seed at higher elevations. Seed produced at an altitude of 2000–2600 m has performed well in the plains under varied agro-climatic conditions. Use of this seed material gives 50–60 percent more yield than seed procured from private growers in the plains. An autumn-to-autumn cycle of seed multiplication, being practiced in Punjab, has been found relatively good for keeping virus content low. The aphid-free period in autumn is good for seed production. Basic seed production at higher elevations can be used for an autumn-to-autumn cycle to minimize dependence on imported stock (Hussain and Said 1967; Kernick 1961; Ware and McCollum 1975).

6.5 Propagation of ornamental plant species

In Pakistan, landscaping and raising ornamental flowering shrubs, trees, and annuals for beautification of the environment and as means of earning a livelihood has not received the attention it deserves. These activities remain limited to public parks, compounds of government institutions, and the four walls of some villas. It is a matter of great satisfaction, however, that the agricultural universities, provincial departments of agriculture, and several horticultural societies in the country have finally been able to show the way by providing teaching, research, and training facilities, and by creating general awareness and interest in this important aspect of horticulture. The demand for quality seed of annuals, shrubs, rose bushes, bulbs, and succulent species, ornamental flowering trees, and house plants has registered considerable increase during recent years. It is encouraging that ornamental plant species can be reproduced readily by easy methods of propagation to meet the ever–increasing demand within the country and for foreign markets.

The following is a brief account of several methods of propagation usually employed in the multiplication of floricultural plants. They are essentially the same as used in the reproduction of fruit species.

6.5.1 Seeding

All annuals and biennials, most perennials, and some ornamental shrubs and trees are reproduced through seed. Winter annuals are sown in autumn, transplanted in November and December, and flower in the subsequent spring season. The seeds of summer annuals are sown in spring, the seedlings are transplanted in April–May, and the plants flower during the summer and autumn months. By staggering sowing and transplanting, the availability of blooms can be extended over a long period. To avoid damage to seedlings from rain and bad weather, seeds can be sown in earthen pots or wooden flats with drainage holes. In a nursery where a large population is required, seed is sown in well–prepared beds of convenient sizes. Seed should always be sown evenly and thinly to avoid overcrowding and associated problems such as damping–off of seedlings. Fine seed like that of begonia and pansy should be mixed with a little sand to facilitate even spreading. Very fine seeds should scarcely be covered; they are spread over the soil surface and firmed in with the palm of the hand or a wooden presser.

Seeds with hard coats like sweet peas benefit from a 24–hour soaking in warm water, and germinate evenly and quickly. Seeds of most species are sown at a depth twice their diameter. They are covered with a thin layer of well–decomposed leaf mold. Covering or mulching by straw, paper, or polythene is helpful in avoiding excessive evaporation and damage to the seed by rodents and birds. The pots or beds should be kept moist, but overwatering should be avoided since it causes the seed to rot. The seedlings are tended for four to six weeks before they are transplanted. During summer, transplanting should be done in the evening and followed by watering.

6.5.2 Division

A number of herbaceous perennials, such as polyanthus and Michaelmas daisies, form clumps. These can be lifted and eased apart to form either smaller clumps or separate plants.

Rhubarb and peonies are two plants which develop a fleshy crown just below the soil level. These crowns can be cut into pieces to form new plants, provided each piece has at least one bud and some roots. In a similar way, rhizomes like those of bearded iris and lily, and tubers such as those of dahlia and begonia can be cut into pieces. Both rhizomes and tubers are types of underground stems which possess leaves, buds, and roots. It follows, therefore, that any section cut to include all these parts should be capable of forming a new plant.

Some plants produce **offsets**, small new plants which grow out from the parent and can be separated as independent plants. A number of house plants such as sansevieria, aechmea, and some cacti produce offsets, as do garden plants like globe artichoke. Offsets should never be separated from the parent plant until they are at least a year old, if success is to be guaranteed. The parent plant should then continue to grow well and produce further offsets.

'**Suckers** are very similar to offsets, but the word generally refers to shoots thrown up from buds on the roots of woody plants. Lilac, *Rhus typhina* (stag's horn sumac), *Kerria japonica*, and raspberry all produce suckers which can be dug up and separated from the parent plant by cutting with secateurs through the root close to the base of the shoot.

In general, plants should be divided in the spring or autumn, and the plantlets should be planted out or potted up immediately.

6.5.3 Cuttings

Shrubs and flowering trees are mostly propagated through soft or hard-wood cuttings. Roses are multiplied by cuttings and by budding. The techniques are essentially the same as employed in the vegetative propagation of fruit trees, hence the details of these methods are not repeated here.

Some annuals, shrubs, and trees propagated by the methods discussed above are listed here (Garner and Chaudhri 1976; Yusuf 1989).

6.5.4 Common plants propagated by various means

SEASONAL ANNUALS AND BIENNIALS – reproduced through seed

Acroclinum	Ageratum	Alyssum
Anchusa	Antirrhinum	Arctotis
Aster	Calendula	Candytuft
Centaurea	Chrysanthemum	Cineraria
Clarkia	Clianthus	Cornflower
Daisy	Dimorphotheca	Echium
Gaillardia	Gazania	Godetia
Gomolopsis	Gypsophila	Helichrysum
Linaria	Larkspur	Linum
Lobelia	Marigold	Mimulus
Nasturtium	Nemesia	Nicotiana

Nigela	Pansy	Petunia
Phlox	Poppy	Salvia
Schizanthus	Statice	Stock
Sweet William	Sweet Sultan	Verbena
Wallflower		

HERBACEOUS PERENNIALS

Begonia	Geranium	Gerbera (single)
Lupines	Pelargonium	Violet
Carnation	Chrysanthemum	Salvia
Dahlia	Gladiolus	Hollyhock
Coleus		

BULBOUS PLANTS – propagated by bulbs, corms, rhizomes

Allium	Amaryllis	Anemone
Zantedeschia	Narcissi (Daffodil)	Hyacinth
Dahlia	Lillium (Easter lily)	Freesia
Iris (bearded)	Iris (Dutch)	Oxalis
Ranunculus	Sparaxis	Tulip
Tuberose	Crocus	Canna
Gladiolus	Muscari	

SUCCULENTS INCLUDING CACTI – propagated by seeds and division

Cereus	Mammillaria	Agave
Liliaceae	Aloe	Gasteria
Haworthia	Yucca	Sansevieria, etc.
Echinocactus, Echinopsis	Euphorbia	Crassula
Kalanchoe	Echevieria group	Opuntia
Epiphyllum		

HOUSE PLANTS AND EVERGREENS – propagated by seeds, division, and cuttings

Ferns	Palms	Dracacna
Ficus	Monstera	Philodendron
Araucaria	Juniper, etc.	Ivies
Vines	Croton	Maranta
Monstera	Sansevieria	Tradescantia

CLIMBERS AND CREEPERS – propagated by cutting and layering

Aristolochia macrophylla	Bougainvillea	Clematis
Common ivy	Chamba	Passion flower
Tecoma grandiflora	*Thunbergia alata*	*Wisteria sinensis*
several varieties of roses	*Antigona indica*	

SHRUBS – mainly propagated by cuttings, occasionally by seed

Abelia	*Tecoma stans*	*Buddleia alternifolia*
Azaleas	Camellia	Queen of the Night
Jerusalem thorn	Nerium	Bottlebrush

Jasmine	Magnolia	Motia
Prunus chealsweeping	Roses	Weeping willow
Ashoka	Poinsettia	Hibiscus
Poinciana pulcherrima		

STUDY QUESTIONS

1. Define *stratification.*
2. Why are certain seeds subjected to low temperatures before sowing?
3. What is the proper time to layer plants?
4. What is the difference between air layering and ground layering?
5. Explain *division*, and name some plants which are propagated by division.
6. What is a hard–wood cutting?
7. Name some fruit trees which are readily propagated by cuttings.
8. When are hard–wood cuttings usually planted?
9. What are the best media in which to plant soft–wood and hard–wood cuttings?
10. What are the advantages of propagating fruit trees from cuttings?
11. Explain *budding* and *grafting*. What is the difference between them?
12. Name the objectives of grafting.
13. Define *stock*, *scion*, and *cambium layer*.
14. Describe the selection and care of scion wood for grafting.
15. What is the proper time for grafting evergreen and deciduous plant species?
16. Distinguish between cleft grafting, veneer grafting, and bark grafting.
17. Explain the method of inarching.
18. Give a brief account of the factors which determine success in grafting.
19. Define the term *incompatibility* as used in grafting fruit trees.
20. Give details of the chief method of reproduction of vegetable crops.

REFERENCES

Ahmed, S. Propagation of mango. Punjab Fruit J., Mango No. 49-53. Lyallpur: The West Pakistan Co–op Fruit Development Board, Ltd.; 1960.

Chandler, W. H. Evergreen orchards. 2nd ed. Philadelphia: Lea and Febiger; 1958.

Childers, N. F. Modern fruit science. 4th ed. Horticultural Publications. New Brunswick, N. J.: Rutgers University; 1969.

Garner, R. J; Chaudhri, S. A. The propagation of tropical fruit trees. Slough: FAO (UN), and Commonwealth Agricultural Bureau; 1976.

Ginai, M. A. A treatise on horticulture. Bureau of Agri. Information (Publication Div.). Dept. of Agriculture, Govt. of West Pakistan; 1968.

Gourley, J. H.; Howlett, F. S. Modern fruit production. New York: Macmillan; 1949.

Hartmann, H. T.; Kester, D. E. Plant propagation; principles and practices. 2nd ed. Englewood Cliffs, N. J.: Prentice-Hall; 1968.

Hussain, A.; Said, M. *Hamari sabzian*. Lahore: Dept. of Agriculture, W. Pakistan; 1967.

Kains, M. G; McQuesten, L. M. Propagation of plants. New York: Orange Judd Publishing Co.; 1952.

Kernick, M. D. Vegetable crops. In: Agricultural and horticultural seeds. Rome: Food and Agricultural Organization of the United Nations; 1961.

Khan, M. D.; Ahmed, R. The mango growing. Lahore: Bureau of Agri. Information (Publication Div.), Dept. of Agriculture, Govt. of Punjab; 1972.

Khan, M.; Ahmed, R.; Shah, A. A.; Iqtidar, M. T. Tissue culture technique. Punjab Fruit J. 40(1-4):70-73; 1987.

Murashige, T. Clonal crops through tissue culture. In: Barz, W. et al. eds. Plant tissue culture and its bio–technological applications. Berlin: Springer-Verlag; 1977.

Qureshi, A.; Tauqeer, S.; Naqvi, A. Plant tissue culture in Pakistan. Islamabad: National Agricultural Research Centre; 1991.

Singh, S.; Krishnamurthi, S. Fruit culture in India. New Delhi: Indian Council of Agricultural Research; 1967.

Street, H. E. Plant tissue and cell culture. 2nd ed. Oxford: Blackwell Scientific Publications; 1977.

Ware, G. W.; McCollum, J. P. Producing vegetable crops. 2nd ed. Danville, Illinois: Interstate; 1975.

Yusuf, S. Practical guide for growing of flowering plants in Islamabad and Rawalpindi. Rawalpindi: Fauji Fertilizer Co., Ltd.; 1989.

Outline of Chapter 7

7. PLANT ENVIRONMENTS

Iqrar Ahmad Khan[1]

LEARNING OBJECTIVES

The fate of a crop in a field situation is determined by the environmental factors which affect it. This chapter should enable a student to:

- Describe the major climatic regions of Pakistan
- Select crops suitable to a specific given set of environmental factors
- Suggest practicable means of controlling the environmental impact of light, temperature, humidity, water supply, air quality, soil, plant–to–plant interaction and competition, insect pests, and diseases

7.1 Light

Sunlight is the ultimate source of energy on the earth. Green plants convert solar energy into chemical energy by photosynthesis. The stored chemical energy of plants is utilized in different forms by non–photosynthetic organisms. In addition to photosynthesis, light affects other plant processes including seed germination, vegetative and reproductive growth, and plant morphology. To understand the effects of light, it is necessary to present some basic terminology and definitions.

Physicists describe the movement of light in terms of both wave motion and of discrete particles, but for most practical purposes, the wave description is used. **Wavelength** is the distance between successive crests or troughs of a wave, and **frequency** is the number of crests passing a given point in a unit of time. Frequency and wavelength are inversely related: the higher the frequency the shorter the wavelength, and conversely.

Shorter waves are emitted at a higher frequency and have higher energy, while longer wavelengths have a lower frequency and lower energy level. Visible light is the portion of the electromagnetic spectrum to which the human eye is sensitive. It ranges between 380 and 760 nm, and the sensitivity response peaks at 555 nm. Photosynthetically active radiation (PAR) also lies within the visible portion of the light spectrum but peaks in different

---

[1]Associate Professor, Department of Horticulture, University of Agriculture, Faisalabad.

ranges. Figure 7.1 depicts the spectrum of solar radiation, also called the electromagnetic spectrum. About 50 percent of the total radiation received on the earth is in the visible wavelengths. The air acts as a filter which admits the visible waves and restricts the penetration of high-energy waves. This is discussed in section 7.4 on air.

7.1.1 Light measurements

Radiant energy is measured and expressed variously. A light meter used to determine the proper exposure time for photography is familiar to all of us. A **foot–candle** (ft–c) is defined as the illuminance from a standard candle received on a one-foot square surface one foot from the candle. This has been replaced by a more reproducible reference unit, called one **lumen** (lm) per square foot. The metric unit of light is the **lux** (lx), which is defined as one lumen per square meter. Thus, one ft–c equals 10.76 lux. A few examples from everyday life are: full sunlight at noon in summer is about 10,000 ft–c or 108,000 lux; full moonlight is about 1/50 ft–c. About 20 ft–c is considered adequate for reading; a 40W florescent tube emits about 25 lux.

The ft–c and lux have been widely used by researchers because of the simplicity and availability of low-cost measuring instruments. The trend in agricultural research is to express radiant energy in terms of PAR, which is the part of the spectrum primarily absorbed by chlorophyll molecules. It is measured by sensitive instruments and expressed in micro-einsteins per square meter per second ($\mu E\ m^{-2}\ sec^{-1}$) in the wavelength range of 400–700 nm. An **einstein** (E) is the energy in one mole, or Avogadro's number (6.02×10^{23}) of photons. In Pakistan, we are fortunate to have ample sunshine throughout the year. Light is never considered to be a limiting factor in the cultivation of horticultural crops in this country.

7.1.2 Plant responses to light

Photosynthesis and photomorphogenesis are the two major categories of plant responses to light. **Chlorophyll *a* and *b*** molecules present in the green pigmented plant parts are responsible for photosynthesis; whereas **phytochrome**, a photoreceptive pigment ubiquitously present in all plant parts, is involved in photomorphogenesis. Phytochrome has two forms, which have different absorption peaks and are photo-reversible. The red-absorbing form (Pr) has an absorption peak at 660 nm and is readily converted to the far-red-absorbing form (Pfr) with an absorption peak at 730 nm.

Photosynthesis is discussed in detail in Chapter 3. A brief description of important photomorphogenic responses is given here.

1. Photoperiodism. Photoperiodism is the developmental response of plants to the relative lengths of light and dark periods. Flowering, tuber and bulb formation, and dormancy/rest periods are important developmental

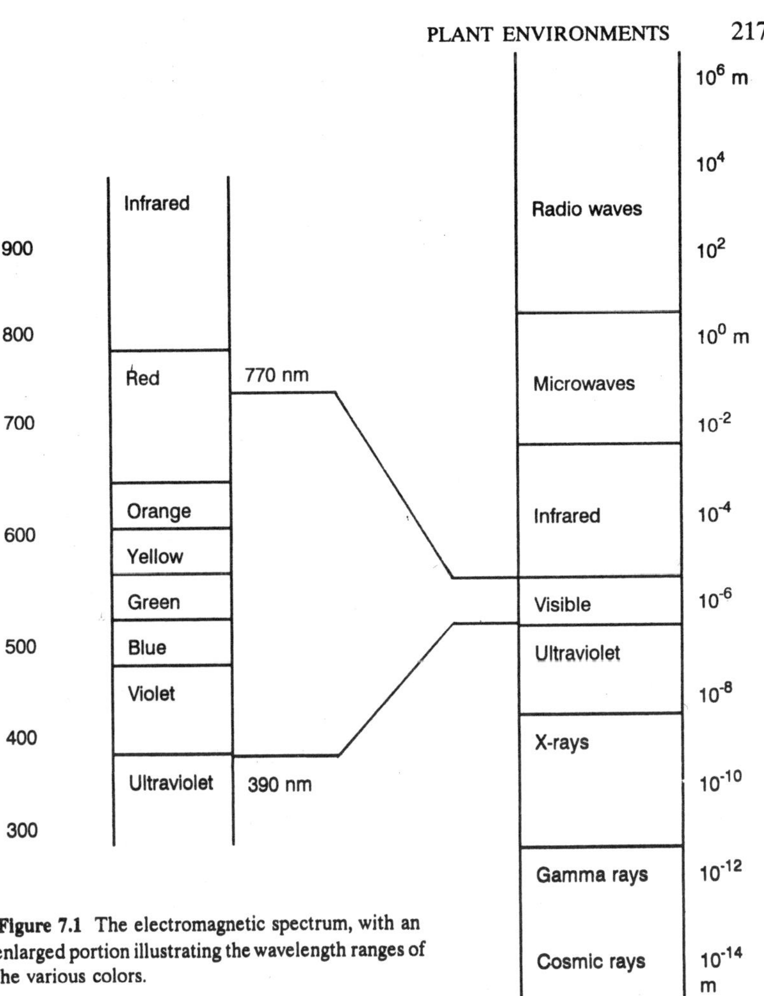

Figure 7.1 The electromagnetic spectrum, with an enlarged portion illustrating the wavelength ranges of the various colors.

processes controlled by photoperiodism.

Photoperiod/daylength varies with latitude (Fig. 7.2).

A. FLOWERING: Three distinct responses are known: short–day, long–day, and day–neutral plants.

Short–day plants flower only when the daylength is less than the critical length, which varies among species and varieties. The plant remains in a vegetative state if the days are longer than the critical length. Some common short–day plants are chrysanthemum, poinsettia, and soybean.

Long–day plants flower only when days are longer than the critical daylength. Some long–day plants are spinach, beets, and radishes.

Day-neutral plants will flower with any daylength, provided the plant is otherwise ready to reproduce. Some species flower after a certain number of nodes have developed on the stem regardless of the light conditions. Common examples of day-neutral plants are tomato, dandelion, rose, and African violet.

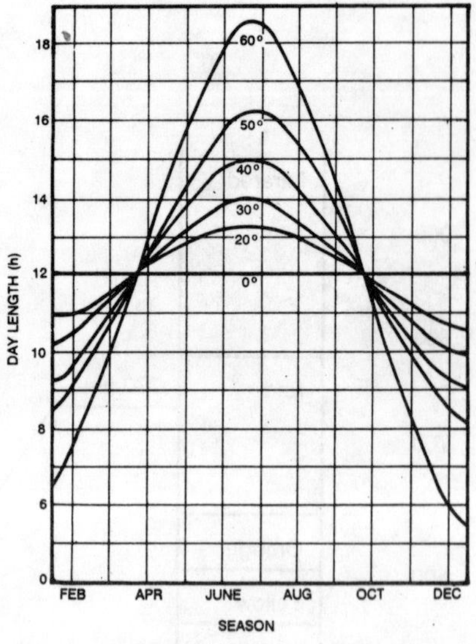

Figure 7.2 As latitude increases, annual variation in day length increases dramatically.

These definitions must be understood precisely. A short-day plant is not necessarily one which requires a shorter daylength than long-day plants. In fact, a short-day plant might require a daylength longer than a long-day plant. It is classified as short-day because it *flowers* below a critical daylength. Similarly, for *flowering* a long-day plant requires a photoperiod longer than the critical daylength. Also, it must be noted that most light-sensitive plants will ultimately flower even if not provided the right kind of photoperiod; however, this will occur after a prolonged period of time. It has been shown through grafting experiments that if bark from a plant induced to flower is grafted to an uninduced plant, the latter will also be induced to flower. Such experiments have led to the hypothesis of a flowering hormone or **florigen**. This means that flowering plants responding to external as well as internal stimuli produce endogenous compounds responsible for the induction of flowering (Halfacre and Barden 1979:263–64).

B. TUBER AND BULB FORMATION: Flowering, tuberization, and bulb formation have evolved for a common function, i.e. reproduction and survival during times which are unfavorable for growth. Bulbs and tubers are storage organs which start development when they receive a light stimulus. Most potato cultivars form tubers under short-day conditions; tuber formation can be stopped by providing light during the night, thus disrupting the short-day conditions. Interestingly, the potato is a long-day plant for flowering purposes. Both of the processes have their own evolutionary significance. The formation of onion bulbs requires long photoperiods, ranging from 12–15 hours depending upon the cultivar. Both bulb formation and flowering are long-day responses in onion.

C. BUD DORMANCY: The induction of dormancy of vegetative buds or the rest period in deciduous trees is directly related to the shortening of days during late summer and before the onset of cold winter weather. As the days lengthen in spring, dormancy is broken and fresh growth begins. This is a protective mechanism which tides the plant over adverse weather conditions. Otherwise, growth might begin in the middle of winter during unseasonal warm weather, and be killed later by the resumption of cold weather. In most deciduous trees, this is controlled precisely by the number of hours of cold after which dormancy will break, technically referred to as **chilling requirement** (Halfacre and Barden 1979:265–66) (See also 7.2.4 on temperature).

7.1.3 Seed germination

The seeds of some plants will not germinate without exposure to light. Many weeds have this property and germinate only when near the soil surface or where the soil has been disturbed; this gives them an advantage over other plants; since their seeds germinate under conditions conducive to their survival and growth. Lettuce is a well–known example of this characteristic. If lettuce seeds absorb water in the dark, only a few will germinate. If the seeds are exposed to red light (660 nm) after absorbing water, germination will be nearly 100 percent. If, however, they are then exposed to far–red light (730 nm), the stimulatory effect of the exposure to red light is reversed. Germination of certain species, e.g. the American elm and some cucurbits, is inhibited by light; and the seeds of many plants are unresponsive to the presence or absence of light (Halfacre and Barden 1979:269).

7.1.4 Phototropism

Plants generally bend in the direction of the most intense light, since the bending of the plant causes it to receive light more evenly on all sides. This response is called (positive) **phototropism.** Shoots and roots have positive and negative phototropism, respectively. Polarity for the phototropic response is at the molecular level and can be found even in embryonic tissues of the radicle and plumule.

7.1.5 Transpiration

Transpiration—loss of water vapor through the stomates—is an important plant function which controls uptake and distribution of water, minerals, and gases. Light affects transpiration both directly and indirectly. In most plants the stomates open in the light and close in the dark, so stomatal transpiration depends directly upon light. However, some of the succulents, such as cacti, open their stomates at night and close them during the day, thus reducing water loss. In these plants, the CO_2–fixing mechanism is different from that in other plants; most of the CO_2 fixation takes place in the dark, and it is mostly in the form of organic acids. This is called the **crassulacean-type metabolism** because it is common in the Crassulaceae and other succulents.

Light affects transpiration indirectly by determining leaf temperature. Leaves efficiently absorb photosynthetically active radiation but tend to transmit and reflect much of the infrared wavelengths which would have strong heating effects. However, of the light energy absorbed, only a fraction is used in photosynthesis; much is converted to heat and re–radiated or dissipated by conduction, convection, or transpirational cooling (Halfacre and Barden 1979:271–72).

7.2 Temperature

The narrow range of temperature in which plants can grow makes temperature one of the most limiting factors in crop cultivation. The optimum temperature for growth of most horticultural plants lies between 15° and 35°C. The tolerance limits for maximum and minimum temperature vary with species. Tomatoes, for example, cannot withstand freezing temperatures, whereas a hardened apple tree will not be harmed at −35°C. Tropical plants like bananas will suffer chilling injury at 4°C. The dramatic increase in the production of vegetables and bananas in the frost–free zone of lower Sindh is a vivid example of the exploitation of plant–temperature relations.

The **temperature** of a substance may be defined as a measure of the relative speed with which its molecules are vibrating. All vibrations cease at absolute zero (−273°C). Temperature reflects the *intensity* of heat, which is a qualitative indicator without any regard for the quantity of heat present in a body of matter. It is necessary to discuss a few important aspects of heat energy at this point.

7.2.1 Heat

Heat is a form of energy that causes an increase in the temperature of matter when transferred into it. A corresponding reduction in temperature takes place when heat is removed, provided the matter does not change states during the process of heat transfer. The changes of state of ice to water, or water to water vapor involve the transfer of heat without a change in temperature.

The number of calories of heat required to change the temperature of 1 g of a substance by 1°C is called **specific heat**. Specific heats for a few common substances are given in table 7.1.

Table 7.1 Specific heat, heat of fusion, and heat of vaporization of some common substances

Substance	Sp. heat (cal/g/°C)	Heat of fus. (cal/g)	Heat of vap. (cal/g)
Water	1.00	80.00	540
Ice	0.50	–	–
Steam	0.48	–	–
Alcohol (ethyl)	0.58	25	204
Wood	0.42	–	–
Glass	0.20	–	–
Steel	0.11	–	–
Oxygen	–	3.3	51

Source: Halfacre and Barden 1979:168.

Substances with a high specific heat undergo a relatively small change in temperature in response to a given amount of heat energy. Because of its high specific heat, water has a strong modifying effect on temperature change. This has great significance for plants, especially for frost protection.

The amount of heat absorbed in changing 1 g of a substance at its melting point from the solid to the liquid state, or released when the substance changes from the liquid to the solid state is its **heat of fusion**. To change 1 g of a substance at its boiling point from the liquid to the vapor state, the heat requirement is called the **heat of vaporization**. Like specific heat, the heat of fusion and heat of vaporization of water are very high compared to those of other substances (Table 7.1).

7.2.2 Temperature measurement

The freezing point of water is 0°C, and its boiling point is 100°C on the Celsius (centigrade/metric) scale. On the Fahrenheit scale, the freezing point is 32° and the boiling point 212°. The Kelvin absolute temperature scale begins at **absolute zero**, at which point matter contains no heat energy. It is equivalent to −273°C or −460.4°F. Ice melts at 273°K and water boils at 373°K. The following equations can be used to convert temperatures from one scale to another.

$$°F = (9/5\ °C) + 32$$
$$°C = 5/9\ (°F - 32)$$
$$°K = °C + 273$$
$$°C = °K - 273$$

7.2.3 Factors influencing temperature

1. Solar radiation. Our main source of heat is the sun. The sun is located at a mean distance of 1.5×10^8 km from the earth, and has a diameter of 1.4×10^6 km, which is about 100 times that of the earth. Its estimated surface temperature is about 6,000°C, and it is constantly radiating energy in all directions. The minute fraction of the sun's radiation which reaches the earth and its atmosphere is enough to sustain all life. Of the incoming solar energy striking the outer atmosphere, 10% is ultraviolet, 40% visible, and 50% infrared radiation. Most of the UV radiation is absorbed by the ozone layer in the upper atmosphere. Living organisms cannot withstand large amounts of UV radiation. Water vapor and CO_2 absorb part of the infrared radiation. Smoke, dust particles, and water droplets in the atmosphere scatter light of differing wavelengths, depending on their size and concentration. It is estimated that 34% of the solar radiation received by earth is reflected back to space by atmospheric clouds, 19% is absorbed by the atmosphere, and 47% reaches the earth directly or as diffused radiation.

2. Latitude. In general, temperature increases with decreasing latitude (Table 7.2). These temperature differences are caused by differences in the amount of radiation received from the sun (**insolation**). The earth's axis of rotation is tilted at an angle of 23½° to its plane of rotation around the sun. At any given time, only half of the earth is illuminated, and the portion illuminated is continually changing. At the spring and fall equinoxes, (March 21 and September 21, respectively), sunlight reaches the entire earth from the North to the South Pole. Thus, all points on the earth have 12 hours of daylight and 12 hours of darkness. However, at all other times of the year insolation striking the earth's surface varies with latitude.

Table 7.2 Temperatures (°C) at five latitudes in the northern hemisphere

North latitude	Mean annual	Mean January	Mean July	Range January–July
90°	−26	−41	−1	40°
60°	3°	−16°	14°	30°
30°	18°	14°	27°	13°
10°	26°	26°	27°	1°
0°	26°	27°	26°	1°

Source: Halfacre and Barden 1979.

The more vertical the sun's rays, the greater will be the concentration of energy per unit area. Thus the intensity of solar energy reaching the earth at the equator, where the sun strikes the earth at an angle of 90°, is greater than at the poles, where the sun's rays strike the earth at a very small angle.

As one moves away from the equator, the same amount of radiant energy is spread over an increasing surface area. As the angle of incidence decreases, the same amount of radiant energy is received by an increasing area, and consequently its intensity is decreased. You can visualize this by shining a flashlight on a dark wall at an angle of 90° and then tilting the flashlight to decrease the angle at which the light strikes the wall. Also, as the angle of the sun decreases, the depth of atmosphere through which its rays must pass increases. At 90°, the depth is 1 atmosphere; at 60° it is 1.2 atmospheres; at 30°, 2.0 atmospheres; and at 10°, 5.7 atmospheres. As the depth of the atmosphere through which it must pass increases, the amount of solar energy reaching the earth decreases because of increased absorption, reflection, and scattering by the atmosphere. The difference in the intensity of sunlight in the early morning and at noon is a result of these factors. Within Pakistan, however, latitude effects on temperature are negligible.

3. Season. Seasonal effects on temperature are closely related to latitude. When one moves away from the equator, temperatures not only decline but vary more between summer and winter. These seasonal effects result largely from the tilt of the earth's axis with respect to its plane of rotation and daylength (Table 7.3). At the equator, daylength is 12 h during the entire year, and solar angle is always relatively close to 90°.

Table 7.3 Effects of latitude and season on daylength

Latitude	Daylength at equinoxes and solstices			
	March 21	**June 21**	**September 21**	**December 21**
0°	12 h	12 h	12 h	12 h
10°	12 h	12 h 35 min	12 h	11 h 25 min
20°	12 h	13 h 12 min	12 h	10 h 48 min
30°	12 h	13 h 56 min	12 h	10 h 4 min
40°	12 h	14 h 52 min	12 h	9 h 8 min
50°	12 h	16 h 18 min	12 h	7 h 42 min
60°	12 h	1 month	12 h	0 h 0 min
80°	12 h	4 months	12 h	0 h 0 min
90°	12 h	6 months	12 h	0 h 0 min

Source: Adapted from Halfacre and Barden 1979:182.

As distance from the equator increases, summer daylength increases and winter daylength decreases. Therefore, in spite of the low solar angle, areas at higher latitudes receive a moderately high amount of solar energy during summer. The average temperature difference between 0° and 90°N latitude in July is 27°C (Table 7.2). During winter, however, far northern latitudes have not only very low solar angles but very short days as well. Thus the mean temperature difference in January between 0° and 90°N latitude is 68°C, much greater than that in July.

All of Pakistan lies between 25°N and 35°N. Average monthly temperatures in the plains of Punjab are graphed in Fig. 7.3.

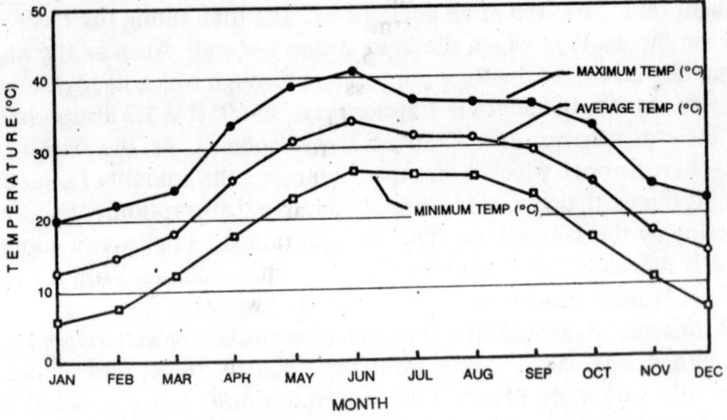

Figure 7.3 Average monthly temperatures, Punjab (1980-8). Source: Department of Meteorology, University of Agriculture, Faisalabad.

4. Elevation. Atmospheric heat is received directly from the earth's surface and only indirectly from the sun. The lower tropospheric air contains more water vapor and dust, therefore absorbs more terrestrial radiation, leading to higher temperatures at lower altitudes. Temperature decreases with elevation; for each 100 m increment in elevation, mean temperatures decline about 0.6°C. A zone of permanent snow exists above 4500 m in the tropics and above 3000 m in the temperate zones.

The effect of elevation on temperature depends on the balance between incoming and outgoing radiation. The most important factor is atmospheric water vapor and its effect on radiation. Approximately half of the total water vapor in the atmosphere is in the lower 1800 m.; thus the lower 0.2% of the atmosphere contains more than 50% of its total water vapour. Since water vapor slows radiation of heat away from the earth more than it blocks incoming solar radiation, atmospheric water vapor has a major impact on temperature at the surface of the earth. Although a site at 1500 m receives somewhat more solar radiation than one at sea level, since there is much less water vapor in the air at the higher elevation, the rate of radiant heat loss is much greater and the effect is a lower temperature at the higher elevation (Halfacre and Barden 1979:183).

Pakistan is a land of contrasts, starting from sea level and rising to the highest mountain peaks. The influence of elevation on temperature plays a tremendous role in the different agro–ecological zones and the horticultural crops grown in specific areas of the country. For annual crops like vegetables and seasonal flowers, the mean and extreme temperature values and the length of the growing season are important. Perennial plants like fruit trees and ornamental shrubs are affected by temperature during the whole year, and frost is a critical component.

5. **Time of day.** The temperature at any given time depends on the balance between incoming and outgoing heat. Plotting the energy received from the sun and that lost by the earth in relation to temperature shows that maximum temperatures occur not at the time of maximum insolation but rather when insolation has declined to where it equals radiational heat loss. Maximum daily temperatures occur in mid-afternoon, well past solar noon, and the daily minimum occurs just before sunrise. For the same reasons, annual maximum temperatures in the Northern Hemisphere occur in late July and August, well after the time of maximum solar radiation; and the annual minimum occurs in late winter, well after the period of minimal insolation. This relationship is called the **lag effect** (Halfacre and Barden 1979:183–85).

AIR TEMPERATURE CHANGES DURING THE DAY AND NIGHT: During the day, radiation from the sun passes through the atmosphere and is absorbed by solid surfaces on the earth's surface. Much of this energy is converted to heat, which raises the surface temperature of the earth. Soon after sunrise, the temperature of surfaces exposed to the sun becomes greater than adjacent air temperature and heat begins to be transmitted to the air by conduction. As the air near the surface of the earth warms up, convection currents gradually begin, and heat is moved upward. As the day progresses, however, the warmest air is close to the ground and temperature declines with increasing distance from the earth's surface. After sunset, the earth no longer receives solar radiation, but radiation of heat from the earth continues. Thus the temperature of many exposed surfaces falls below that of the adjacent air. This effect is particularly important with efficient radiating surfaces like leaves, which have a large surface area in relation to their volume. As the surface of a leaf becomes cooler than the surrounding air, heat is conducted from the air to the leaf, the reverse of the heat flow during the day.

TEMPERATURE INVERSIONS: As the nighttime pattern of conduction continues, the air near the ground becomes cooler than the air above it. Since the cool air is denser than the warm air above it, no convection currents are formed and a **temperature inversion** results; that is, an inversion of the normal daytime situation. Maximum nighttime radiant heat loss occurs when there are long nights, clear skies, cool dry air, and calm conditions. Since clouds block heat loss, temperatures drop much more slowly on cloudy nights. High humidity also restricts heat loss as can be seen by comparing the rapid cooling at night in dry areas like the Thal Desert with the hot nights in humid areas like the irrigated areas of Punjab.

6. Topographic factors. Proximity to large bodies of water has a profound effect on temperature patterns. Since water has a higher specific heat and higher thermal conductivity than soil, it absorbs heat to much deeper levels and warms and cools more slowly than soil. Temperatures vary less over oceans and other large bodies of water than over continents. The moderating effect of oceans and other large bodies of water on climate is called the **oceanic effect**. The differences between the yearly temperature patterns of Karachi and of Multan are largely a result of the oceanic effect.

Slope. The direction in which a slope faces affects its temperature. In general, angles of the slopes being equal, a south slope is warmest, followed by west, east, and north slopes. These temperature differences are the result of differences in the amount of solar radiation received by the soil. South slopes receive the most radiation because they are exposed to sunlight almost all day, even in winter when the sun is low in the sky. A west slope tends to be warmer than an east slope because it is warmed by the sun all morning. This difference is particularly marked in *soil* temperature, and thus low–growing crops, such as strawberries, are more affected than apple trees, which are more influenced by air temperature (Halfacre and Barden 1979:187). Horticulturists should keep slope effects in mind when selecting sites for relatively tender low–growing crops.

Air drainage. During the night the air near the ground is colder and therefore denser than the warm air above. Because it is denser, cold air flows downhill and collects in low areas from which it cannot drain, an effect called **cold–air drainage**. This leads to the formation of **frost pockets**, or depressions which collect cold air; the slope itself remains a relatively warm area, or **thermal belt**.

Plants in frost pockets are in most danger of freeze damage. Therefore, in selecting planting sites the location of the site to be planted in relation to the area into which cold air will drain should be kept in mind. Crops susceptible to freeze damage should be planted on sites which have good air drainage. Thus apple growers often plant orchards on hillsides with less than ideal soil, just to have good elevation and asscciated air drainage. Early–blossoming cultivars may be planted at the top of a slope, and later–blossoming varieties at the lower levels. The size of the area into which cold air drains is also important. The sides of a narrow valley are less desirable than the sides of a hill which rises up from a relatively large area into which cold air can drain (Halfacre and Barden 1979:187–88).

7. Soils. Different types of soil warm up at different rates in the spring. The rate of warming is determined by several factors, including colour, density, rate of heat conduction, air space, and water–holding capacity—the most important factor. Sandy soils with low water–holding capacity tend to warm up first in the spring and are often called **early soils**. Organic soils tend to warm up slowly and are thus classed as **late soils**. Loam, silt, and clay soils fall somewhere between sandy and organic soils, depending on the factors listed above.

7.2.4 Temperature relations of plants

The term **cardinal temperatures** designates several critical temperatures for plants. It includes the **optimum temperature**, at which a plant functions best; the **minimum temperature**, below which a plant cannot grow; and the **maximum temperature**, above which the plant cannot grow. Cardinal temperatures vary greatly among species according to their origin and adaptability. To describe tolerance for *minimum* temperature, plants are categorized as hardy, half-hardy, and tender.

> **Hardy** plants can withstand minimum temperatures of −4 to −2°C. Peas, spinach, turnips, and cabbage are hardy plants.
>
> **Half-hardy** crops can survive minimum temperatures of −1 to 0°C. Some half-hardy plants are carrots, beets, and lettuce.
>
> **Tender** crops cannot tolerate 0°C. Beans, corn, squashes, melons, cucumbers, and tomatoes fall in this category. These crops need a frost-free growing season or frost protection.

A second classification involves the *optimum* growing season temperatures. There are two classes: **cool-season** and **warm-season crops**. The optimum temperature for cool-season crops is 18–24°C. Warm-season crops like temperatures of 25–35°C. Vegetable crops and annual flowering plants are traditionally classified as summer and winter groups according to their temperature requirements (Halfacre and Barden 1979:191). However, most of the winter crops of the plains are grown at higher elevations during the summer months. Trees are classified as temperate, tropical, and subtropical/semi-tropical according to their sensitivity and temperature requirements.

However, some crops can be grown in areas not normally ideal for them by selecting special strains and using management practices to cope with unfavorable weather and temperature conditions. Greenhouses are used for controlled temperature production of high-value ornamental and vegetable crops. Shading and mulching can be used to avoid injury from extreme temperatures. With perennial plants, artificial control of temperature has limited scope.

Length of growing season and degree days. The suitability of an area for any crop is determined by the length of its growing season. The oldest and simplest method of calculating the length of the growing season is the average number of consecutive frost-free days. This varies from 100–120 days in the Northern Areas, and keeps increasing progressively toward the lower parts of the country to almost 365 days a year in the lower tropical/coastal regions of Sindh. Elevation and proximity to large bodies of water will have marked influence on the occurrence of frost.

The length of the growing season determines what *cannot* be grown, but does not tell us what *can* be grown. For example, even though the growing season is long enough, the production of certain tender vegetables is not possible if the temperatures are not sufficiently warm. The concept of **degree days** is a means of correlating temperature with crop development. The calculation of degree days is a method of temperature summation which can be used to predict successive planting and maturation dates.

Degree days are calculated as follows. For each crop, a base (threshold) temperature is determined below which crop growth and development is minimal. For each day of the year, this base temperature (in °F) is subtracted from the maximum for that day. For example, with pea the base temperature is 40°F (4.4°C), so for a day with a maximum of 50°F (10°C), 10 degree days would be accumulated. If the daily maximum is lower than the base temperature, no degree days are accumulated or subtracted. Peas require an average of 2000 degree days. For warm-season crops, the base temperature would obviously be higher than 40°F.

Degree-day calculations yield approximations, as they do not consider the non-linear responses of plants to temperature as well as the effects of other important factors like latitude, illumination, daylength, nutrition, water, and disease and insect problems (Halfacre and Barden 1979:193).

Freezes and frosts. There are two types of freeze. **Radiational freeze** occurs with calm conditions, radiational cooling, and temperature inversion. Its severity is directly dependent on radiant heat loss through temperature inversions. **Advective freezing** occurs as a result of the entry of a large cold air mass. During wintertime in the mid to high latitudes of the temperate zone, large, dry cold air masses move southward from the circumpolar regions causing sudden and drastic temperature drops. No temperature inversion is present; heat is lost directly to the cold air by conduction, and freeze damage may occur.

The terms *freeze* and *frost* are often not discriminated; however, horticulturists often use the words *frost* and *freeze* for radiational and advective freeze, respectively. Thus, *frost* refers to temperatures of 0°C or below associated with a temperature inversion. A **white frost** occurs where the **dew point** (temperature at which relative humidity reaches 100 percent) is above the minimum temperature and is obvious from the deposit of ice or frost (frozen dew) on exposed surfaces. A **black frost** results when the dew point is below the minimum temperature. The visible indication of cold injury is blackening of tender plants.

Other forms of low–temperature injury include chilling injury, desiccation, and heaving. **Chilling injury** is plant tissue damage at the membrane level caused by temperatures slightly above freezing. Desiccation of tissue occurs because of decreased water uptake by the roots. Winterkill takes place in evergreen plants which lose water by transpiration faster than it is replaced through the roots. The **heaving** of soil caused by alternate freezing and thawing damages the plant by ripping its root system. Mulching can effectively prevent heaving. Bark splitting can also occur as a result of low–temperature injury. Limbs can be broken by the weight of ice after heavy snowfalls.

High–temperature injury. Sunburn or 'burning up' of plants takes place during hot weather. High temperature results in excessive water loss, and tissue desiccation. Warm, dry winds aggravate this situation. A significant amount of visible heat injury to fruit, leaves, and other tissues can occur when daily air temperatures reach 40°C or higher. As seen in Figure 7.3, temperatures during the summer months in the plains of Punjab are high enough to cause heat injury. Typically, heat injury occurs on a plant's most exposed south side. Bark splitting, yellowing and browning of leaves, visible spots on the peel of fruits, misshapen fruits, and effects on the quality of produce are common. Soil temperatures can also rise enough to interfere with root growth. Young plants can burn off at the soil line. Extremely high temperatures (46–54°C) may kill the plant.

Vernalization and chilling requirements. Temperature plays a crucial role in flower initiation and development in some plants. Biennial plants typically produce only vegetative growth in the first year; then, after extended exposure to cold during the winter (**vernalization**), flowering takes place in the second year. Some common crops which need low temperatures for flower induction are beets, cabbage, lettuce, radishes, and carrots.

Most winter–hardy deciduous fruits have a specific quantitative chilling requirement for successful flowering. A certain minimum number of hours of exposure to temperatures below 4.4°C (40°F) are counted by an internal–clock mechanism. Any hours of temperature above the critical temperature are to be deducted from the counts. Chilling requirements vary among species and among cultivars. Breeders have developed low–chilling cultivars of apples, peaches, plums, almonds, and apricots which can fruit successfully in relatively warm regions.

Some fruit crops important in Pakistan are listed according to their temperature requirements in Table 7.4.

Table 7.4 Some common fruit crops listed according to their temperature requirements (order of increasing cold–tolerance)

Tropical	**Subtropical**		**Temperature** Mild winter	Severe winter
Coconut Banana Mango Pineapple Papaya				
	Date Fig			
		Citrus Olive Pomegranate		
			Almond Grape (European) Persimmon (Japanese)	
				Peach Cherry Apricot (blossoms tender) Strawberry (very hardy)
				Pear Plum Grape (American) Apple
Low–temperature sensitive		**Slightly frost–tolerant**	**Tender**	**Winter–hardy**
Non cold–requiring			**Cold–requiring**	

* Variation in tolerance depends to a large extent upon species, variety, plant part, and stage of growth.

Source: Adapted from Janick (1986:468), with permission of the publishers.

7.2.5 Temperature management

Plants show a marked response to changing temperatures. Extremes of temperature, even for short periods of time, can lead to irreversible changes, resulting in the death of the whole plant or parts of it. Exhausted fruit–bearing trees are very susceptible to low–temperature injury. Late season growth on trees remains succulent and gets killed. There is no direct control of temperature under field conditions. Adaptations through selection of location, site, choice of plants, and appropriate cultural practices can modify the effects of extreme temperature.

Hardening. Broadly, hardening refers to processes that increase the ability of plants to survive environmental stress. In its restricted meaning, hardening refers to developing the ability to withstand cold injury. Physiologically, cold–hardiness is acquired in response to natural changes in temperature, water availability, and daylength. The progressive onset of cooler weather and shorter days in the fall bring about dormancy and hardiness. Growth slows down and carbohydrates accumulate near the end of the growing season. Water moves from protoplasm to the vacuoles. The changed carbohydrate and water relations leave a larger proportion of water bound to protoplasmic proteins and other colloidal materials, which stabilizes cell structure and imparts hardening.

Nursery plants are hardened before transplanting to enable them to withstand the stress of transplanting. There are several ways to do this. The simplest method is to withhold water and excessive nitrogen fertilization. Sufficient growth during the growing season is necessary to accumulate carbohydrates. Gradual exposure of nursery plants to cold along with withholding water or root pruning can sufficiently harden the plants in about 10 days. Hardened nursery plants look stocky in contrast to tender and 'leggy' unhardened plants. Many vegetable nursery plants develop a dark–green color and waxy covering on the leaves upon hardening. Over–hardening must also be avoided, since prolonged hardening can interfere with the subsequent development of seedlings after transplanting. In cruciferous vegetables (cole crops), over–hardened seedlings bolt without producing heads.

Mulching. Surface covers used as insulating agents to regulate soil temperature are called mulches. Mulching materials include a variety of plant refuses, straw, hay, organic manures, paper, and plastics. Stable soil temperature improves root growth and increases nutrient availability both in summer and winter. It encourages early production of the crop as well. The stable soil temperature also prevents heaving. Additional benefits of mulching are moisture conservation, weed control, erosion control, and clean harvest from vine crops like strawberries. Some problems associated with mulching include the harboring of plant pests, weeds, disease producing microbes, and rodents. Inflammable mulching materials can be a fire hazard.

Frost control. There are three approaches to frost control: (a) avoidance, (b) reduction of heat loss, and (c) addition of heat. For a successful control strategy, the conditions of occurrence of frost and the mechanisms of injury must be thoroughly understood.

Spring frost can be avoided by planting annual crops late. But this approach has its own problems. Predicting the last frost–free date is a matter of statistics, and can be inaccurate. Areas with a short growing season get early fall frost, and late planting will further reduce the growing season. Starting nurseries under protection and transplanting after the frost–free period begins avoids the danger of frost. A futuristic approach to escaping freeze damage involves interfering with the transition of liquid water to ice crystals. The transition of water to ice requires ice–nucleating particles that include certain bacteria (*Pseudomonas* and *Erwinia*). Elimination of ice-nucleating bacteria by bactericides or the use of non–nucleating ('ice–minus') bacteria as antagonists can reduce frost damage. 'Ice–minus' bacteria exist naturally and can also be genetically engineered.

Heat moves up through the soil by conduction, the rate of conduction varying with the soil type. Mineral soils conduct heat better than organic soils. Conditions leading to rapid and prolonged surface cooling are conducive to radiational frost. For example, the introduction of cold polar air followed by clear, dry, calm nights facilitates upward radiation, causing temperature inversions. The absence of wind leaves the coldest air undisturbed next to the ground. Clouds and humidity cause re–radiation back to the earth and prevent frost.

Reduction of heat loss can be achieved by any method that will increase daytime absorption of heat by the soil, or prevent its loss at night. Anything that prevents the accumulation of radiation during the day will increase the chances of frost. Thus clean cultivation, which allows the direct penetration of heat into the soil, is better for frost control than mulched or sodded fields. Covering individual plants and the use of plastic tunnels, cold frames, and a variety of mulches can prevent heat loss at night. Replacing air spaces in the soil by water reduces the danger of frost.

The addition of heat may be accomplished in a number of ways, e.g. by using heaters, flooding, or spray irrigation, and by mixing the air with wind machines. However, the cost of energy for artificial heating and running wind machines may be prohibitive. Smudging (use of smoky heaters), an effective method of frost protection, is now illegal in some parts of the world because of the air pollution that it causes.

7.2.6 Plant–growing structures

A variety of plant–growing structures are in use to protect plants against adverse weather or to provide an ideal growing environment, primarily by temperature control. Systems are available which can provide a completely artificial plant environment. **Greenhouse** is a general name for many structures used for controlling the environment. Commercial production of many vegetables and ornamentals is carried on in greenhouses. A brief description of common plant–growing structures is presented here.

1. Cold frames. A cold frame is an inexpensive form of temperature control. It is an enclosed soil bed covered by a removable, transparent (glass or plastic) frame or sash. Temperature inside the cold frame increases considerably during the day and heat is stored in the soil. Plants can be protected at night due to the excess heat trapped. In warm weather, the sash is raised or removed. Cold frames are commonly used for starting plants from seed or hardening-off seedlings to be transplanted.

2. **Hotbeds.** Hotbeds are essentially cold frames provided with a supplemental source of heat, like hot water, steam, or electric heaters. A cheaper source of heat like fermentation of decaying organic matter can also be used.

3. **Glasshouses.** These are permanent structures, often on concrete foundations, in which temperature and other environmental factors can be regulated (Fig. 7.4). Temperature may be regulated by a combination of heating, cooling, and ventilation. In summer, shading the glass or whitewashing it can improve the efficiency of cooling systems. Ventilation can be used to regulate humidity. Supplemental lights can be used to extend the duration and intensity of light. Plants are usually grown in containers in artificially prepared soil mixes and given well-controlled fertilizer programs. CO_2 enrichments have been shown to improve plant growth and yields in glasshouses. Because of the high cost of glass, other glazing materials like plastics and fiberglass have also been used.

Figure 7.4 A glasshouse

4. **Plastic tunnels.** Sheets of polyethylene are laid down over wooden or metal hoops (Figure 7.5a,b). The increase of temperature within the tunnel encourages plant growth even with low outside temperatures. Tunnels are especially suitable for early production of summer vegetables. The operations may be carried out in a semi-permanent structure (high tunnels)

Figure 7.5 Plastic tunnel. Above: exterior view; below: interior view.

or by spreading the plastic sheets on temporary supports over the field beds (low tunnels). As warm weather approaches, the plastic may be raised or removed during the day and put back at night.

Extensive use of tunnels also has some problems. High humidity encourages fungal pathogens. Proper ventilation must be carried out to lower the humidity. Lack of pollinators may result in no fruit set in crops like cucurbits, and hand pollination must be done to ensure fruit–set in cross–pollinated crops.

5. Shade houses. Shade or lath houses are structures with semi–shade conditions provided by screens, shading nets, or wooden laths. Shade reduces temperature, which in turn reduces water loss by transpiration and evaporation. Shade–loving plants like chrysanthemum, hydrangea, and foliage plants can be propagated and maintained in lath houses. Shade houses are also used for hardening plants, and for transplanting.

7.3 Water

Given certain light and temperature conditions, water is the next most important limiting environmental factor in selecting a planting site and plant materials. Historically, agriculture started along the banks of rivers and lakes, and only later, when man developed a variety of irrigation systems, expanded into arid and semiarid areas. An adequate water supply must be ensured for any horticultural enterprise. The need for water cannot be overemphasized, as it performs many vital functions in plant life.

- It is a necessary constituent of all living plant cells.
- It serves as a solvent for nutrients from soils and CO_2 from air.
- It acts as a raw material in the process of photosynthesis.
- It is a medium of translocation from site of uptake for raw materials and site of production of plant compounds to their sites of utilization.
- It is a reagent/substrate for a variety of chemical reactions, and acts as a pH buffer.
- It maintains the turgidity of plant cells which is required for growth and development, and for vital functions like stomatal opening.
- Because of its high specific heat, transpiration of water dissipates large quantities of heat and suppresses leaf temperature.

7.3.1 Properties of water

Water covers up to 70 percent of the earth's surface. Because of its unique properties, water is ubiquitous and functionally present in all living organisms in large quantities.

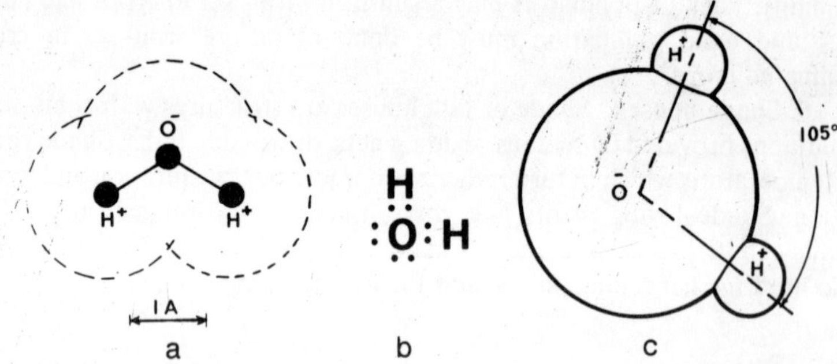

Figure 7.6 The water molecule. (a) angle of separation between the two positive charges (H^+); (b) covalent bonding between the H's and O; (c) position of atoms with effective sizes of isolated atoms.

Most of the unusual properties of water are a result of its molecular structure. Water, consisting of two hydrogen atoms and one oxygen, is a covalently bonded molecule. The bonding angle between hydrogens is 105°, and because of this asymmetry of hydrogen, water has a strong dipole moment, i.e. it is highly polarized, with a strong separation of positive and negative charges (Fig. 7.6). Because of this polarization, it readily shares its hydrogen with the oxygens of other molecules of water and other oxygen–containing substances. Thus water is both very cohesive and adhesive. Its strong dipolar nature results in electrostatic attraction for other dipoles and ions, and in hydrogen–bonding with oxygen–containing molecules. Water will adsorb to substances such as cellulose, but not to polyesters which have few oxygens available for hydrogen bonding.

Many of the physical properties of water play a crucial role in plant–water relations. Ice at 0° has a density of about 0.999 g/cm³ and at 4° is 1.0 g/cm³. Since the densest water sinks to the bottom of a body of water, the coldest water rises to the top, and water freezes from the top down rather than from the bottom up. The frozen water at the surface insulates the water below, preventing large bodies of water from freezing solid and killing aquatic life.

Water has the highest **specific heat** of the common liquids. Water is used as the standard for the **calorie**, the amount of heat required to raise 1 g of water 1°C between 15.5° and 16.5°C. Hence water has specific heat of

1 cal/g per degree. This is important for rates of heating and cooling of plant bodies as well as aqueous bodies. Large bodies of water such as lakes and oceans tend to moderate the temperatures of adjacent land. For the same reason, dry climates tend to have greater temperature extremes than humid climates. The rates of heating and cooling of plant tissue are also affected by the heat capacity of water.

The **heat of vaporization**, the amount of heat required to convert one gram of liquid to vapor, is much higher for water than other liquids. This is due to the cohesiveness of water. Because of its high heat of vaporization, when water evaporates from living surfaces, much heat is lost and a cooling effect is produced.

The **heat of fusion**, the heat required to change a solid to a liquid, is also higher for water (ice) than for other substances. An interesting application of this property of water is in frost protection. Orchards can be protected from frost injury by flooding. When the water freezes, the heat liberated during fusion (i.e. freezing) adds heat to the grove and protects the trees.

Pure water is colorless, tasteless, and odorless. Due to molecular motion, water will tend to scatter short–wavelength light (blue) and transmit the visible long wavelengths (red), with the result that water appears blue. Other properties of interest include high viscosity (resistance to flow), high density (grams per cubic centimeter), and high surface tension (the tendency of the surface to contract and resemble an elastic membrane).

To understand plant–water relations, both atmospheric and soil moisture content must be considered. Water utilization from the soil is strongly influenced by atmospheric humidity, i.e. water in vapor form. It is, therefore, essential to understand various aspects of atmospheric moisture before discussing the soil–plant–water relationship.

7.3.2 Humidity

Absolute humidity is the weight of water vapor per unit *volume* of air, expressed as grams per cubic meter. For a given atmospheric pressure, the volume of air changes with changing temperature. Therefore, absolute humidity will vary with temperature without any change in absolute moisture content.

Specific humidity is the weight of water vapor per unit *weight* of air, expressed as grams per kilogram. Since it is a ratio of weight to weight, temperature and pressure changes do not affect this measure of humidity.

Relative humidity is the ratio of the amount of water vapor present in the air and the amount at saturation for a given temperature and pressure, expressed as a percentage. The water–holding capacity of air depends on temperature. The higher the temperature, the greater the water–holding capacity. The relationship between relative humidity and specific humidity is shown in Table 7.5.

Table 7.5 Effect of increasing temperature on relative humidity with a constant specific humidity

Temperature (°C)	Specific humidity (g/ kg)	Relative humidity (%)
0	3.80	100
5	3.80	70
10	3.80	50
15	3.80	36
20	3.80	26
25	3.80	19
30	3.80	14
35	3.80	11

Source: Halfacre and Barden 1979:216.

Starting from 0°C air temperature and 100% relative humidity, when temperature increases to 30°C, the relative humidity drops to 14%, with no change in moisture content. This shows that relative humidity as a measure of atmospheric moisture is meaningless if compared at different temperatures, and explains the daily variation in relative humidity from morning to evening.

A very useful method of expressing atmospheric moisture is vapor pressure. Water **vapor pressure** is that part of the total atmospheric pressure which is due to water vapor, expressed in millimeters of mercury (mm Hg), or in millibars (1 mm Hg = 1.32 millibars). Since relative humidity is a measure of relative saturation, the actual vapor pressure is calculated by multiplying the saturation vapor pressure by the relative humidity. A range of actual vapor pressures at different temperature and humidity levels is given in Table 7.6. The saturation vapor pressure at 30°C is 31.82; at 50% relative humidity, the actual vapor pressure is 15.91 (31.82 × 0.50).

Table 7.6 Vapor pressure of the atmosphere at selected temperatures and relative humidities

Temp.	Actual vapor pressure (mm Hg) at indicated relative humidity										
°C	0	10%	20%	30%	40%	50%	60%	70%	80%	90%	100%
0	0	0.46	0.92	1.37	1.83	2.29	2.75	3.21	3.66	4.12	4.58
5	0	0.65	1.31	1.96	2.62	3.27	3.92	4.58	5.23	5.89	6.54
10	0	0.92	1.84	2.76	3.68	4.60	5.53	6.45	7.37	8.29	9.21
15	0	1.28	2.56	3.84	5.12	6.40	7.67	8.95	10.23	11.51	12.79
20	0	1.75	3.51	5.26	7.02	8.77	10.52	12.28	14.03	15.79	17.54
25	0	2.38	4.75	7.13	9.50	11.88	14.26	16.63	19.01	21.38	23.76

Temp. °C	Actual vapor pressure (mm Hg) at indicated relative humidity 0	10%	20%	30%	40%	50%	60%	70%	80%	90%	100%
30	0	3.18	6.36	9.55	12.73	15.91	19.09	22.27	25.46	28.64	31.82
35	0	4.22	8.44	12.65	16.87	21.09	25.31	29.53	33.74	37.96	42.18
40	0	5.33	11.60	16.60	22.13	27.66	33.19	38.72	44.25	49.79	55.32

Source: Halfacre and Barden 1979:217.

A parameter commonly used to describe atmospheric water relations is **vapor pressure gradient**. This measure is used to compare the vapor pressure at a relative humidity less than 100% with the saturation vapor pressure. The vapor pressure gradient allows us to compare situations at different temperatures, which is not possible with any of the other parameters described. For example, the vapor pressure gradient at 30°C and 40% relative humidity is 31.82 – 12.733 = 10.09; and at 20°C, 17.54 – 7.02 = 10.52.

For the same relative humidity, a different vapor pressure gradient indicates the relationship of temperature with rate of evaporation. The relative humidity within the stomatal cavity is assumed to be 100%. For a given leaf surface temperature, one can work out the vapor pressure gradient, which will determine the rate of water loss through transpiration. It is important to remember that with a constant amount of water vapor present, vapor pressure does not change with temperature, but the vapor pressure *gradient* changes with temperature (Halfacre and Barden 1979:216–17).

There are several forms of atmospheric moisture or humidity, including rain, snow, vapor, fog, sleet, hail, and dew. Rain forms when the condensation of atmospheric moisture takes place at temperatures above the freezing point. Snow forms when atmospheric moisture condenses at temperatures below the freezing point. When rain falls through a layer of air with temperatures below the freezing point, sleet is formed. If the air currents carry the frozen particles back up so that they accumulate a thicker covering of ice before falling again, hail is formed. Hailstorms cause considerable damage to crops. Dew forms at night when atmospheric moisture condenses on cold surfaces (relative humidity reaches 100%). Fogs and mists are low clouds which do not settle. In addition to providing high moisture in the air, winter fogs help in the accumulation of chilling hours for deciduous fruits.

Atmospheric moisture can be measured by psychrometers, hygrometers, or rain gauges. The **psychrometer** is the simplest instrument to measure the vapor content of air. Typically, it consists of two similar thermometers. The bulb of one thermometer is covered with clean muslin cloth wetted with distilled water, and the bulb of the other is left dry. The two thermometers (wet–bulb and dry–bulb) are mounted side by side. If the relative humidity is below 100%, water evaporates from the wet bulb, and its temperature is lowered proportionately to the surface heat loss. The lower the humidity, the more rapid the evaporation. For a better estimate, the thermometers can be swung in the air until the temperature of the wet bulb no longer decreases. When the temperature of the wet bulb stops declining, the vapor pressures of the water in the wick and the water in the atmosphere are equal. By knowing the difference between the wet and dry–bulb temperatures, one can obtain the relative humidity from psychrometric tables or charts (Figure 7.7).

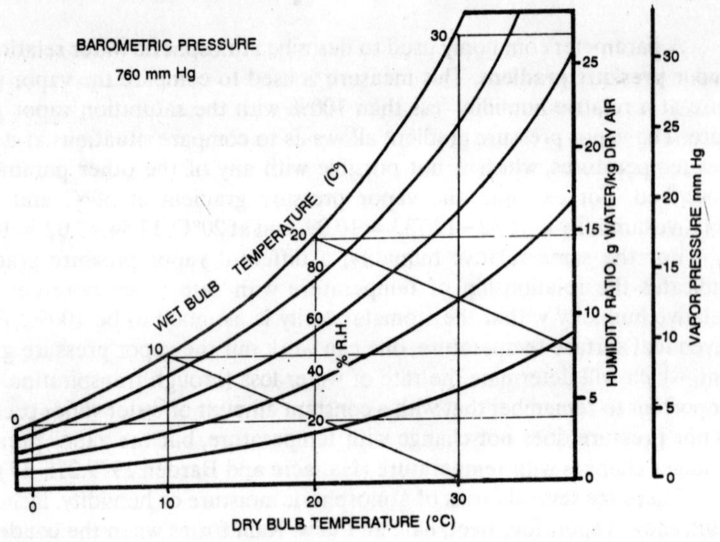

Figure 7.7 Psychrometric chart illustrating relationships between properties of moist air (mm Hg).

7.3.3 Soil moisture

Water in soil is found in several forms. In a saturated soil (after irrigation or heavy rain), water moves through the pore spaces. The proportion of water that drains out/down is **free water** or **gravitational water**. Gravitational water is available to plants, but only for a short time. Most of the water retained by soil is **capillary water**, which is held by its own surface tension (cohesion) and by adhesion to the soil particles. It is the most important component of soil moisture available to plants. A small component of soil moisture, **hygroscopic water**, adheres tightly to soil particles and is not normally available to plants. Water is also present in soils in chemically bound forms, and is also unavailable for plant use. When all free water drains from the soil, it is said to be at **field capacity**. Field capacity differs with the type of soil. The field capacity of a heavy clay soil may be 30 percent, whereas it is only 3 percent in a sand, on a dry–weight basis. After the depletion of all available capillary water, the soil water reaches a level called the **wilting point**. Again, the percentage of water at this point depends on the type of soil. The difference between the wilting point and field capacity is the **available water**.

A number of factors influence the movement of water in soil. Infiltration (movement of water into soil), percolation (movement of water through soil), and surface runoff are all influenced by the initial soil moisture content at the time of rain or irrigation. Soil structure, texture, and vegetation are other factors. The smaller pore size in clay soils, and larger pore sizes in loamy and sandy soils affect the movement of water accordingly. As water is taken up by plants and evaporates from the soil surface, water from lower parts of the soil moves upward. Upward and lateral movement is caused by capillary action, which is a surface tension phenomenon. The finer the soil spaces, the greater the distance water moves by capillary action.

There are several ways of measuring and expressing soil water. The simplest way is as a percentage of the dry weight of soil. A more useful method of assessing soil moisture is to express the tension with which it is bound to the soil particles. Free water is held with very low tension; capillary water is bound tightly; and hygroscopic water is bound extremely tightly. Thus free water has high potential energy (high potential), and bound water (under tension) has low potential energy (potential). The potential of soil water can be expressed in bars (one bar is equivalent to one atmosphere). Available water in most soils is between -0.33 bars and -15 bars.

A **tensiometer** is a device to measure the water tension/potential in soil. A soil tensiometer consists of a porous ceramic cup which is buried in the soil and connected with a tube to a vacuum gauge or mercury manometer situated above the soil surface. The cup and connecting tube are filled with water, which gradually comes to equilibrium with the soil moisture. At equilibrium, the water potential of the cup is the same as that of the soil, and can be directly read from the vacuum gauge or mercury manometer.

7.3.4 Plant water requirements

Determining when and how much water to apply is a major problem in plant water management. Only rarely is natural precipitation enough to meet the water requirements of most horticultural crops. The deficiency must be made up by irrigation. Irrigation timings can be based on soil moisture and meteorological data. The calculation of **consumptive use**, i.e. water lost by evaporation and transpiration, is the best indication of irrigation requirements. Consumptive use varies with temperature, duration of sunshine, humidity, wind, amount of plant cover, the stage of plant development, and available soil moisture. High water consumption takes place during hot, dry, and windy conditions (Table 7.7).

Table 7.7 Average peak moisture use (inches of water per day) for commonly irrigated horticultural crops

	Cool climate		Moderate climate		Hot climate	
Crop	Humid	Dry	Humid	Dry	Humid	Dry
Potato	0.10	0.16	0.12	0.20	0.14	0.24
Tomato	0.14	0.17	0.17	0.22	0.23	0.23
Bean	0.12	0.16	0.16	0.20	0.20	0.25
Vegetables	0.12	0.15	0.15	0.19	0.20	0.23
Deciduous orchard	0.15	0.20	0.20	0.25	0.25	0.30
Deciduous orchard with cover	0.20	0.25	0.25	0.30	0.30	0.35
Citrus orchard	0.10	0.15	0.13	0.19	0.18	0.23

Source: Reproduced from Janick (1986:210), with permission of the publisher.

Crops require most water in the fruiting or seed–forming periods. Plants also differ in their water requirements in relation to their ground–covering ability and root depth (Table 7.8). Except for water–loving plants, an *excess* of soil water can result in problems like root damage, fruit splitting, and infections by pathogenic fungi.

Table 7.8 Normal root–zone depths of mature irrigated crops grown in a deep, permeable, well–drained soil

Crop	Root depth (feet)	Crop	Root depth (feet)
Alfalfa	5–10	Grass pasture	3–4
Artichokes	4	Ladino clover	2
Asparagus	6–10	Lettuce	½
Beans	3–4	Onions	1
Beets (sugar)	4–6	Parsnip	3
Beets (table)	2–3	Peas	3–4
Broccoli	2	Potatoes	3–4
Cabbage	2	Pumpkin	6
Cantaloupe	6	Radishes	1
Carrots	2–3	Spinach	2
Cauliflower	2	Squash	3
Citrus	4–6	Sweet potato	4–6
Corn (sweet)	3	Tomatoes	6–10

Crop	Root depth (feet)	Crop	Root depth (feet)
Corn (field)	4–5	Turnips	3
Cotton	4–6	Strawberries	3–4
Deciduous orchard	6–8	Walnuts	12+
Grain	5	Watermelon	6

Source: Reproduced from Janick 1986:211, with permission of the publisher.

Not all the irrigation water applied is available to the crops. The percentage of applied irrigation water that actually becomes available for use by plants is the **irrigation efficiency**. Water should be applied to bring the soil up to field capacity. Irrigation is best done when the water tension in the depletion zone is above four atmospheres, or when 60% of available water is depleted (Janick 1989:211). Various irrigation methods for vegetable and fruit plants are described in Chapter 10.

7.4 Air

Normal air is composed of 78% nitrogen, 21% oxygen, 0.9% argon, 0.03% carbon dioxide, and 1–3% water vapor. Traces of inorganic and organic compounds are detectable.

Nitrogen is an inert gas, and though an essential element for plants, it is not available unless transformed into nitrate or ammonia ions through the process of nitrogen fixation. Air is the source of this critical element, which is fixed either through biological means or industrial/chemical methods. Abundant oxygen in the air is the basic requirement for life on the earth. Except in flooded soil conditions, which limit its availability to plant roots, oxygen is assumed to be freely available. Despite being a very small fraction of the atmosphere (0.03% or 300 ppm), carbon dioxide is critical to plants, since it is the only source of carbon. Carbon dioxide is never a limiting factor in field conditions; however, it may drop below normal levels in thick plant populations or under greenhouse conditions because of higher than normal rates of photosynthesis. Carbon dioxide enrichment can increase crop yields and improve plant quality in intensive greenhouse production. Atmospheric moisture (water vapor) was discussed in section 7.3.

In preindustrial days, the atmosphere varied relatively little and the normal movement of air kept its constituents in equilibrium. The processes of industrialization and urbanization, however, have caused considerable degradation of air quality. The principal sources of air pollution are automobiles, industries using fossil fuels, heating, refuse burning, and poor disposal of sewage and sludge. Pollutants in the air can interfere with the normal growth of plants, and plant scientists have begun investigations of the effects of air pollution on plant health. The role of horticultural plants in the control of air pollution has also been greatly advocated. A brief description of important air pollutants is presented here.

7.4.1 Air pollutants

Ozone. Combustion engines release nitric oxide (NO^-) which oxidizes quickly in the air to form nitrogen dioxide (NO_2). As a result of different photochemical reactions, nitrogen dioxide releases atoms of oxygen (O), which combines with molecular oxygen (O_2) to form ozone (O_3). Ozone is a highly reactive compound and tends to disappear if it ceases to be produced. Ozone can affect plants by decreasing photosynthesis. Stomata tend to close in the presence of O_3, limiting gas exchange and reduction of carbohydrates. Several plant disorders are known to be caused by ozone. In the upper atmosphere, a deep layer of ozone is present which protects living organisms on the earth from the ultraviolet radiation of the sun.

Sulfur dioxide. Sulfur dioxide is released into the atmosphere from burning fuels and mining operations. It is toxic to plants and threatens human and animal health. It can enter leaves through the stomata and is absorbed on wet cell surfaces. At low concentrations, it interferes with the synthesis of proteins. Larger accumulation results in direct injury to the cells. In the air, it is one of the gases which cause acid rain.

Fluorides. Fluorides are non-degradable hydrocarbons. They are produced in the aluminum, glass, ceramic, and phosphate industries, and are manufactured and used either in gas or particulate forms. Household paints and coolants in refrigerators and air conditioners contain appreciable quantities of fluorides. Once released into the air, fluorides travel to the upper atmosphere and form compounds with the highly reactive ozone. The continuous release of such compounds in the air has been depleting the ozone layer, causing holes in our 'protective blanket'. Fluorides enter plants by diffusing through the stomata. Exactly how fluorides injure plants is not known. They cause mottling and necrosis at the tips and margins of leaves of sensitive broad-leaved plants. Small doses of fluorides have been reported to increase growth of some plants, including grapes and citrus trees.

Other harmful gases in the air include ammonia and chlorine. Ethylene, a natural plant hormone, will injure plants if present in great quantities.

Dust. Dust, or suspended particulate matter, can be injurious to plants. Dust is introduced into the air by soil erosion, agricultural operations, smoke, mining operations, fast moving objects, winds, and industrial processes. When present on leaf surfaces, dust can impair photosynthesis. Wind-blown particles can physically damage the foliage, and finer particles can obstruct stomatal openings. The quality of crops like lettuce and cabbage, and many fruits and ornamentals can be badly damaged by airborne particles. A serious consequence of suspended particles is their ability to act as carriers of agro-chemicals sprayed on crops. Spray-coated dust particles can travel with the winds and land at unwanted destinations causing damage to crops and posing health hazards for man and animals.

Smog. Smog is a word made up from ***sm**oke* + *f**og***. It contains dust, oxides of nitrogen, ozone, sulfur dioxide, hydrocarbons, and other emission products. Emission products from automobiles react in the atmosphere under the influence of sunlight to produce toxic gases which are injurious to plants. The end product of a series of reactions is peroxyacetyl nitrate (PAN). PAN and ozone are toxic to plants at concentrations as low as 5 ppm for 10 minutes.

7.4.2 Acid rain

The pH of normal rain is about 5.6, due to natural carbonic acid formed when atmospheric carbon dioxide dissolves in rain water. However, rains at pH 4.0 or even lower occur in heavily industrialized areas of the world. Acid rain can also be produced naturally by volcanic eruptions. The production of SO_2 and NO_2 by burning fossil fuels is the main cause of this problem. The SO_2 and NO_2 are transformed into dilute sulfuric and nitric acids, respectively, causing acid rain. The acid-causing compounds in rain have been associated with extensive ecological damage, including plant injury and death, the destruction of aquatic life, and extensive damage to buildings. The only control possible is the reduction of sulfur and nitrogen-carrying gases in industrial and vehicular emissions.

7.4.3 The greenhouse effect

The earth's surface and atmosphere remain at temperature equilibrium by absorbing incoming sunlight and radiating an equal amount of infrared energy to outer space. However, the increasing concentration of CO_2 in the air has serious consequences for this ecological balance. Carbon dioxide strongly absorbs and emits radiation in the infrared range. Being transparent to visible light and partially opaque to infrared, CO_2 prevents normal re-radiation of infrared energy to space, thus increasing the temperature of the atmosphere. This result is called the **greenhouse effect**. A rise in the earth's temperature could change precipitation patterns, causing melting of the polar icecap, and a rise in ocean levels, submerging some coastal areas.

Efforts to prevent the increase of CO_2 levels in the atmosphere must be made. An increase in the use of cleaner energy sources with a corresponding decrease in fossil fuel use could reduce CO_2 emission. Another useful approach is to derive energy from renewable carbon sources such as plant debris and plant hydrocarbons instead of carbon fixed in the past. If all energy were obtained from carbon fixed annually, there would be no net change in the amount of CO_2 in the atmosphere.

7.4.4 Wind

The amount, velocity, and seasonal distribution of winds has definite effects on the adaptability and success of particular crops in any region. The major influences are of two types. The first is as an environmental component influencing energy flux. Hot winds tend to increase transpirational water losses and exposed leaves may die of desiccation. Cold winds reduce the influence of radiant energy to or from plant tissues. The second effect is mechanical. In addition to its speed effects, wind may carry sand and dust particles big enough to cause physical damage. Stem and leaf tissues may be bruised and fruits scarred. Tree limbs can be broken, and fruit drop may be extensive. Summer winds in Punjab are one of major causes of damage to our mango crop. In areas with frequent winds, trees will grow lopsided. It is difficult, if not impossible, to protect plants from the effects of wind. In selecting a site for orchards, the wind should be considered as an important factor. Windbreaks can considerably reduce the effect of winds. A discussion of windbreaks follows in Chapter 9.

7.5 Climatic regions

The summation of an area's weather components, including temperature, moisture, and light effects over a period of time is collectively called **climate**. The term **weather** refers to the daily condition of environmental components. Climate is the fundamental factor in the physical environment of plants and determines which plants can be grown in a particular region. On the geological scale, it determines the characteristics of soils, which are formed by the process of weathering. **Microclimate** refers to the climate of a very small area. For example, the climate at ground level will differ considerably from that 500 feet above it. Similarly, the microclimates of an air pocket, or in the vicinity of a body of water, or in low–lying areas are different from the climate of their surroundings. The term **location** identifies a geographic and climatic area, and **site** is related to microclimatic differences within a specific location (Janick 1986:456). The selection of proper location and sites is a major determinant of success in horticultural farming.

Total annual precipitation and its distribution vary from place to place. **Effective precipitation** is the rain water that is not lost by runoff or evaporation and is available to plants. The natural vegetation of an area gives a good indication of effective precipitation, and a variety of vegetation patterns exist in different parts of the world depending upon the amount of effective precipitation. For example, the average rainfall in the Northern Areas and Kashmir is much higher than that in the central parts of the country, and rainfall is negligible in Cholistan and the Thar deserts. There

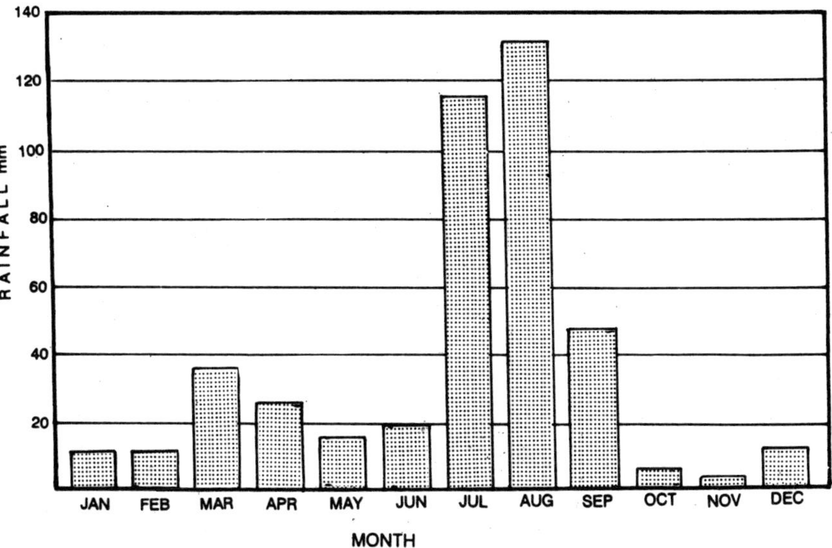

Figure 7.8 Average monthly rainfall in Punjab.

are considerable forest resources in the north, while there is very little life in Cholistan and the Thar deserts. Average monthly rainfall data for the plains of Punjab are presented in Fig. 7.8.

The quantity of light is a critical climatic factor, and plant species native to various regions have evolved to respond to specific photoperiods. Daylength is the most obvious difference between climates. Pakistan is located around 30°N, where the difference between midsummer and midwinter daylengths is about two hours. In addition to daylength, the quantity of light is affected by such atmospheric conditions as the number of sunny versus cloudy days. The average number of sunny days in Pakistan is more than 350 per year.

Climatic regions can be classified in a variety of ways. A common climatic **classification by temperature** divides the earth into three major zones, i.e. tropical, temperate, and polar. The tropical climate can be further subdivided into pure tropical with no distinct winters, and subtropical, where distinct summer and winter exist along with transitional spring and fall seasons. The temperate climates are characterized by harsh winters and mild summers. Another classification is based on **annual and monthly means of temperature and precipitation.**

7.5.1 Tropical rainy climates

This region lies in a band of roughly 20–40° around the equator, and is characterized by the absence of winter. The average minimum monthly temperature is 18°C. Rainfall is abundant, averaging 76 or more cm per year. The distribution of rainfall defines two climatic subtypes. One subtype has rainfall of no less than 6.4 cm per month, distributed more or less evenly throughout the year. The other sub–climate has distinct wet and dry seasons. In the continually rainy areas, the natural vegetation is broad–leaved and evergreen with no dormancy requirements. In the seasonally wet and dry areas, the dominant natural vegetation is coarse grass, which remains dormant during dry periods. Many horticultural crops are native to the tropical regions: mango, papaya, yam, and banana have worldwide significance (Janick 1986:472). Most Southeast Asian and Southeast African countries have a tropical climate.

7.5.2 Tropical dry climates

The dry climates are roughly distributed on either side of the tropical rainy climates. About 26% of the earth's surface lies in this climatic zone. Rainfall is scarce and unpredictable. In this climatic zone, evaporation from the soil surface and transpiration by plants may exceed annual rainfall. The dry climates can also be subdivided on the basis of moisture into arid (desert) and semiarid (steppe). The dry climates may also be subdivided into tropical and subtropical regions on the basis of minimum temperatures. Thus, there are four general types: hot deserts, cold deserts, hot steppes, and cold steppes. The soils in the dry climate, especially the steppes, are very fertile and considered to be the best agricultural lands provided artificial irrigation is available. Subtropical fruit crops like citrus, figs, and melons are native to these areas. Temperate zone crops like carrots and lettuce can be grown during the winter season. Most of the Punjab province and the southern districts of NWFP can be regarded as falling in the steppe zone. Most of Sindh and Balochistan provinces, the Cholistan and Thal areas of Punjab, and parts of eastern NWFP lie in the arid zone.

7.5.3 Humid temperate climates

These climatic regions around the middle latitudes are characterized by distinct winter, spring, summer, and autumn seasons. Two subdivisions can be defined on the basis of an average minimum monthly temperature of 0°C, mild–winter, and severe–winter. The mild–wintered portions of the temperate zones are found in the lower latitudes and on seacoasts on the western side of continents. Within this zone, there are three important climate types: (1) Mediterranean, (2) humid subtropical, and (3) marine.

The **Mediterranean climate** is characterised by dry, warm–to–hot summers, and mild winters with enough rainfall for horticultural crops. One of the best climates for horticultural crops, this zone constitutes less than 2% of the earth's land area. It is found in the Mediterranean region, central and coastal California in the USA, the southernmost part of South Africa, and parts of southern Australia. Parts of Balochistan have climatic features similar to those of the Mediterranean region. Major crops are olive, fig, citrus, grapes, and dates. Its dry summers are ideal for fruit drying, and it is here that this method of fruit preservation originated. This climatic zone provides most of the world's dried figs, dates, prunes, and raisins (grapes).

Humid subtropical climates are usually found on the eastern sides of continents. The annual precipitation is abundant (76–165 cm) and distributed evenly over the entire year. Summers are hot and humid; winters are mild with occasional freezing temperatures which limit the production of warm–season, perennial, and tree crops like citrus. The southeastern USA, many South American countries, and parts of Australia and China have this kind of climate. These are very rich vegetable producing lands. Long growing seasons permit two harvests a year for the same crop. Fruit crops with low chilling requirements (e.g. peaches) are also grown.

Marine climates are found on the western coasts of middle latitude continents. An extensive area in western Europe has marine climate. The marine summers are cool (18–21°C) but long enough for growing most crops. The winters are mild with average monthly minimum temperatures above freezing. Frosts are frequent and winter chilling sufficient to allow a dormant season. Rainfall varies from 76–254 cm; fog and mist are common. About half the days each year are cloudy. Apples, pears, and strawberries are well suited to the marine climate. Cool–season vegetables like peas, lettuce, and crucifers do especially well. The long spring season and mild summers extend the flowering season, making this climate highly suitable for bulbous flowers like tulips and daffodils. This type of climate is not found in Pakistan.

The **severe–winter climates** of the temperate region are characterized by short but distinct summers. These climates are found over the large land masses beneath the poles, and at higher altitudes. This climatic zone can be further subdivided into warm–summer (the prairies of the midwestern USA), cool–summer (western and central Russia), and sub–arctic (Canada, Alaska, Siberia) areas. The high altitude mountainous regions of Murree, AJK, FATA, NWFP, and Balochistan represent severe–winter, warm–summer climates. Murree and AJK receive rains during summer while FATA, NWFP, and Balochistan are outside of the summer monsoon zone. All of them have summer temperatures high enough to allow the cultivation of crops. These climates are ideal for crops like melons, tomatoes, and sweet corn. The cool–summer climates support extensive plantation of hardy deciduous fruits. Also, cool–season vegetables like potatoes and peas are very successful. In sub–arctic regions, there are only 50–75 frost–free days. In these areas, the main horticultural crops are root crops (turnips, beets, carrots) and crucifers (cabbage and cauliflower).

The polar climates (tundra) have mean annual temperatures at or below 0°C, and the average temperature of the warm months is 10°C. There is no horticultural activity in these areas.

STUDY QUESTIONS

1. What is radiant energy? Describe the relationship between visible light and photosynthetically active radiation.
2. What is phytochrome? List various phytochrome–mediated plant responses.
3. Define photomorphogenesis. Discuss the effects of daylength on the reproductive behavior of horticultural crops.
4. What is transpiration and its significance for plants? Explain its relationship to light and temperature.
5. How does water moderate temperature effects on plants? Explain why these moderating effects occur.
6. What is the influence of latitude on temperatures and seasons on the earth? Explain why these effects occur.
7. Describe the relation of temperature to altitude. Explain why these relationships are found.
8. What is a temperature inversion? How do the day and night air temperatures lead to the formation of temperature inversions?
9. Define the concept of *degree days*. How are degree days calculated?
10. Explain what is meant by *chilling requirement*. Compare the concepts of *degree days* and *chilling requirement*.
11. Differentiate between *freeze* and *frost*.
12. Describe some common high and low–temperature injuries to plants.
13. Discuss frost and methods of frost protection.
14. What is a greenhouse? How can a plastic tunnel be used for early production of vegetables?
15. Briefly describe the role of water in plant life.

16. Discuss various aspects of atmospheric humidity. How can you build and use a psychrometer?
17. Differentiate between water potential and tension in soil. How are irrigation requirements determined?
18. What is air pollution? Discuss major causes of air pollution, and describe some common pollutants and their effects.
19. What is the greenhouse effect? Discuss its causes.
20. What is climate? Discuss major variables influencing the world's climatic zones.
21. Distinguish between field capacity, permanent wilting percentage, and available water in soil.
22. A plant with a deep root system withstands drought better than a plant with a shallow root system. Explain why this is so.
23. Define transpiration. In general, the rate of transpiration is greater during summer than winter, on sunny days than on cloudy days within the same season, and during the light period than the dark period of any given 24-hour day. Explain why this is so.
24. When a plant is wilted, it is not making carbohydrates. Explain why.
25. What is meant by the optimum temperature range for plant growth and development?
26. Fruit trees and other woody plants low in carbohydrates at the end of the growing season are more susceptible to winter injury than trees high in carbohydrates. Explain why this is so.
27. Fruit trees and other woody plants with heavy crops are more likely to be injured by low temperatures during the winter than trees which bear moderate crops. Explain why this is so.
28. What is meant by the *rest period*?
29. Explain why there are no citrus orchards in the northern regions of Pakistan.
30. What is light? Discuss the function of light in photosynthesis.
31. Plants grown in dense shade have less extensive root systems than plants grown in the sun. Explain from the standpoint of light supply.
32. When do the equinoxes occur? How long are the light and dark periods on these dates?

33. What is a short–day, long–night plant? A long–day, short–night plant? A day–night neutral plant?

SELECTED REFERENCES

Edmond, J. B.; Senn, T. L.; Andrews, F. S. Fundamentals of horticulture. 3rd ed. New York: McGraw–Hill; 1964.

Gaffney, J. J. Humidity: basic principles and measurement techniques. Hort. Science 13:551–555; 1978.

Government of Pakistan. Hygrometric tables. Karachi: Meteorological Services, Govt. of Pakistan; 1984.

Halfacre, R. G.; Barden, J. A. Horticulture. New York: McGraw–Hill; 1979.

Hartmann, H. T.; Flocker, W. J.; Kofranek, A. M. Plant science: growth, development and utilization of cultivated plants. Englewood Cliffs, New Jersey: Prentice–Hall; 1981.

Janick, J. Horticultural science. New York: W. H. Freeman; 1986.

Janick, J.; Schery, R. W.; Woods, F. W.; Ruttan, V. W. Plant science, an introduction to world crops. 3rd ed. San Francisco: W. H. Freeman; 1981.

Michael, A. M. Irrigation: theory and practice. Shahiabad, India: Vani Educational Books; 1986.

Reuther, W. ed. The citrus industry. Volume III, Production technology. Berkeley: University of California Press; 1973.

Ting, I. P. Plant physiology. Menlo Park, CA: Addison–Wesley; 1983.

U.S. Department of Agriculture. Climâte and man. Yearbook of agriculture. Washington, DC: USDA; 1941.

U.S. Department of Agriculture. Water. Yearbook of agriculture. Washington, DC: USDA; 1955.

Outline of Chapter 8

8. SOILS AND FERTILIZERS

M. Ibrahim Chaudhary[1]

Soil is the foundation of agricultural production. Soils are tremendously heterogeneous, and are put into numerous classes depending upon their physical and chemical characteristics. Each class is best suited for particular types of crops, hence the study of soils is basic to crop management. The continuous drain of nutrients by crop production depletes the soil's fertility, the upkeep of which is essential for profitable agriculture, the major source of livelihood of over 70 percent of Pakistan's population. Thus, soils must be continuously artificially replenished, keeping in view the types and inherent fertility of the soil, and the requirements of the crops to be grown. Organic and inorganic fertilizers are the principal means of maintaining soil fertility; therefore along with the study of soils, study of fertilizers is also necessary. This chapter contains basic information about soils and fertilizers for undergraduate students.

LEARNING OBJECTIVES

Reading this chapter should enable a student to discuss basic concepts of:

- Various types of soil texture with reference to horticultural crops grown in Pakistan
- Soil structure: its formation and importance for crops
- The role of various soil components in crop and soil management
- Essential nutrient elements: their role and deficiency symptoms
- Methods of handling organic manures
- The composition of various organic and inorganic fertilizers
- Methods of determining nutrient requirements, and their appropriate times and methods of application

---

[1] Dr. Ibrahim Chaudhary is Associate Professor, Department of Horticulture, University of Agriculture, Faisalabad.

8.1 Soils

Soil is the outermost weathered layer of the earth's crust which can support a growing plant. It is the main medium of plant growth. Besides providing anchorage to the plant, soil also supplies water and nutrients. Thus for the successful growing of horticultural crops it is of immense importance to understand soil.

There are several types of soils in Pakistan: loamy soils in the northern regions; and clay loam, sandy loam, and sandy soils in central and southern Pakistan. Alluvial deposits constitute the main material for most soils in this country, and there are also some weathered bedrocks and loess soils.

The most important soil components, on which fertility and fitness for crop production depend, are: inorganic minerals, organic matter, water, and air. The proportion of these components to each other determines the agricultural value of a soil. A good soil should contain one–half solid material (organic and inorganic) and the other half equally divided between water and air. Soils are classed as mineral soils and organic soils, depending upon whether inorganic or organic matter predominates. Some main types of mineral soils are: sandy, sandy loam, loamy, silt loam, and clay loam. The organic soils are muck and peat soils.

8.1.1 Mineral soils

Almost all types of mineral soils are used for growing various horticultural crops, each having specific advantages. Some of the important types of mineral soils with their salient characteristics are described here.

Sandy soils. These are generally coarse textured and of low fertility. The sand component is as much as 80–95%, while the remaining 5–20% is silt and clay. These soils are well–aerated and well–drained, and because of speedy decomposition, low in organic matter. These soils are widespread in the south and southwest of Punjab and the deserts of Sindh. Some of them are cultivated. Horticultural crops, especially vegetables, are sometimes grown in sandy soils for early maturity, but yields are low. Frequent irrigation and heavy manuring are important to improve the fertility and texture of these soils. However, they are easily managed. Generally, watermelons, muskmelons, groundnuts, and sweet potatoes are favoured in these soils. After repeated green manuring, cultivation of citrus fruit has been made possible in Leiah, Muzaffargarh, and Bahawalpur Districts.

Sandy loam. These soils are well–aerated, well–drained, and moderately fertile. Most of the soils of the Punjab and Sindh, particularly of the southern irrigated plain, represent this type. They contain 50–80% sand and 20–50% silt and clay, with a little organic matter. The presençe of silt and organic matter increases their water–holding capacity, and hence a wider variety of horticultural crops, e.g. citrus, mango, date palm, cucurbits, root crops, and ornamentals, can be successfully grown.

Silt loam. These soils are common around the Salt Range, the Potohar Plateau in Punjab, in the wet mountains of the northern part of the country, and in the dry western plateau of Balochistan. These are fine–textured, poorly drained, and fertile soils. The proportion of sand is 20–30%, while silt and clay constitute 70-80% of silt loams. Their organic matter content may also stay high due to moderate aeration. Good soil preparation is required for growing apples, pears, and stone fruits. Melons and some vegetables are widely cultivated.

Clay loam. These are quite fertile and moderately well–drained soils; they are found particularly in the northern irrigated plains of Peshawar and Mardan and in the rice belt of Punjab. They are very fine–textured soils, containing 60–80% silt and clay and 20–40% sand. Comparatively more organic matter stays in these soils because they favour slow decomposition. Clay loams require good preparation and are important in growing pears, plums, peaches, and apricots. Vegetables like onions, cabbage, and cauliflower, and other high value horticultural crops are also grown on these soils.

Silty soils. Silty soils are found in the active flood plains of the Indus delta. These soils are fine–textured, and poorly drained and aerated. Mangoes and bananas are successfully grown on these soils in Sindh province.

8.1.2 Organic soils

Organic soils are formed from plant material deposits in and around shallow bodies of water and contain over 20% organic matter. They are of two types: muck and peat soils. **Muck soils** contain 20–70% organic matter and are fine–textured. Plant residue is in an advanced stage of decomposition. These are highly valuable soils, producing high yields of horticultural crops, particularly vegetables and ornamentals. **Peat soils** are composed of coarse and fibrous organic matter. The percentage of organic matter in these soils also ranges from 20–70%. They are important in greenhouse production.

These soils are not found anywhere in Pakistan. Our soils generally have less than 1% organic matter content (Kausar et al. 1979).

8.1.3 Soil moisture

Moisture is the third essential component of soil; without it, life in the soil is impossible, while an excess of it also results in loss of life because of poor aeration. The moisture–holding capacity of agricultural soils is very important. This increases with an increase in organic matter content and with reduced size of mineral particles. The maintenance of organic matter, soil texture, and structure also determine the drainage of the soil. Organic matter and clay particles hold water with a force called soil–moisture tension. The more water in the soil, the looser the bond and the lower the soil–moisture tension, which is zero on saturation. Conversely, the less water in the soil, the stronger the bond and the greater the soil–moisture tension. Generally, water is held more tightly in clay soils than in loamy soils, and least tightly in sandy soils.

Categories of soil moisture. The moisture present in the soil is never pure water but always in solution form called **soil solution**. The concentration of the soil solution keeps on changing with the addition or depletion of fertilizers. Categories of moisture are named after the force by which they are held by the soil particles.

CHEMICALLY COMBINED WATER: This is a component of the soil particles and is not at all available to crops.

HYGROSCOPIC WATER: Held very tightly in the form of a thin layer around the soil particle, none of this water is available to plants.

CAPILLARY WATER: Existing as a film around particles in micropores, capillary water is held relatively loosely. This water moves in the soil and is the water which is most utilized by plants.

GRAVITATIONAL WATER: This water is held very loosely in the macropores and moves downward and is lost quickly from the top of well–drained soils without much benefit to the plants. It is generally rain water and flood irrigation water. Since it is supplied in abundance, the soil is saturated. The water left after the drainage of gravitational water (after 24–48 hours) is called **field capacity**.

8.1.4 Soil air

Though there is a small amount of air in soil water, more air is present in the pore spaces if there is no water there. Air, particularly oxygen, is necessary for a variety of life in the soil, including plant roots. If the soil is so flooded or submerged that all the pore spaces are filled with water, plants will suffocate. If this flooded condition persists, plants die and the soil remains poor for maintaining life. Since aeration is dependent on the number of total pore spaces, compact soils or fine–textured soils are usually poorly aerated. Plant growth is unsatisfactory in these soils because of poor root respiration and low absorption of nutrients by the plants.

8.1.5 Soil reaction

Degree of acidity or alkalinity is called **soil reaction.** Soil reaction is of great significance because it determines the availability and toxicity of plant nutrients and in turn the growth of crops and microorganisms. The preference of a given plant type may be for acidic, alkaline, or neutral soils, **possibly**

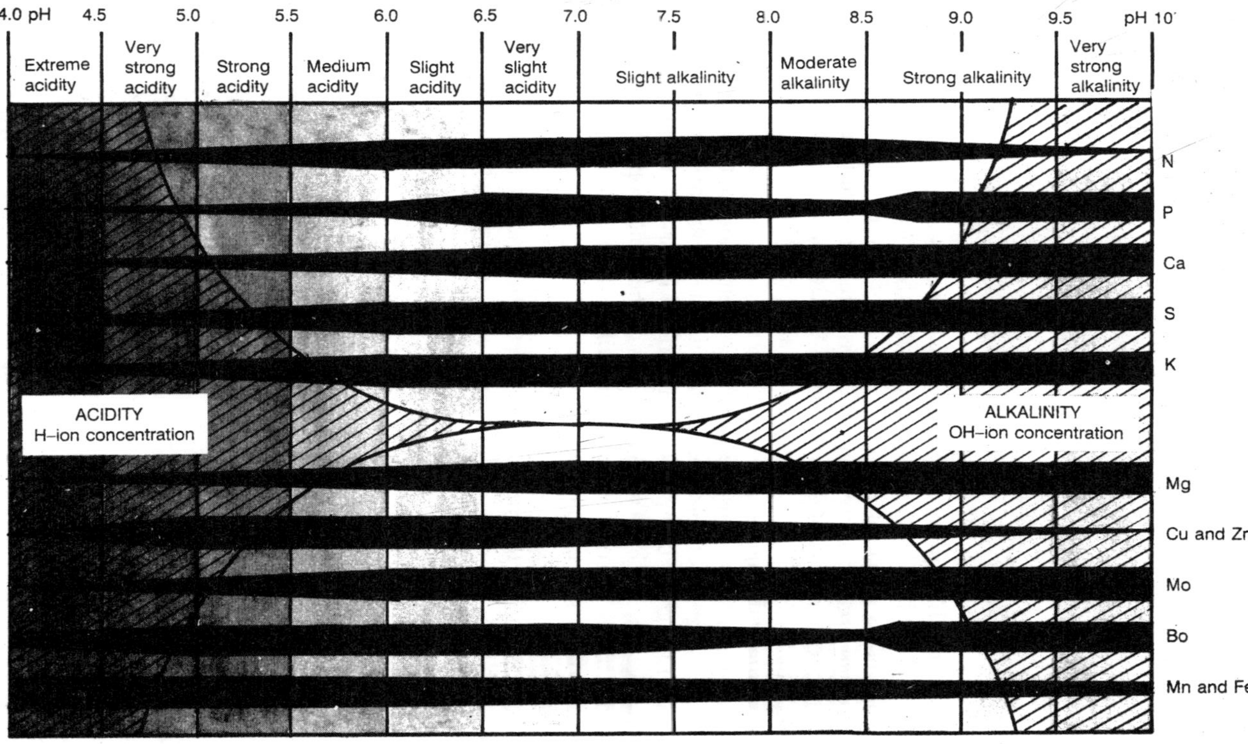

Figure 8.1 General relation of soil pH to availability of plant nutrients. (Adapted from Truog 1946:305; redrawn by Habib–ul–Rahman.)

because of certain physiological characteristics of the plants or their sensitivity to the degree of availability of certain mineral elements.

The numerical expression of soil reaction is called **soil pH**. The pH value reflects the relative quantity of hydrogen ions in the soil solution. The more hydrogen ions present relative to hydroxyl ions, the more acidic the soil solution is, and the lower the pH value. A pH value of 7 is neutral, and above this level is basic. The pH of productive soils ranges between 4.0 and 8.5, but most of the agricultural soils of Pakistan are alkaline in reaction, falling within a range of 7–8.5. Figure 8.1 diagrams the relation of soil pH to the availability of plant nutrients.

Some soils have pH as high as 10, which indicates a problem of sodicity. **Sodic soils** have high levels of alkalis, particularly sodium hydroxide, and are also sometimes called alkali soils. They are black in colour because of accumulated organic matter, and are considered most difficult to reclaim. There is another class of soils known as **saline soils**. These soils contain harmful quantities of chlorides, sulphates, and carbonates of sodium, potassium, calcium, and magnesium.

Acidic and saline soils are suitable for some horticultural plants. The differential degrees of tolerance of various plants to various pH levels is exploited for agricultural purposes. Tolerance is variable within species of the same genus, and even within varieties of the same crop. Date palm and olives are highly tolerant to acidity as well as alkalinity of soils. Ranges of tolerance of some horticultural plants are given in Table 8.1.

Table 8.1 Range of tolerance to soil reaction of some horticultural plants

Type of plant	pH range of tolerance	Optimum pH range
Pome and stone fruits	3.5–8.5	5.5–6.5
Citrus fruits	4.0–9.7	4.5–7.0
Strawberries	4.5–8.5	5.0–6.0
Guava	4.2–8.2	6.5–7.5
Mango	6.5–8.4	6.5–7.5
Cabbage, cauliflower (cole crops)	5.5–8.5	6.0–6.5
Cucumbers	4.5–8.8	5.5–6.8
Muskmelons	5.5–8.5	6.0–7.0
Watermelons	5.0–8.5	5.5–6.8
Potatoes	5.5–8.2	6.0–6.5
Sweet potatoes	4.5–8.9	5.2–6.7
Peppers	4.5–8.8	5.5–7.0
Lettuce	5.2–8.5	5.5–7.0
Onions	4.5–8.7	6.0–6.5

Type of plant	pH range of tolerance	Optimum pH range
Chrysanthemums	5.0–8.5	5.5–6.5
Carnations	5.5–8.2	6.0–7.0

Source: Data from various sources.

8.2 Organic matter

The maintenance of soil fertility is of prime importance in commercial agriculture. The organic matter present in the soil is originally derived almost entirely from green plants grown on the soil. When crop plants are grown, a portion or all of the plant can be returned to the soil, replenishing its organic matter. After a crop is harvested the roots decay in the soil adding organic matter. When the entire plant is returned to the soil, the process is called **green manuring**. Plant material after passing through animals may be returned to the soil as farmyard manure. The organic matter then decomposes and in turn supplies the soil with minerals important for its fertility.

The rate of decomposition of organic matter is generally affected by soil temperature, humidity, and aeration conditions. That is why at the same temperature clayey soils have more organic matter than sandy soils or sandy loams. There is usually more organic matter in the soils of the temperate zone than of the tropical and subtropical regions. The decomposed products of organic matter give soils their dark brown and black colour. The product which gives soil a black colour is called **humus**.

8.2.1 Importance of organic matter

Organic matter is usually a complete food component for plants. It is extremely important for horticultural plants, annuals, perennials, and tree fruits. Since organic matter decomposes rapidly under warm conditions like those in Pakistan, frequent supplements are necessary to maintain and get good yields from orchard soils. Since it has a low specific weight and a very high surface area, a small percentage of organic matter in mineral soils exerts a significant effect on the structure, water absorption, water–holding capacity, cation exchange capacity, and aeration of soil.

8.2.2 Effect on structure

The addition of organic matter provides material from which microorganisms produce vital compounds such as gums and resins, and make minerals available. These products play an important role in building soil structure and increasing the size of soil particles by cementing small particles together to form larger aggregates. This process improves soil structure, enlarging pore spaces and improving the aeration of the soil.

8.2.3 Water–absorption capacity

Organic matter possesses a large surface area in proportion to its weight (low specific weight), and it is hydrophilic. Thus it has great ability to absorb water. Incompletely decomposed organic matter (humus) has about four times the water–holding capacity of a body of clay colloid of the same size. Addition of organic matter to the soil in orchards has been found helpful for fruit trees, particularly during drought conditions. Annual horticultural crops, which have shallow root systems, suffer less water stress if there is a sufficient amount of organic matter in the soil.

8.2.4 Cation exchange capacity

Cation exchange capacity is one of the most important chemical properties of soils, and is usually closely related to soil fertility. It is defined as the quantity of exchangeable cations expressed in milliequivalents per 100 g of soil (Saleem et al. 1986:49). As a general rule, soils with large amounts of clay and organic mater will have higher exchange capacities than sandy soils and soils low in organic matter.

Ion exchange takes place in two stages: cation exchange between soil colloids and soil solution, and cation exchange between soil solution and the plant root. The important cations held on soil particles are H^{+}, Ca^{++}, Mg^{++}, K^{+}, Na^{+}, and in smaller amounts NH^{+}, Mn^{++}, Zn^{++}, Cu^{++}, and Al^{+++} with different binding forces.

Organic matter with colloidal clay will make a humus–clay colloidal complex whose particles have negative charge. The particles have a large surface area; they attract cations and maintain a large quantity of essential ions in available form. Since these ions are not free in the soil solution, they are not subject to leaching.

8.2.5 Sources of organic matter

Farmyard manure (FYM). Farmyard manure includes mainly dung of dairy cattle, droppings of poultry, and animal litter. Until now, it has been the main source of organic matter in Pakistan for the supply of essential minerals needed by plants. Heavy applications of organic matter are required for successful production of fruits and vegetables. Some studies of citrus (Anon. 1937–38) have indicated improved health of the trees, and better fruit quality and yield. Application of 6–8 tons per acre of FYM is recommended to increase vegetable yields (Ali 1962). Now, combined application with commercial fertilizers is preferred in order to assure prompt supply of nutrients and increased production of fruit and vegetables.

The use of FYM is decreasing because sufficient quantities are not easily available. Transport of FYM is another factor which adds to the cost of crop production, while the nutrients available per unit value are far less than in commercial fertilizers. Moreover, with the advent of mechanised agricultural operations, the number of cattle has decreased, thus reducing the production and availability of farmyard manures. Higher cropping intensity and higher cost of FYM have decreased the quantity used per crop.

City sweepings. This source of organic matter is gaining importance, particularly in agricultural lands around the big cities. It is considered a satisfactory replacement of FYM as it is also a rich source of essential minerals. The continuous use of city sweepings enhances the humus content of soils. However, some problems faced by the growers are the numerous plastic bags and broken glass which require additional expense to clear the soil. City sweepings include animal stools, human excreta, remains of fruit and vegetables, and other litter. This material contains a variety of organic matter which is in different phases of decomposition and thus supplies rich and prolonged nutrition to plants.

Sewage manure. This is obtained while cleaning sewers and storm water drains of big cities. The material is rich in major elements but may also contain harmful heavy metals. The use of sewage manure is gaining popularity in vegetable and ornamental plant production. Some research has recently been conducted on citrus fruit using sewage in combination with commercial fertilizers. These experiments have shown good results (Rafiq 1980). The composition of sewage manure is given in Table 8.2.

Sludge water. Sludge water was in use for irrigation and manuring by the Greeks in ancient times, and is still used in many countries of the world. In Pakistan, sludge water has been in use for a long time for the production of vegetables, particularly near the big cities. This water contains human urine, feces, and mud dredged from city drains during cleaning, and other constituents. It has very high manuring value, containing 4 ppm nitrogen and 10 ppm phosphorus. Of the N, 60 percent is in the form of ammonium. It is a cheap source of irrigation water which also supplies manurial components to the plants. Vegetables grown with sludge water are higher in protein and ash content than those grown with canal water. The quality of protein, however, is poor (Yaqub 1986). The presence in sludge water of large quantities of heavy metals is also a matter of concern.

Livestock slurries. Fresh dung and urine combined is called **slurry**. In some countries it is commercially prepared by adding water to animal dung and urine. The P and K content of slurries varies widely, depending upon the type of animal and the feed they eat, but their N content is almost the same. Unlike manures, where most of the N occurs in organic form, slurries contain about half of it as NH_4 – N, which is readily available to plants, and the other half as organic nitrogen. NH_4 – N is susceptible to volatilization, particularly if the slurries are applied on sunny days and not incorporated into the soil promptly. The dairy farmers of Pakistan are planning to prepare slurries on a commercial scale.

Green manures. The value of green manuring was known even by ancient peoples, who used to grow a crop of legumes before other main crops. The ploughing in of legume crops is well understood and practiced. It is the most economical means of increasing organic matter content in the soil. Green manuring is a good tool to improve the texture of sandy and clayey soils and to enhance the productivity of poor soils. Ploughing in a legume crop is often recommended before laying out an orchard, while afterwards cover crops, intercrops, sods, and weeds contribute a great deal to increasing organic matter.

Clean cultivation of commercial orchards is often recommended, but a complete ban on intercrops may cause nutritional problems by reducing organic matter content. Therefore, cover crops or sods are commonly used to prevent erosion and to add organic matter. These cover crops may compete with the trees for food and water, but their restricted cultivation proves useful in the long run. Some crops commonly raised for green manuring are *guar*, sunhemp, *daincha*, *jantar*, vetch, and sometimes *berseem*. These are ploughed in when still tender or before flowering so that they will decompose rapidly. The amount of nitrogen fixed by these green–manuring crops varies from 70-100 kg/ha (Akhtar 1989).

Leaf manures or tree shavings. Leaf manure is an important component in preparing the soil for nursery raising of vegetables and ornamentals and for filling pots. Plant propagation media are also prepared by mixing soil and leaf manures in appropriate proportions. Wood shavings or sawdust are rarely used in propagation media in our country, but the author has seen some people mixing the acacia bark left in tanneries with soil. It appears to be quite useful for indoor permanent plants.

Analysis of organic matter. Many types of organic matter are incorporated into soils. These range from straw to FYM. Straw has a high C:N ratio (80:1), and discharges a lot of CO_2, adding little to the fertility of the soil. FYM has a comparatively lower C:N ratio (10:1) and hence is more useful in enriching the soils. Besides NPK, manures contain all other essential elements. The content of manures may vary depending upon the feed, bedding, and the individual animal. It is known that if all the mineral requirements of the crop are supplied through FYM or other sources of organic matter, horticultural crops are much healthier, and higher and better quality yields are obtained. For this purpose, data on the composition of various organic manures is required. Composition of some organic matter sources is given in Table 8.2.

Table 8.2 Mean composition of manures and crop residues

Material	% (oven–dried basis) N	P	K	Reference
1. Animal and human excreta				
Buffalo dung	1.23	0.55	0.69	Misra and Hesse (1986)
Buffalo urine	2.05	0.01	3.78	"
Cattle dung	1.91	0.56	1.40	"
Cattle urine	9.74	0.05	7.78	"
Sheep dung	1.87	0.79	0.92	"
Sheep urine	9.90	0.10	12.31	"
Human feces	7.24	1.72	2.41	"
Human urine	17.14	1.57	4.86	"
Hen droppings	4.00	1.26	0.68	"
2. Farmyard manure	0.80	0.21	0.68	Benne et al. (1966), Rafiq (1980)
3. Sewage manure	.52			
4. Green manures				
Daincha	3.2	–	–	Akhtar (1989)
Sun hemp	2.4	–	–	"
Guar	4	–	–	
5. Crop residues				
Wheat straw	0.49	0.11	1.06	Misra and Hesse (1986)
Rice straw	0.58	0.10	1.38	"

8.2.6 Application

Pakistan's soils are generally low in organic matter, most of them containing less than 1 percent (Kausar et al. 1979). This low organic matter content is probably due to the high annual temperatures, and repeated application of organic matter to orchards is necessary. These manures are applied during the winter so that the nutrients can become available by the following spring, the blooming and sprouting season. The manure is spread around the trees up to the drip line. Two to three feet around the main stem of fruiting trees is left bare, so that the manure does not damage the stem.

For commercial vegetable production, application of adequate amounts of FYM is emphatically recommended. Since most vegetable crops are annual and have smaller root systems than trees, organic manures ensure against water stress in addition to supplying balanced nutrition. Yield and quality of vegetables are reported to be better with organic manures than with either chemical fertilizers or no treatment because of continuous slow release of the elements. The manure is applied and mixed well during soil preparation.

Annual and perennial flowers like roses supplied with FYM produce large and beautiful blooms.

8.2.7 Handling of organic matter

Fresh FYM should not be applied to fruit trees or other horticultural crops. It contains organically bound elements which are not available to plants until after decomposition. Fresh manures may have a negative impact on the health of trees due to competition for nitrogen between the decomposing bacteria and the plants. Therefore FYM should first be stored for about a year in pits about a metre deep and then transported to the field. The pits should preferably be brick lined to avoid any wastage of minerals through leaching. Addition of an appropriate quantity of nitrogenous fertilizer to the pits will enhance the speed of decomposition. Deep pits are not considered good because there is not enough air in them for speedy decomposition, and in deep pits the manure at the bottom remains undecomposed. The decomposed manures are shifted to the fields, generally during soil preparation for vegetables, and in December and January for orchards.

In Pakistan, much of the animal feces is made into dung cakes for fuel, and few farmers store even the leftover properly. Animal feces, a major component of farmyard manures, is heaped in the outskirts of villages open to wind and rain, which wash away the components of these slowly decomposing manures. City sweepings are handled in about the same way. They are either taken directly to the fields or are heaped in open places outside the cities. Thus the nutrients are wasted and a problem of sanitation and unhygienic conditions is created for the citizens.

8.3 Chemical fertilizers

The rapid increase in human population and depletion of natural resources of cultivated lands has prompted the use of chemical fertilizer to replenish the soils and to increase crop yields per unit area. These chemical fertilizers supply the same essential elements as organic matter, but unlike organic manures, with chemical fertilizers the supply of nutrients to the crop is immediate and effects are observed soon after application. The use of fertilizers in Pakistan was introduced in 1958 with FAO assistance, and it is on the increase every year. NPK is generally applied artificially in the form of fertilizer, nitrogenous fertilizers being the most frequently used.

Sixteen elements have been found essential for the growth and development of plants. These elements are carbon (C), hydrogen (H), oxygen (O), nitrogen (N), phosphorus (P), potassium (K), calcium (Ca), magnesium (Mg), and sulphur (S), which are used by plants in relatively large quantities and hence are called **major** elements; and iron (Fe), zinc (Zn), manganese (Mn), boron (B), copper (Cu), molybdenum (Mo), and chlorine (Cl), which are used by plants in smaller quantities and are called **minor** or **trace** elements. Of these, C and O are available from the atmosphere: C from atmospheric CO_2 through photosynthesis, and O through the process of respiration. Hydrogen is obtained from water as a result of photolysis during photosynthesis. All of the other elements are mostly absorbed by plant roots from the soil unless otherwise applied to the leaves in foliar sprays.

8.3.1 Determination of need of the elements

Scientists have been conducting experiments all over the world to ascertain the nutritional and fertilizer requirements of crops. Some basic methods used to determine plant requirements of various elements are field trials, pot culture, soil analysis, plant analysis, and visual diagnostic methods.

Field trials. These are conducted by applying measured amounts of various nutrients to planted soil and directly observing the harmful or beneficial effects of the treatments. This method is practical and represents the ultimate test to which the findings of other methods are compared. Field trials provide quantitative relationships between nutrients and yields. However, the impact of similar trials may be different in various localities. These trials require considerable time and resources.

Pot culture. This method is an attempt to overcome the wide range of variation in soils, and the other difficulties of field trials. The two common techniques are solution culture, **hydroponics**, and sand or gravel culture. In pot culture, the composition of the nutrient medium is known exactly and can be changed at will. A formulation of nutrients for hydroponics in pots was suggested by Hoagland and Arnon (1950). These methods are used for studying deficiency symptoms and the role of various elements in plants, but the results of these experiments must be verified.

Soil analysis. Soil is the major source of mineral nutrients to plants. The results of chemical analysis may give guidance about potential supplies of nutrients, but they may also be misleading because of factors like solubility, availability, and absorption. Elements may be present in sufficient quantity but unavailable to plants because of the chemical reaction, i.e. pH of the soil. Soil potential is also tested using biological methods, that is by growing plants in a given amount of soil and comparing their nutrient composition to that of the same plants grown in pure sand.

Plant analysis. Since 1920, plant analysis as a source of information about the nutrient status of plants has gained much favour. As a result of many years of pot culture and field experiments, certain guidelines about how the variables involved in sampling are related to the results have evolved. These variables include the organ to be analysed, e.g. leaf blades, leaf petioles, or stem; the age of the organ; and the time and frequency of sampling. If these variables are considered, a reliable guide for nutrient need can be obtained. For example, with apples, mid–terminal leaves sampled in mid–summer give the best nutrient content data. With vines, the third leaf from the base of the terminal shoot is taken. For citrus, four to seven month–old leaves from bearing or non–bearing terminals in early spring and in autumn are helpful in determining the appropriate nutrient status and requirements of the trees. Some other factors which induce variation are rootstocks, cultivar, previous fertilization, rainfall, crop load, and the spray programme. Before making any recommendation, these factors should also be taken into consideration.

Critical nutrient levels for a number of tree fruits have been proposed, and comparing them with the determined nutrient status enables an appropriate fertilizer programme for the particular orchards to be developed.

Visual methods of diagnosis. In this method nutritional requirements are determined by recognizing specific symptoms which appear when one or two elements are insufficient for healthy plant growth. Deficiency of any one element will result in abnormal growth and development, often appearing as chlorosis or necrosis of leaves. It requires skill and experience to identify and differentiate the symptoms of specific deficiencies, because deficiencies of different elements may manifest quite similar symptoms. Sometimes insect or disease attacks further complicate the identification. Also, it is often too late to treat deficient trees when they show deficiency symptoms because by then significant damage has already occurred.

8.3.2 Essential elements

An element is considered essential if deficiency of it makes it impossible for a plant to complete satisfactory vegetative and reproductive growth. The role of each essential element and its deficiency symptoms are specific. Deficiency can only be remedied by addition of the specific element required.

1. Nitrogen. Nitrogen is one of the most widely distributed elements in nature, constituting 78 percent of the atmosphere. Even then, it is the element most often deficient for crop production and the one most frequently applied to plants. There is hardly any plant which does not respond to the addition of this element. Nitrogen is, however, not available to plants unless fixed; available N is mostly in the forms of nitrate (NO_3) and ammonium (NH_4). Biologically fixed N is the main N–source for plants. Biological fixation of atmospheric N is accomplished by bacteria working in the root nodules of legume crops (*Rhizobium* [symbiotic], *Azotobacter* [free living]), and blue–green algae. Gradual decomposition of soil organic matter is another source of N. An important though quantitatively smaller source is the artificial supply of nitrogenous fertilizer to cultivated crops.

ROLE IN PLANTS. Nitrogen is a part of the chlorophyll molecule, amino acids, proteins, nucleic acid, and pigments. Normal metabolic processes can continue only in the presence of an optimum level of N. Leaves, being metabolically most active, contain the highest level of N, which ranges from 2–4.5 percent, followed by roots, stem, and fruit. Fruiting spurs of apple have a higher N–content than non–fruiting, but fruiting terminals of citrus usually contain less N than non–fruiting terminals. Nitrogen is most needed during the blooming period, to mobilise the processes of flower opening, fruit setting, and fruit development. Reserve N moves from other parts like the bark into the leaves during the blooming season.

DEFICIENCY SYMPTOMS. The addition of N enhances vegetative growth, and its deficiency leads to low production, stunted growth, and small yellow leaves. Nitrogen is mobile in the plant, and deficiency symptoms appear in older leaves first. Citrus trees with a low N–level shed their leaves excessively during winter causing a heavy setback to fruiting during the following spring. Other perennials low in nitrogen, particularly deciduous trees, shed their leaves early in autumn.

Heavy application of N promotes luxuriant and succulent growth in some trees, while in others like citrus it sometimes causes heavy production of small, low market–value fruits. Delayed ripening and poor post–harvest life is also reported in apples and other crops. With some leafy vegetables where succulence is important, heavy N–applications increase the value of the crop.

NITROGEN CYCLE. Nitrogen disappears from the soil through leaching, crop removal, volatilization, and runoff; and is added through biological fixation, organic matter, and artificial fertilizers. It is hence in a state of constant change which is referred to as the **nitrogen cycle**.

Rhizobium, the symbiotic bacteria which reside in the root nodules of leguminous crops like peas and beans, fix free elemental nitrogen from the atmosphere into NH_4, which is available to crops after the bacteria die. The process is called biological nitrogen fixation.

Organic matter—manures and plant residues—release N as NH_4 after decomposition. This NH_4 may be absorbed by some plants as such but is generally nitrified through a bacterial process to NO_3, the form generally preferred for uptake. Some chemical fertilizers like urea, ammonium sulphate, and ammonium nitrate are also important sources of additional N. Urea may also be absorbed by plants directly when sprayed on the foliage, but when added to the soil it is nitrified in a series of microbial actions.

Nitrogen is lost from the soil in various ways. It is depleted through uptake by plants and by leaching and erosion. A significant amount of nitrogen is lost when NO_3 is denitrified to gaseous forms as N_2, N_2O, or NO under anaerobic conditions. NH_3 is also lost during hydrolysis of urea if it is applied to bare soils. Urea (NH_2CONH_2) hydrolyses to ammonium carbonate [$(NH_4)_2CO_3$)] very quickly when added to the soil. Ammonium carbonate is an unstable compound and decomposes to ammonia and carbon dioxide. The NH_3 or NH_4 so released is adsorbed by the colloidal fraction of the soil and subsequently nitrified.

2. Phosphorus. Phosphorus is a major essential plant nutrient. It exists in nature mainly as rock phosphate, which is unavailable to plants. It is mined and later converted to available forms of phosphate fertilizer by treatment with sulphuric acid or phosphoric acid.

i. $Ca_3(PO_4)_2 + 2H_2SO_4 \longrightarrow Ca(H_2PO_4)_2 + 2CaSO_4$

Rock Phosphate + Sulphuric acid → Monocalcium Phosphate + Gypsum

ii. $Ca_3(PO_4)_2 + 4H_3PO_4 \longrightarrow 3Ca(H_2PO_4)_2$

Rock Phosphate + Phosphoric acid → Monocalcium Phosphate

In the first reaction single superphosphate is formed, which is a mixture of monocalcium phosphate and calcium sulphate (gypsum). In the second reaction only monocalcium phosphate is formed. The product is called double or triple superphosphate. Plants absorb this element principally in the forms of primary orthophosphate (H_2PO_4) and secondary orthophosphate (HPO_4). Of these ions, H_2PO_4 is absorbed better because of its greater solubility.

The relative amounts of H_2PO_4 and HPO_4 in the soil solution and their absorption by plants is affected by the soil pH. In highly acidic soil (pH lower than 5), P is fixed and not available to plants. In moderately acidic to neutral soil (pH 5–5.7), more H_2PO_4 ions are absorbed, while at soil pH of 7.1-8, absorption of H_2PO_4 and HPO_4 ions is approximately equal. At pH higher than 8, only HPO_4 ions are absorbed by plants. However, the movement of phosphate ions in the soil is very slow and the element stays within the root zone for a long time even if it is not replenished repeatedly.

ROLE IN PLANTS. Phosphorus is an essential element required in the formation of ATP (adenosine triphosphate), and is extremely important in plant metabolism. It is also part of nucleic acids, DNA (deoxyribonucleic acid), RNA (ribonucleic acid), phospholipids, and co–enzyme NADP (nicotinamide adenine dinucleotide phosphate). Phosphorous levels in plant tissue range from .05–1.0%. Phosphorus levels are higher in seeds and lower in leaves and roots. An optimum range of phosphorus in the leaves of horticultural plants is .10–.15%. It hastens maturity in tuber crops like potatoes and seed crops like peas and beans, and improves the quality of citrus fruits and apples. It hastens flowering in tomatoes.

DEFICIENCY SYMPTOMS. Phosphorus is a mobile element, and hence older leaves suffer first from P–deficiency and drop excessively. Growth in phosphorus–deficient plants is retarded. In early summer the leaves look dark–green, but in late autumn in some crops they look purplish because of the accumulation of carbohydrates which are not utilized and transported. Leaves of deciduous trees tend to be narrower and oblong. Branches may terminate to verticillate whorls. Fruit become soft prior to maturity, and yield is reduced. With citrus fruit, the rind is coarser and thicker than normal; acidity is increased and juice is decreased. In apples, fruit ripen early, and are apt to become brown internally. Small–sized fruit and premature fruit drops are common.

3. Potassium. Potassium is the third important major element. However, it is not a part of any organic compound in the plant; hence it is soon released after plant death. Organic matter is therefore a poor source of potassium. In soil, the total content of this element is usually higher than that of N and P, but the forms in which it exists are not readily available to plants. Potassium applied as fertilizer is available slowly and is not easily leached from the soil as its movement downward is very slow, ranging from 3–5 cm per year.

Plants absorb potassium in elemental form K^+. It can move within the plant and travels from regions of excess or older tissues to regions of deficiency or younger tissues. In plants it occurs in larger quantities than phosphorus. The optimum range in horticultural plants is 1.0–2.0% (Forshey 1969; Jones and Embleton 1969). It is expressed usually as K_2O, which can be converted to elemental percentage in a fertilizer by multiplying by .83.

ROLE IN PLANTS. Unlike N, P, and several other elements, K does not form an integral part of organic plant components like protoplasm, fats, proteins, and cellulose. Its functions appear to be catalytic. It is considered essential in photosynthesis, sugar translocation, nitrogen metabolism and protein synthesis, enzyme activation, stomatal movement, water relations, and promotion of the growth of meristematic tissue.

Potassium improves the quality and storage life of fruits like peaches, pineapples, grapes, and apples; ensures uniform ripening in tomatoes and apples; and increases acidity in pineapples and citrus fruit. Adequate potassium improves the size and quality of potato tubers.

DEFICIENCY SYMPTOMS. Since potassium is a mobile element, deficiency symptoms are manifested in older leaves first. Deficient leaves develop chlorosis or spotting on the margins, starting from the tips and advancing towards the centre giving an inverted 'V'–shape appearance. At later stages the leaves develop necrosis. Carbohydrates and nitrogen compounds are accumulated because of failure of translocation. Rolling, cupping, and curling of the leaves and distortion of the mid–rib are common in deciduous fruits. Citrus fruits develop thin skins and increased acidity, and turn colour sooner. K–deficient potatoes produce poor quality potato chips.

4. Calcium. Calcium constitutes from 0.5–5.0% of plant dry matter. These high levels of Ca in tissue are mainly due to its high level in the soil solution rather than to efficiency of root uptake. The Ca concentration of the soil solution is about ten times greater than that of K, while the uptake rate of Ca is usually lower than that of K. Of the inorganic elements, Ca is found in the highest quantity in all plant organs except fruit, where the K level is higher.

ROLE IN PLANTS. Calcium is required for cell division and elongation. It precipitates as calcium pectate in cell–wall structures and in the middle lamella, and makes a cementing layer between the cells. As calcium pectate, it is immobilized and is not available for reuse elsewhere in the plant; hence a continuous supply is essential. It also acts as an activator of enzymes like amylases and ATP–ases.

DEFICIENCY SYMPTOMS. The necessity of Ca for plant growth can be observed by growing plants in solutions without it. Ca–deficient plants show stunted growth. The apical buds disintegrate and secrete a weakly gummy fluid. The root tips become brown and gradually die. In horticultural crops, some physiological disorders like blossom–end rot in tomatoes and capsicum pepper, black–heart in celery, and bitter–pit in apples are also due to Ca deficiency. Yellowing and distortion of leaves often showing a hook at the tip are common, and in severe cases the margins of the leaves turn brown.

5. Magnesium. Magnesium is required by plants in a range from .1-.4% dry weight. The optimum range in the leaves of most horticultural plants is .2–.25%. Magnesium is found in smaller quantities in the soil than Ca and K. There is more exchangeable Mg than K, but much less than Ca. Normally 4–20% of the cation exchange capacity is constituted by Mg, and 80% and 4% by Ca and K, respectively. If a soil is low in Mg and high doses of either K or Ca are added, Mg–deficiency will be induced. Also, at higher pH, Mg is normally unavailable and deficiency is induced. Magnesium is also more often deficient in sandy soils than in fine–textured clay soils because it is rapidly leached.

ROLE IN PLANTS. The main function of Mg in plants is in the formation of chlorophyll. The chlorophyll molecule contains Mg as its central component, however, it also has other functions. Magnesium fertilization increases the sugar content of beets and vitamin content of potatoes. Increase of some vital amino acids in some plants is also observed when Mg is added.

DEFICIENCY SYMPTOMS. When Mg is deficient in plants the symptoms appear on the lower or older leaves. This indicates that Mg is not completely fixed in plants but is mobile and can be reused. Since Mg is a part of the chlorophyll molecule, a deficiency of it induces chlorosis of various appearance on leaves. In cabbage a crimson reddish colour develops; in potatoes, orange–yellow; in cabbage, cauliflower, and turnips, dead spots appear between the veins. Under Mg–deficiency conditions, the germination rate of peas is reduced.

Fruit trees generally need relatively small quantities of Mg, but deficiency causes excessive pre–harvest drop in apples in addition to leaf chlorosis. Peach leaves show marginal chlorosis and interveinal spotting; and grapes and citrus fruits show frenching or interveinal chlorosis. Fruit yields are reduced and alternate bearing is more pronounced, particularly in the seedy varieties of citrus.

6. Sulphur. Sulphur has generally been thought to be adequate in most soils to meet plant requirements. Since SO_2 gas is present in the air, soils repeatedly get some sulphur from rain water. The presence of S in the soil lowers pH and enhances the availability of several elements. Plants absorb it from the soil generally as sulphates, which are mobile in plants. SO_2 is also readily absorbed from the air by plant leaves; this process contributes 30–40% of plant needs, and near factories possibly even more.

ROLE IN PLANTS. Sulphur has not been as extensively studied as other major elements. It appears essential in the synthesis of chlorophyll although it is not part of it. It is a component of the amino acids methionine, cysteine, and cystine, hence plays a part in the synthesis of proteins. It is present in onions as allyl sulphate, a volatile compound which gives pungency and the characteristic flavour to cabbage and other crucifers. It ranges in plants from .1–.8% with a normal range of .2–.4% in mature leaves.

DEFICIENCY SYMPTOMS. There are no sharply defined deficiency symptoms of sulphur in plants. In general the plants turn light–green or yellow, but even under severe deficiency the leaves do not die. The veins of the leaves become greener than the tissue between the veins, but in some plants, e.g. turnips and tomatoes, a marked redness develops in the lower leaves starting in the leaf vein and spreading to the area between the veins. In citrus, besides the leaves yellowing, the fruit fails to develop its normal colour, juice content is reduced, and the rind thickens. Sulphur deficiency normally develops in the leaves of the upper branches, which indicates that it is not readily transferred from one part of the plant to another.

8.3.3 Minor elements

Several mineral elements are essential for a plant to complete both vegetative and reproductive phases successfully, but are required in very small quantities. These are iron, zinc, manganese, boron, copper, molybdenum, and chlorine. Their occurrence, functions, and deficiency symptoms are as follows.

1. Iron. The normal iron content of plants ranges from 100-1500 ppm. It is generally higher in leafy vegetables like spinach and lettuce, where 500–1200 ppm is the norm. In tree fruit it is generally within a range of 40–100 ppm. It is less available under alkaline soil conditions. Deficiency of iron is corrected by soil application of iron chelates like Fe–ethylene diamine tetra–acetic acid (EDTA), through foliar application of iron salts like ferrous sulphate, or even by driving nails into the trunks of trees.

ROLE IN PLANTS. Iron acts as a catalyst in the production of chlorophyll. It is important in respiration, affecting the iron–porphyrin protein complex which acts as an oxygen carrier and activator of oxygen in the plant. Iron is essential in the young growing parts of the plants.

DEFICIENCY SYMPTOMS. Deficiency symptoms first appear on new growing leaves, which indicates that iron is immobile in the plant. The symptoms are specific and can easily be diagnosed. In the beginning the leaf colour turns pale. The next stage is yellowing between the veins of the leaves, leaving the veins green. In severe conditions, new leaves are completely yellow or even pure white. The leaves remain smaller than normal; the growing points may also die; and growth is stunted.

2. Zinc. This element ranges from 5-100 ppm of the dry weight of plants. An optimum range of zinc is reported to be from 20–40 ppm in mature leaves. Zinc is absorbed as Zn^{++} by plants and is an immobile element, hence deficiency symptoms are first manifested by new leaves.

ROLE IN PLANTS. Zinc appears not to be combined in any organic compound, but its presence is observed to affect a number of processes in plants. Optimum amounts of Zn help to maintain a full complement of chloroplasts. It is necessary for the formation of tryptophane, the precursor of indoleacetic acid (IAA). It is a component of enzymes like RNAase, which regulates the formation of RNA, glyco–glycine dipeptidase necessary for the formation of protein, and the dehydrogenase required for glycolysis.

DEFICIENCY SYMPTOMS. Zinc deficiency symptoms are widespread, particularly in citrus and apples. Deficient leaves of tree fruits are easily identified by their characteristic interveinal chlorosis, small leaves, the rosette formation which results from insufficient synthesis of IAA, and short internodes. A higher than normal proportion of male flowers in some citrus cultivars is also attributed to Zn deficiency. In severely deficient cases, peas and beans produce only small seedless pods. Zinc deficiency may be corrected by foliar application of .3–.4% solution of zinc sulphate.

3. Manganese. Manganese is essential, but under acidic soil conditions, where a large amount of this element is present in soil solution, it is frequently toxic. It should have a range of 25–50 ppm in plant leaves for optimum functioning of the plant.

ROLE IN PLANTS. Manganese assists in the formation of chlorophyll. It also functions in enzymes in oxidation–reduction reactions. Hence it is important in photosynthesis, respiration, and also in nitrogen assimilation. Manganese that has already been used by a plant is immobile. Therefore deficiency symptoms first appear on the youngest leaves. In general, the symptoms are yellowing between the veins of leaves while the veins remain green. In severe deficiency cases, complete yellowing and necrosis or dead areas appear on the leaves. Deficiency symptoms can be corrected by foliar spray of soluble manganese salts like manganese sulphate.

4. Copper. Only a small amount of copper is required by plants. In general, a range of 6–10 ppm Cu in plant leaves is considered optimum. Copper in excess of plant requirements is seldom absorbed by the roots. In case less than the required amount is absorbed through the roots under high soil pH conditions, application of copper fungicides to orchard trees is considered sufficient to meet their copper needs.

DEFICIENCY SYMPTOMS. Copper–deficiency symptoms, like those of some other trace elements, appear on fruit and younger leaves first, indicating its immobility in the plant. In fruit trees, gum pockets develop under the bark and stained spots on the bark of terminal twigs, and the leaves drop off. The twigs die back and in citrus the fruit frequently splits open and drops before maturity. Leaf size is reduced and sporadic necrotic areas are common. The leaves of deficient plants develop a dark colour and burnt or drying tips. Onion leaves fade to light green and the tips dry when they are 10–15 cm long. In tomatoes, cupping and rolling of leaves and development of fewer seeds is common.

5. Boron. For most plants the optimum boron level in the leaves is around 25–30 ppm. Boron is seldom deficient in canal–irrigated areas. Boron content of up to 100 ppm has been found in the leaves of citrus fruits. Deficiency is first evidenced in younger leaves, indicating that the element is immobile. If deficiency occurs, it is corrected by soil or foliar applications of borax or boric acid solution.

ROLE IN PLANTS. Boron is needed in cell–wall development, cell division, phloem development, and the movement of sugars. Its role in the germination of pollen is well documented for Italian prunes and some other tree fruits. Boron spraying has increased fruit set and fruit yield many–fold in prunes, almonds, and grapes.

DEFICIENCY SYMPTOMS. If boron is deficient, young growing cells (meristem) are discoloured, disrupted, and die; and the plants are stunted. Young leaves become brittle, thick, and discoloured. Flowering is suppressed, and few fruit are set because of slow pollen–tube growth. In grapes, boron deficiency reduces stigmatic secretion. Corky core of apples, hard fruit of citrus, cracking of plums and avocadoes, gum deposits in the inner peel and excessive dropping in citrus, browning and hollow stem of cabbage and cauliflower, and internal black spots of beets are some commonly reported boron–deficiency symptoms.

6. Molybdenum. Of all the essential elements, the requirement for molybdenum is the smallest. A concentration around 3 ppm is quite enough for healthy growth of plants.

ROLE IN PLANTS. Molybdenum is essential in nitrogen metabolism. It is required for reduction of nitrates to ammonia, and plays an important stimulating role for nitrogen fixation by *Rhizobium* in legumes. It is also reported to enhance vitamin and amino acid content in plants.

DEFICIENCY SYMPTOMS. There are some significant Mo–deficiency symptoms, like whiptail of cauliflower, yellow spot between the veins and gum deposits on the lower leaf surface of citrus, and N–deficiency in peas and beans. Mo–deficient plants show nitrate accumulation in the leaves. Molybdenum deficiency, unlike that of other trace elements, occurs under acidic soil conditions. The symptoms appear on older leaves first, which indicates that the element is mobile and moves from older leaves to younger leaves but does not reverse.

7. Chlorine. Chlorine has recently been added to the list of essential elements. Its frequent presence in the soil, air, water, and fertilizer material usually prevents deficiency of this element. Around 100 ppm of Cl (dry weight of the leaves) suffices for the needs of plants. Excessive concentration of Cl may take place under higher soil–chloride conditions, which in turn may prove toxic. It is abundantly available near the coasts and may be toxic to some plants there. Citrus trees having 280–440 ppm of chloride in their leaves may slowly decline and collapse.

ROLE IN PLANTS. The exact function of chlorine is still unknown, but mango and coconut yields in India have been increased markedly by the application of common chlorine salts. It is quite essential for the growth and development of tomato and sugar beets.

DEFICIENCY SYMPTOMS. Plants which are deficient in chlorine show wilting, chlorosis, and death of tissues with leaf bronzing.

8.3.4 Commercial fertilizers

A substance containing one or more of the nutrient elements that can be absorbed by plants and promote their growth is called a fertilizer. These substances are being used everywhere in the world to increase the yield and quality of crops. The use of fertilizers has started in Pakistan only recently. Nitrogenous fertilizer is the most widely used, followed by phosphatic and potassic. The average amount of fertilizers per acre used in Pakistan is still among the lowest in the world. It is around 63 kg/ha, which is just half of what is used in the USA and only 15% of that used in the Netherlands.

The manufacture of fertilizer in Pakistan is increasing; urea, ammonium sulphates, calcium ammonium nitrate, single superphosphate, triple superphosphate, and nitrophos are now being produced locally. A total of around three million tons of fertilizers per annum are manufactured in Pakistan, and an additional 781,000 tons are imported every year to meet the country's needs. There is, however, tremendous awareness among the growers of the need for more use of fertilizer, and it is anticipated that there will be further increase in its use in the years to come.

Some of the locally manufactured fertilizers and their chemical compositions are listed in Table 8.3.

Table 8.3 Nutrient percentage of some inorganic commercial fertilizers

	Percent nutrient		
Name of fertilizer	N	P_2O_5	K_2O
Urca	46	–	–
Ammonium sulphate	21	–	–
Calcium ammonium nitrate (CAN)	26	–	–
Single superphosphate	–	18–20	–
Triple superphosphate	–	45	–
Nitrophos	23	23	–
Potassium sulphate	–	–	50

Source: Saleem et al. (1986:60).

Method of application. Fertilizers are added to the soil or other growing media to enhance their fertility and provide a prompt supply of nutrients to plants. This in turn improves the health and productivity of the crops. Horticultural crops particularly need additional fertilizer to enhance their production and improve the quality of the produce. It is important to determine when, how, and in what quantity fertilizer should be applied to a particular crop.

The method of application plays an important role in the optimum utilization of nutrients. Effort should always be made to improve the contact of the fertilizer with the plant roots to make it possible for the plant to use the maximum quantity of the fertilizer. Appropriate methods of application are decided keeping in view the type of crop, the season, the nature of the nutrients, and the extent of the plant's requirement. In general, there are two methods of applying fertilizers: to the soil and through the leaves as a foliar spray. There are a number of methods of applying fertilizer to the soil.

1. Banding
   a. With the seed
   b. Beside the seed
   c. Around the trees
2. Broadcasting
3. Along with irrigation (fertigation)
4. Foliar spray

BANDING. Fertilizers are banded with the seed to provide immediate contact with the emerging roots. However, the new roots of some crops, like peas and beans, are sensitive to direct contact, and are easily hurt if fertilizer is closely applied. Directly seeded onions, however, have been observed to respond excellently to this method of application under certain conditions. All N, P, and K may be supplied with beneficial effects using this method.

Banding beside the seed is generally done at a distance of 5 cm from the seed. It gives excellent results with vegetable crops, particularly those raised in the winter season. The author has also observed good results with tomatoes, cucurbits, and okra. Banding of nitrogenous fertilizers provides a boost to young growing seedlings and increases early as well as total yield. With tree fruits, fertilizers are added in bands prepared at a distance of 60 cm from the main trunk under the canopy. With large trees, two to three such bands are prepared at a distance of 30 cm from each other. A band for tree fruits should be about 10 cm wide and 5 cm deep. After the fertilizer is added, it is covered with soil.

An important advantage of banding is that a smaller quantity of soil comes in contact with the fertilizer than in the broadcast method. Thus less fixation of the fertilizer elements takes place. This is particularly important for K and P which in some soils are easily fixed in a form that is unavailable to plants. It is also important for nitrogenous fertilizers, which are easily leached. Application close to the roots enhances the percentage of N absorbed by the plants.

BROADCASTING. In the broadcast method, fertilizer is spread uniformly on the soil surface and later ploughed in. This method raises the general fertility level of the soil; it is very suitable for perennial plants like fruit trees with well established surface root systems. Broadcasting fertilizer under the canopy of trees and then working it in is the general method practiced by tree fruit growers.

FERTIGATION. Fertilizers like the nitrogenous ones, which are needed frequently by plants, are also applied together with irrigation water. This is particularly feasible where water is pumped through a drip irrigation system in the field or in greenhouses. In some instances where the crop is closely planted, like onions, spinach, and lettuce, fertilizer material is placed alongside the source of water, which carries it to the field. Since nitrates immediately leach down, while ammonium rapidly changes to nitrates and is then leached, these irrigations must be light to avoid leaching of nitrogen.

FOLIAR SPRAY. Another method of applying fertilizer is through foliar applications. The sprayed elements are readily absorbed through the stomata. Because of the large amounts of N, P, and K required by plants, foliar application of materials containing these elements cannot be relied on for full fertilization, but can be used as a supplement if needed. With the trace elements, foliar applications are particularly successful, especially when soil applications would result in fixation of the element to an unavailable form. This is frequently true under high soil pH conditions where most of the minor elements except molybdenum are fixed. Under these conditions soil application of these elements is only successful in chelate forms, where the element in question is released slowly and escapes fixation. Foliar applications of iron, zinc, manganese, copper, boron, and molybdenum are generally done in the form of salts, more commonly on tree fruits or vines.

The longer the nutrient solution remains on the leaf surface in the form of a fine film, the more efficient is nutrient uptake by the leaf tissue. Thus on hot clear days when foliar spray evaporates quickly, much of the salts accumulate on the leaf surface without being absorbed. Although dew may be helpful in wetting the material again and again, this may sometimes cause scorching and burning. Such detrimental effects can be avoided by using low concentration solutions, applying neutral solutions, and spraying on cool cloudy days or in the afternoon. The number of sprayings required to correct deficiency depends upon the severity of the problem in the plants.

Time of applications. It is important to decide prudently when to apply fertilizer to get the maximum benefit. It is desirable to apply N at the time when the plant can absorb it best, because otherwise it may leach down. But fertilizers carrying N in organic form (organic fertilizers) must be applied sufficiently early to allow for decomposition to make the nitrogen available to the plants at the time of need. Observations indicate that in orchard trees N appears deficient in the early spring or after the bloom is over. Therefore, for quick response in growth and improved fruit set, nitrogenous fertilizers are most effective when applied to orchard trees starting growth in the spring. With vegetable crops and flowers, these fertilizers produce good results when added at the time of seeding; for seedy vegetables like tomatoes and cucurbits, additional support during the fruit development stage is desirable. If it rains heavily, some additional nitrogenous fertilizer should be given to both orchard trees and vegetable crops to replenish the leached N.

For elements like phosphorus, potassium, and several others which move quite slowly in the soil and remain within the rhizosphere for a long time, the time of application is not so crucial.

STUDY QUESTIONS

1. What is the importance of soil for raising plants, and what characteristics are necessary in agricultural soils?
2. (a) What different types of soil are there? (b) Describe the basis of their nomenclature and the types of regions they are found in.
3. What is the importance of soil structure? How is it improved for agricultural purposes?
4. What is meant by *soil reaction*? What is its significance in raising horticultural crops?
5. Give the soil type and the range of pH optimal for growing citrus, mangoes, stone fruits, cole crops, and cucurbits.
6. What is organic matter, and why is it important in horticulture?
7. (a) Name the various sources of organic matter. (b) How should organic matter be handled?
8. Organic matter is produced from plants, directly or indirectly. Discuss and explain this statement.
9. What is the role of organic matter in improving soil fertility?
10. (a) What is the difference between organic and inorganic fertilizers? (b) Why are both types important in raising horticultural crops profitably?
11. (a) Define *essential element*. (b) Name these elements.
12. (a) What is the role of each of the essential elements? (b) What deficiency symptoms are manifested by various horticultural plants lacking each of them?
13. Describe the different methods of applying fertilizer, pointing out the merits of each.
14. How can we determine the mineral needs of a plant? Describe the merits and demerits of each method.
15. Why is the time of fertilizer application important? Discuss.

PRACTICAL EXERCISES

1. Determine the structures of various types of soils in the field.
2. Compare various types of soils and organic matter for their ability to absorb water.
3. Carry out analysis for NPK of plants and soils.
4. Study the visual symptoms of mineral deficiency, in the field if possible.

REFERENCES

Akhtar, M. J. Integrated use of organic and inorganic nitrogen fertilizers in irrigated lowland rice. Faisalabad: Department of Soil Science, University of Agriculture; 1989. M.Sc. thesis

Ali, Sadiq. Some experiments on the manuring of citrus in the Punjab. Punjab Fruit J. (Citrus number):24–27, 90–99, 108-119; 1962.

Anonymous. Manurial experiments on sweet orange at Fruit Farm Sahiwal; 1930–33. Annual report of Fruit Specialist, Lyallpur 1933–34. unpublished.

Anonymous. Manurial experiments on sweet orange at (Lyallpur) Faisalabad; 1930–37. Annual report of Fruit Specialist, Lyallpur 1937–38. unpublished.

Anonymous. Fertilizer Review 1989. 37, Street 11, F–6/3, Islamabad: National Fertilizer Dev. Centre; 1989.

Benne, E. F.; Hoglund, C. R.; Longnecker, E. C.; Cook, R. L. Animal manures. Michigan Agri. Expt. Sta. Cir. Bul. 231; 1961. Also in: Tisdale, S. D.; Nelson, W. L. Soil fertility and fertilizers. 2nd ed. New York: Macmillan; 1966.

Berger, K. C. Sun, soil and survival. Norman, Oklahoma: University of Oklahoma Press; 1972.

Edmond, J. B.; Senn, T. L.; Andrew, F. S.; Halfacre, R. G. Fundamentals of horticulture. 4th ed. New Delhi: McGraw–Hill Publishing Company Ltd.; 1977.

Forshey C. G. Potassium nutrition of deciduous fruits. Hort. Sci. 4(1):39–41; 1969.

Gardner, V. R. Principles of horticultural production. East Lansing, Michigan: Michigan State University Press; 1966.

Halfacre, R. G.; Barden, J. A. Horticulture. New York: McGraw–Hill; 1979.

Hoagland, D. R.; Arnon, D. I. The water culture method for growing plants without soil. Calif. Agr. Expt. Sta. Circ. 347. Cooperative Extension, Div. of Agricultural Sciences, University of California; 1950.

Jones, W. W.; Embleton, T. W. Potassium nutrition of sub–tropical fruits. Hort. Sci. 4 (1):37–38; 1969.

Kausar, M. A.; Alam, S. M.; Sharif, M.; Pervez, M. I. Micronutrient status of Pakistan soils. Pak. J. Sci. Ind. Res. 22(3):156–161; 1979.

Mengel, K.; Kirkby, E. A. Principles of plant nutrition. 4th ed. Worblaufen–Bern, Switzerland: International Potash Institute; 1987.

Misra, R. V.; Hesse, P. R. Comparative analysis of organic manures. FAO/UNFP Regional project RAS/75/004 Field Document 24. Rome: Food and Agriculture Organization of the United Nations; 1986.

Rafiq, M. Effect of different combinations of organic and inorganic fertilizers on the chemical composition of sweet orange leaves. Faisalabad: Department of Horticulture, University of Agriculture; 1980. M.Sc. thesis.

Saleem, M. T.; Ahmad, N.; Davide, J. G. Fertilizers and their use in Pakistan. Islamabad: National Fertilizer Development Centre, Planning and Development Division, Govt. of Pakistan; 1986.

Truog, E. Soil Sci. Soc. Amer. Proc. 2:305; 1946.

Wilcox, G. E. Potassium needs: diagnosis and use of vegetable crops. Hort. Sci. 4(1):41; 1969.

Yaqub, M. Sewage effect on vegetables. The Daily Pakistan Times (View Point). July 31, 1986.

Outline of Chapter 9

9. ESTABLISHMENT OF GARDENS

Mahmood N. Malik[1]

LEARNING OBJECTIVES

After studying this chapter, a student should be able to:

- Select suitable places for raising desirable horticultural crops
- Select the best layout plans for various horticultural varieties in terms of plant population and longevity for a given plot
- Select the most suitable horticultural variety/cultivar for a given vicinity according to its agro–climatic conditions
- For a given specific situation, select the kind of gardening with better future prospects, i.e. commercial, local, or home gardening
- Plan various key operations for establishing new gardens: preparation of land, layout, planting, and after–care
- Assess capital needs for various operations required in raising different horticultural crops
- Make the best use of available resources: land, capital, labour, and transport for maximum return from various horticultural crops
- Select the best true–to–name plants—budded on recommended rootstocks and of proper age and size

Horticulture is an important branch of agriculture dealing with many fruit, vegetable, and flower crops, and if done scientifically, is a profitable enterprise. Financial returns can be increased many–fold if careful planning is done when selecting the specific crops to establish in a particular region. In most cases, the horticulture industry is a long–term investment, especially the raising of fruit trees. Thus very careful planning is needed while establishing either commercial or home orchards of fruits, vegetables, or ornamentals. Different factors to be considered in planning of gardens and orchards are:

---

[1]Dr. Malik is Professor of Horticulture and Dean, Faculty of Agriculture, University of Agriculture, Faisalabad.

- Selection of proper locality and site
- Provision of suitable drainage
- Adequate preparatory tillage
- Scientific layout of gardens with proper spacing
- Selection of cultivars suitable to the particular region
- Proper planting and post–planting care
- Provision of suitable windbreak protection
- Independent water supply

Other important points include:

- Availability of needed capital
- Availability of labour resources
- Dependable transportation
- Good export as well as domestic markets

9.1 Selecting locality and site

Locality refers to the geographical circumstances of a place in relation to cities, villages, railway stations, roads, etc. It is important for the ease with which horticultural produce can reach the markets of choice in the shortest possible time and with minimum expense. Commercial vegetable and flower gardens should preferably be near daily markets to ensure prompt sale of the produce.

Commercial gardens should be planted in a locality where such crops have already established themselves as commercial enterprises. It is often advantageous to select a site for a certain crop close to successfully established, highly productive plantations of the same or similar crops, so that cooperative marketing of the produce is possible. Also, one can learn much from the experience of the other growers, and needed plant protection measures can be undertaken collectively.

Site refers to a specific place where any given horticultural product is produced: a citrus or mango orchard, a grape vineyard, or a field of tomatoes. Site considerations include two factors—topography and soils. **Topography** refers to the contour of the land, its elevation, and similar features of the terrain. The soil is the home of the plant roots and the reservoir for the plant's essential nutrients and water.

Vegetable crops and herbaceous flowering plants are usually grown on level and slightly sloping sites, as most of these crops need cultivation. Harvesting operations for these crops can also be performed more efficiently on level land than on sloping land. Fruit trees, however, can be raised on steep slopes to a certain extent. The *direction* of the slope **is** also important:

northern slopes are always cooler than southern slopes, so crops grown on them mature and ripen later than those grown on southern slopes.

9.1.1 Climate and soil

Climate refers to meteorological conditions (changes in the atmosphere) in a given region. Temperature is the most important climatic factor influencing the performance of horticultural crops. Temperature determines the success or failure of each specific crop in a given region. Maximum and minimum temperature endurance varies with different species and even varieties. Relative humidity also plays a great role in the performance of plants, and influences the colour and external appearance of the product.

Climatic factors such as temperature, humidity, rainfall, wind, sunlight, frost, and hailstorms; and soil conditions including texture, fertility, depth, alkalinity, salinity, soil reaction, chemical content, drainage, and waterlogging, which influence the growth and development of horticultural crops must also be considered while selecting a suitable site for gardens. Subsoils are more important than surface soils, especially for fruit trees, so examination of subsoils to a depth of 1–2 m is essential for these crops.

There is often a great difference in climatic requirements among varieties of the same species. In most parts of Pakistan, growers have their choice of several excellent fruits, vegetables, and ornamentals. General climatic requirements for horticultural crops are given in Chapter 7, and more specific discussion is found in Chapters 13, 14, and 15.

Soil is the natural resource base for horticultural production, as well as other forms of agriculture. It is an important natural factor which determines the success or failure of a crop. It is, therefore, extremely important that a suitable soil be chosen. Fortunately, many horticultural crops, especially a good number of fruits, may be grown on a wide variety of soil types. Extremes are to be avoided; very heavy soils are difficult to handle, and sandy soils do not hold moisture well and are likely to be infertile. A loam or sandy loam soil is good for most horticultural crops, and such soils are found in many parts of Pakistan. For good performance of a horticultural crop, a deep (especially for fruit plants), well–drained soil free of excess salts, with a fair amount of humus is essential.

Besides providing anchorage to the plant, soil furnishes water and nutrients for its growth and development. The more fertile the soil is the less manuring will be necessary, but manure will still be needed to maintain its fertility. Good drainage is also essential, as most horticultural plants cannot thrive with water standing around their roots. In heavy soils it is sometimes necessary to provide artificial drainage. The soil preference of various horticultural crops is discussed in Chapters 8, 13, 14, and 15.

Unlike fruit crops, most vegetables have a short growing season, which can easily be adjusted to suit the local climate. This fact plus their relative ease and cheapness in production, handling, and shipment of their seeds to distant destinations, and comparatively less hazard of crop failure have facilitated the growing of vegetables outside their natural habitat.

In some areas of the country, the climate permits production of summer vegetables during the winter and spring seasons of other regions when these vegetables cannot be produced there. This objective may also be achieved by **forcing**, production of vegetables out of their normal season by providing protection from cold. Greenhouses, cold frames, and polyethylene tunnels are used for this purpose on a commercial scale in many advanced countries. Polyethylene tunnels are also becoming popular in Pakistan for producing out–of–season vegetables or at least for raising nurseries for early vegetable production. This may also be true for most ornamentals raised commercially. Most of the cut flowers of European countries can be raised during winter in the open in Pakistan and can be exported to markets in Europe, earning high profits and much foreign exchange.

9.1.2 Water supply

Next to climate and soil, a regular and adequate source of irrigation water is an important consideration in site selection for commercial horticulture as well as for home gardens. Good plentiful water should be regularly available all year round especially during the hot and dry season. Horticultural plants in general need more water than field crops. Plants vary in their irrigation requirements, and some plants are more resistant to drought than others. Young plants require much water for their satisfactory growth and bearing. Gardens of plants with high water requirements should ideally be located near canal–fed watercourses. In some countries, regular arrangements are made for sprinkler irrigation of orchards where irrigation through watercourses is not possible because of soil topography or non–availability of canal water.

In most parts of Pakistan, successful gardening is possible only with irrigation. An additional and independent source of irrigation water is needed to supplement canal irrigation during canal closures. This may either be a well, tube well if the ground water is good, or a reasonable water reservoir/tank where canal water can be stored. Consideration should also be given to the quality of water. Excesses of sodium salts and boron are injurious to most horticultural plants.

9.1.3 Market facilities

Horticultural products are usually perishable, and need to be transported and marketed as soon as possible. Under the prevailing conditions in Pakistan, the general outlets for growers are city markets. Therefore, the location should be very close to cities or at least big towns. If the produce is to be sent to distant markets, a railway station in the vicinity is desirable. Joint trailers are now a blessing for marketing produce provided that there are good roads nearby. Although some crops which have good keeping quality can be transported over a long distance without excessive losses, nearness to the market is always advantageous. Products to be exported to foreign markets need a standard packing house nearby. If there is a food processing industry in the vicinity, excess produce can be sold there. All these factors must be kept in mind when selecting sites for establishing new gardens.

9.1.4 Transportation requirements

Improvements in and availability of metalled roads in most of Pakistan, fast motor trollies/trucks, refrigerated trucks, improved rail transportation, and the development of more economical air transportation to distant/foreign markets will stimulate the raising of horticultural crops in areas which may have favourable growing conditions but be far away from markets.

9.1.5 Availability of labour

The growing of horticultural crops is a full–time occupation. The availability of labour in the vicinity must be considered before establishment because most garden operations are still done by hand in Pakistan. Cheap and regular labour, available close to the site, is necessary for carrying out various operations economically and on time. Preparation of land, i.e. cleaning, levelling, and ploughing; installing systems for irrigation and drainage; planting and post–planting care, i.e. pruning, hoeing, and weeding; application of manures and fertilizers; control of insect pests and diseases; picking, packing, transportation, and marketing are all cultural operations which, more or less, are done by manual labour in this country.

9.1.6 Capital needs

Horticultural enterprises require much more investment than other agricultural crops for establishment, maintenance, and continuous profitable returns. Capital requirements must be estimated before deciding to establish a garden.

9.1.7 Personal factors

Success in raising a given horticultural crop depends to a considerable extent on the aptitude of the individual grower. Some farmers adapt themselves easily to vegetable growing, commercial flower production, or fruit growing, while others may be slow to adjust to this type of farming. Economic differences between fruit growing and vegetable or flower raising are important factors influencing the decision to adopt any of these types of farming. Differences between various horticultural enterprises are listed below.

- Fruit production is a long–term investment, while vegetable or flower raising is mostly seasonal.
- Fruit production is sure to produce some crop every year, but this is not necessarily the case with vegetable or flower production.
- Getting into fruit growing is a slow process, and getting out of it may be even slower; this is not true for other horticultural crops.
- Fruit growing is a long–term enterprise, so annual financing and tenanting cannot be organized as with vegetable or flower production.
- Cooperative efforts and organizations can act more easily, and are easier to manage in fruit production than in other horticultural crops.

9.2 Preparation of land

After selecting the site, some preliminary operations are desirable before raising any horticultural crop. The preparation of the soil depends largely upon its present condition, previous history, and the grower's plans. If the land has been under cultivation and well–maintained, nothing further may be required. However, if the site is new and previously uncultivated, it should be thoroughly surveyed for its size, topography, flow of irrigation water, and fertility status. It may be necessary to grade and manure the land, to provide an irrigation system, and to buiid fences. The positioning of main and subsidiary roads, wells, sheds, etc. should be planned clearly. The following operations should be done well in advance, preferably a full season. Any of these operations, if delayed, may cause a considerable loss.

Cleaning. Existing vegetation should be cleared; standing trees should be cut down and their stumps removed. No vegetation should be left, otherwise it may shade the young plantation and compete for water and nutrients. Further, removing it at a later stage may be expensive and risky.

Ploughing. After the site is cleared, it must be given a deep and thorough ploughing twice or thrice in two directions. Then it should be harrowed and cultivated to a fine tilth.

Levelling. The site should be levelled. For convenience in levelling, the field can be partitioned into blocks unless it is already small and manageable. Earth is moved from higher to lower spots, and ditches and low–lying areas are filled up. A uniform gentle slope may be provided, preferably in one direction, to facilitate the flow of irrigation water. Then bunds, channels, and drains may be formed. On hilly slopes, unless contour planting is to be followed, the land should be suitably terraced.

Irrigation system. The irrigation system should be planned and sometimes even installed before establishing the plantation. It may be desirable to put in the permanent water channels before the actual layout of the gardens is done.

Soil enrichment. The soil can be enriched by raising cover crops, preferably legumes. Levelling sometimes upsets the fertility status of the soil. Hayes (1960) advised growing field crops for 2–3 years before planting vegetables to condition the site. A green–manure crop, in addition to FYM, is often the most economical means of increasing the organic matter content of the soil. Growing a green–manure crop is essential not only to improve the fertility of the soil, but also to indicate any unsuitable spots like saline or alkaline patches, poor physical condition of the soil, ill–drained, or rocky patches. All such defects should be corrected before planting.

Fencing. Fencing should be erected all around the site to prevent the entry of animal pests like cattle, goats, or wild pigs, as well as human thieves. Barbed wire fencing is desirable, though it is expensive. It is advantageous to have wire fencing, as it takes less space and neither shades the soil nor takes any nutrients from it. Many types of fencing are used with more or less satisfaction.

9.3 Laying out gardens

Laying out gardens means indicating the actual places for roads, footpaths, irrigation channels, water tanks, manure pits, office or residence blocks, water pump or tube well/water tank, and windbreaks, as well as spacing the plants in the garden. The systematic laying out of gardens is the first step in a successful horticultural enterprise. In this way cultivation and pest management can be carried out more correctly and efficiently, resulting in healthy plantations which ultimately yield more and higher quality produce.

After site preparation, the actual laying out of the orchard can begin. The planting arrangement of the trees in a garden is a crucial operation. In laying out an orchard, a well–considered plan should be followed that provides the maximum number of trees per unit area consistent with sufficient space for the proper development of each tree, convenience in orchard operations, scope for extension, and an aesthetic appearance. Mistakes made at this time are likely to cause losses throughout the life of the orchard, and it is very difficult or sometimes impossible to correct them later.

For economical management, most horticultural plants should be planted in straight rows. With fruit trees and most other horticultural plants, if more than one species are to be planted, each kind should ordinarily be planted in a block. However, with self–incompatible varieties, pollinator plants should be mixed with receptor plants according to a set plan. Varieties of a particular crop ripening at the same time should be grouped together for convenience in harvesting and handling.

Before selecting a layout system, the first thing to consider is optimum planting distance, which differs for fruit trees, vegetables, and ornamentals.

9.3.1 Planting distance

Needs for certain environmental conditions are more crucial for horticultural plants than for other farm crops because of their specific needs for certain climatic and nutritional requirements. In ornamental horticulture, plants are established according to a properly developed landscape plan to enhance their aesthetic value. For this, proper distance, in addition to their other requirements, is essential. In vegetable growing, the number of plants per unit area plays an important role in increasing yields, which cannot be done unless seedlings are planted in properly spaced rows.

Fruit trees. Planting distance depends upon many things, including plant species and variety, climatic conditions, soil fertility, availability of water, and and the training and pruning systems to be followed, especially for fruit and ornamental trees. Methods of cultivation may also influence planting distance. With fruit trees, plant types (seedlings or graftage and rootstocks) have a definite impact on plant size and vigour. When deciding the planting distance of a fruit garden or placing ornamental trees in a landscape, one should bear in mind the ultimate size and spread of the trees at maturity.

Spacing is very important for fruit trees because of their permanency and long bearing life. Fruit trees raised too close together give lower yields of poor quality fruit with reduced size and colour. Such plants will grow taller, be exposed to greater wind injuries, and be difficult to manage. Crowding of their tops will create dense shade beneath the trees, resulting in insufficient light penetration and ventilation. Such plantations are ideal places for the multiplication of disease organisms and insect pests.

With fruit trees, the spacing problem is always difficult and usually results in compromise. Basically, close spacing with a large number of trees per hectare gives heavier early production, while wider spacing gives better late production. Those who want quick production after planting a grove generally lean toward closer spacing, and those who are looking at the grove as a long–term investment generally prefer wide spacing. The Spanish accomplish both ends by doing very heavy hand pruning. They frequently plant their citrus groves at a distance of 10 ft × 10 ft (3 m × 3 m), and keep the trees pruned so that they just touch each other. Such spacing gives a very large plant population and heavy production at an extremely early age, but requires a good deal of handwork to keep production up over a long period.

Double spacing, where plant-to-plant distance is half the distance from row to row, is very satisfactory in the early years, but will prove difficult to manage unless every other tree is removed after about half the bearing life of the tree. Unfortunately, very few growers are willing to take out trees, and the result is a hedgerow effect that is extremely hard to spray, cultivate, and pick. For growers willing to take out trees at the proper time, there are a number of these combinations for different fruit trees that will help to give both high early production and a high late production, with a slight dip for two or three years at the time when the alternate trees are removed.

Calculation of the number of plants per hectare when they are planted by the square/rectangular system is easy. Each tree occupies a plot of land equal to the product of the tree-to-tree distance within rows and the distance between rows. By dividing the area of a hectare (in sq m) by the area occupied by each tree, the approximate number of trees per hectare can be calculated.

$$\text{Total number of plants} = \frac{\text{Area of a hectare (100 m} \times \text{100 m)}}{\text{Row-to-row distance} \times \text{plant-to-plant distance (sq m)}}$$

Examples of the total numbers of plants per hectare of citrus and mango calculated according to the square layout system at different planting distances are given here.

CITRUS

At 8 m distance $= \dfrac{100 \times 100}{8 \times 8} = 156$

At 7 m distance $= \dfrac{100 \times 100}{7 \times 7} = 204$

MANGO

At 10 m distance $= \dfrac{100 \times 100}{10 \times 10} = 100$

At 12 m distance $= \dfrac{100 \times 100}{12 \times 12} = 70$

At 15 m distance $= \dfrac{100 \times 100}{15 \times 15} = 45$

The optimum development of an individual tree requires that it be planted where it will not have to compete with other trees. Normally it is more profitable to plant trees closer together and supply the water and food materials needed for good yields. If trees are too close together, however, no amount of irrigation and manuring will produce as good a crop as properly spaced trees bear. Usually, therefore, the maximum yield per hectare can be had by spacing the trees somewhat closer together than the normal optimal distance, and providing adequate supplementary inputs. Recommended spacings for various fruit trees in Pakistan are given in Table 9.1.

Table 9.1 Recommended spacing for fruit trees in Pakistan

Types of trees	Planting distance (m)
Mango, jaman, ber, amla, fig, walnut	10–15
Guava, mulberry, loquat, litchi, olive	8–10
Sweet orange, mandarin, lemon, grapefruit, kaghzi lime, sweet lime	7–8
Apple, pear, peach, plum, apricot, almond, cherry, pomegranate	6.5–8
Date palm	5–7
Banana, papaya, grapevine	3–4
Falsa	2.5

Vegetables. Some vegetables are first raised in the nursery, and when the seedlings reach a height of about 15–20 cm they are transplanted to the field at species–specific plant–to–plant and line–to–line distances. Some vegetables raised in this way are onions, tomatoes, brinjals, chilies, cabbage, knol khol (kohl rabi), cauliflower, and lettuce. Other vegetables are sown directly in properly prepared beds, maintaining the appropriate distances between beds and between plants. Seeds of coriander, spinach, and fenugreek are planted on both sides of beds in continuous lines or are broadcast. Some information about times of nursery sowing and transplanting, soil requirements, seed rate, and planting distance for important summer and winter vegetables is given in Table 9.2.

Table 9.2 Time of planting, soil type, seed rate, and planting distances of important winter and summer vegetables

Vegetable	Time of sowing and transplanting	Type of soil	Seed rate per hectare	Planting distances
		Winter vegetables		
Potatoes	Spring crop=Jan Autumn crop =Sept–Oct Hill crop=April	Sandy loam and clay loam	1.25–1.5 t Autumn= 2.5–3.0 t	R×R=60 cm (well–irr.) R×R=75 cm (canal–irr.) P×P=18–20 cm
Peas	Sept–Oct April (hills)	Silt loam and clay loam	60–75 kg	B×B=1.25 m P×P=15–20 cm
Carrot	Aug–Nov April (hills)	Rich clay loam	20–25 kg	B×B=60–75 cm P×P=8 cm
Radish	July–Nov April–June (hills)	Sandy loam	8–10 kg	B×B=75 cm P×P=8 cm
Turnip	Aug–Nov April–May (hills)	Deep rich loam	2.5 kg	B×B=.75–1 m P×P=10–12 cm
Sugar beet	Sept–Oct April–May (hills)	Deep rich loam	10–12.5 kg	B×B=40 cm P×P=7–8 cm
Cauliflower	N) July–Aug T) July–Oct N) Feb–Mar (hills) T) April–May (hills)	Clay loam	N=1.25–2.5 kg	L×L=75 cm P×P=30 cm
Cabbage	N) Aug–Oct T) Sept–Nov N) Feb–Mar (hills)	Clay loam	N=1.25 kg	L×L=60 cm P×P=30 cm
Knol khol	N) June–Aug T) Sept–Oct	Clay loam	N=1.25 kg	L×L=45 cm P×P=30 cm
Spinach	Aug–Oct	Heavy loam	25–35 kg	B×B=75 cm Plant seed on both sides in lines as continuous planting.
Fenugreek	Sept–Oct	Loam and clay loam	10–12 kg	Like spinach
Coriander	Aug–Nov and Jan–Feb	Rich clay loam	25–30 kg	Like spinach
Lettuce	Sept–Oct April–May (hills)	Clay loam	N=500 g	B×B=50 cm P×P=15–20 cm
Onion	N) Oct–Nov T) Jan–Feb SP) Oct–Nov	Organic soil	N) 8–10 kg SP) 1500–1750 kg onion bulbs	P×P=8–10 cm L×L=15 cm
Garlic	October	Organic rich soil	750–1000 kg cloves	L×L=15 cm P×P=8–10 cm

Vegetable	Time of sowing and transplanting	Type of soil	Seed rate per hectare	Planting distances
Tomato	N) Nov–Dec T) Feb–Mar N) Feb–Mar (hills) T) April–May (hills)	Clay loam	N=250 g	B×B=1–1.25 m P×P=25–50 cm
Brinjal	N) Feb and June T) Apr and June N) Apr and July	Clay loam	N–400 g	B×B=1 m P×P=45 cm
Chilies	N) Oct and Feb T) Feb and April	Silt loam and clay loam	N=1.25 kg	B×B=1 m P×P=30–45 cm
Sweet pepper	N) Oct–Nov T) Feb–March N) Mar–Apr (hills) T) May–June (hills)	Silt loam to clay loam	N=1.25 kg	B×B=1 m P×P=45 cm
		Summer vegetables		
Muskmelon	Feb–March	Sandy loam	4–5 kg	B×B=2–2.5 m P×P=40–45 cm
Watermelon	March	Sandy loam and loam	4–5 kg	B×B=3–3.5 m P×P=40 cm
Bitter gourd	Feb–April	Clay loam to sandy loam	8–10 kg	B×B=1–1.5 m P×P=30 cm
Tinda gourd	Feb–March	Clay loam	5–6.5 kg	B×B=1–1.5 m P×P=40 cm
Bottle gourd	Feb–Mar and June–July	Clay loam	5–6.5 kg	B×B=2.5–3 m P×P=40–50 cm
Cucumber	Feb–Mar, July Apr–June (hills)	Sandy loam	2.5–4 kg	B×B=2–2.5 m P×P=30 cm
Long melon	Feb–March	Sandy loam	2.5 kg	B×B=2–2.5 m P×P=30–40 cm
Okra	Feb–March	Sandy loam and clay	25–30 kg	B×B=75 cm P×P=20 cm
Sweet potato	April–July	Sandy loam	–	B×B=2–2.5 m P×P=30–40 cm
Turmeric	March–April	Sandy loam	1.7–2.0 tonne	L×L=50–75 cm P×P=20–30 cm
Ginger	March–April	Sandy loam	1.5–1.8 tonne	L×L=40–45 cm P×P=15–20 cm
Ghiya tori	Feb–March	Sandy loam	4 kg	B×B=2.5–3 m P×P=45–60 cm

N = Nursery
T = Transplanting
SP = Seed production
P×P = Plant–to–plant
L×L = Line–to–line
R×R = Row–to–row
B×B = Bed–to–bed

9.3.2 Systems of layout for fruit orchards

A number of different systems for laying out fruit orchards may be employed. The advantages and disadvantages of these systems are described here, and illustrated in Figure 9.1.

i. Square system. The square system is the most common and popular method of laying out fruit orchards on flat ground. In this method plants are set at right angles to each other with equal row–to–row and tree–to–tree distances (Fig. 9.1). This is the most desirable system for long–lived permanent trees of important fruits like mango, citrus, guava, apple, or pear. It is simple and easy to lay out. Important reasons for its popularity are the possibility of intercropping, the ease of two–directional cultivation, and convenience in irrigation, spraying, and harvesting. It forms the basic plan for the quincunx and rectangular layout systems. The only drawback of this system of orchard layout is that some unutilized space is left in the centre.

ii. Rectangular system. In the rectangular system, the row–to–row distance is more than the plant–to–plant distance. This system is particularly suitable for medium to large sized trees. In this case, the trees are set at the corners of the rectangle (Fig. 9.1). In a variant of this system, the plant–to–plant distance is set at half the row–to–row distance, and later, halfway through the bearing life of the trees, every other plant is removed. This procedure is adopted to get maximum returns in the early years of planting. However, the rectangular system is not very good for permanent trees because the land is not fully used and crowding occurs on two sides. Lots of pruning is also needed to allow the plants to grow normally, and only one–way cultivation is feasible in the gardens.

iii. Hexagonal system. This is the best system for garden layout because all plants are equidistant from each other and no unutilized space is left in between. The layout consists of a grid of contiguous equilateral hexagons. Trees are located at each vertex, and in the centre of each hexagon equidistant from each of the vertices. This can also be thought of as a grid of equilateral triangles, in which the length of each edge of the triangles is the desired tree–to–tree distance. A tree is located at each vertex of each triangle (Fig. 9.1). The actual tree desired long–term is set in the centre of the hexagon. About 15 percent more plants can be planted than in the square system. Three–directional cultivation is possible in orchards with this layout system. The only potential problem with this system is that it is more difficult to lay out and harder to understand in the beginning.

iv. Quincunx system. In this system, after laying out the garden according to the square system, an additional tree is set as filler in the centre of every square (Fig. 9.1). It is desirable that quick–bearing filler plants be set in between the long–lived, slow–bearing fruit trees. The filler plants are uprooted at a later stage. In this way 75 percent more plants (fillers) can be

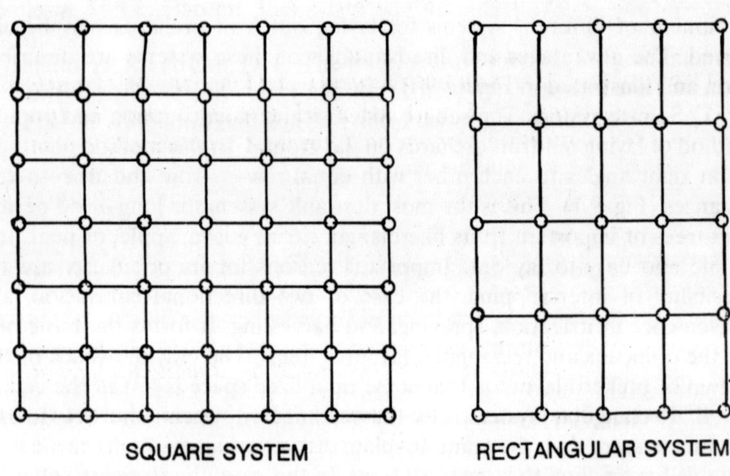

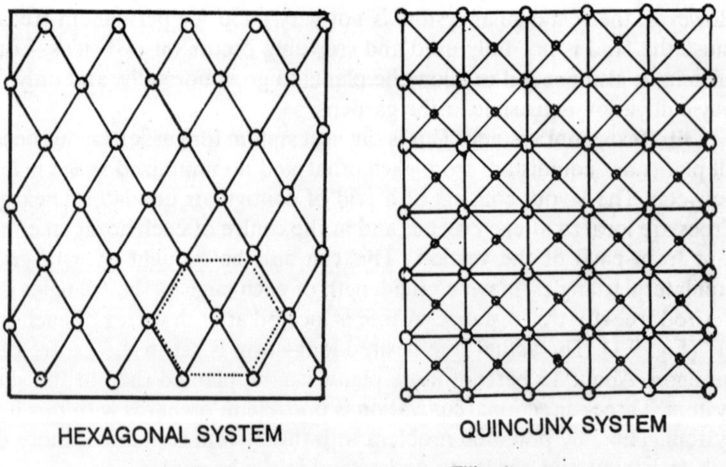

O = Permanent trees , o = Filler trees

Figure 9.1 Orchard layout systems compared

planted than by using the square system. The percentage by which the number of trees can be increased depends on the overall size of the orchard. These fillers can bring better returns than intercrops until the actual fruit trees start bearing normally. However, the fillers may adversely affect the progress of the permanent trees.

v. Triangular system. In the triangular pattern, after laying out a rectangular system one tree is planted at the intersection of the diagonals of each rectangle. This system gives about 50–60 percent more plants than with the rectangular system, the percentage of increase varying with the overall size of the orchard. However, at a later stage this system causes overcrowding, creating problems in cultural operations. This system is not popular and is seldom adopted.

vi. Contour system. All the above systems can be used only on flat or level lands, and not on uneven lands and submountainous areas. On undulating lands and hill slopes, fruit trees have to be set on terraces, and sprinkler irrigation is usually used. To reduce runoff and soil erosion, it is advisable to raise cover crops in the alleys. On soil with only a slight slope, trees can be planted in lines following the contour of the slope. Irrigation and cultivation are then practiced only along thc rows, not across them. The trees will not be equidistant, and the number per hectare will generally be less than with the other systems, but the advantages outweigh these defects. If the land is very steep, each contour may become a terrace.

9.4 Digging and refilling of pits

Though deep soils are usually preferred for raising fruit trees, additional nutrition and organic matter can be provided to the plants at a very early stage by digging pits. This should be done after laying out the garden and marking the actual spots for trees by staking. While digging pits, the upper 30–40 cm of soil should be put aside and kept, and the remaining subsoil scattered in the field or taken away. The pits are left open for 2–3 weeks for exposure to the sun and circulation of air. It is harmful to keep pits open for longer periods because their inner surface gets hardened by prolonged exposure to weather. The main objective of digging pits is to provide a suitable environment for the development of the plant and its roots. Since this is a place to which one cannot have access later, for the whole life of the plant, the preparation of the pit is very important. Also, if there is a hardpan or gravel present in the soil, these problems can be dealt with when the pit is open.

The size of the pit depends upon the type of soil and the species and variety of fruit. It should be larger in hard soils and smaller in sandy loam soils. Generally, for various fruit trees and in most of the orchard soils in this country, a pit of 1 m × 1 m × 1 m in length, width, and depth is most desirable. This is the volume in which most of the root development occurs during the early days of plant growth.

After an appropriate period these pits are refilled with a mixture of equal quantities of the saved surface soil, well–rotted FYM, and silt. The pits should be filled preferably about a fortnight before planting. Unfortunately, adding undecomposed FYM is a widespread practice in Pakistan. This should be discouraged because as manure decomposes in the soil it is injurious to cut roots and attracts white ants. The pits should be filled a little higher (about 10–15 cm) than the surrounding soil. This will allow for soil settling after the first irrigation. The field is then irrigated heavily. After the field dries, the pits are levelled once again. Plants are set in the centres of the pits and in straight rows on all sides.

9.5 Selection and purchase of plants

The selection of a species and the varieties of fruit plants to be raised is an important step in the establishment of a new orchard. Since the earnings a grower will get from his orchard depend on this choice, it should be decided early in the process of planning to enter the field of horticulture. The following important factors govern the selection of fruit plants.

9.5.1 Selection of scion varieties

Scion varieties for an orchard are selected keeping in mind the type of orchard, market demand in the area for a certain species or variety, climatic suitability of the area for a certain fruit or its kinds, resistance to pests and diseases prevalent in that region, and, in the case of some fruit, the requirement for pollinators.

There are several types of orchards, which differ in the markets and the kinds of needs they serve. Consideration of these socio–economic factors is also important when a grower is selecting scion varieties.

1. Commercial gardens. The production of fruits and other horticultural crops on a large scale in localities far from the market is considered commercial gardening. In this type of farming, horticultural crops are even grown in the interior and in places far from the market, where suitable soil, climate, and adequate water supply are available; cost of production is low; and reasonable facilities for shipment to big cities exist. In vegetable growing this type of gardening is known as truck–farming. Horticultural produce may also be raised for export purposes. These commercial gardens contain only a few, good quality, high–yielding varieties, which are raised on large areas to fulfil the demands of major markets in or outside the country.

2. Local market gardens. These produce horticultural crops to meet the requirements of local markets only. Such gardens contain a wide range of crops and varieties which provide produce over a long season. In this case, many varieties with a wide fluctuation in their maturity and ripening times are produced on small areas. A fruit whose popularity is increasing, or one which is being supplied from a distance, offers a good opportunity to growers of local market gardens.

3. Home gardens. Some farmers produce fruits, vegetables, and ornamentals in home gardens for their own needs. When such gardens are very small, they are called kitchen gardens. They include varieties of the owner's own taste and choice to meet his day–to–day requirements.

4. Production for processing. There is an increasing demand for fruits and vegetables for the processing industry. Various fruits are being used for preparing juices, squashes, jams, marmalades, and jellies, and for canning or pickling. Vegetables are also being dried, canned, pickled, and frozen. Sun drying of sliced turnips, onions, brinjals, bitter gourds, fenugreek, cauliflower, and okra is also a very popular practice in Pakistan. Now special kinds of fruits and vegetables are available which can be raised only for processing and are different from the varieties raised for table purposes.

9.5.2 Type of rootstock

When selecting fruit plants for a particular area, careful consideration should be given to the rootstocks on which a particular scion variety or varieties are to be grafted or budded. Rootstocks affect the vigour, productivity, longevity, quality, and resistance to insect pests and diseases of a scion variety. A rootstock should be adaptable to various soil and climatic conditions, resistant to insect pests and diseases, and easily perpetuated and buddable. For rootstock production a high percentage of polyembryony is desirable to produce large numbers of uniform seedlings.

9.5.3 True–to–name plants

When purchasing fruit plants from the nursery, one should first try to visit government nurseries; or, if none are available in the vicinity, one should try to locate nurseries which have a good reputation for past performance. Fruit plants are very problematic, in the sense that they bear fruit after 4–5 years, and the real nature of a fruit tree is usually revealed when it is too late. Sometimes it may not be possible even for an expert to identify the scion variety and the rootstock on which a fruit tree is budded/grafted at the nursery stage. If you want to have a big garden and you can afford the time, it is always advisable to raise your own nursery. The main cause of the present decline in quality and quantity of the fruit industry in Pakistan is the mushroom growth of fruit nurseries. Fruit plants should always be purchased from nurseries which have their own progeny gardens and where the pedigree of the plant is known.

9.5.4 Age, size, and grade

One–year–old, well–grown plants with many leaves, a clean and bright trunk, and abundant roots should be selected. Evergreen plants are transplanted with earth balls, while deciduous plants are transplanted with bare roots and no leaves. One should always check that bud union is at least 15–20 cm above the ground. Higher bud union is now being advocated as a controlling measure against foot rot disease, which can be caused by soil or water touching the scion portion of the trunk. Foot rot does not affect the rootstock portion. With citrus fruits, the trunks of one–year–old plants measure 1–2 cm in diameter about 3 cm above the ground, and two–year–old plants are about 2–3 cm in diameter at the same height.

9.6 Care of plants at arrival

The first thing which should be done on the arrival of nursery plants is to check them for variety, number of plants, and grade. The plants are then sprayed with lime and sulphur, and stored in the shade for about 48 hours. They are kept soaked with water and covered with straw during storage. In the case of deciduous plants, if the roots are dry they should be soaked in water for several hours before planting. Roots broken during lifting of the plants from the nursery soil must be cut off with a clean cut. In order to minimise the unnecessary loss of moisture by transpiration, especially in arid zones, young trees are often stripped of leaves. However, this is not necessary when the roots are packed in balls of soil. Only enough pruning should be done to prevent wilting, which lasts for no more than a couple of days after planting. If more leaves in good condition remain on the plant, growth will start earlier. While pruning at the time of transplanting, the shaping of the tree should also be kept in mind.

9.7 Setting of fruit plants

9.7.1 Time of planting

There are generally two principal planting seasons in Pakistan: spring (Feb–March), and autumn (Sept–Oct). Plants of evergreen fruits can be set in both these seasons, but deciduous fruits only in the spring season. Care should be taken to set evergreen plants wel' before the severe dry and cold season starts, and deciduous plants should be planted when the danger of frost and severe cold is over. Planting season, which differs with locality, is an important factor influencing the successful establishment of an orchard. In areas where winters are not too severe, citrus plants can also be planted throughout the winter, beginning from the rainy season. It is always desirable to plant mangoes in the spring season because an autumn planting needs protection from frost during the winter.

9.7.2 Actual planting

Plants of evergreen fruits should be planted as soon as possible after they are dug from the nursery. If deciduous plants are transplanted while they are dormant, delay does little harm. The plants should be set approximately in the centre of the pit keeping the rows straight on all sides. The use of pegs and a planting board is also recommended to help in keeping the lines of plants straight.

A hole sufficiently large to accommodate the earth ball and/or root system of the plant should be dug in the centre of the refilled pit. The plants should be set about the same height as they were in the nursery, with the bud union quite visible above the ground. If a planting board is used, the plant stem should be in the central notch of the planting board. When the earth ball is present, care should be taken not to break it while planting. The earth around the ball in the hole should be pressed well but not too hard while the plant is held in position. More care is necessary for plants with bare roots to ensure that the roots are well–spaced and no air pockets are left. When placing soil around the earth ball, the topsoil should be put in first, using the subsoil as a top layer.

9.7.3 Post–planting care

After plants are set, they should be irrigated sufficiently to wet all the soil around. This consolidates the soil and helps the roots to establish soil contact and secure a supply of water quickly. A small basin can be made around the plant for this purpose. Sometimes it is advisable to stake the plants immediately after planting to hold the stem erect. The soil level around the plant stem should be made higher than the ground level. Heading back of deciduous plants is also sometimes done. Some young plants are subject to considerable injury from sunburn or frost and severe cold, so shading the plant with *sarkanda* is recommended where needed. Plant stems can also be protected by wrapping them with paper or other materials, or by painting them with whitewash.

9.8 Windbreaks

Strong winds generally damage both foliage and fruit. They damage bearing trees by causing fruit shedding, breaking and blowing off branches, and defoliating the trees. Coastal regions are subject to such high winds and frequent cyclones. Plants have been used for wind protection for centuries. Although the terms *windbreak* and *shelter belt* are used almost interchangeably, *windbreak* is more appropriate for a structure (a fence or a screen) or a single, narrow row of plants, while *shelter belt* refers to several rows of trees or shrubs. A single row of trees can be satisfactory, if they are closely planted, have dense foliage from the ground to the top, and can grow upright in the winds that are characteristic of the area. Wind damage can be minimised by providing such windbreaks around orchards. Some fruit species are more liable than others to shed fruit in high wind. Such species should be planted in the centre of the orchard where wind effect is minimum.

Before planting an orchard, it is necessary to reserve some space for the trees which are to serve as windbreaks on those sides from which hot wind or frost is expected. It is even better to establish windbreaks one or two years before the fruit trees are planted. These should be at an appropriate distance from the orchard plants, which depends upon the type of plants. Usually the recommended distance is about the same as between the fruit trees to be protected. Sometimes a channel 1 m wide and about 1-1.5 m deep is dug about 4–5 m from the windbreak trees. This channel should be cleaned every other year. Doing this can save the fruit trees from competition with the windbreak trees because the channel will not allow the roots of the windbreak plants to come into the root region of the fruit plants.

In some regions or for some crops temporary windbreaks are also made. These are very easy to establish, and may be removed after the extreme conditions are over. For vegetables, mostly tunnels and plastic covers are used and removed when the severe weather is over. Setting of *sarkanda* windbreaks at the northwest side of the rows of vegetables is a usual practice in vegetable forcing. In fruit plants, particularly for small mango plants, temporary windbreaks are very useful during the early years. They are made by pegging *shisham* branches on the north sides of the plants and covering them with sugarcane waste. These are removed every year in late February or early March when the danger of frost and cold winds is over.

It is always advisable to select windbreak plants very carefully. In selecting plant material to be used for windbreaks, some major considerations are: suitability for the soil and climatic conditions of the site, resistance to insect pests and diseases, rapid growth, tallness and dense foliage, mechanical strength to resist damage by winds, and the type of root system. For orchards, windbreak plants should be deep–rooted for support but have limited lateral spread, as trees with widely spreading root systems may compete with the adjoining fruit trees for water and nutrients. Plants also yielding some income should be preferred.

Depending on specific requirements, various types of evergreen or deciduous plants can be selected. In tropical and subtropical regions *shisham*, *kikar*, *jaman*, mulberry, seedling mango, *ber*, eucalyptus, and bamboo are very common. For temperate regions, species which have proven effective as windbreaks are Scotch pine, jack pine, ponderosa pine, Rocky Mountain red cedar, and Australian pine. Where rainfall is sufficient, cone–bearing trees like pines and spruce are best.

Windbreaks are beneficial for fruit trees in many ways. They reduce the destructiveness of strong winds, sunburn, and frost injuries. They improve site conditions by minimising wind erosion and providing protection against sand blown by fast winds. They also suppress transpiration by reducing evaporation caused by hot dry winds. However, there are also some potential negative effects of windbreaks. At certain times they may make a plantation colder. The fruits immediately adjoining the windbreaks are liable to be damaged by insects and diseases, and may be smaller and inferior in colour because of delayed ripening. Also, there may be greater damage from late–spring cold waves in sheltered plantations.

9.9 Orchard planning under various agro–climatic regimes

The response of fruit trees to their environments is complex. A difference of soil and climate which seems very minor may determine the success or failure of a venture. One must either choose the site of an orchard to suit the fruits he wishes to grow, or choose crops suited to his land. The problem becomes more complicated when it comes to varieties of fruits to be grown in a particular region. What fruit will be most suitable to the prevailing agro–climatic conditions? Which of their varieties will be most suitable? What are the possibilities of marketing the produce within and outside the region or country? It is very important to consider all these points before proceeding to the actual planning of the orchard.

Climate is more often a limiting factor than soil in the establishment of an orchard. Some good varieties of apples can only be grown in the regions where elevation is 1500 to 2000 m, where their chilling requirements can be met. Some other varieties of the same fruit do not need so much chilling and can be grown at lower elevations. Tables 9.3–9.7 give some schemes for orchard planning in various regions with different agro–climatic conditions.

Table 9.3 Scheme for mountainous region of high hilly areas with elevation 1500–2000 m (10 hectares)

Fruit	Area (ha)	Varieties
Apple	5	Golden Delicious, Starking, Sky Spur, Kashmir Amri, Amri
Pear	2	Kieffer, Comice, Bartlett
Plum	1	Fazli Manani, Methley, Wilkson
Peach	1	Red French Early, Golden Jubilee, Alberta, No. 6–A
Apricot	1	Charmaghzi, Nuri, White–1, Red Sandian

Table 9.4 Scheme for mountainous region of low hilly areas with elevation 1000–1500 m (10 hectares)

Fruit	Area (ha)	Varieties
Apple	2	Tropical Beauty
Pear	2	Kieffer, Lecontteе
Stone fruits	3	About 1 ha each of plum, peach and apricot; varieties as in Table 9.3.
Persimmon	1	
Loquat and/or strawberry	1	
Minor fruits	1	

Table 9.5 Scheme for sub–mountainous region (10 hectares)

Fruit	Area (ha)	Varieties
Loquat	2	Local
Apple	1	Tropical Beauty
Citrus	2	Kinnow (2 ha), Kaghzi lime and sweet lime (0.5 ha each)
Stone fruits	3	Plum, peach and apricot, about 1 ha each (varieties as in Table 9.3)
Persimmon	1	
Ber	1	Local, Umran 9, 13, 15, and Gohr

Table 9.6 Scheme for plains–subtropical region (25 hectares)

Fruit	Area (ha)	Varieties
Mandarin orange	4–5	Kinnow, Feutrell's Early
Sweet orange	5–6	Musambi, Pineapple, Jaffa, Blood Red, Valencia Late
Kaghzi lime	0.5	
Sweet lime	0.5	
Grapefruit	1	Shamber, Marsh seedless, Foster, Duncan
Mango	8	Malda, Dusehri, Anwar Rataul, Langra, Sindhri, Samar Behisht, Fajri
Guava	2	
Other fruits	3	Ber, falsa, litchi, date palm

Table 9.7 Scheme for semiarid to arid regions (25 hectares)[2]

Fruit	Area (ha)	Varieties
Mango	10	As in Table 9.6
Citrus	5	Kinnow, local sweet orange, sweet lime
Date palm	5	Hilawi, Khadrawi, Zari, Dakki
Other fruits	5	Guava, falsa, pomegranate, etc.

STUDY QUESTIONS

1. Name the areas of Pakistan which can be classified as arid, semiarid, tropical, subtropical, high temperate, and low temperate zones, and their important fruits.
2. What are some important factors which should be considered in planning and establishing new gardens?
3. Climate is considered to be more important than soil in site selection while establishing a new garden. Explain.
4. The texture, fertility, and drainage of the subsoil is particularly important to growers of tree fruits. Explain.

---

[2]This includes the area around Multan, Muzaffargarh, and Bahawalpur.

5. Explain the importance of spacing in tree fruit gardening. Also explain the role of dwarf rootstocks in this regard.

6. Name the important systems for layout of fruit orchards. Explain each briefly, giving its merits and demerits.

7. Explain the importance of windbreaks in gardening. Give examples to illustrate your points.

8. In general, evergreen plants are transplanted with earth balls and deciduous tree without leaves. Explain why this is so.

9. Plan a suitable scheme for an orchard, along with important fruits and their varieties, on 20 hectares of land in your region.

10. Give soil type, time of planting or transplanting, and spacing for important summer and winter vegetables of your region.

REFERENCES

Ali, N.; Malik, M. N.; Gilani, A. H. Practical manual of fruit culture. Faisalabad: University of Agriculture Press; 1979.

Anonymous. Zarai digest: fruit number (*phal numbar* in Urdu). Faisalabad: Agri. Extension, Univ. of Agri.; 1984.

Anonymous. Zarai digest: vegetable number (*sabziat numbar* in Urdu). Faisalabad: Agri. Extension, Univ. of Agri., 1986.

Edmond, J. B.; Senn, T. L.; Andrews, F. S.; Halfacre, R. G. Fundamentals of horticulture. New Delhi: Tata McGraw–Hill; 1977.

Ginai, M. A. A treatise on vegetable culture. Lahore: Bureau of Agri. Information (Publishing Division), Govt. of West Pakistan; 1968.

Hayes, W. B. Fruit growing in India. Allahabad: Kitabistan; 1960.

Hussain, A. Vegetables of Punjab (Urdu). Lahore: Agri. Information Division, Govt. of Punjáb; 1983.

Laurie, A.; Ries, V. H. Floriculture: fundamentals and practices. New York: McGraw–Hill; 1950.

Rajput, C. B. S.; Sri Hari Babu, R. Citriculture. New Delhi, Ludhiana: Kalyani Publishers; 1985.

Ware, G. W.; Mc Collum, J. P. Producing vegetable crops. 3rd ed. Danville, Illinois: Interstate; 1980.

Outline of Chapter 10

10. MANAGEMENT PRACTICES

Mahmood N. Malik[1]

OBJECTIVES

After studying this chapter, a student should be able to:

- Discuss the advantages and disadvantages of the different systems of cultivation
- Given a specific situation (site and crop), decide on the most appropriate cultivation system
- For a given specific situation, decide when and how much water should be applied to a plantation of fruit trees
- Compare the different methods of weed control, telling the advantages and disadvantages of each
- For a given fruit orchard, decide on the most appropriate pruning regime
- Discuss the use of growth–regulating chemicals

10.1 Soil management

Soil management includes various methods of managing the soil to maintain and conserve soil moisture, aeration, organic matter, and nutrition of the plant to its maximum, which is desirable for plant growth and fruitfulness without undue expense. As the soil is the store house of plant nutrients, a good soil management can build up soil fertility and provide ideal conditions for plant growth.

There are many reasons to cultivate garden soils. The most important is to control weeds, which would otherwise compete with the cultivated crop for moisture and nutrients. Cultivation also loosens the soil and helps conserve organic matter by burying it in the soil. This also promotes good root growth and helps soil bacteria to make nutrients available. Proper cultivation prior to planting incorporates air into the soil. These practices turn surface soil and crop residues under, leaving an uneven surface and loose structure

---

[1]Dr. Malik is Professor of Horticulture, University of Agriculture, Faisalabad.

which encourages water penetration and spread of roots. Ploughing and disking are the most effective methods of incorporating organic manures and fertilizers and killing many pests which hibernate in the soil.

Cultivation may also have adverse effects. It may damage soil structure and deplete fertility, cut tree roots, and increase erosion. Regular cultivation also destroys irrigation structures (basins, furrows) in the orchards. It has also been reported that in citrus growing, tillage may not be necessary, or may even be harmful to the health of the trees. Nowadays non–tillage is increasingly popular because the use of weedicides reduces this need for cultivation.

Ploughing is a tillage method still used worldwide. The proper use of ploughing can improve aeration, water penetration, and incorporation of organic matter, but its improper use can create compaction, cause puddling, restrict water penetration, and ruin soil structure. The depth of cultivation has a great effect on crop response. Most vegetables, flowers, and small fruit plants are shallow rooted and have fine feeder roots near the surface. Deep cultivation between rows can do considerable root pruning and drastically reduce total growth and yield. Shallow cultivation can be fully effective in weed removal and does less root pruning of the crop plants. There is less surface–soil evaporation, less soil pulverizing, and reduced power requirements with shallow cultivation. Where annual ploughing is practiced, it is wise to vary the depth to avoid formation of a **plough sole**, a zone of compaction created by the plough bottom taking the same path year after year.

The practice of non–tillage in fruit orchards is being tried in advanced countries. In spite of the rather high prices of herbicides, it is reported to be gaining popularity in controlling weeds of banana, citrus, and avocado orchards. In Australia and California it has been observed that trees produced more when weeds were controlled with weedicides than when tillage was used (Reuther 1973). Similar results may be had by frequent light hoeing or other means of preventing weed growth without greatly disturbing the soil. But the fact that orchards tend to be in better condition and yield more fruit after years without cultivation at least raises questions about the values attributed to cultivation and the necessity for adding other organic material than that contributed by the trees themselves.

10.2 Systems of cultivation

10.2.1 Sod culture

Fruit trees in most of the cooler regions of the world are grown under sod. Under this system natural grasses are allowed to grow under the trees and the soil is not stirred in any way. This prevents erosion, particularly on steep slopes. The sod provides a firm base for machinery and in the orchard it cushions falling fruit and reduces bruising. Because of its competition for nutrients, the sod is not usually started in a young orchard until the trees are well established. The grass is either pastured or mowed from time to time with the mowings preferably left on the ground (sod mulch). Sod is harmful in plains areas as it reduces available soil moisture and nitrogen, but it has a definite advantage on hills and sloping lands as it prevents erosion and also lessens the loss of moisture by runoff and evaporation.

10.2.2 Mulching

Mulches are regularly utilized to conserve moisture, regulate temperature, control weeds, and improve produce quality. Mulching affects organic matter content, microbial activity, nutrient availability, soil compaction, and erosion. Mulches are any type of organic material and contribute to soil humus after decomposition. The adsorptiveness of the mulch enhances water–holding capacity and slows water flow, thus reducing erosion. Soil compaction is reduced by mulches, and earthworm activity, which aids in opening the soil, is increased under mulches. A variety of materials can be used as mulches—grass, straw, hay, dry leaves, wood shavings, weed scrapings, paper, and even polyethylene. For tree mulching, the material should be 20–30 cm deep. Some disadvantages of mulching are the danger of fire and injury to roots and trunks from rats and other rodents (Denisen 1979:87).

10.2.3 Contour planting

Soil conservation techniques are important in soil management. Millions of tons of topsoil are lost every year from fields which are cultivated up and down the slope instead of horizontally along the contours. Contour planting along steep slopes greatly reduces runoff and erosion. Most vegetables, nursery crops, and small fruits can readily be planted along the contours.

10.2.4 Clean cultivation

In clean cultivation, the soil is repeatedly worked throughout the year, keeping the soil free of plant growth. Clean cultivation is helpful in controlling weeds and working farmyard manure and fertilizers into the soil. Clean cultivation is being widely followed in orchards in Pakistan.

There are also disadvantages associated with clean cultivation. Under clean cultivation, with the repeated removal of humus, fertility is also depleted. However, the addition of fertilizers to such soils is not advantageous because the trees, in spite of fertilization, show ill health. It has been reported that clean cultivation reduces tree growth and ultimately tree production due to greater loss of soil moisture during the summer months. Clean cultivation, though otherwise desirable, must not be overdone at the expense of tree growth and health. The advisability of clean cultivation depends on the specific ecological conditions of a region.

10.2.5 Non–tillage

With this system weeds are killed by herbicides. As the ground is left bare, there is good penetration of water. Reduction in cost of cultural and harvesting operations is obvious. However, this system has some disadvantages, as the soil dries quickly and irrigation is needed more frequently.

Although opinions still differ widely about the relative benefits of cultivation, sod culture, and mulching, if one considers their respective merits and demerits for orchards, the inclination is toward non–cultivation practices with the use of chemical weed control.

10.2.6 Cover crops

Cover crops are grown generally in late summer or autumn and ploughed under in late spring. Small grains, small–seeded legumes, and annual grasses make very desirable cover crops in small fruit plantings, vineyards, and nurseries. In the dormant season, they are effective in reducing both wind and water erosion. In cold climates they hold snow which adds moisture to the soil. Cover crops open up the soil with their root growth, add organic matter, and (if they are legumes) build up the nitrogen content of the soil.

10.2.7 Rotation

Crop rotation is an important system of soil management. It is the raising of succeeding crops in the same field following a regular plan so that soil fertility is not adversely affected or may even be improved. It is practiced to equalize depletion of nutrients, avoid toxic effects on the succeeding crop, and control pests. Crops vary in their nutritive requirements in both kind and quantity, so natural adjustments in the soil are made by rotations. For example, tomatoes use a great deal of phosphorus. When a crop of tomatoes is followed by cabbage, which uses much less of this element, the soil is able to recover its balance. Some crops have a toxic effect upon the same crop if they are grown on the same land year after year or even in one succeeding year. Insect pests and diseases become more serious if their host species is planted repeatedly in the same place. Relocating the next season's crop may help in eliminating or reducing the seriousness of the hidden pests in the surrounding vegetation or in the soil.

Since most vegetables are annual crops, they fit well into a four or five–year rotation. On upland soils, for example, the following rotation may prove satisfactory: potatoes–cabbage–onions–legume. Legumes are usually ploughed under for green manuring. A fruit orchard, on the other hand, is a long–term crop lasting about 25–50 years or more. An orchard on level land can be renovated by plowing in green–manure crops, growing fibrous rooted crops which open up the soil, and growing row crops with which good weed–control methods can be practiced. Such a rotation plan may require a generation (Denisen 1979:86–87).

10.2.8 Intercropping

Perennial horticultural crops, particularly fruit orchards, have wide spacing between rows and between plants. Before maturity, these plants are small and do not fully utilize the land, which may permit short–term crops to be interplanted. Vegetables and other small fruits are often interplanted in young orchards. The main objective is to earn some income before the trees begin bearing fruit. Making better use of the land and available nutrients is an important reason for intercropping. However, the main consideration should always be not to let the intercrop compete with the primary crop. Short–term crops are less apt to compete with the primary crop and also easier to eliminate when necessary (Denisen 1979:87).

Intercropping in orchards has been controversial. Farmers want to use available land in the orchard, while experts believe that space between the young trees is necessary to ensure sufficient feeding area for their roots and healthy tree growth. Progressive orchardists and experts agree that crops like wheat, cotton, sorghum, maize, and sugar cane, which exhaust soil nutrients, should be completely avoided as intercrops. Tall growing intercrops, besides exhausting the available food, also spoil the shape of the fruit trees, thus affecting their productivity and health. The same thing happens with some medium–tall crops like cotton. Aside from the nutrient–depleting effects of the cotton crop, orchards intercropped with cotton will tend to grow upward, giving an awkward look. Bearing fruit trees have extensive root systems, so any crop grown between the lines or plants creates nutritional and other problems for the trees.

Moreover, through intercropping there is frequent transmission of insects and diseases to orchard trees, which adds to the cost of plant protection, and in some cases may cause decline and early death of the trees. Cultural requirements like critical times for irrigation and ploughing also differ for fruit plants and field crops. Cotton, for example, requires frequent pesticide spraying (five sprays in the two months from August 15 to October 15), while fruit trees require less. The frequent sprays necessary for cotton may result in serious injuries to fruit trees.

Intercropping with winter crops like berseem was previously recommended, but frequent irrigation to this fodder crop has had serious impact on the longevity and health of many orchards. In Pakistan, one of the main causes of fruit tree decline in general, and citrus decline in particular, is the raising of berseem fodder in orchards. A wet root zone during winter is fatal for citrus orchards, while berseem needs weekly irrigation during this season. The transmission of root rot and other soil-borne diseases to the fruit trees is another very serious problem with intercropping with berseem.

Thus for maximum orchard growth, production, or health, scientific evidence does not favour intercropping. However, farmers with small land holdings and need for diversified income will find it difficult to resist intercropping. Research on the annual biological and economic effects of orchard intercropping is critical, and the results must be made available to farmers.

10.3 Water and nutrient management

Water is an essential environmental factor which should be controlled by drainage, storage, diversion, and irrigation. The wise use of water is obligatory for all farmers, as growing plants have a high demand for water. An optimum supply of soil water is essential for maximum root development. Well developed roots in turn absorb optimal quantities of water and nutrients, producing healthy top growth and higher yields. Excess water, on the other hand, causes unhealthy roots due to poor soil aeration.

Some of the important functions of water in plants are given below.

1. It is a normal constituent of all plant tissues.
2. The rate of formation of carbohydrates and organic nitrogenous compounds is directly proportionate to the amount of water in plant tissues.
3. Water must be present in the soil to encourage fixation and as a medium of transport for nutrient minerals.
4. Water has a moderating effect on temperature.
5. It keeps the plant cell turgid, which is essential for normal growth and cell division.
6. Enzyme activity is adversely affected by a deficiency of water.

Lack of water is the major limiting factor in crop production in vast areas of the world. Much of the productive land is totally dependent on irrigation. In addition to the amount of available irrigation water, it is also essential to know how rapidly this water will be depleted. Probably the most useful measure of water loss is **evapo-transpirational potential**. This measure takes into account both evaporation from the soil and transpiration by plants. Several methods have been used to estimate evapo-transpirational potential which include (a) calculation of incoming, outgoing, and net radiation; and (b) the use of lysimeters. A **lysimeter** is a large container of soil in which a crop is grown; water loss is determined by weight loss of the entire container.

Horticultural plants are very sensitive to their environment. They react adversely to excessive heat, frost, dry soil, excessively wet soil, low soil oxygen, and air pollution. Poor irrigation may result in areas of soil dryness, and excessively wet soil is a strongly contributing cause of low soil oxygen. Damage from excessive heat has been observed to be greater in poorly irrigated garden soil than in soil which has been well irrigated. There is danger in over–irrigation as well as in under–irrigation. There is no simple rule, but factors including specific requirements for kinds of plants, plant size and age, climatic conditions, soil characteristics like water–holding capacity and texture, water already present in the soil, depth of irrigation, interculture practices, and presence or absence of windbreaks and other plants must be considered.

10.3.1 Methods of irrigation

1. VEGETABLES. With vegetable crops, appropriate irrigation practices vary for different species and from area to area, depending on agro–climatic conditions. Vegetable crops are irrigated through surface, sub–surface, and spray systems, which are explained in the following paragraphs.

Surface irrigation. Surface irrigation is the application of water directly to the soil surface. This system is used on deep, compact, and uniformly–textured soils with a gentle slope. Surface irrigation is given to vegetable crops through the border and furrow systems. In the border system, after land–levelling, borders 15–25 cm high are made around the field, which may be divided into sub–blocks for separate irrigation. In the furrow system, the land is thoroughly levelled and 15–20 cm deep furrows are made between the rows. This method is commonly used with vegetables grown in rows in the arid and semi–arid regions of Pakistan.

Although surface irrigation is easy to do, it results in uneven water distribution and wastage of water through leaching on open and porous soils. It cannot be used efficiently on uneven or unlevelled soils. Puddling and baking of the soil also occur with this method of irrigation. Making sub–blocks in the field increases the cost of labour.

Sub–surface irrigation. Water is applied under the soil surface in the root zone of the plants through emitters, which are discussed below under fruit plants. It is continuously available; loss of water is minimized; there is no loss of water from the soil surface; and more soil can be utilized for vegetable raising. On the other hand, it is very costly and difficult to install and operate. It is not useful in porous soils or in those which have hardpan.

Spray irrigation. Spray irrigation is giving water to crops in the form of a spray similar to gentle rain. Irrigation by this method can be used for vegetable growing on all types of soils, and on both levelled and rolling land.

Stationary nozzles, portable sprinklers, and perforated pipes can be used for spray irrigation of vegetable crops and nurseries.

The merits of this method are uniform distribution of water on all types of soils with minimum loss of water and nutrients through leaching, no soil erosion, and full utilization of land. However, its initial and operating costs are very high. It may also not work in areas with continuous strong wind, and may cause excessive loss of water through evaporation. Mechanical difficulties in the working of sprinklers and difficulties in moving pipes in soft vegetable fields are some points which are not in its favour.

2. FRUIT TREES. Success in the fruit industry may depend on how well and economically the trees can be irrigated. Efficient irrigation means providing sufficient water in the root zone of plants at all times, without either excessive penetration below the root zone or excessive waste through surface evaporation and runoff. Planning an efficient irrigation programme requires determining the best answers to the following questions.

1. When and in what quantity should water be applied?
2. What is the most efficient method of using the irrigation water available to plants without undue wastage?

The best method of irrigation in any region should supply a sufficient quantity of good quality water to the plants at the right time with minimum expense. The irrigation schedule should be based on knowledge of water quality, soil and root zone conditions, overall climatic conditions, and variation in the sensitivity of the tree to its environment.

Keeping in view factors such as soil topography, water resources, and availability of capital, the following methods of irrigation are possible under the agro–ecological conditions in Pakistan.

1. Channel system
2. Channel–basin system
3. Modified–basin system
4. Flooding or common method of irrigation
5. Furrow irrigation
6. Sprinkler irrigation
7. Drip or trickle irrigation
8. Alternate–middle irrigation

1. Channel system. In this system, the trees are connected by channels, and water is given to the plants through them (Fig. 10.1–top). Although this is a simple and easy method of irrigation which requires little labour and little technical know–how, it causes much water loss and the passage of infection from one tree trunk to another.

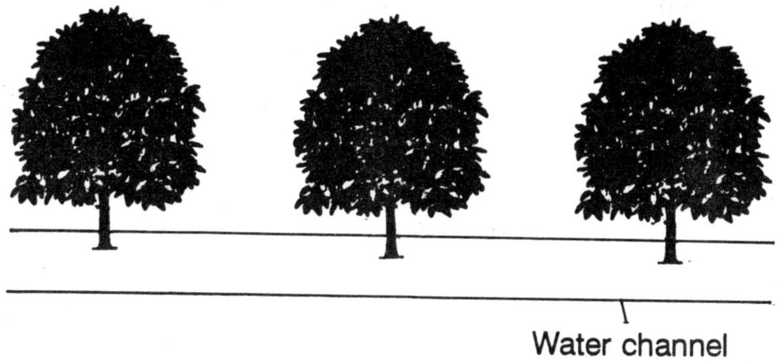

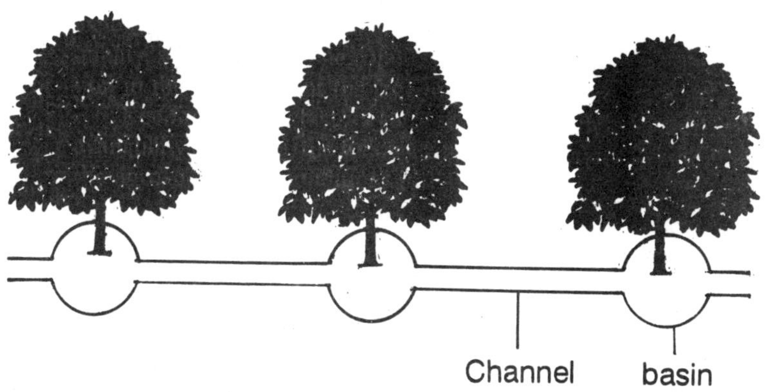

Figure 10.1 Channel system and channel–basin system of irrigation

2. Channel–basin system. Trees are connected with channels, but around each tree a basin is made (Fig. 10.1–bottom). This system has all the merits and demerits of the channel system, except that more of the root zone is wetted, and it requires more labour to make the channels and basins and keep them clean.

3. Modified–basin system. In this system a water channel is made between the tree rows. Basins are made around the tree trunks and then individually connected with the central water channel (Fig. 10.2). Laying out this system is difficult and expensive, and intercropping is impossible, but water loss is somewhat reduced.

4. Flooding or common method of irrigation. Flooding like that done for field crops is also the general practice for irrigating orchards in Pakistan. It is the simplest, easiest, and cheapest method of water application, but causes maximum water losses, leaching of nutrients, and uneven water distribution in the field, which may result in waterlogging. It cannot be used on uneven/unlevelled soils and on soils with non–uniform topography.

5. Furrow irrigation. Furrow irrigation can be done if the slope of the land and size of water source are such that they will not cause erosion. Furrows should be spaced close enough together so that the wetted areas meet and the water is kept in the furrows until it has penetrated to the desired depth. In this method, water loss is minimized and waterlogging does not occur. Also, less labour is needed to set up and maintain the furrows. Sometimes, however, it causes uneven distribution of water, making less water available to the root zone and necessitating more frequent irrigation.

The number of furrows between the rows of trees varies, depending upon circumstances and need. With young trees, at first there may be only one furrow on each side of the row within a foot or two of the row. Later two furrows are made, at a gradually increasing distance from the trees. As the trees grow and their root systems spread, the furrows can be made farther away from the trees and additional ones added as needed. As the trees get larger, three to eight narrow V–shaped furrows are used between each row of trees. Each of these furrows carries a relatively small flow of water, but since several are used and the flow continues for an extended period of time, a large part of the entire soil is usually wetted. This system is most useful in medium to fine–textured soils.

Broad–based furrows are becoming increasingly popular among fruit growers, particularly citrus farmers. A broad, flat furrow wets a larger surface area than a narrow V–shaped one. It also allows the use of a large stream of water in each furrow, so that the water reaches the lower end of the run more quickly, and more uniform irrigation can be achieved. This is advantageous in medium–textured to moderately coarse–textured soils.

6. Sprinkler irrigation. Sprinklers can be used under a variety of conditions, and can be adapted to rolling or steep lands and to soils of

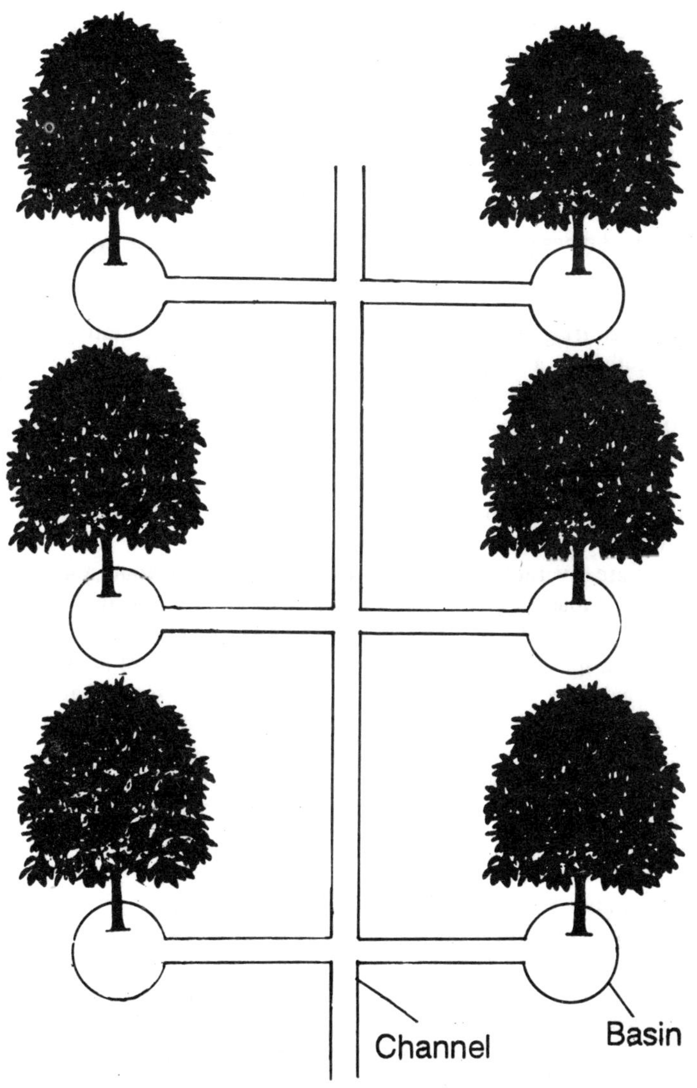

Figure 10.2 Modified-basin system of irrigation

non-uniform topography. Their advantage is that they need no levelling and can make use of small streams of water. Usually sprinklers wet the soil approximately uniformly. The disadvantages of sprinklers are that they have high capital investment and maintenance costs, and they limit the amount of water that can be applied at one setting of the pipelines. Moreover, once the original design of the system is adopted, its operation is inflexible. Low-hanging branches of some fruit trees may interfere with the spray from the sprinklers and cause uneven water distribution. Salts present in water also get deposited on the leaves, which may injure them.

Different types of sprinklers are available. The choice of type should be based on the soil and wind conditions of the area, and also on the quality of water. Some of these types are explained in the following paragraphs.

Overhead irrigation systems have a medium to large, rotating sprinkler head mounted on a tall riser extending above the tree tops. The spacing between sprinklers depends upon their size but usually it is 15–25 m. This method of irrigation may prove unsatisfactory in a strong wind, but it has the additional advantage of increasing humidity and modifying the climate within an orchard on hot days. It may also be used for frost protection. Overhead irrigation should not be used with poor-quality water or in climatic conditions that could produce leaf burn.

There are several types of **under-tree sprinkler irrigation** systems. One type has portable, lateral aluminum pipes to which low-angle, rotating sprinkler heads are attached by means of a short, 15 cm riser. The sprinkler heads may be spaced at distances of 6, 9, or 12 m. Distance between pipes is usually the width of two tree rows. The trees interfere somewhat with uniform water distribution. *Perforated*, portable, lateral aluminum pipelines attached to a pressurized water source can also be used. The application rate tends to be higher than that with rotating sprinkler heads, but sandy soils are capable of absorbing a high rate of application.

Individual tree sprinklers having a very low rate of output are a popular modification of the under-tree system. The orchard is equipped with plastic distribution lines 5–10 cm in diameter, which are usually buried with risers and valves at every other row of trees. To the valve is connected a 1.25 cm diameter flexible plastic hose like a garden hose. Another modification used for young trees is a water line with fixed-jet or micro-sprinklers opposite each tree. As the trees grow, the line is gradually moved away from the tree row. When it gets too far away for the fixed jet to irrigate sufficiently, the jet is replaced by a rotating sprinkler.

7. Drip or trickle irrigation. Sprinkler irrigation wets a large land surface, whereas drip or trickle irrigation only wets a specific area surrounding the plant. The rate of application is so slow that little or no flow of water occurs on the surface. The system discharges the water onto the ground through one or more **emitters** adjacent to each tree, which are usually connected to a narrow lateral plastic line that extends parallel to the tree row. It may either be buried slightly between trees or lie on the surface.

Since in this system there are a limited number of emitters, a large root area cannot be wetted. Frequent irrigations are needed, each of limited volume to avoid over–wetting the soil. Irrigation is generally done daily or every other day depending upon the needs of the tree.

Though there are several advantages of drip irrigation, namely water savings, restriction of weed growth to wetted areas, utilization of problem soils, and savings in labour, there are also many disadvantages. It is very costly to install; the continuity of the flow of water through the emitters is unreliable; and salts accumulate around the root zones of the plants. As the root area becomes restricted to the wetted region only, problems of weak tree anchorage and unsatisfactory yield may arise because of the small root zone and small root volume. Therefore, drip irrigation must be tested under specific field conditions before its installation on a large scale in any region of the country is recommended.

8. Alternate–middle irrigation. In alternate–middle irrigation every other space (middle) between rows is watered with under–tree sprinklers, leaving half the between–row spaces unwatered. At the next irrigation, the previously unwatered middles are irrigated (Fig. 10.3). Thus one side of every tree in the orchard receives water at each irrigation. This method is based on the fact that water absorbed by any portion of the root system is freely translocated to any part of the branches and leaves of the tree.

Alternate–middle irrigation is very helpful in overcoming the effects of water shortage. It provides a means of covering the entire plantation in a short period, so that the intervals between irrigation are not so long that trees become stressed. However, this system cannot be easily adopted on a permanent basis. It is generally used just to get plants through a dry period during times of water shortage. Much labour is required to move the sprinklers every time; the person supervising the operation must also be able to run all the operations smoothly and accurately.

10.3.2 Timing of Irrigation

The aim of irrigation should be to wet the soil to an even depth around the root zone. Water which penetrates below the root zone cannot be used by the roots and is, therefore, wasted. The depth of rooting can markedly affect how long a crop can thrive without the addition of water. Lettuce, strawberry, and radish plants are shallow–rooted and deplete the root zone of available moisture more rapidly than do deep–rooted crops such as tomatoes and tree crops. Thus frequent but light irrigation is needed for shallow–rooted crops, and heavy but less frequent irrigation for deep–rooted plants.

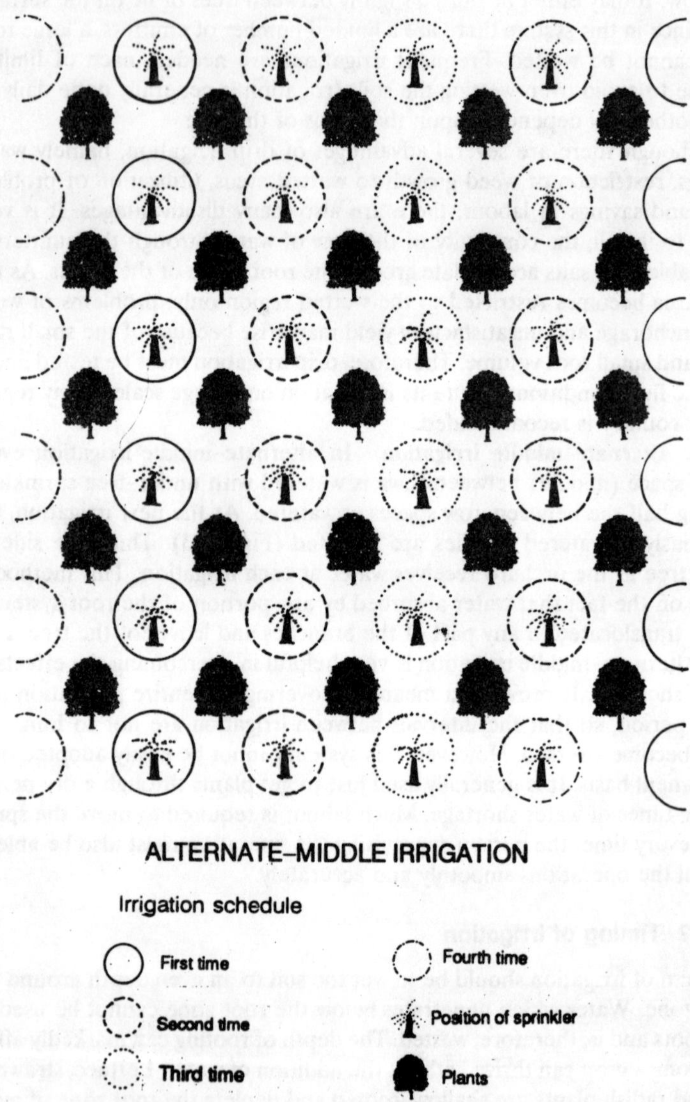

Figure 10.3 Alternate–middle irrigation

Irrigation frequency also depends upon the soil type. Fine–textured soils hold moisture longer than coarse–textured soils, and deeper soils hold more water than shallow ones. Organic matter increases the water–holding capacity of light soils and improves aeration of heavy soils.

Irrigation needs vary from species to species and from region to region. For example, with citrus fruits 30 hectare–cm/ha irrigation is needed near the coast, and this requirement increases to 90 or more hectare–cm/ha in the interior. As much as 1.8 hectare–m/ha are used in some desert plantations. In California, where citrus orchards are extensively irrigated, and where water is carefully conserved, the total water provided by rainfall and irrigation is usually from 90–100 hectare–cm annually. If rainfall provides ⅓ to ½ of this quantity, the remaining ⅔ to ½ has to be provided by irrigation. In seasons of extreme drought and scarcity of water, even less than 60 hectare–cm is sufficient to carry trees through; but if this situation persists year after year, failure will inevitably result.

With fruit orchards, in order to determine when to start irrigation, systematic soil examination should be done between two and three weeks after the last good spring rain and/or 10 days to two weeks after each irrigation. It is also important to check the depth to which water has penetrated two to four days after each irrigation. The interval between irrigations for fruit trees generally ranges between 10 and 20 days. The proper interval is not determined by the calendar or by the soil type, but should be based on the rate of use by the trees. Water should be given to the plants when most of the root zone is nearly dry. A good rule is "always irrigate dry soil, never irrigate wet soil." In the climatic conditions prevailing in most parts of Pakistan, after March until the beginning of winter orchards should be irrigated fortnightly, and during winter irrigation is done every month depending on precipitation.

10.3.3 Critical periods

There are critical periods for water supply in the life span of every vegetable. During this critical period irrigation is essential. If water is not supplied during this period, yield will be badly affected. Critical periods differ for various vegetables; some important ones are given here.

Critical periods	Vegetables
Germination	All species, particularly summer or fall vegetables
Pollination	Lima beans
Pod enlargement	Peas, lima and snap beans
Head development	Cauliflower, cabbage, broccoli
Root, bulb, and tuber enlargement	Carrots, radishes, turnips, onions, potatoes

Critical periods	Vegetables
Flowering: ear, fruit, and seed development	Cucumbers, peas, squashes
Fruit set and development	Melons
Flowering to harvesting	Eggplant, peppers, tomatoes

10.3.4 Water quality

Most horticultural plants are very sensitive to the presence of salts in irrigation water as well as in the soil. It is, therefore, important to know the quality of water used for irrigating various horticultural crops. These crops vary in their tolerance of various salts, so a water analysis is advisable. The water should not contain amounts of dissolved salts excessive for the crop to be raised. As a general rule, water which is hard (less sodium, more calcium and magnesium salts) is better for irrigation than water which is soft (less calcium, excessive sodium salts). This is because Na^+ present in soft water gets attached at the exchangeable sites of the soil particles, which leads to an undesirable soil structure.

Irrigation water may come from canals, wells, tube wells, ponds, or lakes, which may contain salts in variable quantities. Water containing salts at 2000 ppm or above causes injury to most horticultural plants, and even lower concentrations may prove dangerous unless they are diluted or drained away by rain or liberal irrigation. Chlorides and sodium salts are most harmful. Chlorides are said to be more injurious than sulphates, while carbonates are considered to be the most injurious of all the salts. If there is any sign of salt accumulation in the soil, an irrigation of at least 15 hectare–cm is recommended. When the natural salt content of the soil or irrigation water is high, water should be applied in excess of the needs of the plants in order to reduce the concentration of soluble salts in the root zone.

10.3.5 Nutrient management

Pakistan's soils are mostly deficient in N, P, Zn, Mn, Fe, and B. Deficiency of any of these elements can severely affect plant growth and development. Deficiencies invite diseases, hinder uptake of certain other elements, and also disturb the physiological processes of plants. Amounts of various nutrients and/or fertilizers to be applied depend on the fertility status of the soil, and also on various environmental factors. N, P, Zn, M, Fe, and B should be included in the normal maintenance fertilizer programmes of horticultural crops. A combination of organic manures with inorganic ones gives better results than either one used singly.

In applying different fertilizers, not only the deficient nutrients but also small amounts of some other elements should be applied. This, however, must be done very carefully, otherwise the excess of some of these elements may cause problems with the others. The major elements can be applied by any method, but because of their immobility minor elements should be applied only through foliar sprays. Deficiency symptoms and nutritional management for various horticultural crops are given in Chapter 8 in general, and in Chapters 13, 14, and 15 in particular.

10.4 Weed and pest management

It is difficult to define *weed* precisely, but one of the most useful is 'a plant growing where it is not wanted'. In general, any plant which is out of place is a weed. Plants are classified as weeds because they compete for moisture, soil nutrients, and light. It is quite possible for a plant to be considered a weed in one situation but a desirable plant in another. Certain plants such as pigweed are essentially always weeds, but others such as Bermuda grass, which is a very undesirable weed in a vegetable field, can be cultivated as a turf and pasture crop.

10.4.1 Damage from weeds

Weeds can cause damage in various ways, but the damage can be grouped into the following five main categories.

1. Reduction in crop yield. Because of their great number and rapid growth rate, weeds effectively compete with crop plants for moisture, nutrients, and light, causing costly direct damage in the form of reduced crop yields. A troublesome feature of weeds is that they typically produce large numbers of seeds which remain viable for a long time, making them very hard to eradicate. The general notion that one year of seeding is equal to seven years of weeding, is true for most weeds.

2. Crop contamination. The leaves or seeds of weeds can contaminate food crops. Contamination is minimal with tree fruits but is particularly serious in leafy vegetables. Contamination of grass seed with weed seeds can be troublesome when grass is raised from seed.

3. Hosts for pests. Weeds also harbour insect pests and disease–causing organisms, encouraging their reproduction.

4. Poisonous weeds. Sometimes poisonous species of weeds cause problems. With horticultural crops, poison ivy can cause serious discomfort to people who are allergic to it. Some species of mushrooms are also deadly poison.

5. Lack of aesthetic value. Weeds are very undesirable from an aesthetic point of view because they detract from the appearance of areas such as lawns, gardens, and golf courses. In addition, when weeds become larger they interfere with gardening operations. With trees, sometimes climbing weeds shade the leaves and cause great damage.

10.4.2 Types of weeds

Weeds can be grouped into annuals, winter annuals or biennials, and perennials. Those which are very difficult to control and are extremely serious pests are called **noxious** weeds. These types of weeds are explained below.

Annual weeds. These weeds are propagated by seeds and can easily be controlled by cultivation and specific herbicides during their early growth. They become more difficult to eliminate if permitted to seed. They generally produce many seeds and unless controlled will disperse seeds many–fold for succeeding seasons. In Pakistan, they fall into two groups.

A. RABI SEASON WEEDS: *Poa annua* (Poa grass), *Phalaris minor* (*dumbi sitti*), *Avena fatua* (*jangli jawi*), *Chenopodium album* (*bathu*), *Chenopodium murale* (*krund*), *Fumaria indica* (*shahtra*), *Rumex acutus* (*jangli palak*), *Asphodelus tenuifolius* (*piazi*), *Anagallis arvensis* (*billi booti*), *Convolvulus arvensis* (*lehli*), *Cornopus didymus* (*jangli haloon*), and *Sisymbrium irio* (*khub kalan*).

B. KHARIF SEASON WEEDS: *Tribulus terrestris* (bhakra), *Cyperus rotundus* (*deela*), *Euphorbia hirta* (*dodhak*), *Trianthema porthlacastrum* (*It–sit*), *Dactyloctenium aegyptica* (madhana grass), *Solanum nigrum* (makku), *Ehinochloa colonum* Swank, *Heliotropium supinum* (*oant chara*), *Datura indica* (*dhatura*), and *Euphorbia pilulifera* (*hazardani*).

Biennial weeds. These weeds grow for two seasons. They germinate in late summer or fall, overwinter, and then resume growth and flower the next spring. They are an especially serious weed problem in the following spring. Typical examples are *Capsella bursapastoris* (shepherd's purse), *Lepidium virginucum* (peppergrass), *Stellaria media* (chickweed), and *Hordeum jubatum* (squirrel–tail grass).

Perennial weeds. Perennial weeds live for many years and are a serious problem for many horticultural crops, especially perennials. Many perennial weeds can propagate themselves both by seed and vegetatively. They are often deep–rooted, very persistent, and able to withstand many efforts to eliminate them. In some cases an entire crop season is lost reclaiming land infested by perennial weeds. Some of them become extremely competitive even as individual plants. They include *Sorghum halepense* (*baru* grass), *Desmostachya bipinnata* (*dhabb*), *Achryanthes aspera* (*puth kanda*), *Calotropis procera* (*ak*), *Cynodon dactylon* (*khabal* grass) and *Cyperus rotundus* (*deela*).

10.4.3 Weed control

Weeds can be controlled mechanically, competitively, biologically, and chemically. A combination of one or more of these methods can often be more effective than a single one in eradicating weeds in horticultural crops.

Mechanical. Pulling by hand, hoeing, ploughing, and mowing are methods of mechanical weed control. With many crops, hoeing is the standard practice, but where possible it is being replaced by ploughing, either by draft animals or recently with garden tractors. Mulches such as straw, sawdust, and dried sugar cane leaves not only provide excellent weed control but offer other benefits such as stabilizing soil temperature and conserving moisture. Black plastic sheets (polyethylene) are also very effective in weed control.

Competitive. This is the least expensive method of weed control, but it is only possible in a few circumstances. For example, by maintaining a good stand of grass in a lawn, weed problems are reduced because the germinating weeds are not able to compete with the established grass.

Biological. As the use of chemical pesticides is very controversial, suitable biological control is very much needed. In some instances it has been found very successful. For example, prickly pear was a very serious weed problem in Australia, but was controlled by the introduction of the Argentine moth. Great care must be taken that the insect or the pathogen introduced will not harm desirable plants, animals, or the environment.

Chemical. The most widely used technique of modern weed control involves chemicals called herbicides, or weed killers. They may kill by contact or systemic action. **Contact herbicides** kill the tissue on contact, while **systemic herbicides**, which are readily translocated, affect the entire plant. With small weeds the entire plant dies; with large weeds, the foliage dies but the stem usually remains and new shoots may arise later from axillary buds. Discovery of selective herbicides was a milestone because they can be applied in a mixed stand of plants and kill the undesirable plants without harming the others.

The introduction of 2,4–D (dichlorophenoxyacetic acid) in 1944 began the era of modern weed control. This was the first of the growth–regulator type of herbicides. It is not only effective at very low dosages but is selective and readily translocated within the plant. Other phenoxy–herbicides are 2,4,5–T (2,4,5–trichlorophenoxyacetic acid) and 2,4,5–TP (2,4,5–trichlorophenoxypropionic acid). These are highly toxic to broad–leaved weeds, but not to grasses, and so are widely used for weed control in grasses such as turf, corn, and other cereals. Another group of organic herbicides includes the amides, aliphatics, benzoic acid derivatives, carbamates, and urea derivatives.

Contact or systemic (translocated) herbicides either kill the tissue on contact or affect the entire plant, since they are readily translocated. They include inorganics, petroleum oils, and certain organics like paraquat or diquat. Paraquat is a non-selective herbicide and is often used to kill cover crops. It is deactivated as soon as it comes in contact with the soil, so it has no effect on the germination or growth of the following crop. The translocated types are applied to foliage or to the soil, but in either case are absorbed and translocated. Contact herbicides have residual characteristics which make them effective over a long period as pre-emergence herbicides. They may be applied before crop emergence but not necessarily before weed emergence.

Herbicides which dissipate after application include the petroleum compounds such as Stoddard solvent (a dry-cleaning fluid), fuel oil, and kerosene. They are also used on weeds which are present before crop emergence. On carrots, these petroleum herbicides are selective and can be applied post-emergence without injury to the carrots. They give a complete kill of nearly all annual weed species.

Pre-planting, pre-emergence, and post-emergence herbicides are various soil applied herbicides. Pre-planting herbicides are effective against annual weeds, and are applied a few days or weeks before seeding. Pre-emergence herbicides kill germinating seeds; they are put on after seeding but before crops and weeds emerge. They must, therefore, be either highly selective or applied in such a way as to avoid contact with the crop plants. A second and third application at three to four-week intervals are needed for continued control of late germinating seeds. Pre-emergence herbicides include Sasone, Alanap 1, and methyl arsenite. Post-emergence herbicides are most effective at the young seedling stage and may cause some burning of perennial grasses. Some post-emergence herbicides are potassium cyanate (KOCN) and phenylmercuric acetate (PMA).

Systemic herbicides do not kill immediately. The translocated types are applied to foliage or to the soil and are absorbed and translocated to other parts of the plant. Plants are eventually killed completely, including the root system and storage organs. Major herbicides applied to the foliage include 2,4–D, 2,4,5–T, Dalapon, Dicamba, and Glyphosate. Dalapon, 2,4–D, MCP, and amino triazole are systemic herbicides. 2,4–D produces a hormonal action; it causes proliferated growth and in light doses is sometimes a growth stimulant. On turf grasses, 2,4–D is selective; it kills most broad-leaved weeds without apparent injury to the established grass, but grassy weeds are harder to control since 2,4–D is generally ineffective on them.

Systemic herbicides include: Stomp and Dowpon as pre-planting; Arelon, Tolkan, Graminon, and Kenoron as pre-emergence and post-emergence herbicides; Tibunil, Treflan, Topogard, and Ordram as pre-emergence; and Dicuran, Cotogard, Karmex, Saturn, and Brimonal–M as post-emergence herbicides. The most popular contact type herbicide in this country is Gramoxone 20 EC, which is usually used as a pre–planting as well as post–emergence herbicide. Dropp is a contact type which is used as a post emergence herbicide only.

Precautionary measures. All herbicides are poisonous, so they should be handled carefully to avoid injury to man, animals, and desirable plants. In many cases, drifting spray from 2,4–D and other chemicals has caused serious losses of vegetables, shrubs, and flowers. Some herbicides are absorbed through the stem, so prolonged contact with the plant stem should be avoided. It is essential to read and follow the precautions on the label.

Concluding remarks. In spite of tremendous advances made in recent years, the control of weeds continues to be a major worldwide problem. Chemicals have provided us with the means to increase food production with decreasing labour inputs, but they bring their own problems. One of these is the potential damage to human beings, animals, and the environment. It seems likely, though, that with adequate precautionary measures, chemicals will continue to be a major weapon against weeds. Certainly, however, alternative approaches to weed control such as biological control must also be more fully explored.

10.4.4 Pest management

Growers of horticultural crops in Pakistan generally have little knowledge of common diseases and insect pests. Also, they usually consult the experts only when an attack has become incurable; in many cases the problem could have been controlled had it been reported earlier. A grower should keep a close eye on the health of his plants. He should be in regular touch with spraying and plant–protection programmes of the Agriculture Department. Sometimes the reason for the declining health of a crop is misunderstood, and an insect attack is thought to be an elemental deficiency. If it is diagnosed correctly, the cause of the decline can be overcome by controlling the insects. Insect and disease–management programmes for various horticultural crops, and the control of various individual diseases and insect pests are discussed in detail in Chapter 11.

10.5 Pruning and training

The management of plant structure and fruiting wood is called **pruning**. It involves removing parts of a plant's top or root system to increase its usefulness. Limbs, branches, twigs, shoots, or roots can be removed. Pruning also includes the **training** of plants, or shaping them to forms that function more efficiently. Pruning is important for the successful production of fruits, nuts, grapes, and many flowers and ornamentals. Horticulturists look at the yield, size, colour, shape, or quality of flowers and fruit in terms of potential profit, while for amateurs factors like size, beauty, or quality are ends in themselves. Pruning helps both to achieve their goals more effectively.

10.5.1 Principles of pruning

Some important principles of pruning are summarized here, following Denisen (1979:202–3).

Modification of apical dominance. Apical dominance occurs when hormones produced in the stem apices travel down the stem and inhibit or reduce branching and growth of lateral buds. When the terminal growing point is removed, the production and flow of these hormones to lateral buds is stopped and the initiation rate of lateral growth of branches is increased.

Balance of roots and top. Plant growth, development, and reproduction are significantly influenced by the ratio of roots to top. Reduction of leaf area in the growing season or reduction of the number of buds in the dormant season has little effect on root area, but a reduced number of growing points results in stronger shoots with larger leaves. Increased leaf area increases transpiration and photosynthesis, and puts more demands on the roots. However, root pruning reduces the absorbing area, slowing top growth. Stored food is utilized to replace roots, and top growth does not resume immediately after root replacement since manufactured food must first be stored in the stem. Proper pruning effects a balance of top and roots.

Altering growth phases. Regular annual pruning of a growing tree stimulates shoot growth. Heavy annual pruning of young fruit trees delays early fruit production, therefore pruning should be minimal from the juvenile to the productive stage. For maximum flower and fruit production, however, a plant should show good annual shoot growth. If annual shoot growth decreases, as happens with older trees, pruning will stimulate growth, and production is usually increased. Nitrogen fertilizer produces a similar stimulating effect of increased shoot growth. Both fertilizing and pruning are useful in maintaining good flower and fruit production. However, excessive fertilizing or pruning can cause the plant to revert to a vegetative state.

Environmental factors. Desirable pruning and training practices are influenced by several environmental factors (Denisen 1979:203–4).

1. Trees grown in heavy shade are pale coloured, have fewer flowers, and are usually smaller. Frequent pruning to maintain good form allows light to strike the leaves and produces dense foliage in hedges.
2. Pruning can also influence air movement. In spreading foliage, air movement is increased. Open structures allow better spray penetration for controlling insect pests and diseases.

3. Excess moisture tends to produce water sprouts on trunks and primary branches. If excess moisture is preceded or accompanied by severe pruning, the increased water sprout production wastes plant growth.
4. Since pruning reduces transpiration, it is useful during drought periods.
5. Temperature should be considered when deciding whether to prune. Soft, succulent growth resulting from over–pruning or late summer pruning is more susceptible to winter injury, because there is less time for hardening and storing food before the cold weather.
6. Crown and crotch injuries due to cold are more likely on trees which have not been trained or pruned to desirable branching angles.
7. The trunks of trees trained to a low–headed shape receive less intense light, and the bark may be protected from sun scald.

10.5.2 Objectives of pruning

Major objectives of pruning are summarized below (Denisen 1979:205–8).

a. Controlling the direction of growth. The natural form of a plant can be modified to induce it to branch and spread more profusely. Low–branching types can be trained to branch higher. Branches can also be trained to grow away from utility wires or buildings.

b. Developing a strong framework. Some trees have naturally narrow crotch angles (40° or less from the vertical). Narrow crotches result in a greater loss of limbs from wind storms and heavy loads of fruit than crotches with larger angles. The strongest crotches are those in which branches grow up from the trunk at angles ranging from 40–90°. Scaffold branches should be evenly spaced around the tree, with each branch at least 90° from the next one.

c. Controlling the amount of growth. Pruning can either dwarf or invigorate a tree. The type, manner, and time of pruning is determined by the objective. A combination of dormant and summer pruning promotes dwarfness. It is the frequency of pruning rather than its severity that is critical in promoting dwarfness. Increased vigour is produced by dormant pruning of older wood.

d. Improving productiveness. Yield can be either increased or decreased by pruning. The type of pruning depends on the fruit–bearing habits or the vegetative response of each species. Decreasing the number of fruit buds will usually give fewer but larger fruits and may increase the percentage of desirable fruit. Severe pruning can stimulate excess vegetative growth of scion wood, but suppress fruit bearing.

e. Improving quality of product. Fruit has better colour and flavour with adequate light. Improving product quality is one of the main reasons for annual dormant pruning of many fruit trees.

f. Utilizing space efficiently. Training, staking, and pruning are done simultaneously when tomatoes or roses, for example, are grown for the most efficient utilization of space. Pruning facilitates cultural operations, for example good spray penetration and the use of maintenance equipment, and also makes fruit picking easier. Fruit trees with very high tops are difficult to harvest.

g. Increasing the usefulness of plants. Pruning to modify growth increases plant utility. A properly pruned shade tree can provide heavier shade and a regularly clipped hedge can become almost impenetrable. Such trees and shrubs provide better screening and wind protection. The aesthetic qualities of ornamentals can be enhanced: roses can produce larger blooms; vines can produce more concealing foliage, and large or spreading plants can be miniaturized by training and pruning.

10.5.3 Training

Training is pruning management done to develop a tree framework strong enough to bear large fruit crops without the branches breaking. There are three main training systems: (a) central leader, (b) modified leader, and (c) open centre or vase system. These three structures are schematized in Figure 10.4.

a. Central leader system. This system resembles the natural growth pattern of most trees. Trees are trained to a main stem and a series of well–spaced, subordinate lateral branches. Apical growth is encouraged, resulting in taller trees than with other systems. The advantage of this system is the development of strong crotches, and the disadvantage is internal shading, which may weaken the central leader and thus reduce the life of the tree (Denisen 1979:205).

b. Modified leader system. This system reduces the height of the main trunk and encourages scaffold branches to become larger and longer, thus lowering the top of the tree. It combines features of the central leader and open centre systems. The central leader is cut back slightly so that it does not become dominant, and laterals are cut back and selected repeatedly until an appropriate number and distribution of branches is reached. The central leader is then cut and the tree is left with a rounded open top, low, well–spaced limbs, a strong framework, and well–distributed fruiting wood. It is low enough to facilitate various orchard operations. This is the most desirable pruning system for many fruits (Edmond et al. 1977:308).

c. Open centre or vase system. This training system develops a series of well–spaced, coordinate lateral branches rather than a main or central

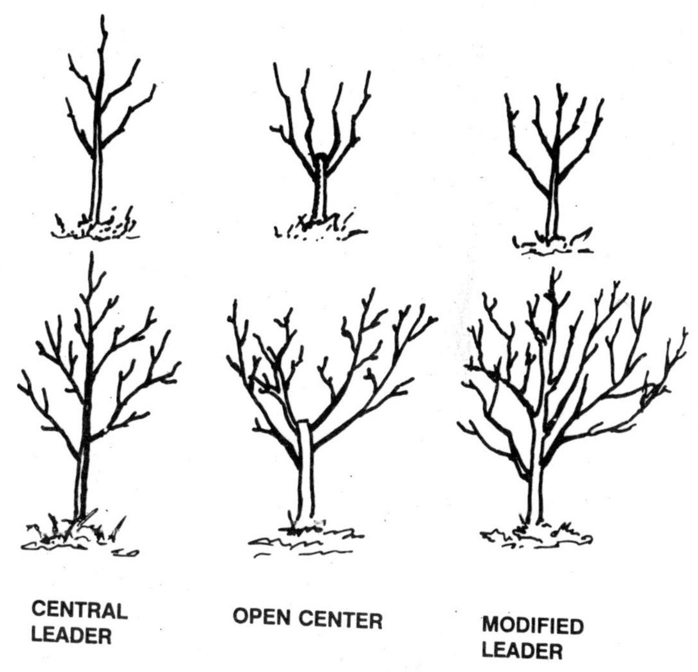

Figure 10.4 Training systems schematized (re-drawn following Edmond et al. 1977:308).

trunk. These branches are cut back equally each year, which gives them equal dominance. The advantages of this system are sufficient light penetration for the fruiting of inner branches, and a low–headed tree that facilitates pruning, thinning, spraying, and picking. Its main disadvantage is that the tree becomes weak with crowded crotches which often break under a heavy load of fruit (Edmond et al. 1977:307).

10.5.4 Kinds of pruning

There are several different kinds of pruning, involving both top and root. These are: (a) heading back and thinning out; (b) fine and coarse (bulk) pruning; and (c) root pruning. There are also a number of special pruning practices. Fruit thinning is also considered to be a type of pruning. Desirable results are often accomplished through a combination of these methods, which are explained in detail below and illustrated in Figure 10.5.

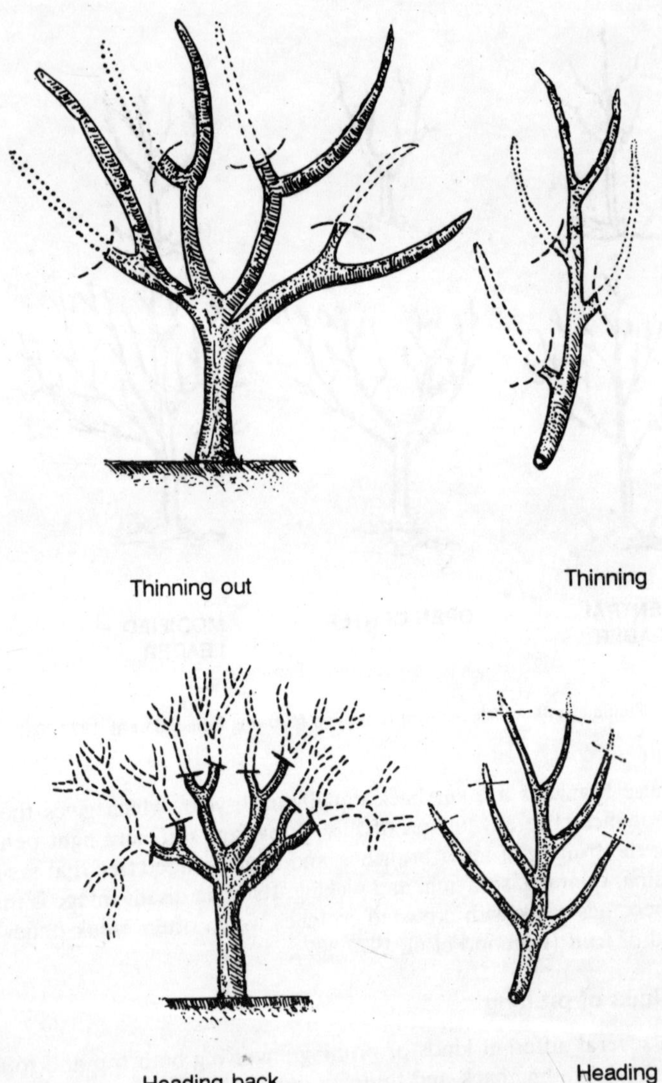

Figure 10.5 Types of pruning for fruit trees

a. Thinning out and heading back. **Thinning out** means to remove certain shoots, canes, spurs, or branches completely from the base. **Heading back** removes only terminal portions of these parts. Thinning out may remove the same percentage of a plant's structure as heading back, but there is substantial difference in what is removed, and an even greater difference in the response of the plant. Heading back leaves the shoots more crowded than does a corresponding thinning. Thinning out usually does not effect much reduction in height or spread, but heading back of comparable severity effects substantial reduction in height, spread, or both, and leaves a less open plant. Thinned plants become taller, and more spreading and open; while headed plants become more compact. In general, heading back stimulates the development of more growing points by overcoming apical dominance, while thinning out does not have this effect.

b. Fine and coarse (bulk) pruning. As bearing branches become older and weaker, they set fewer flowers and fruit of poor quality, and die within a few months. Removing these branches earlier will result in vigorous new wood. Removal involves a large number of small cuts; it is fine or thin-wood pruning. The term applies to shoots, canes, spurs, and older growth. Coarse or bulk pruning is done from the upper and older portions of the top which invigorates the lower, interior, shaded, weaker portions.

c. Root pruning. **Root pruning** is mostly removal of a part of the root system by cutting or heading back the ends of the roots, but there is usually some thinning out too. Like top pruning, root pruning affects both the total growth of the plant and its vegetative–reproductive balance. In general, pruning the roots reduces the area for the absorption of nitrogen and other essential elements, and water. Growing-point cells are reduced as are cell division and enlargement, and there is reduced utilization and accumulation of carbohydrates. Reduction of these processes favours reproductive over vegetative processes (Edmond et al. 1977:325).

Major effects of root pruning are: (i) reduced absorption, (ii) more branched main roots and feeder roots, and (iii) reduced top growth. Since reduced absorption promotes dwarfness, root pruning was a standard practice for dwarfing plants. Since carbohydrates are stored in the primary and secondary roots of woody plants, root pruning removes some of the carbohydrates which would otherwise be used for growth.

Root pruning can be done on cabbage plants to prevent the heads from cracking following an excess of moisture. A spade inserted around the leaf area will cut many of the roots. Undercutting of the tap roots of nursery plants is also done to produce plants with more compact root systems for ease in handling, shipping, and replanting. Root pruning is especially helpful in transplanting evergreen plants like citrus or palm trees. It is also employed in the field with established plants to check over-vigorous growth, which may hasten bearing and increase productivity, but also may produce undesirable results (Denisen 1979:215–16).

10.5.5 Special pruning practices

Dehorning. Severe cutting back of trees and shrubs to keep them from becoming too tall is called **dehorning**. A somewhat less drastic heading back is employed with too–tall fruit trees to make spraying and harvesting easier.

Pinching. Removal of a plant part that is soft enough to be readily broken with the fingers is called **pinching**. One may remove a shoot or spur at its base (thinning) or remove its terminal portion (heading back) by pinching. If pinching is done early in the season it stops the growth of the pinched shoot and causes it to branch; if done late in the season it stops lengthening without stimulating branching.

Disbudding. Removal of buds is easier after they have started to break. Disbudding is useful in preventing the development of laterals on the trunks of recently set trees and sometimes in limiting the number of laterals on their main branches. However, the labour involved may often be greater than the benefits derived.

There are a number of other specialized pruning practices such as **stripping**, **notching**, **binding**, **ringing**, and **girdling** (scoring). They are limited to treatment of the bark of the trunk or main branches and serve to restrict the downward flow of food from leaves to roots, resulting in the accumulation of food reserves above the point of restriction. This encourages fruit–bud differentiation, flowering, and fruiting. Tying knots in the flexible stems of young seedlings has an effect similar to that of girdling. Bending and tying down long vigorous shoots on a young tree in its formative years also has a similar effect. These operations should be regarded as special practices rather than ones to be regularly employed.

Fruit thinning. Removal of blossoms, tiny fruits, buds, and spurs are ways of accomplishing fruit thinning. Pruning some fruiting branches prior to bloom will decrease the number of fruit. More shoot growth will occur, more carbohydrates will be manufactured, and the remaining fruit will be larger. Also, if adequate nutrients are available, increased food production will result in an abundance of fruit buds for the following year's production. Thus fruit thinning in years when an excess of fruit buds are produced will help to overcome the biennial bearing habit of fruit and may force a biennial bearer to become annual.

With all fruit thinning, whether mechanical or chemical, it is important to remove the excess fruit while it is very small, less than 1.5 cm in diameter. If it is thinned after it gets larger than 2.5 cm in diameter, the pattern of fruit development has already been well established, and there may be little beneficial effect on the current crop (Denisen 1979:211).

The fruit industry in foreign countries, especially the apple industry, uses chemicals for flower and fruit thinning. The general purpose of fruit thinning with chemical sprays is to encourage annual bearing and to increase fruit size. The apple tree regulates itself naturally to some degree, since less than 5–20% of the flowers develop into fruit. However, chemical thinning is still necessary. Dinitro (DN) sprays are used to reduce fruit set. They are sprayed on clear days with moderate temperature right after the bloom stage to kill the blossoms. If frost or rain follows its application, the spray may greatly over–thin the blossoms, so DN sprays are very risky in rainy weather. If a spray of naphthalene acetic acid (NAA) is used with polyoxyethylene sorbitan monolaurate or Tween 20 as a wetting agent, the amount of NAA needed for thinning is greatly reduced. Experimentation with these types of sprays on important commercial fruits is still needed in Pakistan.

10.5.6 Chemical pruning

The use of chemicals in the regulation of plant growth is becoming more widespread because manual pruning is time–consuming and also very expensive. Chemicals are used for the following purposes.

- To increase the number of new shoots per plant
- To control the shape of the plant
- To increase the number of flowers
- To control the time of flowering

Pruning with chemicals has become an important part of growing ornamental plants. For chemical pinching, fatty acids, methyl esters, and alcohols of chain length C_8–C_{12} are effective in killing or inhibiting axillary bud growth. Chemical pinching of azaleas and chrysanthemums increases the number of flowers per plant. Chemicals such as methyl caproate, methyl caprylate, methyl laureate, and methyl stearate are effective as chemical pinching agents. The commercial production of chrysanthemums would require hours of tedious manual disbudding, but the commercial product HAN (C_{10}–C_{13} alklapthalenes) has proven very effective for this process, with varying response in different chrysanthemum cultivars.

Root pruning is also done with chemicals. Often a dominant tap root develops if the root is not pruned to induce strong, fibrous growth. Since hand pruning takes much time, chemicals are used to treat layers in the seed beds. When the roots reach these layers, they are chemically pinched. For example, when cork oak acorns are planted 8 cm above layers of Osmocote and Perlite soaked in copper naphthenate, the tap roots are unable to penetrate the treated layers, and a more compact fibrous root system develops.

Plant height can also be controlled chemically. In floral crops these chemicals alter plant growth so that the crop can be grown with less labour. Chemicals used for this purpose have the desirable effect of reducing internode length without reducing the number of leaves. Amo 1618, Phosphon, Cycocel, and Daminozide are used to retard the vertical growth of various flowering plants like chrysanthemums, lilies, bedding plants, poinsettias, azaleas, and hydrangeas. One of the newest growth retardants is A–Rest (cyclopropyl–a–(4–methoxphenyl)–5–pyrimidine methanol). It controls plant height and is active at a low concentration.

10.5.7 Pruning of herbaceous and ornamental plants

Herbaceous plants. Although most pruning is done on woody plants, it is also important for production of herbaceous plants, in which the balance between reproduction and vegetation can be controlled or altered by pruning. Vegetables and flowering crops for which pruning is important are tomatoes, cucumbers, chrysanthemums, camellias, dahlias, zinnias, calendula, mignonette, and herbaceous fruits like strawberries (Denisen 1979:237).

Tomato pruning is carried out both in greenhouse and outdoor production. In greenhouses, individual plants are usually trained to a single stem which is supported by staking or string. After the optimum height is reached, the top of the plant is removed, and some side shoots are removed to reduce the number of growing points and encourage fruit production. Pruning in the greenhouse allows plants to be grown close together and encourages fruit to develop above the soil. It also makes spraying and fruit picking easier. In field production, tomato plants are trained to one, two, or three stems and the fourth, fifth, sixth or succeeding flower clusters are topped. Stakes are usually used for support. This allows close spacing of plants, and leads to higher early yields, and fruit ripening off the ground. However, pruning increases the total cost of production, as the cost of stakes and labour must be included (Edmond et al. 1977:334).

Cucumbers grown in greenhouses are regularly trained and pruned. Individual plants are trained to a single stem and supported by strings to a wire trellis 1–2 m above the ground. For almost two months after plant set, the primary laterals are cut back to the first or second female blossom and the secondary laterals are removed. This involves both heading back and thinning out, and allows plants to be set close together and fruit to develop above the ground (Edmond et al. 1977:334).

Chrysanthemum culture involves two types of pruning, depending on the objective. Pinching back terminal shoots in early summer promotes branching, contributing to a mass bloom effect useful in border varieties. If large individual flowers are wanted, the plants are trained to a single stem. Disbudding also produces large single blooms. Depending on the variety, either crown buds or terminal buds are removed, but in either case only one flower is allowed to develop on each plant. Removal of all the axillary buds results in one large bloom per stem (Edmond et al. 1977:324–25).

Ornamental plants. Two types of pruning are done on ornamental trees, shrubs, and vines: (1) removing dead, diseased, or injured branches; and (2) trimming to make plants proportionate to the size of their growing area.

Severity of pruning depends upon the species and the size of flower desired. Severe pruning is practiced in roses which are grown for cut flowers. Weak stems are thinned and sturdy stems are headed back. In heading back, all cuts are made to an outer bud to promote spreading, and from 6–10 buds are usually left on each stem. Flower stems are cut so as to leave one to three nodes of the current season's growth on the stem. Climbing roses are pruned soon after flowering and the flowers are removed. This stimulates the development of new shoots which produce flowers in the following year. Rambler roses need a small amount of heading back, and in regula roses only weak, dead, and diseased wood is removed (Edmond et al. 1977:320–21).

For hedge plants, new lateral growth is necessary to produce dense compact growth from base to top. Thus the heading–back techniquc of pruning is used. In general, plants are headed back to a uniform height at planting, and as the new shoots reach a length of 15–30 cm they are headed back to 7–15 cm. This practice is continued until the desired height is obtained (Edmond et al. 1977:321).

10.5.8 Amount of pruning related to bearing habits

The amount of pruning varies with plants of different ages. In general, after planting the aim should be to build up a proper framework. After the framework is built up, the next objective is to produce fruiting wood. Thus at first when plants are young, severe pruning is done to build up a strong framework. As the plant approaches bearing age, pruning is comparatively less severe to permit production of leaves and fruiting wood. In some cases even no pruning is desirable. With fruit crops like apples and pears, heavy pruning at this time will greatly delay fruiting. The severity of pruning increases as the tree increases its bearing area. In older trees, there is often too much bearing wood for the roots and leaf area to support. At this time old unproductive wood is pruned away to make the plant bear regularly.

The severity of pruning also varies for different fruit trees depending on their habits of fruiting. For example, grapes bear only on current season wood, so for them pruning is really an annual rejuvenating process. It removes all the vine except enough one–year–old wood to produce a sufficient number of fruiting shoots and leaves the minimum amount of old wood that will make a satisfactory framework for the vine. Thus each year about 90% of the vine is pruned out. Falsa is also quite severely pruned to get a good crop on the current season's growth. If it is not pruned, loose and vigorous plants result, which yield poorly.

With peaches, fruit is borne laterally on long vigorous one–year–old wood. Thus one–year–old wood must be developed for fruit production, and healthy older wood maintained for the production of satisfactory new fruiting wood. For regular production it is necessary to provide for an adequate annual supply of suitable shoots. Moderately vigorous trees are necessary to obtain such relatively thin fruit–bearing twigs. Pruning also serves as a way of fruit thinning. Often ⅓ or more of the peach tree is pruned out.

Apple trees bear their fruit chiefly on old spurs distributed throughout the tree. The object of pruning is to maintain an adequate supply of sufficiently vigorous spurs. This is accomplished by a light thinning throughout the tree. A heavy pruning like that appropriate for peach trees would force these spurs into long vegetative growth and fruiting wood would be decreased. With apple trees, about ⅓ or less of the fruiting wood is removed. Fruit–bearing habits and the amount of pruning desirable for various fruit plants are given in Table 10.1.

Table 10.1 Fruit–bearing habits and amount of pruning in different fruit plants

Fruit	Bearing habits	Amount of pruning
Apple	Fruit buds are mixed and on opening produce flowering, fruiting shoots with terminal inflorescence. They are borne on spurs of varying ages and occasionally on long shoots.	About ⅓ or less fruiting wood is removed after harvesting the fruit.
Pear	Same as for apples	
Peach	Almost all fruit buds have flower parts only and are borne laterally on one–year–old shoots.	About ⅓ or more of peach tree is pruned every year.
Sweet cherry	Fruit buds are usually borne laterally on short spurs. The terminal bud is a leaf bud by which growth of the spur is continued each year.	About ⅓ or less of tree is removed up to 6–7 years, and very little after that.
Almond	Fruit buds have flower parts only and are present laterally both on spurs and shoots.	Less than ⅓ of the fruiting wood is removed.
Plum	As for almond	

Fruit	Bearing habits	Amount of pruning
Apricot	As for almond	About ⅓ of fruiting wood is removed.
Sour cherry	As for almond	More than ⅓ of fruiting wood is removed.
Grapes	Flowers are borne laterally on new growth of the current season which arises from lateral buds on the canes.	Severe pruning is done every year.
Falsa	Same as for grapes	
Mango	Fruit buds are borne terminally on long shoots.	Some pruning is done to control alternate bearing and to remove old, undesirable and diseased dry branches.
Loquat	As for mango	
Citrus	Flower buds are present on new growth appearing in spring, laterally or terminally.	Regular yearly pruning is essential only for lemons. For other citrus fruits only undesirable and diseased dry branches are removed.
Banana	Inflorescence is lateral and once in the life of the plant.	Whole plant is removed after harvesting the fruit.
Pomegranate	Flowers are borne terminally on spurs located on older peripheral wood. Spurs usually stop bearing after 2–3 years and new spurs are formed.	Light thinning out is practiced and no heading back is done.
Persimmon	Fruit buds are mixed and produce fruiting shoots on which flowers are borne laterally. Fruit buds appear on two–year or older wood.	Only undesirable, dried, and diseased parts of the plant are removed.
Walnut	The tree is monoecious. The pistillate flowers are borne terminally on spring shoots. The staminate catkins appear laterally on spring shoots.	Less than ⅓ of the fruiting wood is removed.
Pecan	As for walnuts	

Fruit	Bearing habits	Amount of pruning
Guava	Fruit buds are mixed and are terminal on one–year–old shoots. The buds on opening produce leafy shoots on which flowers are borne in the axils of the outermost leaves.	Only dry diseased undesirable parts are removed.
Jaman	As for guava	"
Olive	The bearing habit is similar to guava but the inflorescence is in the axils of the lower instead of the outermost leaves. The flowering shoots appear from lateral or terminal buds.	"
Jujube (*ber*)	The lateral mixed fruit buds produce flowering shoots, on which solitary flowers are borne laterally.	After 3–4 years severe pruning is done.
Fig	Bears lateral fruit buds.	"

10.6 Grafting for control of growth

Future fruit orchards will likely have trees smaller than those produced by seedling rootstocks. More trees can be planted per unit area, and ultimately the yield per unit area can be increased. Rootstocks determine the size and growth habit of a tree. About 27 different rootstocks for apples, which vary from dwarf to vigorous, have been released, and increasing proportions of apple trees are being planted on them. Among these the Malling and Malling–Merton rootstocks released by the East Malling Research Station in England are increasing in popularity. The vigour of citrus trees also depends on the growth habits of the rootstocks selected.

Sweet orange and grapefruit grafted on *Poncirus trifoliata* (var. Flying Dragon) rootstock are hardly 2 m tall. Experimental plantations of grapefruit and sweet oranges of about 1000 plants per acre (.4 hectare) have been made at the Citrus Research Station, University of California at Landcove, in the USA. Dwarf apple or citrus trees may yield one–half or even less of what plants on conventional rootstocks do, but the increased number of plants per unit area increases the total yield. However, it may be that the tree life is also decreased. Thus the long–term value of this practice remains to be established. To answer such questions, experimentation with grafting on these rootstocks on larger areas for a sufficiently long time is required.

10.7 Growth regulators and their application

The relationship between the growth of a plant and its organs has been known since the late seventeenth century. Tropistic response of a grass tip coleoptile was noted by Darwin in 1897, but it was Went in 1928 who found a specific diffusible substance in the tip of *Avena coleoptile*. The substance has become known as **auxin**, and is necessary for growth.

In the 1930's, study of an unusual fungal disease of rice which caused excessive growth of the plant identified a new growth–producing substance produced by that fungus (*Gibberella fugikuroi*). The new substance became known as **gibberellin**. Gibberellin–like compounds have been isolated by many researchers, mostly from fungi but also from higher plants. The most important was called GA_3. Another versatile compound was identified in 1955, and was called **kinetin**; it is an active cell–division stimulant. Research on growth regulators has dealt with not only the development of growth–promoting substances, but also with toxic substances and growth inhibitors, many of which have been isolated from plants.

The term *growth hormone* refers only to naturally occurring substances including auxins, gibberellins, kinins, inhibitors, and retardants. The term *plant growth regulator* includes not only naturally occurring hormones but also synthetic compounds. These compounds, when applied to plants even in very small quantities, can modify their characters, improve quality, increase economic yield, and facilitate harvesting. Descriptions of some of these substances are given in the glossary of this book. See the entries for **hormone, phytohormone, auxin, growth regulator, auxin synergists, auxin antagonists, anti–auxins, epinastic agent, gibberellins**, and **kinins**.

10.7.1 Practical applications of growth regulators

Plant growth regulators were first used in 1932 to stimulate root development. Since then numerous compounds and growth regulators have been developed and tried at varying concentrations for specialized purposes with several plants. Although these substances have not yet been used on a large scale, their outstanding results and the world's growing problems like the high population growth rate, high cost of labour, and limitations in some farm operations (like harvesting) have made their application more useful and economical.

Some of the practical applications of growth regulators for specific purposes are summarized below.

a. Rooting of cuttings. Some growth regulators and other substances that can be tried for the rooting of cuttings are indolebutyric acid (IBA), indoleacetic acid (IAA), naphthalene acetic acid (NAA), 2,4,5–trichlorophenoxypropionic acid (2,4,5–TP), vitamins, sugars, and amino acids.

The usual method of application is to dip the base of the cutting to a depth of one inch in the prepared solution: for 24 hours with dilute solutions (50 ppm or so), and for up to 30 seconds with a concentrated solution (10,000 ppm or so). Application by dusting or in a lanolin paste is also common. Table 10.2 summarizes information about growth-regulating substances, dosages, and application times for rooting some major crops.

Table 10.2 Growth regulators: dosage and timing for rooting major crops

Fruit crop	Growth regulator and dosage	Time of application and remarks
Grapes	IAA: 100–200 ppm	Soaking before callusing, good response
Guava	IAA: 50 ppm	Early spring treatment, good response
Pear	IBA: 50–100 ppm	Mid-Nov. treatment before callusing, good response
Mango	IBA: 10,000 ppm IAA: 10,000 ppm NAA: 10,000 ppm	No response
Olive	IAA: 15–50 ppm IBA: 15–50 ppm	No response
Sweet lime	Seradix A	Spring treatment, good response

b. Fruit setting and development. It is now well established that the growth of fruit after fertilization results from a hormonal stimulus produced from within the ovarian tissues. The sources of these hormones may be the pollen tube, ovary, the young embryo, or the developing seed. In nature also the parthenocarpic production of fruit is brought about by auxins, gibberellins, and kinins. Table 10.3 lists growth-regulating substances used to promote fruit setting with major crops.

Table 10.3 Growth-regulating substances used to enhance fruit setting

Growth-regulating substance	Dosage	Fruit crops tested with good response
4–chlorophenoxyacetic acid (4–CPA)	25–250 ppm	Tomato, grapes
Cycocel (CCC)	100 ppm	Grapes
Naphthoxyacetic acid (BNOA)	50–60 ppm and 5–10 ppm	Tomato, grapes

Growth–regulating substance	Dosage	Fruit crops tested with good response
2,4–dichlorophenoxyacetic acid (2,4–D)	.001 mg/litre (=.001 ppm)	Pear
2,4,5–TP	2–7.5 ppm	Fig, grapes
Gibberellic acid (GA_3)	25 ppm	Fig, grapes
Gibberellic acid (GA_3)	10–20 ppm	Pear

Sprays should be applied at the full–bloom stage.

c. Flower and fruit thinning. The most outstanding feature of chemical thinning sprays is their effect on alternate bearing. These sprays reduce fruit set early in the season, which lets the tree form more fruit buds for the next year's crop. In addition, thinning improves the size, colour, and quality of fruit. In many countries chemical thinning has replaced hand thinning and has provided an easy method to overcome the problem of alternate–year bearing. Table 10.4 lists important chemical substances used for fruit thinning, along with their dosage and application times.

Table 10.4 Chemicals used for fruit thinning: dosage and time of application

Name of growth regulator	Dosage	Time of application – remarks
Sodium dinitro–orthocresylate (Elgetol)	½–1⅓ pint in 100 gallons	At full bloom in apple – good response
Naphthalene acetamide (Amide–Thin)	50 ppm	14–21 days after full bloom in apple – good response
2,3,5–tri–iodobenzoic acid (TIBA)	50–75 ppm	21 days after full bloom in apple – good response
Ethephon	20–200 ppm	4–8 weeks after full bloom – effective in peach, plum, and apple
Gibberellic acid (GA_3)	10–20 ppm	15–25 days after full bloom – good results in apple and pear
Naphthalene acetic acid (NAA)	400–500 ppm	14 days after fruit set in citrus – good response

d. Pre–harvest drop. Fruit drop causing reduction in crop yield occurs at various stages of fruit development. The first and the heaviest drop is of flowers, which occurs because of structural defects, lack of pollination, adverse weather, or nutritive conditions. The second drop is of young fruit, which takes place during March or June. It may be due to embryo abortion, low seed content, low moisture and nitrogen supply, poor tree health, or adverse weather conditions. Another drop is sometimes experienced in the month of September, which may be due to stem-end rot, a fungal disease, nutritional deficiencies, or adverse environmental conditions.

The pre-harvest drop, which is the most drastic, occurs at the time the fruit is mature but before it is harvested. Depending on the variety of citrus, this drop occurs from November through January or even in February–March with late varieties. Factors contributing to this drop are low auxin levels across the abscission zone, high nitrogen content in soil, high temperature, winds, pest and disease attack, and delayed picking. Table 10.5 lists important chemicals used to control pre-harvest drop.

Table 10.5 Chemical control of pre-harvest drop: dosage and timing

Chemicals	Dosages	Remarks
Naphthalene acetic acid (NAA)	10–20 ppm for apple, 15 ppm for citrus	Controls drop 2–3 days after spray and remains effective for 7–14 days.
2,4–Dichlorophenoxyacetic acid (2,4–D)	5–7 ppm for apple, 18–20 ppm for citrus	Effective for 3–4 weeks; needs 10 days to become effective.
2,4,5–trichlorophenoxypropionic acid (2,4,5–TP)	10–20 ppm	Very effective for McIntosh and other apple varieties.

e. Regulation of dormancy in plants. In nature chilling is the main agent for breaking dormancy and rest conditions in bud and seeds. Chilling inactivates growth-inhibiting substances and stimulates the production of growth-promoting substances, eventually breaking dormancy. Growth promoters found at the end of dormancy are indole-acetic acid, indole–3–acetonitrile, indole–3–pyruvic acid, gibberellic acid, and many other gibberellin-like and unknown compounds. Growth inhibitors like acid inhibitor B-Type, coumarin, parascorbic acid, naringenin, and other unknown substances have been found during deep dormancy. Chemical regulators are also effective in breaking dormancy.

Dormancy can also be prolonged by some chemical regulators. These include methyl esters of naphthalene acetic acid, which were tried for root and tuber crops, and to prevent sprouting of nursery plants; maleic hydrazide (MH), which prevents sprouting of potatoes and onions; and salt of NAA, which prolongs dormancy of fruit trees. MH–30, used at approximately one gallon per acre or 2000 ppm as a pre-harvest foliar spray two weeks before harvest, has revolutionized the handling and storage of onions and potatoes for domestic and foreign trade. Table 10.6 lists some chemical substances used to regulate dormancy.

Table 10.6 Chemicals used to regulate dormancy

Growth regulator	Dose	Remarks
Ethephon	10–500 ppm	Prolongs dormancy (delays sprouting)
Thiourea	.5–3% solution	Soaking seeds for 24 hours stimulates seed germination
Kinetin	100 ppm	Soaking seeds for 3 minutes stimulates germination of some high–temperature dormant seeds like lettuce
Gibberellic acid (GA_3)	200–500 ppm	Overcomes physiological dormancy in various kinds of seeds.

f. Regulation of time of maturity of fruits. Some auxins have been found to hasten maturity. 2,4,5–trichlorophenoxypropionic acid (2,4,5–TP), when applied as a spray at the beginning of pith hardening, has hastened maturity of apricot fruit; this effect is further enhanced by irrigation. Application of 2,4–D in 1% carbowax to Bartlett pears at the immature stage resulted in an increase in the rate of ripening. Gibberellic acid (GA_3) at 20–40 ppm, on the other hand, delayed the maturity of lemon, lime, grapefruit, Valencia orange, and Satsuma mandarin fruit, thus providing greater flexibility in harvesting and marketing.

g. Widening crotch angles in fruit trees. The crotch angle in fruit trees seems to be determined by the amount of hormones produced in the growing parts and reaching the crotch by downward movement. It has been shown for Delicious apples in the USA and for Victoria plums in England that indolebutyric acid (IBA, 250–4000 ppm) applied to the terminal part of decapitated plants in the nursery increased the crotch angle proportionately to the concentration of IBA.

h. Effect on storage life of fruits. Application of gibberellic acid to foliage after the fruit attains its full size and mature colour has increased the storage life of persimmons. Post–harvest storage life can be increased by 11 days with 200 ppm, 9–10 days with 100 ppm, and 6–8 days with 50 ppm. The quality of the sprayed fruit is not adversely affected. Moreover, the use of GA not only increases the post–harvest storage life but also extends the harvest period.

i. Effect on flowering and sex expression. Gibberellin may replace either the cold or the photoperiod requirements for flowering and seed production in many long–day vegetable and flower crops like radishes, lettuce, cabbage, spinach, petunias, and chrysanthemums. The initiation of flowering without a cold treatment is also done with GA application for biennials like celery, cabbage, turnips, sugar beets, and daisies.

In cucumbers, GA (1000–1500 ppm) stimulates formation of male flowers on all-female (gynoecious) cucumber plants, thus making possible the production of hybrids. A single application of 500–1000 ppm of SADH 30–40 days after full bloom increases flowering in pears and apples. Use of 1000 ppm Cycocel 40–50 days after full bloom has a similar effect on pears.

j. Controlling dwarfism. Sprays of 25–50 ppm gibberellin can partially control the dwarfing effects of virus yellows on red tart cherries. Fruit quality is also improved. Growth retardants like Cycocel and Alar have been found effective in controlling the size of some ornamental flowering plants.

k. Regreening of oranges and lemons. The appearance of the yellow colour associated with aging in Washington Navel oranges and lemons may be delayed and green colour prolonged by spraying them with 15–20 ppm gibberellin. The rind can be kept in a juvenile state, thus making it possible to shift the harvest date of lemons from early to late spring according to market demand.

10.7.2 Preparation of spray solutions and pastes

The most common method of applying growth regulators is in dilute solutions, either used as foliar spray or as a dip solution. There are also other methods such as dusting or mixing in lanolin paste. The practical problem is to prepare the solutions in the correct concentrations, as parts per million or as percentages. The simplest way to do these calculations is given here.

Parts per million (ppm). A 1 ppm concentration solution is prepared by dissolving 1 mg of the substance in 1000 cc (one litre) of distilled water. The same proportion is used for any required dosage. For example, to prepare a 25 ppm solution of IAA, 25 mg of IAA is dissolved in 1000 cc of distilled water to get the required strength.

On percentage basis. A 1% solution means 1 gram of the material dissolved in 100 ml of solvent.

The relationship between solution strengths given in ppm and percentage is shown below.

1%	= 10,000 ppm	= 10,000 mg/litre	or 10 g/litre
0.1%	= 1,000 ppm	= 1,000 mg/litre	or 1 g/litre
0.01%	= 100 ppm	= 100 mg/litre	or 0.1 g/litre
0.001%	= 10 ppm	= 10 mg/litre	or 0.01 g/litre
0.001%	= 1 ppm	= 1 mg/litre	or 0.001 g/litre

STUDY QUESTIONS

1. Explain the different systems of cultivating gardens. Discuss the advantages and disadvantages of each system.
2. It is not recommended to cultivate berseem in orchards. Why? Recommend other winter and summer intercrops, if any are appropriate, for orchards of different fruits.
3. Explain the importance of water and its functions in the life of horticultural plants.
4. What are the critical periods for water supply during the life span of important summer and winter vegetables?
5. What important management practices should be followed with vegetables from after sowing until harvesting?
6. Pruning the top promotes vegetative processes and retards reproductive processes. Explain why this is so, and give examples.
7. How does thinning out differ from heading back? Explain each, discussing its advantages and disadvantages.
8. Discuss the fruit–bearing habits and amount of annual pruning desirable for important fruits of your region.
9. What are the different forms to which fruit trees are trained? Explain each with its merits and demerits.
10. Modified–leader type trees produce strong crotches and a more substantial frame than the open centre type. Explain why this is so.
11. Explain the important objectives of disbudding/pinching of certain flower crops. Give appropriate examples for each point you discuss.
12. How does pruning of mature citrus tree differ from that of peach trees? Which citrus fruits need annual pruning, and why?
13. What are the objectives of pruning in ornamental horticulture? Explain with appropriate examples.
14. How does root pruning reduce the total amount of plant growth?
15. Define a weed. Explain why weeds markedly decrease yield.
16. What are herbicides? State the main difference between non–selective and selective herbicides.
17. Write a comprehensive note comparing chemical control with conventional control of weeds.

18. Distinguish between hormones, growth regulators, auxins, gibberellins, and kinins.
19. Explain the role of growth regulators in pre–harvest fruit drop. What do you recommended to control pre–harvest fruit drop?
20. How do growth regulators control biennial bearing and dormancy in fruit plants?
21. What are the merits and demerits of various methods of irrigation in vegetable crops?
22. Explain briefly various systems of irrigation of fruit plants with appropriate diagrams, where possible.

REFERENCES

Denisen, E. L. Principles of horticulture. 2nd ed. New York: Macmillan Publishing Co.; 1979.

Edmond, J. B.; Senn, T. L.; Andrews, F. S.; Halfacre, R. G. Fundamentals of horticulture. New Delhi: Tata McGraw–Hill; 1977.

Gardner, V. R. Principles of horticultural production. East Lansing: Michigan State Univ. Press; 1986.

Ginai, M. A. Floriculture and landscape gardening. Lahore: G. A. Ginai; 1980.

Halfacre, R. G.; Barden, J. A. Horticulture. New York: McGraw–Hill; 1979.

Hayes, W. B. Fruit growing in India. Allahabad: Kitabistan; 1960.

Hussain, A. Vegetables of Punjab (Urdu). Lahore: Agri. Information Division, Govt. of Punjab; 1983.

Laurie, A.; Ries, V. H. Floriculture: fundamentals and practices. 3rd ed. New York: McGraw–Hill; 1975.

Leopold, A. C.; Kriedemann, P. E. Plant growth and development. 2nd ed. New York: McGraw–Hill; 1975.

Nickell, L. G. Plant growth regulators: agricultural uses. New York: Springer Verlag; 1982.

Rajput, C. B. S.; Sri Hari Babu, R. Citriculture. New Delhi, Ludhiana: Kalyani Publishers; 1985.

Reuther, Walter, ed. Weed control in citrus. Citrus industry, Volume 3. Berkeley: University of California Press; 1973.

Ware, G. W.; McCollum, J. P. Producing vegetable crops. 3rd ed. Danville, Illinois: Interstate; 1980.

Outline of Chapter 11

11. INSECT PEST AND DISEASE MANAGEMENT

Ali Asghar Hashmi[1]

OBJECTIVES

A student who has studied this chapter should be able to:

- Recognize the symptoms of the most common and serious diseases and pest infestations and know the most effective control methods
- Use this chapter as a reference to diagnose other disease and pest problems and decide on the best treatment

11.1 Introduction

Research efforts in the production technologies of horticultural crops can be highly rewarding. If the harvest is free from insect pests and diseases, the produce not only appeals to the eye but also fetches high prices in the market. However, because of the severe incidence of pests and diseases in the developing countries, many fruit and vegetable crops cannot be cultivated on a commercial scale. Production of fruits and vegetables in Pakistan is also limited by this problem.

In the case of multiple–organism invasion, the problem of diagnosing pest and disease attacks is complicated. While it may be relatively easy to diagnose damage by an insect because of its visibility, the damage done by diseases is difficult to diagnose because the pathogen is not visible. Also, there are many pathogen species which cause differing pathological responses in different varieties.

Familiar diagnostic categories of disease symptoms in plants include:

- Necrotic symptoms: rot, canker, spot, blight, yellowing, and wilting
- Hypoblastic symptoms: dwarfing, rosetting, mosaic, and chlorosis
- Hyperplastic symptoms: galls, blast, and witch's broom.

The ease with which a disease can be diagnosed depends upon the pathogen and plant part infected, plus several other factors. For example,

---

[1]Dr. Hashmi is Principal Scientific Officer, National Agricultural Research Centre, Islamabad.

many fungal diseases are easier to diagnose than viral diseases because viral diseases may mimic the symptoms produced by nutrient imbalances, pesticide damage, pollutants, or other causes. Phloem–based diseases are more difficult to diagnose than diseases which are manifested on the plant surface.

Disease diagnosis in Pakistan is even more difficult because of lack of sophisticated laboratory equipment and personnel trained in the diagnosis of microbial diseases. There is also less interest in funding basic research on diagnostics than in research for chemical control.

Another problem in Pakistan is the lack of an organized surveillance and forecasting system for pests and diseases. Fixed–plot and roving pest and disease surveys are being carried out on a variety of commodities, primarily with the aim of monitoring the situation so that farmers can be advised to adopt suitable control measures. However, only visible damage signs have been consistently recorded to gauge the various pests. No plan for studies on recurrent patterns of insect or pathogen attack for diagnostic purposes has been made. Research work on crops has mainly been oriented towards the testing of chemicals, hence very little information about integrated pest management is found.

Although a large number of insects and diseases are described in the literature, little is known about the extent of the losses actually caused by them. Thus the status of many as pests is hard to determine.

The purpose of this chapter is to familiarise the reader with some of the major insect pests and diseases of fruits and vegetables in Pakistan. Since all agriculture students take basic courses in entomology and plant pathology, only damage or symptoms and control options have been listed for each organism.

11.2 Fruits

11.2.1 Insect pests

11.2.1.1 Citrus

1. Citrus psylla, *Diaphorina citri*

DAMAGE: Citrus psylla nymphs suck sap through their sharp stylets. Millions of them may infest the leaves and twigs of a single citrus plant. In addition, they also produce honeydew, promoting the growth of black fungus that retards photosynthesis. The attack also stops the growth of leaves and twigs. In addition, flower and leaf buds may wither and die, and fruits remain small and juiceless.

CONTROL: (1) Cultivate plants properly. Do not grow citrus hedges around citrus orchards as they act as reservoirs of the pest. (2) Encourage natural enemies of the pest: *Tetrastichus radiatus* W., which is parasitic on its nymphs; larvae of *Chrysopa* sp., maggots of *Syrphus balteatus* (Degean), lady-bird beetles, and certain mites and spiders which feed on its adults and nymphs. (3) Spray with Malathion/Phosphamidon/Methylparathion/Dimethoate 0.05%.

2. Cloudy–winged whitefly, *Aleurolobus citrifolii*

DAMAGE: Leaves infested by sap–sucking nymphs and adults curl over and fall. Adults avoid light and stay on the underside of leaves and the north side of trees. A sooty mould develops on the honeydew secreted by the nymphs. Infested trees produce scanty blossoms, most of which are shed. The few fruits produced are small and of poor quality.

CONTROL: (1) Cultural practices recommended are the same as for citrus psylla. (2) Encourage predaceous beetles which feed on eggs and nymphs: *Brumus suturalis, Catana parcesetoa, Menochilus sexmaculatus*, and *Scymnus* spp. (3) Spray with any of the following: Lime–sulphur wash (1:2:16) before flowering; Phosphamidon/Methyl Demeton/Dimethoate 0.03%; Malathion 0.05%; resin compound diluted with 6 parts water; or resin soap, dissolved in 50 parts water.

3. Citrus/grape mealybug, *Planococcus citri*

DAMAGE: Nymphs and females extract cell sap from the plants, and excrete honeydew on which fungus develops. In addition to reduced general vigour, necrotic spots 2–3 mm long form on plant parts and interfere with their functioning. In severe infestations, no fruits are set, or they drop prematurely.

CONTROL: (1) Wrap polythene sheeting around the main trunk of the tree, since the insect overwinters on the lower portion of the trunks. (2) For effective control, spray with Phosphamidon/Ethion 225 ml active ingredient (a.i.) mixed with one gallon summer oil in 100 gallons of water.

4. Citrus mealybug, *Pseudococcus filamentosus*

DAMAGE: The citrus mealybug feeds on the cell sap, thereby weakening the plant. It also excretes honeydew which promotes the growth of black mould and thus reduces photosynthesis. Leaves become pale, wilt, and eventually die. No fruit are set when infestation is heavy.

CONTROL: Spray with Diazinon/Parathion/Phosphamidon 0.03% or Malathion/Dimethoate 0.05%.

5. Citrus red scale, *Aonidiella aurantii*

DAMAGE: Scale insects weaken plants by feeding on their sap and injecting a toxic substance in their tissues. They infest leaves, branches, and fruits, appearing as small dots because of their many shields, and encrusting them completely in heavy infestations. Leaves become pale and eventually dry along with twigs.

CONTROL: Remove and burn infested leaves and twigs. Spray with Phosphamidon/Ethion 225 ml a.i. mixed with one gallon summer oil in 100 gallons of water. If summer oil is not available, repeat sprays several times.

6. Citrus leaf miner, *Phyllocnistis citrella*

DAMAGE: Larvae of citrus leaf miner feed on the leaf parenchyma of citrus making zigzag galleries inside the leaves, which appear silvery from outside. Brownish patches form on older leaves, permitting citrus canker to invade the tissues. Infested leaves twist or fold but remain on the plant for a considerable time, allowing the infestation to spread to fresh leaves. Leaf miners reduce photosynthesis, thereby reducing plant vigour. Nurseries are most susceptible to leaf miner attack.

CONTROL: (1) Remove and burn infested leaves and twigs. (2) Chemical control is same as for citrus yellow and red scale when infestation is severe.

7. Lemon butterfly, *Papilio demoleus*

DAMAGE: Larvae feed on citrus leaves and terminal shoots, leaving only the mid-ribs. Defoliation reduces photosynthesis, thus heavily infested trees do not set fruit. Nurseries are sometimes completely defoliated. It is considered the worst leaf-eating pest of citrus nursery plants in the Punjab.

CONTROL: (1) In small orchards, pick and kill the caterpillars. (2) Spray with Malathion/Methylparathion 0.05%.

8. Fruit-piercing moth, *Ophideres* spp.

DAMAGE: These moths suck juice from fruit. Also, bacteria and fungi invade through the punctures made by them. Areas around punctures become pale, and ultimately the fruit turn prematurely yellow.

CONTROL: (1) Remove weeds and natural vegetation around the orchard. (2) Unripe fruit: attract and kill moths with a bait containing 1 kg crude sugar (*gur*), 60 g vinegar, and 60 g lead arsenate in 10 litres of water. Ripe fruit: ward off moths by producing smoke for two hours after sunset.

9. Citrus mite, *Paratetranychus citri*

DAMAGE: Nymphs and adults feed on leaves, green fruit, and green bark of citrus, giving them a speckled appearance. Heavy infestation may completely defoliate the plants, especially at the nursery stage. Infested fruits become prematurely yellow and remain small.

CONTROL: (1) Encourage predatory mites such as *Amblydromella plebius*, *Gnorimus tabella*, and *Euseius mediocris*. (2) When infestation is severe, spray with acaricides: Lime–sulphur wash (1:2:16); 1% wettable sulphur; Dimethoate 0.03%; Dicofol/Malathion 0.05%.

10. Citrus nematode, *Tylenchulus semipenetrans*

DAMAGE: Citrus nematode reduces terminal growth and vigour, causes leaves to yellow and dry, reduces fruit yield, and degrades its quality.

CONTROL: Just before flowering, and also after fruit picking, fumigate soil with Nemagon at 30 litres/ha, after diluting it with water.

11. Lance nematode, *Hoplolaimus indicus*

DAMAGE: In addition to citrus, this nematode also attacks tomatoes, brinjal, maize, and sugar beet. Higher populations of the nematode occur in citrus trees 10 years and older. The population peaks in April.

CONTROL: For new plantations, use nematode–resistant rootstock. e.g. *Poncirus trifoliata.*

11.2.1.2 Mango

1. Mango mealybug, *Drosicha stebbingii*

DAMAGE: Nymphs and wingless females suck the cell sap. Tender shoots and flowers dry, and young fruits lose their juice and eventually drop off. No fruits are set when infestation is serious.

CONTROL: (1) Destroy eggs laid under infested trees after scraping the soil 15 cm deep to expose the eggs. Burn all dry leaves and twigs in May–June. (2) To prevent nymphs from crawling up the stem, apply 8–cm–wide sticky bands, or polythene sheets, about a metre above ground, towards mid–December. Burn nymphs congregated below the bands, or shake them into kerosene mixed with water. (3) Spray nymphs below bands with Methylparathion 0.05%; and those which crawl over the bands, with Methylparathion/-Phosphamidon 0.1%. For full–grown females use 0.1–0.2% Methylparathion. Spray bugs on soil or descending tree trunks with kerosene.

2. Mango fruit fly, *Dacus dorsalis*

DAMAGE: Mango fruit flies lay eggs inside the fruit. The eggs hatch into maggots which feed on the pulp and make it unfit to eat. At first the infestation is not visible on the fruit surface; later brown spots appear on the yielding surface of the infested fruit.

CONTROL: Destroy all infested fruit to reduce future pest populations. Spray frequently with a mixture of 0.05% Malathion and 1% molasses, or Dimethoate 0.03%, until a fortnight before picking.

3. Mango bud mite, *Aceria mangiferae*

DAMAGE: Mango bud mites suck the sap of buds, killing them when the mite population exceeds the threshold. The mites cause the greatest damage to young trees, attacking the terminals of their branches and feeding on bracts, producing brown to black patches. Damage to buds causes severely distorted new growth manifested in bushy appearance of seedlings, vegetative distortion, or dieback of blossoms. On older trees, the mite may produce sporadic coarse galls in the branch forks containing large numbers of mites. All varieties of mango are susceptible to the bud mite.

CONTROL: (1) Never use bud wood from infested trees. To avoid overlooking the microscopic mite, dip bud wood in a solution of Omit 0.04%. (2) Replace extensively damaged trees. (3) Cut out and burn galls on older trees, sealing the wounds to prevent infection. (4) Apply full cover spray with Dicrotophos/Dimethoate/Formothion 0.1–.2% emulsion in summer.

4. Mango scale, *Phenacaspis cockerelli*

DAMAGE: This scale is usually dispersed by birds and bats in their feathers and fur. Severe infestation completely coats the underside of leaves, and they become yellow and die. Flower spikes and fruits may also be infested.

CONTROL: For minor infestations, remove and burn infested leaves and twigs. For severe infestations, spray infested plants with Phosphamidon/ Ethion 225 ml a.i. with one gallon summer oil in 100 gallons of water; or with Metasystox/Phosphamidon 0.05–0.07%.

5. Mango stem borer, *Batocera rufomaculatus*

DAMAGE: Adults of mango stem borer feed on leaf petioles and bark of young shoots; and the grubs on the wood. When they are confined to the heartwood the tree usually survives, but when they traverse the sapwood and phloem, the trees always dry up and die due to a disruption of their water and food transport systems. In severe attacks, numerous large circular holes appear on the trunk or branches.

CONTROL: Plug the borer holes with cotton soaked in kerosene or petroleum, and plaster with mud. Or, inject Methylparathion 0.2% into the holes and plug with mud.

6. Mango hoppers, *Idioscopus clypealis* and *Amritodus atkinsoni*

DAMAGE: Nymphs and adults suck sap from the inflorescence and young shoots, causing the inflorescence to wither, turn brown, and fall off. Infested trees, therefore, do not bear much fruit. Their growth is also retarded.

CONTROL: Spray with Carbaryl 0.1% or Methylparathion/Phosphamidon/Endosulfan/Dimethoate/Malathion 0.05% in the morning when the adults are less active.

11.2.1.3 Date palm

1. Date palm scale, *Parlatoria blanchardii*

DAMAGE: Larvae feed on leaves and fruits by sucking their sap through their stylets. Leaves may be killed if infestation is heavy and fruit rendered unsightly thereby reducing its marketability.

CONTROL: Spray with Phosphamidon/Ethion 225 ml mixed with one gallon summer oil in 100 gallons of water.

2. Red palm weevil, *Rhynchophorus ferrugineus*

DAMAGE: Grubs attack date palm crowns. Outer leaves of infested trees become chlorotic and die first, followed by inner leaves. The weevils later tunnel into the trunk, weakening it and rendering it liable to break.

CONTROL: To reduce pest population, destroy stumps and date-palm debris in orchards.

3. Rhinoceros beetle, *Oryctes rhinoceros*

DAMAGE: Adults bore into unopened leaves and feed on their sap. V–shaped cuts on mature leaves are signs of rhinoceros beetle attack. The beetles themselves are not visible during the day since they are night and early morning feeders.

CONTROL: (1) Cut, split, dry, and burn all dead palm trunks. (2) Remove all dead palms at not more than a metre above ground. (3) Dig out, split, and burn stumps more than a metre in height. (4) Burn all other debris in orchards. (5) Encourage natural enemies, e.g. *Seolia procer*, a grub–parasite. (6) Spray BHC.

11.2.1.4 Grapes

1. Grapevine leaf hopper, *Erythroneura* spp.

DAMAGE: Adults and nymphs of grapevine leaf hopper suck sap from leaves, which, under heavy infestation, die after becoming chlorotic or brown. Loss of foliage reduces photosynthesis thereby decreasing the size and quantity of fruit. Tiny white spots on leaves indicate the attack of this pest.

CONTROL: Find insect parasites and predators on this pest, e.g. the fungus *Entomophthora sphaerosperma*, and encourage them.

2. Grapevine thrips, *Rhipiphorothrips cruentatus*

DAMAGE: Adults and nymphs generally feed on the underside of grapevine leaves, scraping the leaves with their stylets and sucking the oozing cell sap. Infested leaves turn whitish, appear withered, turn brown, curl up, and drop. Infested vines do not bear, or the fruit drops prematurely. Any fruit produced is inferior.

CONTROL: Spray with Dimethoate/Endosulfan/Phosphamidon 0.03%, Nicotine sulphate 0.04%, or Malathion 0.05%.

3. Grapevine leaf roller, *Sylepta lunalis*

DAMAGE: First to third instar larvae skeletonize leaves by eating the lower epidermis. Each larva of the fourth and fifth instars rolls a leaf around it.

CONTROL: Spray Malathion 0.05%, Dimethoate 0.04%, or Phosphamidon 0.04%.

4. Convolvulus hawk moth, *Herse convolvuli*

DAMAGE: Larvae feed on grapevine leaves, each consuming 10–20 leaves daily, defoliating the vines completely.

CONTROL: If infestation is light, pick and destroy larvae. To reduce pest population, use light traps to destroy adults. For severe infestation, spray with Malathion 0.05%, Dimethoate/Endosulfan 0.03%, or Carbaryl 0.01%.

5. Hawk moth, *Theretra alecto*

DAMAGE: In severe infestations, larvae may defoliate entire vineyards, each larva eating 10–15 leaves a day.

CONTROL: Same as for convolvulus hawk moth.

6. Grapevine beetle, *Sinoxylon anale*

DAMAGE: Adults bore into dormant vines about a foot above ground level. Their galleries radiate from the centre, ending in several exit holes at the stem periphery. Both adults and grubs consume the xylem and phloem tissues of the vine, killing the part above the attacked zone by disrupting the flow of water and nutrients.

CONTROL: Carefully prune and burn infested parts of the vine, also removing loose bark. Clean cultivate the orchard.

7. Grapevine girdler, *Sthenias grisator*

DAMAGE: Adults girdle green branches 1.25–2.5 cm thick, about 15 cm to 3 m above ground level, before laying their eggs. This kills the vines above the girdle.

CONTROL: (1) Remove girdled branches to suppress the pest population. (2) Pick adults at night using torches, and destroy. (3) Spray plants at ground level with BHC 0.2%.

11.2.1.5 Banana

1. Banana weevil, *Cosmopolites sordidus*

DAMAGE: Larvae destroy the corm tissue of the plants, killing seedlings when they reach the growing point.

CONTROL: (1) To destroy feeding and sheltering places, chop up pseudostomas from which banana bunches have been cut, and scatter them to rot. Cut pseudostomas close to the ground. (2) Plant only clean and uninfested suckers. (3) Cut old stems from rhizomes at ground level, chop into small strips and use as mulch. (4) Cover cut ends of rhizomes with compacted soil. (5) To trap adult beetles, cut old pseudostomas and place in strategic locations with cut surfaces downwards. Collect and destroy once a week. (6) Introduce the predacious beetles *Plaesius javanus* from Indonesia, and *Dactylosternum hydrophiloides* from Malaysia. (7) Apply Aldrin and Dieldrin as dust around pseudostoma bases and to cut surfaces of rhizomes before covering with soil. (8) Dip suckers suspected of being infested into Dieldrin solution before planting.

11.2.1.6 Ber

1. Ber fruit fly, *Carpomyia vesuviana*

DAMAGE: These flies puncture the fruit and lay their eggs inside it. Fruit growth slows around the puncture and the fruit is deformed. Infested fruits rot around the stones and emit a putrid smell. The flies prefer the fleshy varieties of ber. Late–maturing fruits in particular are riddled with maggots.

CONTROL: (1) Destroy wild ber bushes in the vicinity of grafted ber trees to keep down the pest population. (2) Spray trees with a mixture of Malathion 0.05% and molasses 1%, or Dimethoate 0.03%.

2. Ber beetle, *Adoretus pallens* and *A. nitidus*

DAMAGE: Ber beetles feed on leaves, cutting round holes in them at night. In severe infestations all the trees may be completely defoliated. Such trees do not set fruit due to the cessation of photosynthesis.

CONTROL: Attract beetles by light and drown them in kerosenized water.

3. Ber mite, *Larvacarus transitans*

DAMAGE: These mites may infest any part of ber twigs. The first manifestation of attack is the appearance of minute galls on the twigs. These grow and harden with time, attaining a size of about 9 × 4.5 mm. Each gall houses a mite, busy sucking sap from the host with its chelicerae. The fruits set by infested trees are small and crumpled.

CONTROL: Spray ber trees with Dicrotophos/Dimethoate/Formothion 0.1–0.2% emulsion, or Omit 0.04% when the mite is migrating to new sites.

11.2.1.7 Apple

1. Apple lacebug, *Stephanitis pyrioides*

DAMAGE: Nymphs and adults feed on apple leaves which become discoloured and lose their photosynthetic efficiency.

CONTROL: (1) Use coccinellid adults and larvae as predators. (2) Spray with Gusathion–M, Lorsban, Malathion, Mitac, Novathion, Perfekthion, Sevin, or Sumithion.

2. Apple wooly aphid, *Eriosoma lanigerum*

DAMAGE: These aphids attack stems, twigs, and roots near the surface. Attacked apple trees are weakened as the aphids suck their sap, and nursery stock often succumb. Infested trees can be recognised by white wooly tufts on the stems and twigs, and cracks in the bark in summer. Fruit from heavily infested trees are small, malformed, and without flavour.

CONTROL: (1) Use parasitoid *Aphelinus mali* and predators of the families Chrysopidae, Hemerobidae, Syrphidae, and Coccinellidae, in biological control. (2) Remove aphids early in the season. (3) As a prophylactic measure carefully examine nursery stock and fumigate before transplanting. Seal pruning and other wounds on the trees. (4) For light infestations spray any one of the following: Lorsban, Malathion, Metasystox–R, Perfekthion, Phosdrin, or Primor. (5) For severe attacks, spray Kilnal.

3. Aphids, *Aphis spiraecola* and *Brachycaudus helichrysi*

DAMAGE: Aphids attack leaves of apple, peach, plum, and almond. Infested leaves curl to various degrees, and fruits are frequently deformed.

CONTROL: (1) Several coccinellid, syrphid, and chamaemyiid predators keep the pest population low. (2) Spray with Anthio, Gusathion–M, Lorsban, Malathion, Metasystox–R, Perfekthion, Phosdrin, Pirimor, or Zolone.

4. San Jose scale, *Quadraspidiotus perniciosus*

DAMAGE: San José scale infests stems, branches, leaves, and fruits of apple, plum, pear, and peach. Scale–infested fruit exhibit small reddish discolourations, infested branches, and grayish deposits of various sized scales. In heavy infestations, overlapping scales may cover the bark completely. Because nymphs and adult scales suck cell sap, infested trees are weakened and may often be killed, especially if they are young.

CONTROL: (1) Encourage the following parasites and predators: *Chilocorus infernales, C. nigritus, C. rubidus, C. bipustulatus, Menochilus sexmaculatus, Prospaltella perniciosi* (Aphelinidae), encyrtids, *Encarsia perniciosi*, and other coccinellid spp. (2) In winter, thoroughly spray with 27 litres diesel oil + 6 kg copper sulphate + 1.5 kg lime, emulsified and diluted 5–6 times with water. Alternatively, spray with 3% ESSO tree spray oil emulsion, or DNOC 0.2%, as soon as tree buds start swelling. During spring and summer, preferably at nymph swarming, spray with Phosphamidon/Ethion 225 ml mixed with one gallon summer oil in 100 gallons of water. (3) The scale can also be controlled with Metasystox/Phosphamidon 0.05–0.07% spray; or suppressed with Diazinon or Methylparathion 0.05%, or Methyl Demeton 0.03% spray.

5. Apple codling moth, *Cydia pomonella*

DAMAGE: Larvae tunnel through the fruit and eat the seed. Infested fruit is small, shrivelled, and often falls even in moderately strong wind. A small hole covered with frass may be seen on its surface. The affected part of the fruit is black and rotten.

CONTROL: (1) At the moth's egg–laying time, release the egg parasite *Trichogramma chilonis*. (2) Use 6 inch–wide cloth or gunny–bag bands, impregnated with Gusathion–M, Supracide, Zolone, Lorsban, or Novathion.

6. Tent caterpillar, *Malacosoma indicum*

DAMAGE: This caterpillar defoliates apple, pear, apricot, and walnut trees. In severe infestations it even feeds on the soft bark of twigs after the leaves are exhausted.

CONTROL: (1) Destroy egg clusters in winter. (2) Mop up caterpillar tents with rags dipped in kerosene oil, tied to a pole. (3) Spray Malathion 0.5%.

7. Mite, *Cenopalpus pulcher*

DAMAGE: Infested leaves are speckled. Heavily infested ones become dull and brittle, and many are shed prematurely.

CONTROL: (1) *Typhlodromus* spp. are effective predatory mites. (2) Spray with Anthio, Danital, Ethion, Gusathion–M, Hostathion, Kelthane, Lorsban, Malathion, Metasystox–R, Mitac, Nevron, Omit, Perfekthion, or Supracide.

11.2.1.8 Peaches and pears

1. Black peach aphid, *Pteroclorus persica*

DAMAGE: Since black peach aphid is a heavy sap feeder, it attacks tree trunks or large branches, and sucks sap from the bark of the host through its stylet. Honeydew drips from the infested branches and even trickles down the branches and tree trunks, giving them an oily appearance. Since the insects are very prolific and have a short life cycle, they cause sap depletion in the infested trees.

CONTROL: (1) Destroy the colonies and egg masses by crushing with rags tied on sticks. (2) In serious infestations spray with contact insecticides such as Primicarb, Malathion, or Dichlorvos.

2. Green peach aphid, *Myzus persicae*

DAMAGE: In severe infestations this aphid weakens the trees by sucking its sap, resulting in poor fruit set and early fruit drop.

CONTROL: (1) Spray with Primicarb. If fruit is near ripening, use short residual insecticides such as Dichlorvos and Movinphos.

3. Peach–curl aphid, *Anuraphis helichrysis* and several unidentified nymphs and adults

DAMAGE: Nymphs and females suck sap from tender leaves, causing them to curl. Severely infested plants set little fruit, or the fruit shrivels and drops.

CONTROL: (1) Encourage predacious insects such as *Coccinella*, *Adalia decempunctata* L., *Menochilus sexmaculatus* F., larvae of *Syrphus* spp., larvae and adults of *Chrysopa* spp. (2) Spray Phosphamidon or Metasystox 0.03% in early spring, when the aphids hatch from overwintering eggs.

4. Pear psylla, *Cacopsylla bidens*

DAMAGE: Nymphs feed on pear leaves, arresting their growth and also injuring fruit buds, causing loss of fruit. A distinguishing feature of pear psylla attack is sooty leaves, shoot tips, and fruitlets. Pear sucker infestation is characterised by blackened shoots and foliage.

CONTROL: (1) Encourage predators such as *Chrysopa* spp., or coccinellids such as *Menochilus sexmaculatus*, *Harmonia dimidiata*. (2) Spray Gusathion–M, Lorsban, Malathion, Mitac, Novathion, Perfekthion, or Sumithion.

5. Plum/pear/apricot/olive scale, *Parlatoria oleae*

DAMAGE: The scale attacks all above–ground parts simultaneously–trunks, branches, leaves, and fruits. Infested trees weaken and die because the insects not only suck their sap but also inject toxic substances into their tissues through their saliva. Heavy infestation can be recognised by a whitish–grey coating on the branches, pinkish circular spots on leaves and fruits, and the nymphs congregated on the mid–ribs of leaves.

CONTROL: Apply winter oil in winter. Methylparathion and synthetic pyrethroids should be used during their active season.

6. Fruit flies, *Dacus zonatus* and *D. dorsalis*

DAMAGE: Larvae feed on the fruit pulp. Egg deposition punctures cause chlorotic spots on the fruit, which later becomes mushy, turns brown, and usually drops.

CONTROL: (1) Use methyl eugenol lures to attract and kill adults. Put a small cotton swab soaked with lure in a trap measuring 10" × 6" fitted with two tubes measuring 2.5" × 1.25", one on each side of the trap. Hang the trap 1–2 metres from the ground on the branches of the tree. (2) Spray Diptrex, Nexag, Nogos, or Zolone.

7. **Buprestid borer,** ***Sphenoptera lafertei*** (flat–headed peach borers, or stone–fruit borers)

DAMAGE: Young larvae bore through the bark and start feeding in the phloem and cambium, making minute irregular galleries and leaving a powdery mass. Entry points of larvae are marked by gum globules produced by the host as a reaction to the injury. The outer bark of infested trees gets loosened and the brownish frass shows through its splits. The leaves become chlorotic and brown at the tips and along the margins. Tree branches are killed as a result of girdling by the pest. Adult beetles emerge through oval holes they make in the bark and also in the wood.

CONTROL: (1) Prune and destroy infested branches. (2) Insert cotton soaked in petroleum, kerosene, or ECDT mixture into borer holes; or inject Methyl parathion 0.2%, and plaster with mud. (3) Alternatively, spray trunks and branches with Lorsban, Folidol–M or Gusathion–M in March, April, or May.

11.2.1.9 Apricots

1. **Apricot chalcid,** ***Eurytoma*** **spp.**

DAMAGE: Not visible externally.

CONTROL: (1) Collect and burn dropped fruits. (2) Spray with appropriate pesticide at emergence of adults in spring.

11.2.1.10 Walnuts

1. **Walnut weevil,** ***Alcidodes porrectirostris***

DAMAGE: Weevils feed on green twigs, leaf petioles, flowers, young fruits, and kernels, but grubs feed only on the kernels, reducing them to a useless black mass.

CONTROL: Reduce pest population by destroying fallen fruit.

11.2.1.11 Almonds

1. **Almond bark beetle,** ***Scolytus amygdali***

DAMAGE: Adults bore into new wood at the base of buds and fruit spurs. Adult females tunnel into the sapwood to make their brood chambers. Heavy infestations weaken the trees, making them susceptible to other pests.

CONTROL: Prune and burn infested wood.

2. Cherry tip borer, *Quettania coeruleipennis*

DAMAGE: The cherry tip borer infests almond, apricot, plum, and pomegranate. Grubs bore into shoots or branches and work downwards, consuming the tissues en route. The residual fluid oozes out of the holes the grub makes on the sides of the branches as it moves down. The hollowed–out branches ultimately die.

CONTROL: (1) Destroy infested shoots and branches. (2) Plug holes made by grubs with cotton soaked in kerosene, petroleum, or EDCT mixture; or, inject holes with Methylparathion 0.2%. Plaster with mud. (3) As soon as beetles emerge in May, spray Malathion 0.05%, Dimethoate/Endosulfan 0.03%, or Carbaryl 0.1%, at 3–4 week intervals.

3. Termites

DAMAGE: Termites feed on cellulose which they normally obtain from dead plant parts. However, they also occasionally attack living plant parts, when they are near their normal food.

CONTROL: Remove all dead woody material from the orchard. Do not use green manure or raw farmyard manure. Irrigate regularly.

4. Fruit flies

DAMAGE: Maggots feed on fruit pulp. At first, infested fruit does not show any sign of attack, but later circular and depressed round patches on the skin signal infestation. The attacked fruit turns orange prematurely and gradually begins to fall.

CONTROL: (1) Collect all fallen infested fruit and bury 1.5 m deep. Hoe or plough soil under infested trees frequently to expose pupae to weather. Irrigate after mixing a suitable insecticide, e.g. sanitary fluid, in water. (2) Encourage parasites of immature stages and pupae, e.g. *Splalangia grotuisi* and *Opius longicaudatus*. (3) Spray with Trichlorphon/Dichlorvos 0.1% or Phosphamidon 0.057%.

5. Muskmelon melon fly, *Myiopardalis pardalina*

DAMAGE: The female pierces young fruit with her sharp ovipositor and lays eggs inside the fruits. The fruit is deformed and its growth arrested.

CONTROL: Collect all infested fruits from the soil, or from the trees after the crop is over, and destroy them.

6. Hairy caterpillars, *Euproctis fraterna, Euproctis lunata*

DAMAGE: Young caterpillars feed gregariously on the underside of leaves; adults feed singly. Infested trees have sieve–like dry leaves with jagged edges.

CONTROL: Spray Malathion/Endosulphon 0.03% on young caterpillars, and Methylparathion 0.05% on mature ones.

7. Red spider mite, *Tetranychus telarius*

DAMAGE: These mites spin webs and feed on the underside of leaves. The webs retard leaf growth and the feeding of the mites gradually kills them. Consequently fruit set is poor and yield low.

CONTROL: Spray lime–sulphur wash (1:2:16) Dimethoate 0.02%, Malathion, Dicofol 0.05%, or Omit 0.03%.

11.2.1.12 Pomegranates

1. Anar butterfly, *Virachola isocrates*

DAMAGE: Larvae feed inside the fruit. Butterfly attack can be recognised by a hole on the rind of the fruit, surrounded by a discoloured area.

CONTROL: (1) Destroy all infested fruit, both fallen and on the trees. (2) To deal with heavy infestation, spray twice with Methylparathion 0.05% at 10-12 day intervals.

11.2.1.13 Miscellaneous pests

1. Quetta borer, *Acolesthes sarta*

DAMAGE: Mainly a pest of forest trees, the Quetta borer also infests apple, almond, cherry, quince, walnut, and peach. Grubs feed inside the wood and make numerous galleries inside it. The bark loosens from the infested parts and falls to the ground. It is lined with frass.

CONTROL: (1) Locate galleries and clean them with a long wire. Insert cotton soaked with a persistent insecticide deep into each gallery on the trunk surface.

2. Spotted spider mite, *Tetranychus urticae*

DAMAGE: Spotted spider mites attack fruits, vegetables, and ornamentals. They suck cell sap from leaves and spin fine webs on them, giving them an unhealthy mottled appearance. Apple leaves turn rust coloured and dry up. Since buds are also attacked, severely infested trees do not flower normally the following spring.

CONTROL: (1) Encourage predaceous mites and other insects which normally keep spotted spider mite population low. (2) Control serious infestations with miticides, Chlorbenside, Chlorfenson, Tetradifor, Tetrasul, Dicofol, Chlorbenzilate, Binapyeryl, Dimethoate, Phosphamidon, or Thiometon.

3. Red spider mite, *Panonychus ulmi*

DAMAGE: This mite attacks pome and stone fruits, piercing the leaves and sucking leaf sap. Infested leaves appear speckled and turn bronze in colour; consequently their photosynthetic efficiency greatly decreases. Reduced photosynthesis decreases both flower bud formation and fruit size.

CONTROL: Apply Tetradifor, Triadimefon, or Cyheatin.

11.2.2 Diseases

11.2.2.1 Citrus

1. Withertip (CAUSAL ORGANISM: *Colletotrichum gloeosporioides*)

SYMPTOMS: Withertip causes drying up of twigs from the top downwards. The disease is therefore also known as dieback. The dry twigs later become silvery and are studded with the fruiting bodies of the fungus which appear as black dots. The disease progresses from small twigs to branches, causing their death. If left unattended, it may kill the host.

CONTROL: (1) Prune and burn dry twigs and branches. (2) Apply 1 kg ammonium sulphate with 10 kg well–rotted FYM per tree. (3) In addition, spray trees with 4:4:50 Bordeaux mixture. Cupravit, Copper A, and Perenox 0.2% sprays are also useful.

2. Wilt (CAUSAL ORGANISM: *Fusarium* spp.)

SYMPTOMS: Characteristic damping-off symptoms in seedlings; in older plants there is general decline in vigour resulting in leaf chlorosis and premature fruit drop.

CONTROL: Grow seedlings in seed beds which have tested negative for the presence of *Fusarium*. Otherwise, disinfect soil before sowing seed by applying 1 part formaline in 320 parts water to soil around trees to an area equal to the area of their crown. Then cover with moist burlap for two hours.

3. Root rot (CAUSAL ORGANISM: *Rhizoctonia bataticola*, *Rhizoctonia solani*)

SYMPTOMS: Rotting of roots. Cracking of bark near soil level. Exudation of gum. Chlorosis and leaf drop from infected branches. Sweet lime, lemon, and lime are highly susceptible; sweet orange and mandarin moderately so; and sour orange is resistant.

CONTROL: (1) Avoid predisposing factors: excessive irrigation, inadequate drainage, planting on saline soils, heaping soil around plant base. (2) Raise seedlings in disinfected nursery beds. (3) Do not grow fodder in citrus orchards.

4. Sooty mould (CAUSAL ORGANISM: *Capnodium citri*)

SYMPTOMS: Velvety black coating of mould on leaves, hindering photosynthesis; and on fruits, decreasing their market appeal.

CONTROL: Control sucking insects since the disease develops only on honeydew produced by them.

5. Foot rot (CAUSAL ORGANISM: *Phytophthora parasitica; P. citrophthora*)

SYMPTOMS: *P. parasitica* causes gummosis, foot rot, and root rot, but seldom afflicts tissues higher up the trunk. *P. citrophthora* causes gummosis and root rot. It afflicts the aerial parts of the trees more often than *P. parasitica* and causes brown rot more often. *Phytophthora* spp. cause damping-off of emerging seedlings. They may also attack leaves, shoots, and fruits near the ground which get splashed by infected soil. Under environmental conditions favourable to the fungus, the disease may spread to all the fruits causing them to drop. If they don't drop and are harvested, the fungi present in them may cause post-harvest decay. *Phytophthora* may also cause the rot of feeder roots which can be serious in susceptible varieties. Feeder-root rot of highly susceptible rootstocks may cause tree decline and reduction of fruit yields in mature orchards.

SUSCEPTIBILITY: The susceptibility rating of various kinds of citrus is roughly as follows: Highly susceptible: lemons, limes, sweet oranges, and grapefruit. Susceptible: Navel and other early and mid-season oranges. Tolerant: Valencia orange. Very tolerant: Tangerines and their hybrids.

The various rootstocks can be arranged on a descending scale of susceptibility as follows: Most susceptible: sweet orange and some sources of rough lemon. Tolerant to bark infection: Troy and Carrizo citranges, Rangpur lime, and most selections of rough lemon. High degree of resistance: Swingle citrumelo, alemow, and to a lesser extent, Cleopatra mandarin and sour orange. Nearly immune: Trifoliate orange.

CONTROL: Plant only *Phytophthora*-free nursery stock. To obtain such stock, citrus seed should be treated with hot water at 52°C for 10 minutes to eliminate *Phytophthora* spp. Locate nursery beds on virgin soil if possible; otherwise, steam nursery beds or fumigate with methyl bromide or metam–sodium. To prevent foot rot and root rot in the nurseries, drench soil with Metalaxyl; or spray foliage with Fosetyl–Al.

To avoid infecting *Phytophthora*-free nurseries:

- Do not plant material from infested sites.
- Do not till with implements used on such sites.
- Do not allow people or livestock from infested sites.
- Divert runoff coming from such sites.
- Do not bring infected soil into *Phytophthora*-free nurseries.

6. Greasy spot (CAUSAL ORGANISM: *Mycosphaerella citri*)

SYMPTOMS: Greasy spot infection can be diagnosed at the early stage by yellow mottled appearance of the upper side of the leaf, and a pale-orange to yellowish brown blister on the lower. Later such leaves turn dark-brown or even black and appear greasy. But most of the infected leaves drop long before this stage is reached.

CONTROL: (1) Plough in leaf litter to reduce the intensity of the disease. (2) Spray fungicides such as copper compounds, Benomyl, and Zineb on infected leaves.

7. Green mould (post-harvest) (CAUSAL ORGANISM: *Penicillium digitatum*)

SYMPTOMS: Green mould decay first appears as a soft, slightly discoloured spot. This enlarges to 2–4 cm in 24–36 hours, and soon involves the juice vesicles. White mycelia appear on the surface of the rind, and produce olive-green spores after they reach a diameter of about 2.5 cm. Infected fruit rot in a humid environment, and shrink, wrinkle, and dry in a dry environment.

CONTROL: (1) Pick and handle fruit carefully to reduce injury to the rind and consequent infection by green mould. (2) Spray orchard with Benomyl up to 3 weeks before harvest; or the fruits after they have been picked. Several other chemicals can also be used for this purpose. (3) Careful picking and handling of fruit minimises injuries to the rind and the risk of green mould. Disinfectants such as chlorine, quaternary ammonium compounds, formaldehyde, and alcohol are useful for preventing inoculum buildup.

8. Blue mould (post–harvest) (CAUSAL ORGANISM: *Penicillium italicum*)

SYMPTOMS: Tissues infected by blue mould become mushy. White powdery mycelium develops on the surface of the lesion and blue spores form on it, which may later turn brownish–olive.

CONTROL: Same as for green mould.

9. Pink disease (CAUSAL ORGANISM: *Corticium salmonicolor*)

SYMPTOMS: Pink disease primarily attacks the bark of mature trees causing necrosis and gumming. The bark gets covered with a smooth crust of pink hyphae which eventually penetrate the bark and kill it. Entire trees or their large branches may be killed by such girdling.

CONTROL: Prune and burn infested limbs. Then spray with copper fungicides or lime sulphur.

10. Tristeza (CAUSAL ORGANISM: citrus *Tristeza* virus (CTV))

SYMPTOMS: Almost all species, cultivars, and intergeneric hybrids of citrus are susceptible to CTV. Symptoms developed by the infected host depend on its species, the environment, and the severity of the virus isolate. In sweet orange, mandarin, and grapefruit on sour orange rootstock, the virus in the scion causes girdling of the rootstock, thus disrupting the flow of food to the roots. Some virus strains may cause severe stem pitting, resulting in the disruption of the stem tissue.

CONTROL: No effective control methods have been applied against this virus on a field scale in Pakistan. Quarantine and bud wood certification can help to prevent the entry of the virus into virus–free areas.

11. Citrus canker (CAUSAL ORGANISM: *Xanthomonas campestris*)

SYMPTOMS: Citrus canker bacteria cause lesions on the shoots, leaves, and fruit. Lesions on the bark girdle the shoot, killing the part above the lesion. Leaf lesions begin as small, yellowish–brown pinhead eruptions. As they age, the under-surface becomes raised and they turn into spongy light-brown pustules with depressed centres. The water-soaked margin that develops around the necrotic tissue of the spot is the most reliable diagnostic sign. Lesions on the fruit are firm, woody, and split by large cracks, making the fruit unmarketable. Trifoliate orange rootstock is very susceptible to this disease. Sour citrus is moderately susceptible. Young trees are more susceptible than older trees.

CONTROL: (1) Select rootstock and bud wood only from disease-free trees. (2) Prune and burn infected shoots. Collect, burn, or bury infected tree parts by soil cultivation. (3) Uproot and burn badly attacked trees. (4) During active growth apply copper fungicide to all rootstock, grafted plants, and mother trees.

12. Dieback (CAUSAL ORGANISM: Not known; perhaps fungi, bacteria, viruses, malnutrition; or a combination of some or all of these.)

SYMPTOMS: Shoot dieback, leaf loss, small leaves with marginal and interveinal chlorosis, reduced fruit set, and very dry dead shoots without fungal fruiting bodies are characteristic of this disease.

CONTROL: Prune and burn dead shoots. Collect debris and dead leaves and twigs from the orchard floor and burn.

11.2.2.2 Mango

1. Fruit rot/Mango anthracnose (CAUSAL ORGANISM: *Colletotrichum gloeosporioides* and *Glomerella cingulata*)

SYMPTOMS: Anthracnose attacks leaves, petioles, twigs, and fruits. Tender branch tips are the first casualty, and they start dying from the top. Oval brown spots may appear on leaves, turning necrotic under humid conditions and sometimes rupturing the attacked tissues. Petioles turn gray or black, causing the leaves to droop and dry. Twigs exhibit elongated black necrotic areas. Black spots appear on fruit which also start rotting.

CONTROL: Prune and burn diseased twigs and branches. Seal cut ends of large branches with coaltar or sulphur paste. Collect fallen leaves and twigs from the orchard floor, and burn. In February–April and September, spray orchards with 6:6:100 Bordeaux mixture; or alternatively Cupravit 1.2%, Melprex, or Cobox.

2. Root rot (CAUSAL ORGANISMS: *Rhizoctonia solani* and *Fusarium oxysporum*)

SYMPTOMS: Mango root rot first attacks the rootlets, which rot. It then spreads to the smaller roots and then the larger. As a result of the destruction of their root systems, the trees start dying from the top downward.

CONTROL: One month before blossoming, or two months after it, apply antagonistic organisms *Trichoderma* sp. and *Arachniotus* sp. (commercial names Aspergopak and Trichopak). Add these to wheat straw and mix with the soil in the root zone of infected trees. Then apply nitrogenous and phosphatic fertilizers.

3. Dieback (CAUSAL ORGANISM: *Diplodia natalensis*)

SYMPTOMS: The first indication of dieback is discolouration and darkening of the bark some distance from the tip. The dark area advances and young green twigs start withering from their bases. Their leaves turn dark brown and their margins roll upwards. The twigs and branches shrivel and die. The disease is favoured by high summer temperature (25–31°C), high relative humidity (above 80%), and rains.

CONTROL: (1) Prune and destroy infected twigs and branches. (2) Apply requisite quantity of balanced fertilizer. (3) Give three consecutive sprays to infected trees (pre–blossom, and before and after rains). Topsin–M or Daconil 2787 can be used for this purpose. If these are not available, use Bordeaux mixture (4:4:50) or Mancozeb–based chemicals.

4. Sooty mould (CAUSAL ORGANISM: *Tripospermum acerinum*, *Capnodium mangiferae*, and *Meliola mangiferae*)

SYMPTOMS: Sooty mould appears as a black film on leaves, and in severe cases, on the whole of the plant. The fungus does not enter the tree tissues, and feeds on honeydew secretions of insects on the trees. But it hampers photosynthesis and therefore weakens the trees.

CONTROL: (1) Control honeydew secreting insects. (2) In severe infection, spray 0.2% wettable sulphur.

5. Malformation of inflorescence (CAUSAL ORGANISM: not known, perhaps virus)

SYMPTOMS: The panicle is shortened and thickened, and if the attack is severe the floral branches get bunched and later turn into a black mass. In less severe attacks, the panicles are moderately reduced, causing some of the floral branches to separate out, and several small cones to form. In other cases vegetative leaves are intermingled with small cones on the panicle. The proportion of staminate and hermaphrodite flowers is greatly disturbed. In many cases only staminate flowers are found.

CONTROL: (1) Prune and burn branches with malformed inflorescences. (2) Do not use diseased branches for budding or grafting. (3) Fertilize, irrigate, and weed approriately; control other diseases. Research is underway to develop biological control methods using antagonistic organisms such as *Triochoderma harzianum*, *Aspergillus flavus*, or *Arachnites* spp., with organic soil amendments.

6. Powdery mildew (CAUSAL ORGANISM: *Oidium mangiferae*)

SYMPTOMS: Powdery mildew appears in February and March as whitish–grey mycelium on the inflorescence and young leaves. Infected flowers fail to open, consequently fruit set is greatly reduced. In severe attacks, the entire crop may fail. The fungus is favoured by warm cloudy weather, high humidity, and low night temperature.

CONTROL: (1) Collect and burn infected leaves and twigs. (2) Spray with Topsin–M or Daconil.

11.2.2.3 Grapes

1. Powdery mildew (CAUSAL ORGANISM: *Uncinula necator*)

SYMPTOMS: Powdery mildew first appears as a grayish–white powdery growth on growing tips. Later it infects shoots, stems, flowers, and berries, causing malformation and cracking. Under favourable conditions it forms dark–coloured spore bodies. Some vineyards may be attacked by the disease every year.

CONTROL: (1) Cut and burn diseased parts of the vines. Collect and burn debris on the ground. (2) Spray with Benlate (4–6 oz/100 gallons of water), Afugan, Bordeaux mixture, Perenox, Fermate, Orthophalten, Polyram, or Cobox before blossoming, after berry formation, and three weeks later.

2. Black rot (CAUSAL ORGANISM: *Phyllosticta ampelicida*)

SYMPTOMS: *Phyllosticta* attacks new growth, shoots, tendrils, petioles, and peduncles. On new growth, groups of red necrotic lesions appear. On leaves, circular, tan–coloured spots (2–10 mm in diameter) appear 1–2 weeks after infection. On young shoots, black cankers showing pycnidia appear throughout the season.

CONTROL: For Muscadine grapes, spray with Maneb or Captan after blooming, every two weeks until August.

3. Anthracnose or Bird's–eye rot (CAUSAL ORGANISM: *Elsinoe ampulina*)

SYMPTOMS: Bird's–eye rot disease produces lesions on leaves and young shoots. Leaf lesions are circular with brown to black margins. Their necrotic centres eventually drop out, giving the leaves a shot–hole appearance. On the shoots the lesions are small and isolated. They may crack and make the shoots brittle.

CONTROL: (1) Do not plant highly susceptible cultivars, especially in poorly drained soils. (2) During the dormant season, spray DNOC, DNBP, Bordeaux mixture, or lime–sulphur. (3) During the growing season, spray Captan, Folpet, or Maneb at two–week intervals from the time the shoots are 5–10 cm long. Spray within 24 hours of a hailstorm.

4. Downy mildew (CAUSAL ORGANISM: *Plasmopara viticola*)

SYMPTOMS: *Plasmopara* attacks leaves and all other green parts of the vines, forming yellowish to reddish–brown lesions. White cottony spore masses grow on the lower sides of leaves. Shoot tips thicken, curl into hooks, and bear white masses of spores, ultimately turning brown before they die.

CONTROL: (1) Drain wet soil. (2) Prune ends of infected shoots and burn. Burn infected leaves and other ground litter. (3) In vineyards susceptible to downy mildew, spray fungicides, in particular Cymoxanil, a non–systemic penetrating fungicide specific to mildew.

5. Leaf spot (CAUSAL ORGANISM: *Phomopsis viticola*, *Fusicoceum viticola*, *Cryptosporella viticola*)

SYMPTOMS: Light–green or chlorotic spots with dark centres develop on diseased leaf laminae. These may drop out creating holes in the leaves. Severely infected leaves turn yellow, then brown, and then fall. Young shoots, rachises, and petioles develop yellow spots with dark centres. These enlarge turning the infected tissues dark–brown to black, appearing as streaks and blotches.

CONTROL: (1) While planting new vineyards use healthy planting stock. (2) In diseased orchards, remove maximum possible dead and diseased material in pruning and burn it. (3) Spray Captan, Folpet, or Maneb when shoots are 1–3 cm long, and again when they grow to 6–12 cm.

11.2.2.4 Banana

1. Black tip or fingertip (CAUSAL ORGANISM: *Botryodiplodia* sp., *Gloeosporium masarum*)

SYMPTOMS: Fruit tips turn black and the fruit rots wholly or partially.

CONTROL: (1) Destroy diseased plant parts. (2) Spray 4:4:50 Bordeaux mixture or Zerlate @ 2 lb/100 gallons.

2. Pseudostem rot (CAUSAL ORGANISM: *Gloeosporium sp.*, *Botryodiplodia sp.*, *Erwinia carotovora*)

SYMPTOMS: Affected plants are discoloured and rhizomes and suckers rot. They have weak roots with dark–brown lesions and necrotic tips. The pseudostem often tips over and breaks. Badly affected plants remain stunted, and mature plants may produce small fruits which fail to emerge from the shoot tip.

CONTROL: Plant orchards using young, healthy secondary suckers or disease–free corms treated with 2% copper sulphate solution for 10 minutes, and then dry before planting.

3. Bunchy top (CAUSAL ORGANISM: A virus transmitted by the aphis *Pentalonia nigronervosa*)

SYMPTOMS: Acute primary infection of the bunchy–top virus reduces height growth to about 9–24 inches. Young infected plants are therefore stunted, and the leaves of severely infected older plants get bunched at the top. Also, the leaves are more erect. Though such plants may live for a year or two, they either do not mature at all, or do not produce fruit. Secondary infection can be diagnosed by nodular dark–green dots and dashes along secondary veins and the mid–rib on the underside of the basal part of leaf laminae, or along the leaf stalk.

CONTROL: When planting new orchards, use disease–free planting stock. Tend plantations properly, and along with tending operations, inspect each tree carefully. Mark any tree which shows symptoms of the disease. Thoroughly spray such trees as well as all trees around them, up to 2–3 rows. The diseased stool and a ring of surrounding stools must be completely destroyed by killing with kerosene, gas oil, or herbicides; by digging up and cutting into small pieces; or by burning.

4. Leaf spot (Sigatoka disease) (CAUSAL ORGANISM: *Mycosphaerella musicola*)

SYMPTOMS: The disease first appears as 1–2 mm long linear markings on the third or fourth leaf from the centre. As the spots increase in size, they coalesce, destroying a considerable part of the leaf. In some areas, petioles and mid–ribs rot due to the attack of secondary fungi and bacteria. In severe attacks, fruit remain undersized.

CONTROL: Same as for bunchy–top disease.

11.2.2.5 Papaya

1. Damping off, stem rot (CAUSAL ORGANISM: *Pythium aphanidermatum*)

SYMPTOMS: *Pythium*–attacked plants fall down and die because their stems at ground level become soft and spongy, and their terminal leaves become chlorotic, wither, and die.

CONTROL: (1) Uproot and burn diseased plants. (2) Apply Bordeaux mixture to affected parts of lightly attacked trees. (3) Drain soil. Avoid water stagnation around healthy plants.

2. Charcoal rot (CAUSAL ORGANISM: *Macrophomina phaseoli*)

SYMPTOMS: Charcoal rot destroys root and stem parenchyma. Attacked plant parts look like charcoal due to the development of carbonaceous sclerotia and pycnidia. Leaves wither.

CONTROL: (1) Burn attacked parts. (2) Apply Bordeaux mixture to attacked areas on the plants.

3. Anthracnose and dieback (CAUSAL ORGANISM: *Colletotrichum gloeosporioides* and *Glomerella cingulata*)

SYMPTOMS: Anthracnose causes blight of seedlings, leaves, shoots, and blossoms; also leaf spots, fruit rot, cankers, and dieback of twigs and branches. It may also kill the entire plant. Dieback attacks in early spring. Dark–brown bands appear usually at the bases of new shoots. The pathogen kills plant tissues in these bands, and young shoots therefore get cut off from water and nutrients in the plant body, and wither and die. Twigs and branches are girdled in the same way.

CONTROL: Control: Spray Bordeaux mixture or Cupravit at 15–20 day intervals.

4. Fruit rot (CAUSAL ORGANISM: *Rhizopus oryzae*)

SYMPTOMS: Fruit rot first appears as necrotic spots on the fruit surface. The spots enlarge and join together and whitish hyphae appear. The fruit becomes soft and inedible.

CONTROL: Store fruit at 10–15°C to increase its shelf life.

11.2.2.6 Apple

1. Ripe rot (CAUSAL ORGANISM: *Rhizopus arrhizus*)

SYMPTOMS: Characteristic symptoms are canker on fruit stalks, and the discolouration, disfigurement, rotting, and shedding of fruit. In some cases mummified fruit remains hanging on the trees, providing a primary source of infection for the perpetuation and spread of the disease.

CONTROL: (1) Prune diseased parts. Pick and burn unshed infected fruit, pruned material, and leaves, twigs, branches, and dead fruit on the ground. (2) Spray Fermate mixed with Malathion in February. Repeat at 15-day intervals from July to September.

2. Powdery mildew (CAUSAL ORGANISM: *Podosphaera leucotricha*)

SYMPTOMS: Powdery mildew attacks plants of all ages. Small whitish or grayish cobweb-like patches appear on shoots, flowers, fruits, and leaves. At first they are confined to the underside but later spread to all sides. Infected leaves curl, get brown, and fall. Infected buds fail to open. Infected fruit get distorted and covered with white mycelia.

CONTROL: (1) During the winter, prune and burn diseased plant parts and all plant debris on the ground. (2) Spray with Sulforon (2 lb/100 gallons of water), Lime–sulphur (1:40), or Benlate (4 oz/100 gallons of water): in February before buds open; after flower petals fall; three weeks after this; and after another three weeks.

3. Sooty blotch and fly speck (CAUSAL ORGANISM: *Gloeodes pomigena*, and *Schizothyrium pomi*; *Microthyriella rubi*)

SYMPTOMS: Attacked fruit have sooty black spots which can be rubbed off, revealing pale brown marks. Do not mistake sooty blotch for sooty moulds, which stand out more clearly, being usually blacker and sticky.

4. Apple scab (CAUSAL ORGANISM: *Venturia inaequalis*)

SYMPTOMS: Apple scab attacks petioles, leaves, blossoms, and fruit, but its symptoms are most visible on the leaves and fruit. Diseased leaves may have lesions on both sides. They may also be distorted, dwarfed, die, and fall. Trees may be weakened by defoliations over several consecutive years, and rendered susceptible to cold injury. On the fruit, lesions first develop around the spur end of the fruit, later spreading to other parts of its surface. Scab infection early in the season causes uneven growth, and cracking of the skin and flesh of the fruit. When infection occurs in late summer or early autumn, the fruit may develop rough, black, circular lesions during storage. Such 'pin–point' scab consists of small lesions up to 1 cm in diameter. Apple scab is more corky than sooty blotch (#3), and cannot be rubbed off.

CONTROL: (1) Select resistant varieties. (2) Spray systemic fungicides at the green–tip stage of the buds.

5. Fire blight of pear and apple (CAUSAL ORGANISM: *Erwinia amylovora*)

SYMPTOMS: The bacteria first attack the blossoms, which become water–soaked and dark–green, and later turn dark–brown to black on pear, and brown to dark–brown on apple. On movement into the leaves through their petioles, the bacteria first discolour the mid–rib, then darken the lateral veins and the tissue around them. When the terminals are infected, they wilt from the top, usually developing a crook. Initially the infected leaf and shoot tissues are water–soaked and dark–green. Later they turn brown to black.

CONTROL: (1) Inspect orchards frequently and break infected spurs or terminals. If pruning tools are used, sterilize the tools with Clorox 1:10. (2) Control sucking insects starting before the bloom stage and continuing throughout summer since they create wounds for the entry of bacteria.

6. Root rot (CAUSAL ORGANISM: *Phytophthora cactorum*)

SYMPTOMS: Trees infected by root rot either do not leaf, or their leafing is delayed. Their bases also exhibit red–brown rot. Mature trees die over several years and young trees die suddenly. Nursery stock develop red–brown rot at their bases, and their leaves become pale and drop.

CONTROL: (1) Do not plant on poorly drained soil. Plant trees on ridges, to improve drainage. Do not over–irrigate. (2) Remove and burn any affected plant in the nursery. When planting apple on sites where root rot has occurred before, drench the base of each tree with Captan at planting.

7. White root rot (CAUSAL ORGANISM: *Rosellinia necatrix*)

SYMPTOMS: Roots and stem bases of trees infected with white rot are covered with white wooly mycelium. Attacked trees may die gradually, or suddenly, after a shock like drought.

CONTROL: (1) Use resistant rootstock. (2) Do not plant apple on infected or poorly drained soils. Plant trees on ridges and irrigate and fertilize moderately. (3) Avoid intercropping. (4) Remove and burn infected material. (5) To control the disease, drench soil around young trees with Derosal or Benlate.

11.2.2.7 Peach and almond

1. Leaf rust (CAUSAL ORGANISM: *Tranzschelia pruni–spinosae* Var. *discolor* in advanced stages of infection)

SYMPTOMS: Leaf rust appears in spring. Infected leaves fall prematurely. Cankers appear on young wood and fruit become distorted.

CONTROL: Captan, Thiram, and Zineb may be used to control the disease.

2. Leaf curl (CAUSAL ORGANISM: *Taphrina deformans*)

SYMPTOMS: Attacked leaves pucker, swell, turn yellow, die, and fall. Attacked twigs swell, blossoms wither, and fruit is deformed.

CONTROL: Spray appropriate fungicide before leaf fall, and again before bud swell.

3. Bacterial spot (CAUSAL ORGANISM: *Xanthomonas pruni*)

SYMPTOMS: Bacterial spot infects leaves, fruit, and young shoots. Lesions mainly appear on the lower side of leaves as areas soaked with water. They later appear as brown to black spots and their centres often fall out and their margins develop a reddish colour. The incidence is generally most severe at the tip, killing about an inch or more of it. Heavily attacked leaves become yellow and fall. Severe defoliation early in the season weakens the trees and reduces fruit size. If fruit are infected early in the season, they develop cracks on their surface.

CONTROL: (1) Plant only resistant varieties. (2) Tend and fertilize properly.

4. Brown rot (CAUSAL ORGANISM: *Monilinia fructicola* and *M. laxa*)

SYMPTOMS: The disease infects blossoms, spurs, shoots, and fruit. Infection travels from blossoms, which wilt and become brown, to the flower cluster base, and to the spur. Mature fruit are more affected than immature ones. On fruit the disease first appears as small light–brown circular spots on the skin which expand rapidly under favourable conditions. Rotten fruit may fall down or dry on the tree.

CONTROL: (1) Spray fungicides when flowers open, and repeat at 4–5 day intervals during period conducive to the spread of the disease. (2) Spray pre–harvest fungicides three weeks before fruit ripening to prevent brown rot on ripening fruit. (3) During packing, pick and handle fruit carefully to prevent decay in storage and transit. Use clean packing crates, and remove ripe and rotting fruit.

5. **Peach scab** (CAUSAL ORGANISM: *Cladosporium carpophilum*)

SYMPTOMS: Peach scab is most often observed on fruit, though it also infests twigs and leaves. Small, circular, greenish spots appear mostly on half–grown fruit. The lesions turn black as spores are formed. If infestation is severe, the skin and flesh may crack.

CONTROL: Spray Anthracol, Captan, Dithane M–45, Liromanzeb, or Perenox after flower bud development, at 10–14 day intervals until 40 days before harvest.

6. **Shot hole** (CAUSAL ORGANISM: *Stigmina carpophila*)

SYMPTOMS: Shot hole attacks all stone fruits, symptoms of the disease varying with the host: leaf spots on cherry and plum; twig and bud blight on peach, producing cankers up to 10 mm in diameter; on apricots, bud blight and spots on leaves and fruit; on almond, leaf spot and blossom and fruit infection.

CONTROL: (1) Prune in winter and burn the pruned material. (2) The disease can easily be controlled by appropriate fungicides sprayed at the correct time. Recommended fungicides are Captan, Cobox, Cupravit, Dithane M–45, Liromanzeb, Perenox, and Tri–Miltox.

7. **Powdery mildew** (CAUSAL ORGANISM: *Sphaerotheca pannosa*)

SYMPTOMS: Powdery mildew attacks leaves, shoots, and fruit. White patches on the underside or the entire leaf, and on the shoots, and white powdery spots on the fruit are characteristic of the disease.

CONTROL: (1) Prune primary infected shoots and buds in spring and early summer and burn before fruit formation. (2) Varieties vary in the susceptibility of their fruit to powdery mildew, hence most apricot and peach varieties need treatment. Several fungicides, e.g. Bayleton 25, Antracol, Captan, or Afugan, can be sprayed to protect nursery stock and fruit from powdery mildew attack.

8. **Root rots** (CAUSAL ORGANISM: *Phytophthora cactorum, Rosellinia necatrix, Corticium rolfsii*)
   a. White rot of mature orchards
   b. Rot of fruit tree nurseries

SYMPTOMS: These organisms cause crown and root rot of mature trees which exhibit characteristic red-brown rot at their bases as well as below ground. Infected trees may die suddenly, especially if they are young, or over several years. The leaves of infected nursery stock turn pale and drop, and red-brown rot develops at the bases of their stems.

CONTROL: (1) Plant disease-free stock. (2) Do not plant orchards on poorly drained soil. (3) Plant trees on ridges for improved drainage and water moderately. (4) Remove infected trees and burn them. (5) Drench trees planted on infected sites with Captan or a copper fungicide.

11.2.2.8 Pomegranate

1. **Fruit rot** (CAUSAL ORGANISM: *Zythia versoniana*)

SYMPTOMS: Infested fruit start rotting and later the fruit crack and dry up.

CONTROL: (1) Collect and burn infected twigs, fruit, and fruit stalks from trees and the orchard floor in January–February. (2) Spray with a suitable copper-based fungicide, e.g. Perenox @ 1.5–2 lbs/100 gallons of water in July, August, and September.

11.3 Vegetables

11.3.1 Insect pests

11.3.1.1 Potato

1. **Potato tuber moth, *Phthorimaea operculella***

DAMAGE: The potato tuber moth attacks at two stages. In the early stages of plant growth, caterpillars mine or bore potato leaves and shoots, causing leaf blotches and killing the shoots. Premature leaf withering, and caterpillar excreta on leaf axils are indicative of the attack of this pest. The second stage of its attack is at tuber digging. Moths lay their eggs on the tuber, and on hatching the larvae penetrate the tuber skin and make galleries through the flesh, leaving their excreta near the entrance holes.

CONTROL: (1) Do not leave harvested potatoes in the field overnight. (2) Store in sacks. (3) Thoroughly clean storage places and spray with 1% Malathion or 0.6% Diazinon before storing potatoes. (4) Plant disease-free potato seed. Plant seed tubers at least 20 cm deep and cover with about 25 cm fine, loose soil. (5) Encourage specific parasitoids, e.g. *Copidosoma hochleri* and *Apanteles subandinus*.

2. Golden nematode, *Heterodera rostochiensis*

DAMAGE: Golden nematode-infested plants look sickly and stunted. Their lower leaves yellow and drop. Tuber size is significantly reduced.

CONTROL: (1) Do not plant successive crops of potato in the same fields; start crop rotation. (2) Do not plant potato in infested fields.

3. Mole cricket, *Gryllotalpa africana*

DAMAGE: Mole cricket is polyphagous. It lives in tunnels 2–3 feet deep in agricultural fields. It attacks seedlings of herbaceous crops, buds of sugarcane sets, and roots of crops.

4. Green stink bug, *Nezara viridula*

DAMAGE: Green stink bug nymphs and adults suck sap from leaves, causing them to shrivel.

CONTROL: (1) Adopt crop sanitation. (2) If infestation is serious, spray with carbamate or organophosphorus compounds.

5. Painted bug, *Bagrada cruciferarum*

DAMAGE: Damage and control similar to those for the green stink bug.

11.3.1.2 Cucurbits

1. Melon aphid, cotton aphid, *Aphis gossypii*

DAMAGE: Melon aphids suck leaf juice, stunting and distorting the leaves which become chlorotic and fall prematurely. Ants are attracted to the aphids to feed on honeydew, and indicate their presence.

CONTROL: (1) Do not use very high doses of nitrogenous fertilizers. (2) Encourage the predatory insects *Aphidius* sp. (Braconidae) and *Aphelinus* sp. (Chalcididae), *Entomophthora exitialis,* and *Verticillium lecanii*, which feed on the aphids. (3) In severe infestation, spray with insecticides appropriate to the crop, using recommended concentrations.

2. Red pumpkin beetle, *Aulacophora africana*

DAMAGE: Adults as well as grubs of the beetle feed on young cucurbit vegetables. Grubs bore into roots and stems causing their death. Seedling stages are particularly vulnerable. Grubs also bore into fruits which touch the ground, causing holes on the surface. Adults feed on leaves and ovaries, causing holes in leaves and withering of flowers. Withered flowers are symptomatic of attack by adult beetle.

CONTROL: (1) Dust plants with cow dung ashes. (2) For severe infestations, spray with Methylparathion Dimethoate 0.01% or Malathion 0.05%.

3. Red pumpkin beetle, *Aulocophora foveicollis*

DAMAGE: This insect attacks pumpkin, *tinda*, *ghia tori*, cucumber, and melon. Grubs bore into roots, underground stems, and fruits touching the soil. Adults bite holes in cotyledons, flowers, and foliage. Infested seedlings are killed. Growth of larger plants is retarded, which may cause their death.

CONTROL: (1) Collect and destroy beetles. (2) Clean cultivate. (3) In case of severe attack, dust with carbaryl or pyrethrum 5% when leaves are without dew. Methylparathion, Dimethoate 0.02% or Malathion 0.05% will also control the attack.

4. Hadda beetles, *Epilachna duodecastigma* and *E. vigintioctopunctata*

DAMAGE: These beetles attack brinjal, tomato, potato, and cucurbitaceous vegetables. Adults as well as grubs feed selectively on upper leaf tissues, giving the leaf a lace-like appearance. Infested leaves become brown, dry up, and fall, defoliating the plants completely.

CONTROL: Spray Malathion 0.05% or Methylparathion 0.02%.

5. Melon fruit fly, *Dacus cucurbitae*

DAMAGE: *Dacus cucurbitae* infests muskmelon and cucurbits. Female flies puncture the fruit epidermis with their sharp ovipositor and lay eggs in cavities, sealing them with a gummy secretion. On hatching, the grubs feed on the pulp, making the fruit inedible. Infested fruit also remains small and becomes yellow. Damage is most serious in melons.

CONTROL: (1) Use resistant varieties. (2) Regularly collect infested fruit and bury about a metre deep. After the harvest destroy pupae in the soil by ploughing and heavy irrigation. (3) Search for effective pupal parasitoids. (4) Employ male sterilization or methyl eugenol traps to reduce pest population. (5) Poison baiting with 0.5 kg protein hydrozylate plus 2.5 kg Malathion in

100 litres of water will attract adult flies and kill them. Spraying melon fields with Malathion 0.05% plus molasses 1% is also effective.

6. Cutworms, *Agrotis* sp.

DAMAGE: Young cutworm larvae feed on leaf epidermis. They are night feeders, cutting seedlings at or below the ground and hiding in soil cracks or holes during the day.

CONTROL: (1) Collect and destroy moths by light traps. (2) Weed and hoe frequently. (3) Irrigate.

7. Red vegetable mite, *Tetranychus telarius*

DAMAGE: This mite attacks cucurbits, brinjal, and *bhindi*. Active stages of red vegetable mite suck cell sap from the underside of leaves. Infested leaves die, and fruit set is poor.

CONTROL: Spray crop with lime–sulphur wash (1:2:16), or dust with 1:5 sulphur and lime at about 25.5 kg per hectare.

11.3.1.3 Onions and garlic

1. Onion thrips, *Thrips tabaci*

DAMAGE: Thrips usually collect at the leaf base or flowers of onion and garlic. Adults and larvae suck leaf sap through their piercing mouth parts. Infested leaves curl, wrinkle, and gradually die. Infested plants neither set seed nor develop bulbs.

CONTROL: (1) After harvest, remove and destroy plant debris on the ground. (2) Plough and harrow the fields to expose the pupae to weather. (3) Use early, and quick–maturing varieties. (4) Flood standing crop to keep pest in check. (5) Eradicate weeds. (6) Employ hymenopterous parasitoids of the family Eulophidae for biological control. (7) Control serious infestations with Lindane 0.1% or Malathion 0.05%.

11.3.1.4 Chilies

1. Tobacco caterpillar, *Spodoptera litura*

DAMAGE: Young larvae feed gregariously on the lower surfaces of leaf laminae. At the early stage of infestation, attacked areas of the leaf blades dry up. Eventually leaf blades are consumed entirely, leaving only the mid-ribs, thereby killing the plants. Older caterpillars cut seedlings at ground level after dark. During daytime they hide in the soil near the plants.

CONTROL: (1) Collect infested leaves with masses of young larvae and destroy. Collect older larvae and adults with hand nets and kill them. (2) After harvest, deep–plough the soil to expose the hibernating pupae to weather. (3) Encourage the protozoa, viruses, fungi, and bacteria which parasitize the larvae. Towards this end, avoid using insecticides unless the crop is seriously threatened. In such cases, as a last resort, spray Endosulfon 0.05%, Carbaryl 0.22%, Phosphamidon 0.03%, or methylparathion 0.025%.

11.3.1.5 Peas and beans

1. Cowpea aphid, *Aphis craccivora*

DAMAGE: Infestation is evident by the black colour of the apical parts of the plants. The surface of attacked plants appears greasy because of the honeydew secreted by the aphids. The aphid feeds on sap and therefore weakens the host plants. It is also a vector of about 30 disease–causing viruses, called the 'rosette virus complex'.

CONTROL: (1) Plant fields densely with lucerne or groundnuts to discourage aphid attack. (2) Several parasites and predators attack this pest, e.g. *Aphidius absinthii* Marshall, *Aphelinus basalis* Westwood, *Aphidencyrtus aphidivorus* (Magr.), *Ephedrus* sp. nr. *Cerasicola* Stary, *Trioxys* sp., *Sinensis* Mackauer, and *Lysiphlebus fabarum* Marshall. Among the predators, two anthocorids, one chamaemyiid, one chrysopid, nine coccinellids, and nine syrphid species are recorded in Pakistan. (3) Use of insecticides should therefore be avoided as far as possible to encourage the biological control agents to propagate. Chemical control may be attempted with Nicotine or Pyrethrum if necessary. When absolutely inevitable, use the recommended insecticide as directed.

2. Tobacco budworm, gram caterpillar, gram pod borer, *Heliothis armigera*

DAMAGE: The larvae of this pest attack at flowering or fruiting because they preferentially feed on floral parts or fruits of tomatoes, peas, and okra. On cotton and gram the larvae bore into the bolls and pods.

CONTROL: (1) Destroy alternate host plants and any plant remaining after the harvest. (2) Deep plough after harvest to destroy the hibernating stages in the soil. (3) Time sowing to minimise the duration of flowering. To control serious infestation, spray Carbaryl 0.1%, Endosulfan 0.25–0.5% ULV.

11.3.1.6 Okra

1. Cotton whitefly, *Bemisia tabaci*

DAMAGE: Nymphs and adults of cotton whitefly suck cell sap and thus weaken the hosts. Chlorotic spots appear on the leaves, which enlarge causing irregular yellowing, extending from veins to leaf margins. Infested leaves dry and fall. Juvenile stages of the pest secrete honeydew which gives the infested plants a dark hue.

CONTROL: (1) Remove weeds. (2) To keep the pest population in check, encourage hymenopterous parasites, e.g. *Encarsia lutea* and *Eretmocerus mundus*; the phytosid mites *Amblyseius aleyrodis* and *Typhlodromus sudanicus* which prey on the eggs and young larvae of the whiteflies; and the insect pathogenic fungi *Aschersonia aleyrodis* and *Verticillium lecanii*. (3) If unavoidable, spray Basodin 0.1–0.5%, or other recommended insecticides.

2. Jassid, *Amrasca devastans*

DAMAGE: These pests affect both okra and cotton. Jassids suck sap from the underside of leaves and inject toxins. Leaf edges turn upward (okra) or downward (cotton), and leaves turn pale then fall. Defoliation weakens the plant, causing loss of cotton bolls.

3. Spotted bollworms, *Earias insulana* and *Earias vitella*

DAMAGE: Larvae of spotted bollworm bore into growing shoots, flower buds, and flowers. Fruits of okra get distorted and become unfit to eat. The plants are weakened and may die.

11.3.1.7 Brinjal (eggplant)

1. Brinjal lacewing bug, *Urentius sentis*

DAMAGE: Both adults and nymphs of the lacewing bug suck leaf sap. Infested leaves look mottled because of yellow spots and scale-like excreta deposited by the pest. Heavy infestation causes the leaves to turn yellow, shrivel, and fall.

CONTROL: (1) Uproot and burn withered plants. (2) Spray Diazinon 0.02%, Dimethoate 0.05%, or any other suitable insecticide.

2. Hadda beetle, *Epilachna demurili*

DAMAGE: Young larvae feed only on the underside of the leaf, giving it a lacelike appearance. Adults and older larvae eat holes in the leaves.

CONTROL: (1) Burn plant debris after harvest. (2) Clean-cultivate the soil. (3) Hoe regularly; pick and destroy adults, egg clusters, larvae, and pupae. (4) Utilize the fungus *Verticillium lecanii* to kill the larvae and pupae.

(5) To control serious infestation, spray Carbaryl 0.01 lb a.i., Methylparathion 0.20–0.25 lb, or any other suitable insecticide.

3. Brinjal fruit borer, *Leucinodes orbonalis*

DAMAGE: Early generation larvae bore into growing shoots which die and hang down. Later generations bore preferentially into flower buds and fruit. The whole crop may be destroyed if damage to fruit is severe.

CONTROL: (1) Collect and destroy or bury infested fruits and dead shoots. Remove and destroy plant debris regularly from the fields. After harvest, pull out the plants and destroy. (2) Deep plough to destroy any developmental stage left in the soil. (3) A large number of insecticides can be used to destroy the pest, should its infestation become serious, including Sevin 85 S.P., Malathion 57 E.C., Nogos/Vapena 100 E.C., DDVP or Dipterex 80, Decis/Cymbush/Sumicidin/Ripcord/Arrivo. These should be used according to the directions on the package.

4. Brinjal shoot borer, *Euzophera perticella*

DAMAGE: The larvae feed on pith of the stem, causing the apical shoots to die and hang down. Larvae in advanced stages of development bore into the stem at ground level, killing stray plants in the field.

CONTROL: (1) Remove and destroy all infested plants from the nursery and the field. (2) After harvest, uproot and destroy all the plants. (3) If infestation is severe, spray with Sevin, Malathion, Elsan, or any other suitable insecticide.

5. Root–knot nematode, *Meloidogyne incognita*

DAMAGE: This pest attacks brinjal, *bhindi,* and tomatoes. Attack is indicated by discolouration and decreased size of leaves and stunting of the plants.

CONTROL: The nematode can be controlled by the application of 34 litres of Nemagon 60 per ha. Intercropping of okra with sesame reduces the incidence of nematode attack on okra.

11.3.1.8 Turnip, radish, mustard

1. Mustard sawfly, *Athalia lugens proxima*

DAMAGE: Mustard sawfly larvae skeletonize the leaves. Young plants are totally defoliated, leaving only the thicker veins. Infested older plants do not bear seed.

CONTROL: Several insecticides can control this infestation, e.g. Malathion 57 E.C., Rogar 40 E.C., Curacron 50 E.C.

2. Mustard aphid, *Rhopalosiphum (Lipaphis) erysimi*

DAMAGE: Nymphs as well as adults suck cell sap from several parts of the plants, greatly reducing their vigour. Their leaves curl up, and they fail to produce healthy seed.

CONTROL: The infestation can be controlled by spraying Malathion 0.05% or Nogos.

11.3.1.9 Cruciferous crops

1. Cabbage butterfly, *Pieris brassicae*

DAMAGE: The first instar larvae scrape the leaf surface. Successive instars feed on tender leaves from the margins inward, leaving only the harder veins. In severe infestations the plants are completely defoliated.

CONTROL: (1) Collect egg masses, larvae, and pupae by hand and the butterflies with hand nets; then kill them in water to which kerosene has been added. (2) Encourage *Apanteles glomeratus*, a parasitoid on the larvae. (3) In serious infestation spray Malathion 0.05% or Diazinon 0.02% up to three weeks before harvest. After that, use Carbaryl 0.2%.

2. Diamondback moth, *Plutella xylostella* (= *P. maculipennis*)

DAMAGE: Newly hatched larvae bore into the underside of the cabbage and feed in the head or the heart, and sometimes in the central leaves.

CONTROL: (1) In Malaysia and the Philippines, interplanting tomatoes with cabbage reduced the infestation of diamondback moth by 35%. (2) Mass release of *Trichogramma* wasps e.g. *T. evanescens* or *T. pretiosa*, predators on the eggs of lepidopterous species, can reduce the population of diamondback moth. (3) Serious infestation can be controlled by spraying Malathion 0.05%, Endosulfan 0.05%, or Carbaryl 0.15%.

11.3.2 Diseases

11.3.2.1 Potato

1. Brown rot (tuber rot) (CAUSAL ORGANISM: *Pseudomonas solanacearum*)

SYMPTOMS: Brown rot causes foliage wilting, yellowing, and marginal necrosis; reduction of leaf size, plant height, and dry weight of tubers; and exudation of a foul–smelling liquid in the advanced stages of the disease.

CONTROL: Disinfect seed potatoes.

2. **Wet rot and storage rot (tuber rots)** (CAUSAL ORGANISM: *Trichurus spiralis*)

SYMPTOMS: Infected tubers turn brown to black, and then rupture.

CONTROL: Select seed tubers from a healthy crop and keep in cold storage.

3. **Dry rot (tuber rot)** (CAUSAL ORGANISMS: *Fusarium angustum*, *Fusarium oxysporum*)

SYMPTOMS: Dry–rot attack is indicated by sunken, shrivelled, or wrinkled areas on the tuber surface, which are brown or black in colour, with fungus on them. On the inside, the infected tubers exhibit cavities with white fungal mycelium.

CONTROL: (1) Select cultivars tolerant to tuber rot. (2) From a disease–free crop, select healthy, uninjured, mature tubers, and surface sterilize with formaline. (3) Disinfect storeroom and maintain appropriate temperature in store.

4. **Wilt** (CAUSAL ORGANISM: *Fusarium oxysporum*, *Fusarium radicola*)

SYMPTOMS: The wilt organism causes leaf chlorosis, wilting, drying, and death of leaves within a fortnight.

CONTROL: (1) Select wilt–tolerant cultivars. (2) For seed, select a healthy autumn crop, harvest carefully, pick out uninjured tubers, disinfect their surface, and store carefully in cold storage at appropriate temperature. (3) In soils infested with wilt, adopt crop rotation including non–susceptible crops.

5. **Early blight** (CAUSAL ORGANISM: *Alternaria solani* (= *Macrosporium solani*))

SYMPTOMS: Early blight first appears as brown to black lesions on leaf margins, increasing in size from tiny dots and causing curling, blighting, and drying of leaves.

CONTROL: (1) Remove and destroy all infected debris. (2) Select seed tubers from healthy crop. (3) To control serious infection, spray Bordeaux mixture, Kocide, Cobox, Cupravit, or Dithane–M 45.

6. Leaf spot (CAUSAL ORGANISM: *Cercospora concors*)

SYMPTOMS: This disease can be diagnosed by dark fungal growth around yellow spots on the dorsal surface of the leaf, and fluffy grey colour of the ventral side.

CONTROL: (1) Select healthy seed tubers for planting. (2) If necessary, use fungicides for controlling the infestation.

11.3.2.2 Cucurbits (Muskmelon, watermelon, long melon, bitter gourd, bottle gourd, ash gourd, cucumber, and squashes)

1. Downy mildew (CAUSAL ORGANISM: *Pseudoperonospora cubensis*)

SYMPTOMS: At first pale green spots appear on the dorsal surfaces of leaves giving them a mottled appearance. The spots increase in size, turn yellow, and coalesce to cover large areas of the leaves. Then they start turning necrotic brown from the centre. The ventral side of the leaves is covered with light purple mycelium, bearing large lemon–shaped sporangia. Downy mildew decreases flower set and fruit development by killing the foliage.

CONTROL: (1) Destroy diseased plant debris. (2) Destroy cucurbit weeds. (3) Rotate crops. (4) Control serious infestations with copper–based fungicides or Dithane M–45.

2. Powdery mildew (CAUSAL ORGANISM: *Erysiphe cichoracearum* DC)

SYMPTOMS: Powdery mildew infects seedlings, and leaves and stems of older plants. It first appears as separate white lesions which enlarge and join enveloping the entire plant in the powdery mycelium. Infected leaves curl up, turn chlorotic, then brown, and finally fall. Loss of leaves reduces plant growth, fruit–set, and fruit development. Due to lack of foliage, the fruit also becomes malformed and sunburnt.

CONTROL: (1) Destroy diseased plant debris. (2) Dust with a sulphur–based fungicide such as Sulforon.

11.3.2.3 Tomato

1. Fruit rot (CAUSAL ORGANISM: *Alternaria tenuis*)

SYMPTOMS: Fruit rot can generally be diagnosed by the appearance of small brown lesions on leaves and dark–brown to black concentric lesions on the fruit. As the disease develops, the lesions crack and the fruit drops.

CONTROL: (1) Until resistant varieties are developed, use copper–based fungicidal sprays to control the disease.

2. Wilt (CAUSAL ORGANISM: *Fusarium oxysporum, F. lycopersici*)

SYMPTOMS: Wilt initially causes the drooping of petioles, followed by the chlorosis, wilting, and finally the death of foliage. The disease progresses towards the top of the plant, turning the vascular system brown, causing the leaves to wilt, and stunting growth.

CONTROL: Apply Dexon (2% solution) to soil or mix Brassicol dry with the soil at the rate of 1.5 kg/ha.

11.3.2.4 Chilies

1. Dieback (CAUSAL ORGANISM: *Vermicularia capsici, Gloeosporium piperatum*)

SYMPTOMS: Dieback causes the plant to die from the top downwards.

CONTROL: 1) Use healthy seeds. (2) Destroy diseased plant debris.

2. Fruit rot (CAUSAL ORGANISM: *Alternaria tenuissima*)

SYMPTOMS: Fruit rot first appears on leaves and mature stems as elevated or depressed concentric spots which grow in size, turn brown, and eventually cause rotting.

CONTROL: (1) Collect and destroy diseased fruit lying on the ground. (2) Control disease with 2–3 sprays of a copper–based fungicide such as Bordeaux mixture, repeated biweekly.

3. Powdery mildew (CAUSAL ORGANISM: *Leveillula taurica*)

SYMPTOMS: Powdery mildew first appears as white powdery spots on the ventral surface of the foliage, which in time, covers it entirely. On the dorsal side of the foliage, yellow lesions appear. Due to loss of foliage, fruit size and number is reduced, causing a drop in yield. Infection is favoured by a cool, humid environment.

CONTROL: (1) Destroy diseased plant debris. (2) Dust the crop with Sulforon or spray with Thiovit or Cosan.

11.3.2.5 Peas

1. Powdery mildew (CAUSAL ORGANISM: *Erysiphe polygoni*)

SYMPTOMS: White powdery spots appear on infected leaves in January. By March stems and pods are also affected. Ultimately all the foliage is covered and the plant becomes pale brown, withers, and loses leaves. Smaller and fewer pods are formed, and in severe infestations yield is considerably reduced.

CONTROL: (1) Plant resistant varieties. (2) To control serious attack, use fungicides such as Sulforon or Kumulus (0.2%).

11.3.2.6 Okra

1. Root rot (CAUSAL ORGANISM: *Macrophomina phaseoli*)

SYMPTOMS: Plants infected by root rot become yellow, then wilt, droop, and die. Their roots and rootlets get shredded, and rot as the disease advances. **CONTROL:** (1) Use resistant varieties, e.g. TYP 15. (2) Adopt crop rotation, excluding alternate host crops like tobacco, sesamum, mung, mash, groundnuts, potatoes, and tomatoes from the rotation, since these too are susceptible to root rot.

11.3.2.7 Cabbage and cauliflower

1. Downy mildew (CAUSAL ORGANISM: *Peronospora parasitica*)

SYMPTOMS: Downy mildew attack is first indicated by the appearance of separate areas of yellow cells on the dorsal surface of the leaves. The ventral side of the leaves exhibit white sporangiophores. With the spread of the disease, the lesions increase in size between the major veins.

CONTROL: Use Dithane M–45 or similar fungicide.

2. Black rot (CAUSAL ORGANISM: *Xanthomonas campestris*)

SYMPTOMS: Black rot affects the inflorescence of cauliflower, and causes the veins of cabbage leaves to turn black.

CONTROL: (1) Improve drainage and field sanitation. (2) Adopt effective crop rotation, excluding crucifers for at least five years. (3) Immerse seeds in water at 50°C for 25 minutes for cabbage, and 20 minutes for cauliflower.

3. Club rot (CAUSAL ORGANISM: *Plasmodiophora brassicae*)

SYMPTOMS: Club rot causes gall formation and bruising of the root and hypocotyl. The severity of the disease increases in wet soils, and at temperatures from 20–26°C.

CONTROL: For cabbage: (1) Sterilize the soil with methyl bromide or other soil sterilants. (2) In acid soils, apply hydrated lime to raise soil pH above 7.0. (3) Before planting dip roots in 4% Calomel or Benomyl slurry.

For cauliflower, (1) select disease-free and well-drained soil. (2) Eradicate crucifer weeds. (3) Adjust soil pH to 7.2 by the application of hydrated lime. (4) Use crop rotations with no crucifer for at least seven years. (5) Sterilize infected soil with methyl bromide before planting. (6) Plant disease-free seedlings.

4. Damping-off (CAUSAL ORGANISM: *Thanatephorus cucumeris*)

SYMPTOMS: Damping-off causes the seedlings to collapse due to decay of their stems at ground level.

CONTROL: Treat seed with Captan before sowing.

5. Fusarium wilt (CAUSAL ORGANISM: *Fusarium oxysporum, F. congluti nans*)

SYMPTOMS: Leaves turn dull yellow-green and may also curl. Their veins become yellow or dark-brown. They may also die and be shed prematurely.

CONTROL: Use varieties resistant to *Fusarium* wilt.

6. Black leg (CAUSAL ORGANISM: *Leptosphaeria maculans* [= *Phoma lin gam*])

SYMPTOMS: Black leg attack starts with papery brown lesions on juvenile leaves. They become dry and are covered with black pycnidia. The disease may also cause stem decay at ground level. In this case, plants die because their stems and roots are destroyed.

CONTROL: (1) Select well-drained soil. (2) Adopt crop rotation free of crucifers for at least four years. (3) Eradicate crucifer weeds. (4) Before sowing, soak cabbage seed for 25 minutes, and cauliflower seed for 20 minutes in water at 50°C. Transplant only disease-free seedlings.

STUDY QUESTIONS

1. List the symptoms of at least three diseases which occur on cucurbits.
2. How can you differentiate between the damage caused by fruit flies and that caused by fruit borers?
3. Prepare a calendar of fruit and vegetable insect pests for your area from the information provided in this chapter.

REFERENCES

Anonymous. *Zarait name, tahafzay nibatat* number. Lahore: Publication Department, Department of Agriculture, Govt. of Punjab; 1977. 118 p.

Atwal, A. S. Agricultural pests of India and South–East Asia. 2nd ed. New Delhi: Kalyani Publishers; 1986.

Berrie, A. M.; Mohyuddin, A. I. Integrated pest and disease control of fruit crops in Pakistan. Saidu Sharif, Swat: Malakand Fruit and Vegetable Development Project; 1990.

Catava, A. Compendium of citrus diseases and disorders in Pakistan: a preliminary survey. A Pak/Italian Fruits, Vegetables and Olives Project. Saidu Sharif, Swat: Publication Agrotee (S.P.A.); 1987.

Chadha, K. L. Pest control in horticultural crops. Extension Bulletin 25:1–33. Bangalore: Indian Institute of Horticultural Research; 1985.

Dixon, G. R. Vegetable crop diseases. London: Macmillan; 1981.

Fadeez, Iu. N.; Novozhilov, K. V. ed. Integrated plant protection. New Delhi: Oxonian Press Pvt. Ltd.; 1987.

Frazier, N. W.; Fulton, J. P.; Thresh, J. M.; Converse, R. H.; Varney, E. H.; Hewitt, W. B. Virus diseases of small fruits and grapevines. Berkeley, California: University of California Press; 1970.

Geddes, A. M. W.; Iles, M. The relative importance of crop pests in South Asia. Natural Resources Institute Bulletin No. 39. Chatham, England: Overseas Development Administration; 1991.

Guan–Soon, L.; Yuan–Ba, D. eds. Status and management of major vegetable pests in the Asia–Pacific Region with special focus towards integrated pest management. RAPA Publication 1990(3). Bangkok, Thailand: FAO; 1990.

Hafeez, A. Plant diseases. Islamabad: Pakistan Agricultural Research Council; 1986.

Hashmi, A. A.; Hassall, K. A. Pests and pesticides in Pakistan. Islamabad: Pakistan Agricultural Research Council; 1989.

Hashmi, A. A. Integrated pest management. Islamabad: Pakistan Agricultural Research Council; 1993.

Indian Council of Agricultural Research. Handbook of agriculture. New Delhi: Indian Council of Agricultural Research; 1978. 1303 p.

Parsons, D. The production of citrus in Dir District. Saidu Sharif, Swat: Malakand Fruit and Vegetable Development Project; 1990. 30 p.

Pathak, V. N. Diseases of fruit crops. New Delhi: Oxford and IBH Publishing Co. Pvt. Ltd.; 1989.

Pradhan, S. Insect pests of vegetables. In: Srivastava et al. Agricultural entomology, Vol. II. New Delhi: All India Scientific Writers Society; 1983.

Rangaswamy, G. Diseases of crop plants in India. 2nd ed. New Delhi: Prentice–Hall; 1984.

Sherf, A. F.; Macnab, A. A. Vegetable diseases and their control. New York: John Wiley and Sons; 1986.

Subramanian, T. R. Diseases of fruit and vegetable crops and their control. Extension Bull. 17:1–36. Bangalore: IIHAR; 1987.

Westcott, C. Westcott's plant disease handbook. 4th ed., revised by R. K. Horst. New York: Van Nostrand–Reinhold Co.; 1979.

Outline of Chapter 12

12. POST–HARVEST HANDLING

Wasim A. Farooqi[1]

LEARNING OBJECTIVES

After studying this chapter, the student should be able to:

- Discuss various post–harvest techniques and related problems
- Identify problems in handling fruit and vegetables in the orchard or market
- Demonstrate how to handle fresh produce in a manner that minimizes post–harvest losses
- Organize technical and administrative resources for the fresh fruit and vegetable trade
- Anticipate problems which could occur under undesirable environmental circumstances
- Manage handling of fresh produce for distribution to distant markets
- Disseminate basic technical know–how to the end users in the horticultural trade and industry

12.1 Introduction

Like other developing countries, Pakistan needs to increase the quantity of food available for its rapidly growing population. There are various ways to approach this problem. One is to produce more food; another is to conserve whatever is produced. In the past, much emphasis has been put on growing more food, while the post–harvest aspect (i.e. conservation of food after harvest) has been generally ignored. Undoubtedly we have to grow more food for the fast–growing population of Pakistan— not only by increasing the area under cultivation, but also by applying science and technology in all agro–operations. Horticultural produce supplies essential food items of daily use. There has been significant progress in increasing the area under these

[1]Dr. Farooqi is Principal Scientific Officer/Post-harvest Technologist, Nuclear Institute for Agriculture and Biology, Faisalabad.

crops since Independence. Also, exotic, high–yielding, and good–quality varieties have been introduced in parts of Pakistan. The area under cultivation and production has increased for citrus (especially kinnow), grafted mango varieties, bananas, and deciduous fruits (especially apples).

A major and often neglected step towards offering a greater volume of nutritious food is to prevent its loss between the time of harvesting and consumption. The Pakistan National Commission on Agriculture estimated that defects and inadequate facilities in post–harvest handling, transport, storage, and marketing may cause up to 20–40 percent loss of fruit and vegetables. The value of this loss amounts to millions of rupees annually.

Post–harvest losses of horticultural crops affect both the nutritional status of the population and the economy of the country.

a. Nutrition. Fruit and vegetables are a rich source of the vitamins and minerals essential for human nutrition. The fruit and vegetables wasted in transit from harvest to consumer represent a loss in the quantity of available food. This is important not only in quantitative terms, but also from the point of view of quality nutrition.

b. Economy. Careless harvesting and rough handling of perishables bruise and scar the skin, thus reducing quality and market price. Such damaged produce also fails to attract the international buyer, and brings the exporting country less profit and a bad name. This ultimately results in huge economic losses to the country.

If we review the present situation in Pakistan, it is quite evident that harvesting operations, post–harvest handling in the orchard or field, storage and transport conditions, and marketing are scientifically defective. There is tremendous need for improvement in all these field and market operations. Unless existing conditions are improved, we cannot minimize post–harvest losses, extend the storage shelf life of these perishables, or increase export earnings. Some of our produce like citrus (especially kinnow) and mangoes are of high quality and can compete in international markets, but because of poor post–harvest handling operations and management, the quality of this produce deteriorates before it reaches its destination. Although large quantities of our fruit and vegetables are exported, they face tough competition in international trade and do not bring premium prices.

To improve the situation, it is essential to create awareness among growers, farm workers, managers, traders, and exporters about the extent of the losses being incurred and their economic consequences. These groups of people involved in the fruit industry also need to learn the basic principles of fruit handling and storage. In addition, the government needs to provide basic infrastructure like storage, handling, grading, packing, transport and marketing facilities, and technical expertise. This could be carried out jointly by the public and private sectors.

12.2 Causes of post-harvest losses

Produce may be injured mechanically by the picker if he/she has long nails, handles the produce roughly, or throws it on the ground. Even when harvesting is done carefully, the following factors cause post-harvest spoilage of horticultural produce (Farooqi et al. 1991).

1. Loss of moisture from the surface of produce through evaporation, transpiration, and respiration. Loss of moisture from the produce results in shrivelling and loss of market value.
2. When produce is stored at higher than optimum temperatures, increased metabolic rate, especially respiration, results in the onset of early senescence (aging) and ultimately the death of tissue.
3. Interruption in the normal metabolic activity of the produce, due to either extremely low or high temperature. At sub-optimal temperatures, the produce may exhibit the physiological disorder **chilling injury**. This happens with tropical fruits like bananas, mangoes, and avocados. At high temperatures, above 30°C, the produce gets 'boiled'. In this case, its external appearance deteriorates and it spoils quickly.
4. Invasion of the produce by various pathogens results in fruit rot. Injured and senescent tissue is more sensitive to the attack of decay-causing microorganisms.

12.3 Physiology and biochemistry of fruits and vegetables

12.3.1 Structure

Fruit and vegetables are plant organs made of millions of cells bounded by more or less rigid cell walls. The cell wall is composed of cellulose, pectic substances, and lignins. It is permeable to water and solutes. Each cell consists of the cytoplasm and usually one or more vacuoles, fluid reservoirs containing various solutes like sugars, salts, and amino and organic acids. They are surrounded by a semi-permeable membrane known as the **tonoplast**. The crispness of a fruit or vegetable is related to the turgidity of the cells comprising it.

Important cell organelles are the **plastids**, which are involved in energy fixation (**chloroplasts**), colour change (**chromoplasts**), and starch storage (**amyloplasts**).

12.3.2 Composition

Fruit and vegetables are composed of water, carbohydrates, protein, lipids, organic acids, vitamins, minerals, and flavouring compounds in volatile form. The quantities of these components depend upon the species or variety of fruit or vegetable, the agro–climatic conditions where they are grown, plant–nutritional factors, and stage of maturity and ripening. Tables 12.1 and 12.2 give the approximate amounts of sugar and some essential vitamins in selected fruits and vegetables.

Table 12.1 Sugar content of some ripe fruits

	Sugar (g/100 g fresh weight)		
Fruit	**Glucose**	**Fructose**	**Sucrose**
Apple	2	6	4
Banana	6	4	7
Cherry	5	7	0
Date	32	24	8
Grape	8	8	0
Orange (juice)	2	2	5
Peach	1	1	7
Pear	2	7	1
Pineapple	2	1	8
Tomato	2	1	0

Source: Wills et al. (1989:9).

Table 12.2 Approximate levels of vitamin C, vitamin A, and folic acid in some fruits and vegetables

Commodity	vitamin C (mg/100 g)
Black currant, guava	200
Broccoli, Brussels sprout	100
Papaya	80
Citrus, strawberry	40
Cabbage, lettuce	35
Mango, carrot	30
Pineapple, banana, potato, tomato, bean, cassava	20
Apple, peach	10
Beet, onion	5

	Vitamin A (β–carotene equiv. mg/100 g)
Carrot	10.0
Sweet potato (red)	6.8
Spinach	2.3
Mango	2.4
Tomato	0.3
Apricot	0.1
Banana	0.1
Potato	0.0
	Folate (μg/100 g)
Spinach	80
Broccoli	50
Brussels sprout, pulses	30
Cabbage, lettuce	20
Banana	10
Most fruits	<5

Source: Wills et al. (1989:13).

The nutritional constituents listed in Table 12.2 are essential for normal human health. The daily dietary need depends upon age, health, condition, and the amount of physical work performed. Post–harvest deterioration results in loss of vitamin content.

12.3.3 Ripening and respiration

Ripening. After maturation, senescence begins with ripening, which may occur while the fruit or vegetable is attached to the plant or after harvest. The ripening fruit undergoes many physico–chemical changes after harvest, which determine fruit quality.

Ripening is a complex process with many interdependent steps. The major metabolic process which takes place in the harvested produce is **respiration**. This is the oxidative breakdown of more complex material normally present in the cells such as starch, sugars, and organic acids, into simple molecules such as CO_2 and H_2O with the concurrent production of energy and other molecules which can be used by the cell for sympathetic reactions. Respiration occurs both in the presence of O_2 (aerobic) and in its absence (anaerobic). (See Chapter 3 of this volume, section 3.2, for a detailed discussion of respiration.) Anaerobic respiration is known as **fermentation** and results in the development of an off flavour in the produce. Rate of respiration is a good indicator of the metabolic activity of fruit tissues and can serve as a commercial indicator of the potential storage life of produce. Such information can help in better and scientifically organized marketing.

During ripening, visible and invisible changes occur in a fruit.

- Change of ground colour from green to yellow (with the exception of a few varieties which have green–coloured skin)
- Change of starch into sugar (development of sweet taste)
- Softening of pulp
- Development of aroma (production of aromatic compounds)

These changes make the produce edible and acceptable to the consumer. But from the commercial point of view these processes should be controlled so that the produce has a prolonged storage/shelf life.

Flavour is one of the most important aspects of quality which develops under normal conditions. Some of the components of aroma of several fruits and vegetables are given in Table 12.3. The flavour of Kinnow mandarin is unique among the citrus fruits, and has been studied by Asi et al. (1989).

Table 12.3 Distinctive components of the aroma of some fruits and vegetables

Product	**Compound**
Apple, ripe	Ethyl 2–methylbutyrate
green	hexenal, 2–hexenal
Banana, green	2–hexenal
ripe	Eugenol
overripe	Isopentanol
Grapefruit	Nootakatone
Lemon	Citral
Orange	Valencene
Raspberry	1–(*p*–Hydroxyphenyl)–3–butanone
Cucumber	2, 6–Nonadienal
Cabbage, raw	Allyl isothiocyanate
cooked	Dimethyl disulphide
Mushroom	1–Octen–e–ol, lenthionine
Potato	2–Methoxy–3–ethyl pyrazine, 2.5–dimethyl pyrazine
Radish	4–Methylthio–*trans*–3–butenyl isothiocyanate

Source: Wills et al. (1989:14)

Respiration. As described before, respiration is the most important process involved in maturation and ripening, and can be used as an indicator of the stage of fruit development as well as of potential storage life.

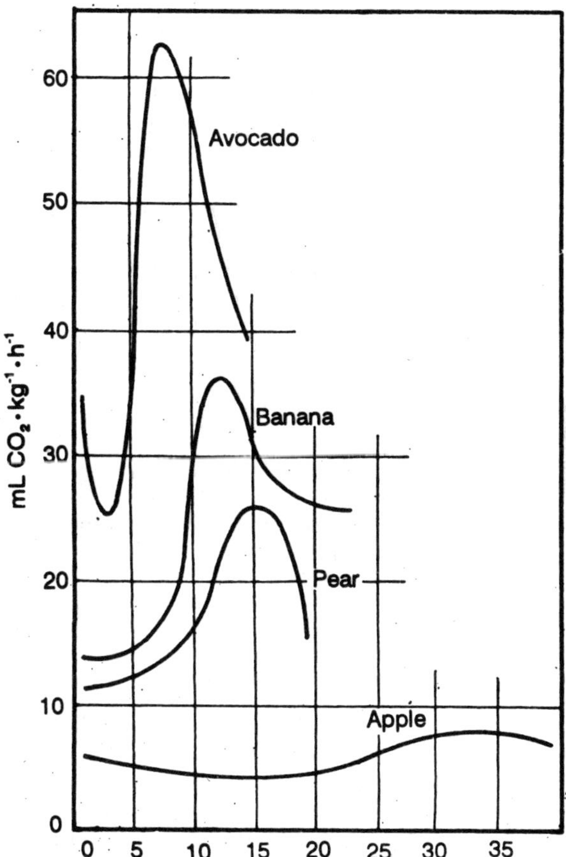

Figure 12.1 Respiratory patterns of some climacteric fruits stored at 15–20°C. Source: Farooqi (1970); Wills et al. (1989).

Respiration involves the oxidative breakdown of certain organic substrates. Respiratory peaks of apples, pears, bananas, and avocados are shown in Fig. 12.1.

Post–harvest respiration. The basic principle in post–harvest handling is that horticultural produce (fruit and vegetables) are living, not only while they are on the tree/plant but also after they are detached from the plant.

They continue their metabolic reactions and maintain their physiological functions. They respire through the uptake of O_2, produce CO_2 and heat, transpire, and lose water. The major difference between the pre– and post–harvest stages is that when the fruit is attached to the plant it replaces consumed energy through the flow of cell sap, but after it is harvested it has to depend entirely on its stored food energy. This is the point where deterioration begins.

In the physiological development of any fruit or vegetable, there are three stages: growth, maturation, and senescence. Growth involves cell division and enlargement, while maturation begins when growth stops. Both these stages are developmental phases. After maturation, senescence starts. This is the stage when a fruit/vegetable should be harvested (before ripening). In senescence, anabolic (synthetic) biochemical processes give way to catabolic (degradative) processes, leading to aging and finally death of the tissue. Changes which may occur during the ripening of fresh fruit are:

- Seed maturation
- Colour changes
- Abscission (detachment from parent plant)
- Changes in respiration rate
- Changes in rate of ethylene production
- Changes in tissue permeability
- Softening: changes in composition of pectic substances
- Changes in carbohydrate composition
- Organic acid changes
- Protein changes
- Production of flavour volatiles
- Development of wax on skin

On the basis of respiratory pattern, horticultural produce can be divided into two major groups: **climacteric** and **non–climacteric** fruits. The climacteric is a period in the development of certain fruits during which a series of biochemical changes results in increased respiration, leading to ripening and marking the change from growth to senescence. A simple classification on this basis is given in Table 12.4.

Table 12.4 Classification of some edible fruits according to their respiratory behaviour during ripening

Climacteric fruits	Non–climacteric fruits
Apple (*Malus sylvestris*)	Cherry, sweet (*Prunus avium*)
Apricot (*Prunus armenica*)	sour (*Prunus cerasus*)
Avocado (*Persea americana*)	Cucumber (*Cucumis sativus*)
Banana (*Musa* sp.)	Grape (*Vitis vinifera*)

Climacteric fruits	Non–climacteric fruits
Blueberry (*Vaccinium corymbosum*)	Lemon (*Citrus limonia*)
Cherimoya (*Annona cherimola*)	Pineapple (*Ananas comosa*)
Chinese gooseberry (*Actinidia chinensis*)	Satsuma mandarin (*Citrus reticulata*)
Feijoa (*Feijoa sellowiana*)	Strawberry (*Fragaria* spp.)
Fig (*Ficus carica*)	Sweet orange (*Citrus sinensis*)
Mango (*Mangifera indica*)	Tamarillo (tree tomato) (*Cyphomandra betacea*)
Muskmelon (*Cucumis melo*)	
Papaya (*Carica papaya*)	
Passion fruit (*Passiflora edulis*)	
Peach (*Prunus persica*)	
Pear (*Pyrus communis*)	
Persimmon (*Diospyros kaki*)	
Plum (*Prunus* spp.)	
Tomato (*Lycopersicon esculentum*)	
Watermelon (*Citrullus lanatus*)	

Source: Wills et al. (1989:24).

12.4 Harvesting

Harvesting at the proper stage of maturity, the method of harvesting, and the handling operations are crucial aspects which determine the shelf life and quality of produce. In Pakistan, harvesting of fruit and vegetables is generally done by hand, by people who are not aware of the principles upon which the best harvesting dates should be determined. This factor adds to post–harvest losses. Prediction of correct harvesting dates may depend upon the number of days from flowering (apples), temperature–time values called heat units, physiological criteria like pressure testing of fruit, ground colour of the fruit peel, or percent total soluble solids (TSS) of the pulp juice.

The concept of **heat unit** provides a way to obtain an objective estimate of the time required for the development of a fruit to maturity. A certain number of heat units (**degree–days**) is required to bring any given crop to maturity. The number of degree–days required by a given crop is determined over a period of several years by calculating the sum of the differences between the daily mean temperatures and a fixed base temperature (commonly the minimum temperature at which growth occurs). Based on weather conditions in a given year, the characteristic number of degree–days required for a given crop to mature can be used to forecast the probable date of maturity for the current year. As maturity approaches, this forecast can be confirmed by other means.

12.4.1 Hand harvesting

Harvesting of fruit and vegetables is generally done by hand, either by pulling, plucking, twisting (twist–pick), or with clippers. If the fruit is harvested by pulling, a portion of the peel is detached along with the stalk, or the button end is ruptured. This damaged portion of the harvested produce becomes more susceptible to microbial invasion, resulting in early spoilage.

Two methods of harvesting tested, pulling and clipping, produced differences in the post–harvest spoilage of oranges due to stem–end and blue mould rots. Clipping was found to be superior to pulling. Also, twist–picking of citrus fruit results in less post–harvest loss from brown stem–end rot than pulling. Vegetables like okra, peas, and tomatoes are hand–picked and placed in a cloth bag or basket lying nearby.

Mangoes are generally harvested when they begin to change colour from green to yellow, or when a few ripe fruit start dropping from the tree (*tapka*). If fruit is to be consumed locally, it should be harvested when ripe or partially ripe, but if it is to be dispatched to distant markets, it should be harvested at the hard, green stage. Picking should be done by hand or with a special mango–picker made of a bamboo pole with an attached knife and a cloth bag. Harvested fruit should be collected in cloth bags and should not be allowed to fall on the ground, otherwise they will suffer mechanical injury and will spoil quickly. Harvested fruit should be kept in field boxes and stored temporarily in the shade for despatch to the grading unit. Other fruits like citrus and apples should also be harvested by clipping or twisting and not by pulling (Farooqi 1991).

12.4.2 Mechanized harvesting

The use of machines for harvesting of horticultural crops has advantages as well as disadvantages. The **advantages** are:

- A greater quantity of produce can be harvested in a shorter time.
- The handling operation and possible damage is known and can be controlled.
- Multi–step operations like harvesting, grading, containerization, and storage can be manipulated at one platform.
- The cost of harvesting per unit is less.
- The field operation (harvesting) will not be badly influenced by social or administrative problems like strikes, curfew, etc.

Disadvantages are:

- The initial cost of machines is very high.
- Some farm labourers will become surplus and will add to the social problem of unemployment.

- Orchardists with small holdings cannot afford to buy the machines, nor would it be economical for them to use them if they could.
- No single type of harvester can be used for all kinds of fruits and vegetables, therefore its utilization will be uneconomical.

12.4.3 Chemical harvesting

Some chemicals are used to help in harvesting. These chemicals are sprayed on the tree prior to harvesting to loosen the attachment of fruit to the plant — the connection between the stalk and the button of the fruit. The tree is then shaken and the fruit drops. Such a practice can only be used with firm or hard fruit like apples or oranges.

Local research on the use of chemicals. The effect of certain chemicals like ethylene dibromide (EDB), 2,4–dichlorophenoxyacetic acid (2,4–D) and maleic hydrazide (MH) was studied on citrus and mangoes (Farooqi 1983; Rashid et al. 1984). 2,4–D was successfully used to keep the buttons of citrus fruit intact and green during storage, which resulted in delaying its aging. Concentrations of 500 and 200 ppm for Kinnow mandarin and Eureka lemons, respectively were found optimum. MH at a concentration of 1000 ppm delayed the ripening of hard green mangoes (Samar Bahisht) during storage at temperatures of 23–30°C. Effects of EDB on bananas during ripening were also studied (Farooqi and Hall 1972).

12.5 Post–harvest techniques

12.5.1 Curing

After harvest, some horticultural products, especially tuber and bulb vegetables like potatoes, onions, and garlic, are conditioned by exposing them to certain temperature and humidity conditions. This technique is called **curing**. It helps to evaporate excess moisture from the produce, develop a waxy layer on the fruit, and reduce the danger of microbial attack and resultant rot. Thus this treatment enhances the shelf and storage life of the vegetables. According to Janick (1979:429–30), potatoes are cured at about 13–16°C during two to three weeks of storage, and sweet potatoes are cured at a temperature around 27°C.

12.5.2 Gamma irradiation

It is possible to extend the shelf life of fresh fruit and vegetables at ambient as well as refrigerated temperatures by gamma irradiation (Goldblith 1970). Irradiation helps to extend the shelf life of horticultural produce in the following manner (Maxie et al. 1966; Farooqi et al. 1967, 1974).

1. It inhibits the sprouting of tubers and bulbs like potatoes, onions, and garlic (Hamid et al. 1965; Farooqi and Donini 1976).
2. It delays ripening and senescence by interfering with basic metabolic processes, as in mango, papaya, banana, and guava (Hatton et al. 1961).
3. It disinfects the produce of fruit flies and weevils, as in mangoes and dried fruits, and helps to solve storage and quarantine problems.
4. It pasteurizes the fruit surface, as in citrus.

Complete sterilization of fresh fruits is not practicable, because the high gamma radiation doses required would damage important quality attributes such as appearance, texture, taste, and flavour. Most attention has, therefore, been centred on the use of sub–sterilization doses (pasteurization) which inhibit physiological processes such as sprouting, ripening, and senescence. Optimum radiation doses for pasteurization have been standardized for various kind of fruits and vegetables. A list of fruits and vegetables for which gamma irradiation has been cleared by the WHO for different countries, and the tolerable radiation doses for various fruit and vegetables is given in Table 12.5.

Table 12.5 Tolerable radiation doses for various fruit and vegetables

Commodity	Objective	Maximum tolerable dose (kGy)	Minimum radiation dose required (kGy)	Limitations
		FRUITS		
Apples	Control of scald and brown core	1.0–1.5	No effect below 1.50	Tissue softening, cheaper and effective alternatives available.
Apricots	Inhibition of brown rot	0.50	2.00	Tissue softening
Avocados	Ripening delay	0.25	2.00	Cheaper and effective alternatives available
Bananas	Ripening delay	0.50	0.30–0.35	
Cantaloupes	Ripening delay	2.00	No effect below 2.00	
Grapes	Inhibition of grey mould	0.25–0.50	10.00	Tissue softening, development of off–flavour, cheaper and effective alternatives available
Lemons	Inhibition of *Penicillium* rots	0.25	1.50–2.00	Skin injury to fruit
Limes		0.25	1.50–2.00	Cheaper and effective alternatives available. Development of off–flavour

Commodity	Objective	Maximum tolerable dose (kGy)	Minimum radiation dose required (kGy)	Limitations
Nectarines	Inhibition of brown rot	1.00	2.00	Tissue softening
Papayas	Disinfection	0.75–1.00	0.25	
Oranges	Inhibition of *Penicillium* rots	2.00	2.00	Skin injury to fruit. Cheaper and effective alternatives available
Peaches	Inhibition of brown rot	1.00	2.00	Tissue softening
Pears	Ripening delay	1.00	2.00	Abnormal ripening; cheaper and effective alternatives available
Raspberries	Inhibition of grey mould	1.00	2.00	Tissue softening
Strawberries	"	2.00	2.00	Cheaper and effective alternatives available
		VEGETABLES		
Asparagus	Inhibition of growth	0.15	0.05–0.10	
Onions	Inhibition of sprouting	0.04–0.06	0.10	
Potatoes	"	0.20	0.08–0.15	
Tomatoes	Inhibition of *Alternaria* rot	1.00–1.50	3.00	

Source: van Kooji (1984); Farooqi (1986).

There is widespread interest in the use of gamma irradiation for extending the shelf life of citrus fruits. This technology offers promise for reducing spoilage from harvest until it reaches the consumer. While the decay organisms can be inactivated by adequate exposure to gamma rays, it is necessary to limit the dosage so that no significant changes occur in the physical and chemical properties of the fruit. Sensitivity of the rot-causing organisms varies depending on several factors including differences among species, the stage of infection on the fruit, and the extent of the infection which influences the amount of radiation required for inactivation.

Changes in fruit respiration, colour, texture, acidity, and vitamin C content are usually intensified as the radiation dose is increased. However, no single minimum radiation dose required to protect fruit from decay can be stated. A number of factors determine the required radiation dose. Higher doses are needed to retard old infection established in the fields than for young infections which develop after harvest. Several decay-causing organisms vary in radiation resistance. The flux as well as dose of gamma radiation apparently influence the control of infections. Several important species causing decay in citrus are known to have radiation resistance in the following increasing order: *Trichoderma viride*, *Phomopsis citri*, *Penicillium italicum*, *Penicillium digitatum*, *Geotrichum candidum*, *Diploidia natalensis*, *Alternaria citri*. *Alternaria citri* is by far the most resistant to radiation of these fungal organisms. A radiation dose may be sufficient to prevent development of the organism most commonly causing spoilage, but secondary decay may result from the more radiation–resistant species. Beraha et al. (1959) reported that the flux or intensity as well as the administered dose of gamma radiation influenced both rate of development and total incidence of blue mould of oranges and green mould of lemons. Decay on oranges was not completely controlled by a 1.82 kGy dose delivered at 30 Gy/min. However, effective control was achieved with radiation given at 2200 Gy/min over a dose range of 1.57–1.82 kGy. At a flux rate of 400 Gy/min, control of infection is effective when the dose range is 1.25–1.37 kGy.

The radiation dose required for sufficient inactivation of organisms to provide adequate protection for fruits during marketing is quite different from the dose required for complete destruction of the organisms. Minimum lethal radiatior doses for fungi (causing citrus fruit decay) vary between 20 kGy (*Penicillium*) and 60 kGy (*Alternaria*). These doses would be excessive for treatment of fresh fruits as they are not only phytotoxic, causing peel injury, but also damage texture and flavour.

Research work has been carried out by various scientists on the effect of gamma irradiation on physical and chemical changes in many fruits. The work on citrus is summarized here.

Physical changes in citrus resulting from gamma irradiation.

A. SKIN INJURY. Dennison et al. (1966) and Farooqi (1966) reported the development of skin injury as a result of gamma irradiation of Washington Navel, Pineapple, Temple, and Valencia oranges. Pineapple and Valencia were found more susceptible to skin injury than the Navel and Temple cultivars. The injury was manifested by the development of dry sunken areas turning brownish in colour as the storage period increased. The depth and the area of the peel exhibiting the injury varied proportionally with the irradiation dose. Radiation injury to the peel could, however, be reduced by manipulating the storage temperature following irradiation.

B. GAS POCKETS. Maxie et al. (1963) found that lemons irradiated with dosages above 2 kGy developed severe cavities (gas pockets) when stored for three or more weeks at 15°C but not at a temperature of 20°C. Development of cavities along the segment membrane was thought to be due to pectin degradation in these areas, or to the differential pressure of the gases of the internal atmosphere present at different locations in lemons. Dennison et al. (1966) reported the development of gas pockets in Avon and Lesion lemons grown under Florida conditions which were irradiated at a dose of 2 kGy or above, when they were stored at 10°C for four weeks.

Chemical changes in citrus.

A. RESPIRATION. Respiration rates have been used as an index of the metabolic activities of fruit during ripening and senescence, mechanical injury, and chilling injury. An increase in respiration of Shamouti-type Jaffa oranges may occur for a few days after irradiation, followed by a decline in respiration. High respiratory activity proportional to the radiation dosage was obtained by Dennison et al. (1966) working on Temple oranges.

Citrus fruits, being non-climacteric, do not ripen after removal from the tree; however, they undergo many metabolic changes which result in senescence. Some of these changes include varied levels of respiration (i.e. evolution of CO_2 or uptake of O_2), organic acids, sugars, pectic substances, and skin pigments. Keeping the fruit at low temperatures has been recommended to slow down these changes and extend shelf life. Research has been carried out by various workers on the effects of radiation on metabolic changes which occur after harvest.

B. ASCORBIC AND CITRIC ACIDS. The major chemical of nutritional importance in citrus fruits is ascorbic acid (vitamin C) which is present in reduced form. Effects of radiation on ascorbic acid content vary with the species and variety of citrus fruit as well as radiation dose level. Radiation causes immediate loss in the reduced form of the acid which is followed by recovery within 24 hours. At higher doses like 4 kGy, the loss may be as much as 15 percent. Change in citric acid level upon irradiation is followed by ascorbic acid loss. Decline in the percentage of citric acid, as an effect of radiation, is variable within varieties of citrus fruits.

C. ORGANOLEPTIC PROPERTIES. The effects of gamma irradiation on the appearance, texture, taste, and flavour of oranges and grapefruit are modified by storage temperature and duration. Generally, these fruit may be irradiated with dosages up to 2 kGy without any appreciable damaging effects on organoleptic characteristics. Maxie et al. (1963) found (except at 3 kGy) no significant difference in the taste of pulp or juice from Navel oranges stored at 0°C for 10 weeks. Overall flavour scores, especially of fruit treated at the lower level of irradiation, generally improved with time of irradiation. Dennison et al. (1966) irradiated Florida grown Navel oranges to 0–3 kGy and stored them at 10°C. One day after irradiation, irradiated oranges were rated less acceptable than unirradiated oranges. However, no significant differences were found after 13 and 28 days of storage. Navel oranges irradiated with 2–3 kGy were rated less acceptable after 2, 11, and 25 days of storage at 2°C. Temple oranges irradiated with dosages above 1 kGy and stored at 2°C for 7 and 13 days, were rated lowest of all. Similar trends were reported for irradiated Duncan grapefruit when stored at 15°C and 18°C.

Maxie et al. (1963) state that California Navel oranges subject to 2–3 kGy radiation doses and stored for two months at 0°C were decidedly more orange in colour. However, irradiated Florida Navel and Temple oranges behaved differently, exhibiting a bleaching of colour with a resultant yellow–orange colour after storage for 30–35 days at 10°C. Maxie et al. (1963) reported that the flesh of both Eureka and Lisbon irradiated lemons developed a translucent appearance followed by a brownish colour within one to two weeks after irradiation. This condition was particularly noticeable in fruits subjected to 2 and 3 kGy radiation doses.

Local research on gamma irradiation. The effects of gamma irradiation (from ^{60}Co) on the inhibition of sprouting in potato tubers (Ultimus and Dacca local) and onions (Punjab Red) were studied during storage (Hamid et al. 1965; Farooqi et al. 1967; Ali et al. 1970). A dose of 60–100 Gy was found to be optimum for the inhibition of sprouting. No significant changes in the rate of respiration (O_2 uptake), biochemical constituents (ascorbic acid, carbohydrates, and phosphorylase and invertase activity), and cooking qualities were recorded.

Low–dose gamma irradiation at 300 Gy to hard, green, pre–climacteric Desi and Samar Bahisht mangoes (Rashid and Farooqi 1984) and guava (Sufeda) significantly ($p<0.05$) delayed the ripening of the fruit. No phytotoxic effect of radiation at this dose was observed. The vitamin C content, acidity, reducing, non–reducing, and total sugars, and sugar–acid ratio of the irradiated fruit remained normal.

Radio–pasteurization of citrus fruits was also done using 1–3 kGy radiation doses. Skin damage to the fruit in the form of pitting or dark brown patches was observed. The extent of damage was dependent upon radiation dose and storage time. Radiation caused an increase in the respiration (CO_2 and C_2H_4 production) of citrus fruit. However, no significant change in biochemical constituents like vitamin C, acidity, reducing, non–reducing, or total sugars was observed except at a dose of 3 kGy.

Shelf life extension of some minor fruits like *falsa* and *ber* was also observed at doses below 0.5 kGy. There were no deleterious effects on the acidity, TSS, vitamin C content, glucose, fructose, or sucrose contents of *falsa*; nor were there any on the vitamin C, carotenoids, pectines, sugars, and acidity of *ber* during storage. The effect of gamma irradiation on shelf–life extension and processing quality of dates (CVS. Hillawi and Khudrawi) was also studied under non–refrigerated storage conditions. A dose of 1 kGy proved useful for this purpose and had no deleterious effect on the eating quality of the fruit (Siddique et al. 1987).

12.5.3 Wax coatings

As described earlier, post-harvest deterioration of fresh horticultural produce is due to water loss, senescence, and development of physiological disorders and fungal attack. While refrigeration is the principal means of retarding this deterioration, and external atmosphere control is finding wider application, certain ancillary treatments are of value (Hall and Trout 1944). Waxing the fruit surface adds shine and reduces water loss and rate of respiration, thus improving external appearance and extending storage life. The effect of wax coating on certain physiological processes which particularly reduce edible quality have been studied by various workers. Platenius (1939) studied the effect of waxing in reducing the shrinkage of potatoes, while Jones and Richey (1939), Nelson (1939), Trout et al. (1953), and Schomer and Pearson (1968) studied the effect of waxing on deciduous fruits like apples. Such a study had been previously carried out by Mack and Janer (1942) on cucumbers. All these workers observed considerable reduction in weight loss of the produce as compared to control samples.

Since research on the waxing of apples gave very encouraging results, research was continued by many workers like Hall and Trout (1944) and Eaks and Ludi (1960). Other fruits for which this treatment showed economic benefits were citrus, for which it became a common commercial treatment to maintain freshness.

Several wax formulations have been developed and tested experimentally. Suitable emulsifiers are used to produce a wax emulsion. Synthetic or natural resins are added to the wax emulsions to give more gloss to the treated produce. Water–wax emulsions are safer to use than solvent waxes, which are highly flammable. Water–wax emulsions can be used without drying the fruit prior to the treatment, therefore, washing and waxing treatments can be combined. Carnauba wax, sugarcane wax, thermoplastic terpene resin, and shellac are common substances which are used commercially; however, waxes of plant origin are preferred.

Developments in wax emulsions were not limited to the formulation of new types of wax, but also included the addition of chemical fungicides, which gave waxing treatment an additional function. For example, diphenylamine added to the wax emulsion not only gave additional shine and reduced weight loss and respiration rate of apples and pears, but also reduced the occurrence of scald. Addition of post-harvest fungicides to the wax emulsion to control *Penicillium* rot in stored citrus fruits also proved helpful.

Waxes may be applied by dipping, spraying, foaming, or brushing. Foaming is a better method of application since it leaves a very thin layer of wax on the fruit. If a thicker layer of wax is applied to the fruit surface, it becomes an undesirable barrier between the external and internal atmospheres and restricts exchange of respiratory gases (CO_2 and O_2). This may result in anaerobic respiration, resulting in fermentation and development of an off flavour. It is, therefore, necessary to adjust the thickness of the wax coat according to the variety and storage and marketing temperatures.

Local research on the effect of wax coatings. The effect of the wax emulsions Fruitex, Britex–561, and SB 65 on oranges (Pineapple and Valencia), mandarins (Feutrell's Early and Kinnow), lemons (Eureka), and grapefruit (Marsh) was studied during refrigerated and non–refrigerated storage (Farooqi et al. 1988). Wax coating of fruit improved its external appearance by giving additional shine to the fruit surface. It reduced weight loss and kept the fruit firmer, thus maintaining its fresh look, while it reduced rates of respiration and ethylene production, thus delaying senescence. No significant changes in constituents like vitamin C, acidity, and sugar content (in citrus fruit) were recorded due to the treatment. However, decrease in vitamin C content and acidity, and an increase in sugar content were observed as a result of storage. Higher concentrations of wax resulted in the development of off flavour in citrus fruits particularly during holding at 20°C, making this treatment unsuitable for this fruit.

Some preliminary work on the waxing of mangoes (Samar Bahisht) was also carried out during storage at temperatures ranging from 30–35°C. Weight loss of wax–coated mangoes was significantly reduced ($p<0.01$) and the appearance of the treated fruit was maintained. Ripening of the wax–coated mangoes was delayed ($p<0.05$). It was, however, found necessary to tailor the thickness of the wax coat according to storage temperature to avoid development of off flavour.

12.5.4 Antifungal agents

Continued metabolic activities of the fruit result in senescence and ultimately death of the tissue. The attack of microorganisms at any stage (pre– or post–harvest) results in decay. Huelin (1942) in Australia reviewed the storage and export problems of citrus fruits in the period between 1935 and 1942. According to Hall (1938), borax was found the most effective of several materials used commercially as post–harvest dips. It was used to a limited extent until the mid–1950's; later on, a combination of borax and boric acid was preferred. Long and Roberts (1958) reported the replacement of borax and boric acid with sodium orthophenylphenate (SOPP) in citrus–packing houses. SOPP was preferred as it was found more effective, did not leave any residue, and caused less fruit shrinkage.

Supplementary use of diphenyl wraps gave good protection against moulds and stem–end rot, especially in export packs. Diphenyl wraps were an effective fungicide, but they left taints in fruit flavour. Although dibromotetrachloroethane and esters of orthophenyl phenol were found relatively better in keeping fruit flavour normal, these fungicides were not approved by health authorities. Eckert and Kolbezen (1962) reported that 2–aminobutane (2–AB) was very effective against green mould (*Penicillium digitatum*). The use of 2–AB was extensively tested in California. In Gosford (Australia), 2–AB was found as effective as SOPP in controlling green mould in stored citrus fruits, and did not cause skin injury. In the USA, Harding (1976) reported the effectiveness of Imazalil against post–harvest decay of citrus. Pierson and Wright (1977) discovered that Imazalil was effective against Benzimidazole–resistant strains of *Penicillium expansum*. Kaplan and Dave (1979) reported the use of Imazalil with aqueous carriers such as spray, in wax emulsions, and applied by drenching and dipping. Kuramoto and Yamada (1975) reported the effective use of Guazatine in controlling rot in citrus fruits. The use of this fungicide for the control of sour rot in lemons, oranges, and tangors was also tested by Rippon and Morris (1981). According to these workers, the use of Guazatine gave better results than SOPP and 2–AB. Tridemorph, RH 2161, and benomyl are also used to control rot in stored citrus with beneficial effects.

Thiabendazole, 2–(Thiazol–4–yl) benzimidazole (TBZ), is a well–known anthelmintic agent and was reported by Staron et al. (1964) as a plant fungicide. Since then, a number of research workers have reported its effective use as post–harvest fungicide. Crivelli (1966) reported complete inhibition of citrus moulds *Penicillium digitatum* and *Diplodia natalensis*. The use of TBZ for the control of citrus moulds was reported in Australia by Seberry and Baldwin (1968). Merck Sharp and Dohme introduced Tecto–60 in Pakistan in 1971, which was replaced by Tecto–40 in 1972. Research on the use of TBZ for the control of mould rot was then undertaken, and its effectiveness for Kinnow mandarin was reported by Farooqi et al. in 1975.

Pathogens which can develop resistance to fungicides are becoming more common with the advent of chemicals that inhibit specific metabolic processes. Wild (1974) has reported this problem for citrus fungicides. The information that citrus green and blue moulds can develop resistance to the Benzimidazole fungicides which control these moulds most effectively has caused the citrus industry great concern (Wild and Eckert 1982). If the resistance of pathogens to certain fungicides is known, scientists should explore the possibilities of using other fungicides. The development of new antifungal agents from time to time is therefore needed to combat this problem. The above mentioned fungistatic chemicals were found effective for *Penicillium* rots and sour rots which develop at any stage during storage.

During extended storage in both under–refrigerated and non–refrigerated conditions, a disease known as black rot or internal blackening causes damage to stored citrus fruit. The fruit appears wholesome externally but is found black when cut open. This disease has been reported to be caused by *Alternaria citri* which enters through dried buttons (Farooqi et al. 1979). This disease creates a practical problem during extended storage and marketing, and its incidence has been reported from time to time in several countries. However, no single antifungal agent has been found effective in controlling this infection of *A. citri* in stored citrus fruits. In Pakistan, Malik et al. (1972) reported a broad–spectrum antibiotic substance from a local strain of *Bacillus subtilis* AECL–69 which was effective against many types of pathogens including *Alternaria* spp. This antifungal complex (AFC) was therefore tested on Valencia oranges, and application of it to the cut stem ends prior to storage was found effective in controlling black rot in stored citrus fruit.

As evident from the above cited literature, TBZ is the most effective fungicide for controlling citrus fruit diseases caused by *Penicillium* spp. Being an anthelmintic agent, it is also safe and recommended by medical specialists as a treatment for roundworms in human beings at a dose of 50 mg/kg/day. This fungicide has no phytotoxic effect on the fruit skin. The post–harvest residue of TBZ in oranges is far less than the normal dose given to a child, and hence is safe.

Acceptable post–harvest residues of several fungicides for several countries are given in Table 12.6.

Table 12.6 Acceptable post–harvest fungicide residue for citrus

	Fungicides (ppm)						
Country	**TBZ**	**SOPP**	**Imazalil**	**Diphenyl**	**2,4–D**	**Benomyl**	**2–A,B**
Australia	10	20	–	110	5	10[x]	30
Austria	6	12	–	70	–	7[v]	–
Belgium	6	12	5	70	–	–	–
Canada	10	10	–	110	2	10	–
Denmark	6	12	5	70	–	10	–
Finland	6	10	5	70	–	–	–
France	6	12	5	70	–	1.5[u]	–
W. Germany	6	12	5	70	2	7[v]	30
Hong Kong	–	10	–	100	–	–	–
Ireland	6	12	5	70	–	–	–
Italy	6[s]	12	–	70	0.5[t]	–	–
Japan	10	10	–	70	–	–	–
Luxembourg	6	12	–	70	–	–	–
Netherlands	6	12	5	70	–	1.5[v]	30
New Zealand	3	10	5	110	5	5	30
Norway	6	10	5	70	–	–	–
Poland	10	10	–	70	–	–	–
Portugal	6	12	–	–	–	–	–
S. Africa	6	–	–	–	–	5	5

Country	TBZ	SOPP	Imazalil	Fungicides (ppm) Diphenyl	2,4–D	Benomyl	2–A,B
Sweden	6	10	5	110	2	10	30
Switzerland	6	12	5	70	–	7[v]	–
UK	10	12	–	70	2	–	–
USA	10	10	10[x]	110	5	10	30
FAO/WHO	10	10	5[x]	110	2	10	30[x]

x Temporary tolerance
u If treated with benomyl, treatment with another fungicide prohibited.
t Registered as pre-harvest use only.
s Maximum of 3 ppm allowed if used with other fungicides.
v Acceptable daily intake.
Source: Farooqi (1986)

Pakistani research on use of antifungal treatments. TBZ, Benlate, Captan, Antracal, and hot water dip, alone and in combination with gamma irradiation (300 Gy), were used to control the post-harvest problem of anthracnose and stem-end rot in mangoes. Use of fungicides had no significant effect on the control of decay in mangoes; however, hot water dip (55°C) alone highly significantly ($p<0.01$) controlled this disease.

Dipping of citrus fruit in TBZ prior to storage highly significantly ($p<0.01$) reduced fruit rot in stored citrus fruit. Use of antifungal wax emulsions like Fruitex, Britex–561, and SB–65 also helped to minimize post-harvest decay in stored citrus fruits; however, like other post-harvest applications, they failed to control black rot. However, application of antibiotic F (ABF) derived from local strain of *Bacillus subtilis* AECL–69 on the cut stem ends prior to storage highly significantly ($p<0.01$) controlled the incidence of this disease. When TBZ and ABF were mixed in water and fruits dipped for two minutes, a synergism was observed; such an application gave the best results for stored citrus fruits (Farooqi et al. 1981, 1985). No phytotoxic effect of ABF was recorded on Kinnow and this process was patented (Farooqi et al. 1988, 1991).

12.5.5 Packaging

Improvements in packaging can lead to consumers receiving produce in the market in a fresh condition with less damage, longer potential shelf life, and better eye appeal. Modern packaging has contributed to the improved handling of fruits and vegetables from farm gate to consumer in technically advanced countries; however, it needs urgent attention in underdeveloped countries.

For many centuries, widespread distribution of most perishables was restricted to their producing areas (Hale et al. 1982). In the developed countries, farmers have modern warehouse facilities to wash, grade, and pack produce before distribution. However, the pace of such progress may be slow in some countries due to poor policy and management. Since there is tough competition in the international market, good packaging needs to be instituted, otherwise the exporter is likely to lose buyers.

Good packaging maintains but does not improve the quality of the fruit or vegetables packed in it, therefore the best possible produce should be packaged (Ahmad et al. 1979, 1989). Moreover, packaging is only one of the post–harvest procedures and not a substitute for other methods of conservation. Some of the benefits of improved packing are listed here.

- It protects against mechanical damage, moisture loss, and theft.
- It provides beneficial modified atmosphere (MA) conditions, which can extend shelf life.
- It serves as an efficient handling unit.
- It reduces the cost of handling, transport, and marketing.
- It facilitates the use of new modes of transportation.
- It provides service and sales motivation.

As described earlier, proper packaging protects fresh produce from the external environment, especially bruising and abrasion during transit. Cuts caused by sharp edges may cause serious quality losses from subsequent microbial decay. To get full protection, the containers should be lined or padded. Trays or tissue wraps also prevent damage from contact with rough surfaces or adjacent produce. Packaging reduces moisture loss and thus prevents shrivelling, especially when moisture/vapour barrier materials are used. By packing topped carrots or other root crops in polyethylene bags, weight loss is significantly reduced. Similar results have also been reported for mango and citrus fruits. Prevention of rapid wilting in vegetables is also important to retard loss of vitamin C (ascorbic acid) and carotene.

Beneficial MA may develop in film–wrapped packages. This retards ripening and extends storage life for produce like tomatoes and bananas, however, such sealed packages must be perforated or opened before marketing to allow normal ripening and prevent injury from high levels of CO_2 or low levels of O_2.

Types of containers and packaging material. The following types of packaging containers and materials are used for fresh fruit and vegetables.

FIELD CONTAINERS: Field containers are of many types, depending upon the crop and availability of raw materials in the country. Fruits should generally be harvested in canvas or cloth picking bags and transferred to large wooden boxes in the orchard.

SHIPPING CONTAINERS: Packaging for shipping and handling requires containers strong enough to protect produce from bruising, vibration, and the weight of other stacked containers. Containers strong enough to be stacked allow maximum utilization of space in storage and transport. In developed countries, many shipping containers are used only once and are not returned to the shipper, but in less–developed countries baskets or boxes are often returned or sold for multiple use. Fibreboard (corrugated) cartons are becoming popular for shipping tropical and subtropical produce because of their light weight and low cost. Absorbing moisture and losing strength may be disadvantages; therefore, quality material should be used. In some cases fibreboard cartons may be wax coated against moisture.

BULK BOXES OR PALLET BINS: When the containers of fruit and vegetables are going to terminal market warehouses, pallet bins are prepared. That is, the boxes of produce are stacked in container–sized lots and stored to await further transport.

CONSUMER PACKAGES: Small consumer–sized packages for horticultural produce are used in retail shops. The produce is washed, cleaned, and graded and then packed in nylon net, plastic, or foamed plastic bags. In advanced countries, produce is often kept in the rear of the shop before sale.

Utilization of packaging in tropical regions. With the increase in export of horticultural produce from the tropical countries, improved packaging needs to be introduced. It is estimated that 25% or more of perishables go to waste in such countries due to improper transport, lack of storage, and inadequate packaging.

Local research on packing techniques. The effect of certain lining materials like tissue paper, wax paper, newsprint paper, thin polyethylene, and cellophane films on the storage life of mango and citrus fruits was studied (Farooqi et al. 1979). Weight loss of the fruit was found to be least when the containers were lined with thin cellophane or polyethylene (0.03 mm), followed by wax paper and newsprint paper. Prevention of weight loss was least with tissue paper lining. The effect of packaging on grapes and citrus fruits has been studied by Beattie and Outhred (1970) and Ahmad et al. (1979), who found that polyethylene and cellophane films helped to reduce weight loss and maintain the fresh look of citrus fruits.

12.5.6 Storage techniques

Refrigeration. Refrigeration is the principal means of extending the shelf life of most fruits and vegetables after harvest, as it slows down metabolic activity (Fidler 1968). Fruits of tropical and subtropical origin show a physiological disorder called **chilling injury** if exposed to temperatures below 10°C but above their freezing point. Severity of injury generally increases as storage temperature decreases and the duration of storage increases. Chilling symptoms may develop during or after exposure. The injury in many fruits and vegetables is manifested as discolouration, susceptibility to decay, or failure to ripen; it is progressive with time, indicating a general degeneration of metabolism. Low–temperature injury is at two levels.

Group A: Subtropical and tropical fruit and vegetables such as bananas, mangoes, tomatoes, cucumbers, and sweet potatoes are subject to injury after short exposures to temperature below 10°C.

Group B: Temperate climate fruits such as plums, peaches, and apples, may suffer chilling injury when stored for long periods below 5°C.

Chilling injury, particularly to produce in group A, is an economic problem and constitutes a major obstacle in the storage, handling, and transport of fresh produce. Bananas are a typical example of highly chilling–sensitive fruits of this group. They may be injured by a 12–hour exposure below 8°C and even by temperatures as high as 12°C or 14°C at longer exposures, the degree of injury varying with variety and other factors. Research has been carried out to find optimum storage conditions for various fruits. Some of these results are given in Table 12.7.

Table 12.7 Optimum storage conditions for various fruits

Fruit	Temperature (°C)	Relative humidity (%)	App. storage life (weeks)
Avocado	12/8	85–90	2
Banana, green	12.8--15.6	85–90	4
Banana, ripe	12.8	85–90	1.5
CITRUS			
Mandarin	5.6–7.2	85–90	6
Orange CV. Valencia	4.4–6.1	88–92	5–6
Lime, yellow	11.1–12.8	85–90	8
Lime, green	11.1–12.8	85–90	7
Lemon	5.6–7.2	85–90	6
Grapefruit	5.6–7.2	85–90	8–12
Pummelo	7.2–8.9	85–90	12

Fruit	Temperature (°C)	Relative humidity (%)	App. storage life (weeks)
Date	6.7	85–90	2
Fig	0.0–1.7	85–90	7
Guava	8.3–10.0	85–90	2–5
Litchi	1.7	85–90	8–12
Mango	7.2–10.0	85–90	3–6
Papaya, green	10.0	85–90	3–4
Papaya, turning	8.3	85–90	2–3
Passion fruit	5.6–7.2	85–90	3
Persimmon	0.0–1.7	85–90	7
Pineapple, green	8.3–10.0	85–90	4–6
Pineapple, turning	4.4–6.7	85–90	1.2
Plum	0.0–1.7	85–90	2
Pomegranate	0.0–1.7	85–90	11

Post–harvest chilling sensitivity of fruit is influenced by many factors such as variety, climate in which a variety is being grown, maturity at harvest, stage of ripeness, fruit composition, and composition of the storage atmosphere. Useful research on fruits has been carried out by many workers (Hatton et al. 1980, 1982; Farooqi 1983, 1987).

Local research on refrigerated storage. Studies on the chilling sensitivity of mangoes and citrus were carried out. It was observed that the skins of both mango and citrus fruits exhibited damage due to chilling, while the physiology of the whole fruit was less affected. Chilling sensitivity of Sensation variety of mango was found different than that of Samar Bahisht (Farooqi et al. 1985). A set of recommendations for mangoes based on the experimental results at NIAB was prepared, and successful export trials were recently carried out (Farooqi and Qureshi 1991; Langer 1991). In studies of the chilling sensitivity of citrus fruits, Marsh grapefruit and Eureka lemons were found significantly ($p<0.05$) more sensitive to chilling during storage than Kinnow mandarins (Farooqi et al. 1987).

Controlled–atmosphere storage. The composition of gases in the storage environment can affect the storage life of produce. An atmosphere with reduced concentration of O_2 or increased concentration of CO_2 or both (decrease in O_2 and increase in CO_2) is called a **controlled–atmosphere** (CA) or **gas storage**, or **modified atmosphere** (MA). Under this type of storage, metabolic activities are slowed down, which results in shelf life extension.

Many volatile compounds evolved in the stored produce may accumulate in the storage atmosphere. Ethylene gas (C_2H_4) is the most important of these compounds. The accumulation of C_2H_4 may enhance ripening, thus reducing shelf life of the produce. It is therefore necessary to remove C_2H_4 gas from the CA storage.

Pakistani research on controlled-atmosphere storage. The storage life of fresh fruits and vegetables has been extended by controlled-atmosphere storage. Langra mangoes remained in a hard green condition for two weeks during storage in nitrogen atmosphere at 25–30°C. However, when they were transferred to normal air, they failed to ripen normally due to the damage which had been done during storage at sub-critical O_2 level. In Pakistan there are no commercial CA stores. Further research on the use of CA storage needs to be conducted, particularly for climacteric fruits of tropical origin. The benefits of this technology can only be realized if practical research is carried out and the findings utilized for the benefit of farmers.

Hypobaric storage. Hypobaric or low-pressure storage is a form of CA storage in which the produce in stored in a partial vacuum. The vacuum chamber is flushed continuously with water-saturated air to maintain the level of O_2 and minimize water loss. Reduction in the partial pressure of O_2 and reduction in C_2H_4 levels in hypobaric storage result in retarding ripening of fruits. A reduction in air pressure to 10 kilopascals (0.1 atmosphere) is equivalent to reducing O_2 concentration to about 2% at normal atmospheric pressure (76 cm Hg).

Research on hypobaric technology has been carried out by Salunkhe and Wu (1973) on deciduous fruits, by Burg (1975) on various fruits and vegetables, and by Spalding and Reeder (1976a, 1976b, 1977) on avocados, limes, and mangoes, respectively. Hypobaric stores are expensive to construct because of the low pressure required inside the store. Its high cost is therefore one of the factors limiting the practical expansion of this technology. Moreover, for its economical use, more research is required on how it affects different kinds of horticultural produce.

12.6 Transportation

Road. Fresh horticultural produce is generally transported by road in Pakistan—using animal-drawn carts, trolleys, trucks, or trailers. None of these are refrigerated. In most cases roads are uneven and damaged, which causes mechanical damage to the produce due to vibration during transit. Moreover, the way the produce is packed and stacked is very haphazard. Under such conditions the shelf life of produce is significantly reduced.

Rail. Produce is also sent to distant markets by rail. The arrangements are not adequate because there are no special facilities for produce; special compartments where temperature can be controlled are generally not available.

Sea. For export purposes, sea transport by launches or ships is available. In addition to the Pakistan National Shipping Corporation (PNSC), many shipping companies offer this facility and provide reefer containers.

Air. Air transport of highly perishable items is very expensive and possible only in the export business. Air-freight charges per kg of produce from one destination to another are three to four times those for sea freight. However, not all horticultural produce can be transported by ship because of the time factor. An aircraft takes only a few hours to carry produce from one destination to another, whereas a ship takes days or weeks. The exporter has to decide between cost and time. If shelf life of horticultural produce could be increased, surface shipment would become more feasible and growers/dealers would have more choice in shipping methods.

12.7 Marketing

Since horticultural products are living plant organs and do not cease their metabolic activities even after harvest, their handling, storage, distribution, and marketing (both domestic and export) is more complicated than for other agricultural produce. Economic benefit to a grower depends not only upon increased production but also on the profitable sale of the produce, which is dependent on planning and careful marketing. Present day marketing is a highly specialized field, in which both qualitative and quantitative information about product demand is gathered through market research, often utilizing sophisticated, modern electronic media. On the basis of such information, supplies of produce can be directed toward markets where there is more demand at a specific time. This can help to balance the supply and demand situation.

International marketing is much more difficult and specialized than domestic, as the demand for a product in each importing country is very specific as regards variety, quality, and acceptable post-harvest chemical residues. Therefore, the exporting country has to tailor the produce in accordance with the demands of the importing country. In some cases when a new or exotic variety is to be introduced in a country where the consumers are not aware of its attractive taste or nutritive value, sales promotion work has to be carried out so that the new product can be introduced to that segment of the export market.

In Pakistan much improvement is needed in the marketing of horticultural produce. Surveying markets in the four provinces of Pakistan, one can easily identify the following major problems.

- Poor hygienic conditions of markets
- Improper packing and grading of produce

- Rough handling during marketing
- Undesirable role of the middleman

Each step of the marketing process is an important aspect of post–harvest handling and demands the attention of all those who are directly or indirectly involved in the marketing of fresh horticultural produce. Such steps will increase the amount of nutritious food available to people and increase both domestic and foreign–exchange earnings.

STUDY QUESTIONS

1. Why is handling horticultural produce different from handling other agricultural produce like grains?
2. Write a note on the precautions that should be followed in the harvesting of fruit and vegetables.
3. Discuss the causes of post–harvest deterioration of horticultural produce.
4. Outline the classification of fruits according to respiratory behaviour.
5. Draw the respiratory curve of any two of the following fruits: banana, avocado, apple, or pear.
6. Describe some of the post–harvest techniques which help to extend the shelf life of horticultural produce.
7. What is the role of radiation technology in the storage behaviour of fruit and vegetables?
8. What are the maximum tolerable radiation doses for vegetables like potatoes, onions, and tomatoes?
9. What is wax coating? Discuss the advantages and disadvantages of this technology. Name one fruit on which waxing is used commercially in Pakistan.
10. Comment on the relationship of antifungal treatment and prolonged shelf life. Is such treatment necessary? If so, under what circumstances, and for what types of produce?
11. Discuss post–harvest fungicide residues, giving the minimum concentration of residue allowed by the WHO for TBZ, SOPP, and Diphenyl in Australia, Germany, and the UK.

12. Discuss the role of packing in the post–harvest treatment of horticultural produce. What constraints does a grower in Pakistan face in packing his produce?

13. Discuss the use of refrigeration technology in post–harvest storage. How are various fruits and vegetables grouped with respect to optimum storage temperatures?

14. What is chilling injury? Which fruits are more susceptible to it?

15. Give the optimum storage temperatures for the following fruits: persimmon, pomegranate, guava, lemon, litchi.

16. Discuss the advantages and disadvantages of controlled–atmosphere and hypobaric storage.

17. Marketing strategies and technology are critical for perishable fruits and vegetables. (a) Describe the practices currently followed in marketing these commodities in your local area. (b) Suggest feasible improvements in marketing of these commodities for the growers of your area.

18. The business of exporting horticultural produce has been slow in the past. How would you suggest using recently developed technology to improve the volume of export of these commodities?

19. Describe research on post–harvest technology and handling that has been done in Pakistan to date.

20. Compare and contrast the storage needs of apples, citrus fruits, and mangoes.

21. (a) Describe post–harvest handling, storage, and distribution practices currently being followed in Pakistan. (b) Formulate a comprehensive plan for improvement of this sector of agriculture in the country.

REFERENCES

Ahmad, M.; Shah, Z.; Durrani, Z.; Chaudhry, M. A.; Khan, I. Effect of film packaging on physico–chemical characteristics of Blood–Red oranges. Pak. J. Agri. Res. 10:66–73; 1989.

Ali, M.; Farooqi, W. A.; Sattar, A.; Mahammed, A. Inhibition of sprouting in onions by gamma radiation. Food Irradiat. 10:10–14; 1970.

Asi, M. R.; Nawaz, R.; Hussain, A.; Farooqi, W. A. Some studies on the volatile components of Kinnow (*Citrus reticulata* Blanco). Proc. 1st Natl. Chem. Conf. Peshawar; 1989:p. 645–58.

Beattie, B. B.; Outhred, N. L. Packaging treatments for the storage and export of Australian grapes. Aust. J. Exper. Agric. 10:124–128; 1970.

Beraha, L.; Ramsey, G. B.; Smith, M. A.; Wright, W. R. Factors influencing the use of gamma irradiation to control decay of lemons and oranges. Phytopathol. 49:91–96; 1959.

Burg, S. P. Hypobaric storage and transportation of fresh fruits and vegetables. In: Postharvest biology and handling of fruits and vegetables. Westport: AVI Pub. Co.; 1975.

Bussel, J.; Maxie, E. C. Gas exchange in bartlett pears in relation to gamma irradiation. Proc. Amer. Soc. Hort. Sci. 88:151–159; 1966.

Crivelli, G. Ricerche sulla utilizzione commercializazone degli agermi. Nota V: Prov sul potere fangistatico del 2–(4–triazoli) benzimidazole sui *Penicillium* delle aranac. I! Freddo 20, 25; 1966.

Dennison, R. A.; Grierson, W.; Ahmed, E. M. Irradiation of Duncan grapefruit, pineapple and Valencia oranges and Temples. Proc. Fla. State Hortic. Soc. 79:285; 1966.

Eaks, I.L.; Ludi, W. A. Effects of temperature, washing and waxing on the composition of the internal atmosphere of orange fruits. Proc. Am. Soc. Hort. Sci. 76:220; 1960.

Eckert, J. W.; Kolbezen, M. J. Control of *Penicillium* decay in citrus fruit with 2 aminobutane. Nature 194:888; 1962.

Farooqi, H. Phyto–toxic effects of some hormones and chemicals on citrus plant. Faisalabad: Dept. of Botany, University of Agriculture, Faisalabad; 1988. M.Sc. thesis, 57 pp.

Farooqi, W. A. Studies on the radiation preservation of oranges (*Citrus sinensis* L). Proc. Sem. Rad. Preserv. of Food. Report No. 8. Lahore: Atomic Energy Centre, Radiobiology Division; 1966:p. 19–20.

Farooqi, W. A. Some physiological studies on banana, apple and pear during storage and ripening. Macquarie University; 1970. M.Sc. Thesis, 149 p.

Farooqi, W. A. Technical aspects of fruit storage. Punjab Fruit J. 29:72–77; 1981.

Farooqi, W. A. Biochemical and physiological basis of shelf life extension of citrus fruits by refrigeration and C.A. storage. NIAB/PL–480/FS2 (A USDA Res. Project Report); 1983, 103 p.

Farooqi, W. A. Physiological, biochemical and microbiological studies on the shelflife extension of citrus fruit. University of the Panjab, Lahore; 1986. Ph.D. Thesis, 152 p.

Farooqi, W. A. Post–harvest technological studies on citrus fruits. In: Citrus. Islamabad: Horticultural Foundation of Pakistan; 1991, brochure, p. 33–36.

Farooqi, W. A. Some practical steps to minimize postharvest losses of mango fruit. In: Mango. Islamabad: Horticultural Foundation of Pakistan; 1991, brochure, p. 25–27.

Farooqi, W. A.; Ahmad, M. Postharvest studies on citrus and mangoes. Proc. Natl. Sem. on Dehydration and Preserv. of Veg. and Fruits, organized by ATDO, held at AARI, Faisalabad, Nov. 23, 1980. Faisalabad: AARI; 1980:p.40–46.

Farooqi, W. A.; Ahmad, M. Process for the preservation of Kinnows. Pak. Patent No. 127370; 1981.

Farooqi, W. A.; Ahmad, M.; Hussain, A.; Naqvi, M. H. Effect of gamma irradiation on Kinnow mandarin during storage. Nucl. 11:25–29; 1974.

Farooqi, W. A.; Ahmad, M.; Hussain, A. M.; Qureshi, M. J. Effect of thiabendazole and packaging material on Kinnow mandarin during storage. Nucl. 12:25–29; 1975.

Farooqi, W. A.; Ahmad, M.; Khalid, Z. M. Effect of polyethylene and some lining materials on the shelf–life extension of citrus fruits. Citrus Ind. 10: 14–20; 1979.

Farooqi, W. A.; Ahmad, M.; Khanum, S. A.; Hatton, T. T. Studies on the shelf life extension of citrus fruits. Proc. Int'l. Citri. Soc., Tokyo 2:760–61; 1981.

Farooqi, W. A.; Ahmad, M. S.; Zain–ul–Abidin. Effect of wax–coating on the physiological and biochemical aspects of Kinnow fruit. Pak J. Sci. Ind. Res. 31:142–145; 1988.

Farooqi, W. A.; Ali, M.; Muhammed, A. Effect of gamma irradiation on guavas (*Psidium guajava* L.). Food Irradiat. 8:37–40; 1967.

Farooqi, W. A.; Daud, K.; Hussain, M.; Sattar. A., Jr. Studies on the postharvest chilling sensitivity of mango fruit (*Mangifera indica* L.). Proc. Fla. State Hortic. Soc. 98:220–21; 1985.

Farooqi, W. A.; Donini, B. Effect of gamma irradiation on onion sprouting. I. Influence of dose rate and postharvest period of treatment. Riv. Sci. Tecn. Um. 6:301–303; 1976.

Farooqi, W. A.; Hall, E. G. Effect of ethylene dibromide on the respiration and ripening of bananas (*Musa cavendishii* L.). Nucl. 9:22; 1972.

Farooqi, W. A.; Mohyuddin; Hamid, M. A. Irradiation of potatoes in Pakistan. Food Irradiat. 7:41–44; 1967.

Farooqi, W. A.; Qureshi, M. J. Commercial trials on the export of mango to Germany. In: Mango. Islamabad: Horticultural Foundation of Pakistan; 1991, brochure, p. 23–24.

Farooqi, W. A.; Sattar, A., Jr.; Qureshi, M. J. Improvement and maintenance of fruit quality by the application of post–harvest techniques. Proc. Natl. Sem. Food Safety and Nutrition. PCSIR: Lahore; 1991.

Farooqi, W. A.; Shaukat, G. A.; Malik, M. A.; Ahmad, M. S. A study on the control of internal blackening in oranges. Punjab Fruit J. 37:51–55; 1979.

Farooqi, W. A.; Shaukat, G. A.; Malik, M. A.; Ahmad, M. S. Studies on the inhibition of *Alternaria citri* in stored citrus fruit. Proc. Fla. State Hortic. Soc. 94:279–80; 1981.

Farooqi, W. A.; Shaukat, G. A.; Malik, M. A.; Ahmad, M. S. Control of black–rot in inoculated citrus fruit. Citrus Ind. 66:62–67; 1985.

Farooqi, W. A.; Shaukat, G. A.; Malik, M. A.; Ahmad, M. S. Influence of pH on the growth of *Alternaria citri* in citrus fruit juice. Proc. Fla. State Hortic. Soc. 98:214–15; 1985.

Farooqi, W. A.; Shaukat, G. A.; Malik, M. A.; Ahmad, M. S. Process for the control of black rot in stored citrus fruits. Pak. Patent No. 130437; 1988.

Farooqi, W. A.; Zain–ul–Abidin; Ahmad, M. S. Physiological and biochemical studies on chilling sensitivity of citrus fruit. Punjab Fruit J. 40:10–17; 1987.

Fidler, J. C. Low temperature injury to fruits and vegetables. In Recent Advances in Food Science 4:271– 283. Sydney: Pergamon Press; 1968.

Goldblith, S. A. Radiation preservation of food: The current status. J. Fd. Technol. 5:103–110; 1970.

Government of Pakistan. National Comm. on Agri. Islamabad: Ministry of Food and Agriculture, Government of Pakistan. n.d.

Hale, P. W.; Davis, P. L.; Marousky, F. J.; Bongers, A. J. Evaluation of a heat shrinkable polymer film to maintain quality of Florida grapefruit during export. Citrus Veg. Mag. 46:39–48; 1982.

Hall, E. G. The storage of oranges. Agri. Gaz. of N.S.W. 49:558; 1938.

Hall, E. G.; Trout, S. A. Some effects of waxing on weight loss from oranges and certain vegetables. J. Aust. Inst. Agr. Sci. 10:80; 1944.

Hamid, M. A.; Mohyuddin, M.; Islam, F.; Ifzal, S. M.; Bhatti, A. S.; Faiz–ur–Rehman, A. T. M. Preservation of potatoes by the application of gamma irradiations, RB/FPI. Lahore: Atomic Energy Centre Lahore; 1965, 65 p.

Harding, P. R., Jr. R. 23979, a new imidazole derivative effective against postharvest decay of citrus by molds resistant to thiabendazole benzol and 2 amino butane. Plant Dis. Rep. 60;643; 1976.

Hatton, T. T. Jr.; Beraha, L.; Wright, W. R. Preliminary trials of gamma irradiation on mature Irwin and Sensation mangoes. Proc. Fla. Mango Forum 21st Annual Meet. Homestead, FLA: Florida Mango Forum; 1961:p. 15–17.

Hatton, T. T.; Cubedge, R. H. Pre–conditioning Florida grapefruit to prevent or reduce chilling injury in low temperature storage. Hort. Sci. 15:375–76; 1980.

Hatton, T. T.; Cubedge, R. H. Conditioning of Florida grapefruit to reduce chilling injury during low temperature. J. Amer Soc. Hort. Sci. 107:57–60; 1982.

Huelin, F. E. The handling and storage of Australian oranges, mandarins and grapefruit. Commonwealth Scientific and Industrial Research Organization (CSIRO), Australia Bull. No. 154; 1942.

Janick, J. Horticultural science. 3rd ed. San Francisco: W. H. Freeman; 1979:p. 418–452.

Jones, S. E.; Richey, H. W. The use of wax emulsions in reducing desiccation of transplanted tomato plants and apples in storage. Proc. Amer. Soc. Hort. Sci. 36:751; 1939.

Kaplan, H. J.; Dave, B. A. The current status of Imazalil, a postharvest fungicide for citrus. Proc. Fla. State Hortic. Soc. 92:37; 1979.

Kurumoto, T.; Yamada, S. Occurrence of sour rot in Satsuma mandarins. Bull. Fruit Tree Res. Stn. Japan. Ser. B2 87; 1975.

Langer, C. Exporting mangoes from Pakistan: a handbook. Karachi: Export Promotion Bureau; 1991, 179 p.

Long, J. K.; Roberts, R. E. The phyto–toxic and fungicidal effects of sodium orthophenyle phenate in controlling green mould wastage in oranges. Aust. J. Agric. Res. 9:609; 1958.

Mack, W. B.; Janer, J. R. Effects of waxing on certain physiological processes of cucumbers under different storage conditions. Food Res. 7:38; 1942.

Malik, M. A.; Shaukat, G. A.; Ahmad, M. S.; Muhammed, A. Isolation and crystallization of an antibiotic substance from a local strain of *B. subtilis.* Nucl. 9:111; 1972.

Maxie, E.C.; Sommer, N. F.; Eaks, I. L. Radiation technology in conjunction with postharvest procedures as a means of extending shelf–life of fruits and vegetables. U.S. Atomic Energy Commission Report TID 1920–26; 1963.

Maxie, E. C.; Eaks, H. L.; Sommer, N. F. Studies on radiation induced ethylene production by lemon fruits. Radiat. Bot. 6:445–455; 1966.

Maxie, E. C.; Amezqiota, R.; Hassan, B. M.; Johnson, F. Effect of gamma irradiation on the ripening of banana fruits. Proc. Am. Soc. Hort. Sci. 92:235–254; 1968.

Morris, S.C. Synergism of *Geotrichum candidum* and *Penicillium digitatum* in infected citrus fruit. Phytopath. 72:1336–1339; 1982.

Nelson, R. E. Dry–bright wax for apples. Min. Hort. 67:131; 1939.

Pierson, C. F.; Wright, T. R. Imazalil for the control of a benzimidazole tolerant strain of *Penicillium expansum* from Anjone pears. Proc. Am. Phytopathol. Soc. 4:94; 1977.

Plantenius, H. Wax emulsions for vegetables. NY Agri. Exp. Stn. (Cornell) Bull. No. 723. Ithaca: Cornell University; 1939.

Rashid, A.; Farooqi, W. A. Studies on the effect of gamma irradiation and maleic hydrazide on the shelflife of mangoes. J. Agric. Res. 22:151–58; 1984.

Rippon, L. E.; Morris, S. C. Guazatine control of sour rot in lemons, oranges and tangors under various storage conditions. Scient. Hort. 14:245; 1981.

Salunkhe, D. K.; Wu, M. T. Effects of sub–atmospheric on the ripening behaviour of same chemical changes on certain deciduous fruits. J. Am. Soc. Hort. Sci. 98:113; 1973.

Schomer, H. A.; Pierson, C. F. The use of wax on apples and pears. Proc. 63rd Ann. Meeting of Wash. St. Hort. Soc. 1967. Pullman, WA: Washington State Horticultural Society; 1968:p. 198–200.

Seberry, J. A.; Baldwin, R. A. Thiabendazole and 2–amino butane as post-harvest fungicides for citrus. Aust. J. Exper. Agric. 8:440–443; 1968.

Spalding, D. H.; Reeder, W. F. Low pressure (hypobaric storage) of avocados. Hort. Sci. 11:491–492; 1976.

Spalding, D. H.; Reeder, W. F. Low pressure (hypobaric) storage of limes. J. Am. Soc. Hort. Sci. 101:367–370; 1976.

Spalding, D. H.; Reeder, W. F. Low pressure (hypobaric) storage of mangoes. J. Am. Soc. Hort. Sci. 102:367; 1977.

Siddique, M. M.; Cheema, A.; Farooqi, W. A.; Randhawa, M. A. Effect of difference doses of chemicals and irradiation on the physiological changes and storage behaviour of date fruit. J. Agric. Res. 25:207–212; 1987.

Staron, T.; Allard, C.; Gug, M. On the anti–fungal properties of 2–(4–thia–zolyl) benzimidazole or thiabendazole. Bulletin French Soc. of Practicing Vet. Oct. 1, 1964.

Trout, S. A.; Hall, E. G.; Sykes, S. M. Effect of skin coatings on the behaviour of apples in storage. Aust. J. Agric. Res. 4:57–81; 1953.

van Kooji, J. Food irradiation makes progress. Bull. No. 26. Vienna, Austria: International Atomic Energy Agency (IAEA); 1984.

Wild, B. L. Pathogen resistance to citrus postharvest fungicides. Food Tech. Aust. 26:505–508; 1974.

Wild, B. L.; Eckert, J. W. Synergy between a benzimidazole sensitive and benzimidazole–resistant isolates of *Penicillium digitatum*. Phytopathol. 72:1329–1332; 1982.

Wills, R. B. H.; McGlasson, W. B.; Graham, D.; Lee, T. H.; Hall, E. G. Postharvest: an introduction to the physiology and handling of fruit and vegetables. Oxford, London, Edinburgh, Boston, Melbourne: BSP Professional Books; 1989.

Outline of Chapter 13

13. FRUIT CROPS

Muhammad Ibrahim Chaudhary[1]

This chapter includes description of the botany and morphology, soil and climatic requirements, propagation, cultural practices, varieties, and important problems associated with major evergreen and deciduous fruits of Pakistan. The fruits are discussed with emphasis on their commercial importance.

LEARNING OBJECTIVES

A student who studies this chapter should be able to:

- Tell the soil and climate requirements of the major fruit crops of Pakistan
- Rank the major fruit crops of Pakistan by their commercial importance
- Given a specified location in Pakistan, tell which fruit crops can be successfully grown there, and what special measures may be needed

13.1 Citrus fruits

13.1.1 Botany

The classification of citrus fruits is controversial among botanists as well as horticulturists. The unique tendency of citrus to cross between species and also between genera has further complicated classification. Swingle (1943) divided the genus *Citrus* into three independent genera: *Citrus*, *Poncirus*, and *Fortunella*. He put 16 species under *Citrus*, four under *Fortunella*, and only one species under *Poncirus*. Hodgson (1961) divided *Citrus* into 36 species. Swingle's classification is usually followed, with some modifications as suggested by Hodgson.

All three genera belong to the sub–family Aurantioideae of the family Rutaceae. The sub–family consists of two tribes, Clausenae and Citreae. Citreae has 33 genera including *Citrus*, *Poncirus,* and *Fortunella*. The taxonomic relationship of the common citrus fruits is as follows.

---

[1]Dr. Chaudhary is Associate Professor, Department of Horticulture, University of Agriculture, Faisalabad.

Family	Rutaceae
Sub–family	Aurantioideae
Tribe	Citreae
Genus	1. *Citrus* (sweet orange, mandarin, grapefruit, lime, and lemon)
	2. *Fortunella* (kumquat)
	3. *Poncirus* (trifoliate orange)

These three genera can be distinguished in general from each other by their leaf and ovary characters. *Poncirus* has 3–lobed (trifoliate) leaves and is deciduous. *Fortunella* and *Citrus* are both unifoliate and evergreen, but the ovary of *Fortunella* is 3–6 celled while that of *Citrus* has 8 or more cells.

Genus *Citrus*. The genus *Citrus* has evergreen trees and shrubs. These are normally spiny with thick, leathery leaves and winged petioles. Flowers are white or in some species pink on the outside, pentamerous, axillary or as terminal lax cymes, and scented. They have small calyxes, hard sepals, and thick petals with densely spaced oil cells. The stamens are numerous (15–60), and the ovary superior with 8–15 carpels containing few to several ovules. The fruit has a thick leathery rind; botanically it is a special type of berry called a *hesperidium*. The fruit is globose, sub–globose, or elliptical and filled with juice sacs or vesicles. The fruit contains few to many white or light–green seeds, which are generally polyembryonic.

The genus *Citrus* is divided into two sub–genera, *Eucitrus* and *Papeda*. All the edible citrus cultivars are species of *Eucitrus*. *Papeda* species contain acrid oil droplets in the juice vesicles and are inedible. A few of the important commercial species are:

Citrus sinensis (L) Osbeck	Sweet orange
Citrus reticulata Blanco	Mandarin
Citrus paradisi Macf.	Grapefruit (pomelo)
Citrus grandis (L) Osbeck	Chakotra (pummelo)
Citrus limettioides Tan.	Sweet lime
Citrus aurantifolia Swing.	Kaghzi lime
Citrus limonia (L)	Lemon
Citrus medica (L)	Citron
Citrus aurantium (L)	Sour orange
Citrus jambheri Lush	Rough lemon

Genus *Poncirus*. The genus *Poncirus* has only one species, *P. trifoliate* Raf., which has compound leaves with three leaflets and is deciduous. The tree is small and spiny. Flowers are sessile, and borne on previous–year wood in the axils of the spines. They open wide and flat, and are creamy white. The inedible, pubescent fruit has 6–8 segments, is filled with large smooth seeds, and is orange–yellow at maturity. *Poncirus* is generally used as a rootstock in colder regions and for growing dwarf trees.

Genus *Fortunella*. The genus *Fortunella* has four species, of which the commercially important ones are *F. margarita* Swing. (oval kumquat) and *F. japonica* Swing. (round kumquat). The trees are small with small green leaves paler on the underside. The flowers are white and smaller than those of *Citrus*. The fruit is small, orange colored, 3–6 celled, acidic and juicy, but with a sweet and edible rind. The other species are *F. crassifolia* Swing. and *F. hindsii* Swing.

Hybrid species. A unique characteristic of the citrus group is that it has intergeneric hybrids. *Poncirus* and *Fortunella* can be crossed with *Citrus* and vice versa. Some successful hybrids reported are:

Citranges	Poncirus × sweet orange
Citrangequats	Citrange × Fortunella
Tangelos	Mandarin × grapefruit
Limequats	Lime × Fortunella
Citrumellos	Poncirus × grapefruit
Citrandrins	Poncirus × mandarins

Distribution and importance. Citrus cultivars are grown in varying quantities in countries with tropical or subtropical climates. Cultivation in each country is determined by the minimum temperature occurring in the region, the temperature tolerance of the specific cultivar, latitude, altitude, proximity to large bodies of water, ocean currents, air drainage, and local conditions. Citrus stands first in area and production among the world's tree fruits. A world production target for 1990 was fixed at 17 million metric tons (MMT) (Wardowski et al. 1986). Important citrus–growing countries are China, India, Israel, Japan, Pakistan, Palestine, Syria, Turkey, Italy, Spain, France, Greece, Russia, South Africa, Algeria, Tunis, Egypt, Libya, Sudan, Argentina, Brazil, Jamaica, Mexico, the USA, and Australia.

In Pakistan also, citrus fruits are the most important fruit crops. They are grown on the most area—160,000 hectares—and production is 1.5 MMT annually. Citrus is grown in all four provinces of Pakistan, but Punjab produces over 95% of the crop. Citrus cultivation in Pakistan has recently made great strides, particularly from the 1960's onward. Introduction of Kinnow mandarin has given a great boost to the citrus industry. Feutrell's Early, another mandarin, is cultivated as an early variety. Other varieties of sweet oranges are Blood Red, Pineapple, and Mosambi. Grapefruits include Duncan, Marsh Seedless, Foster, and Shamber. Other commercial cultivars are Eureka lemon, Kaghzi lime, and sweet lime. The progressive increase in citrus production is evident from Table 13.1.

Table 13.1 Production (thousand metric tons) and area (thousand ha) of evergreen fruits. (Figures are 5–year averages; numbers in parentheses indicate area.)

Year	Punjab	Sindh	NWFP	Bal.	Pak.
		CITRUS FRUITS			
1975–80	668.2	30.9	22.5	1.0	722.6
	(69.6)	(2.7)	(2.7)	(0.3)	(75.3)+
1980–85	1137.4	34.2	27.4	1.9	1200.9
	(116.2)	(3.6)	(3.3)	(0.5)	(123.6)+
1985–90	1422.6	34.8	29.8	3.6	1490.8
	(152.6)	(3.9)	(3.6)	(0.7)	(160.7)
		MANGOES			
1975–80	305.5	255.0	1.2	3.2	564.9
	(24.2)	(33.2)	(0.1)	(0.5)	(58.0)
1980–85	383.0	260.6	1.4*	4.1	649.1+
	(32.7)	(33.5)	(0.1)	(0.6)	(66.9)
1985–90	465.7	260.4	1.7	5.0	732.8
	(44.2)	(33.9)	(0.2)	(0.7)	(79.0)
		BANANAS			
1975–80	6.6	102.7	7.5	0.6	124.4
	(1.3)	(12.3)	(0.4)	(0.1)	(14.1)
1980–85	9.2	114.2	9.4	0.8	133.6
	(1.8)	(12.8)	(0.5)	(0.1)	(15.3)
1985–90	11.5	170.9	9.2	.9	192.5
	(2.3)	(18.8)	(0.5)	(0.1)	(21.7)
		GUAVAS			
1975–80	86.7	19.5	13.0	1.1	120.3
	(12.7)	(2.4)	(1.5)	(0.2)	(16.8)
1980–85	193.2	14.9	18.0	1.3	227.4
	(26.3)	(2.4)	(1.9)	(0.2)	(30.8)+
1985–90	288.9	16.4	22.4	1.7	329.4
	(39.2)	(2.7)	(2.3)	(0.3)	(44.5)
		DATES			
1975–80	42.4	64.3	4.5	81.7	192.9
	(6.5)	(7.2)	(0.6)	(8.8)	(23.1)
1980–85	63.5	67.5	6.5	82.0	219.5+
	(9.4)	(10.2)	(0.9)	(9.1)	(29.6)
1985–90	92.1	93.2	6.5	85.6	277.4
	(14.0)	(16.1)	(0.9)	(9.5)	(40.5)

Source: Data from *Agricultural Statistics of Pakistan, 1989–90*:89–100.

* The original document has 0.4, which is a typographical error.

\+ Totals for Pakistan vary by .1 from those in *ASP, 1989–90* due to rounding differences.

13.1.2 Soil and climatic requirements

Soil. Citrus trees can be grown on a wide range of soils, including deep sandy loam, loam, and clay loam. Citrus trees, however, do not grow well in very heavy, clayey, sandy, alkaline, or waterlogged soils, or when a hardpan is present. In general, sandy loam or loamy soils are preferable. Citrus trees tolerate soil pH from 5.5–8.5. Soil requirements depend upon the type of rootstock used for the various species and varieties. Rough lemon is a good rootstock for the dry, sandy loam soils of Punjab, whereas sour orange performs better on the moist and heavy soil of NWFP.

Climate. Citrus fruits are grown in tropical and subtropical climates. In these regions citrus groves are found from sea level to 450–750 m elevation. In Pakistan, the central divisions of Punjab—Sargodha, Faisalabad, Lahore, and Multan—produce excellent citrus. In NWFP, Peshawar, D.I. Khan, and Dir are important citrus-producing areas.

Temperature is the main factor affecting the production and quality of citrus fruits. In cool regions, vegetative growth is less, fruit growth is slow, ripening is delayed, and the fruit is apt to be acidic. The smaller-sized CVS. become still smaller in these regions and are therefore not recommended. In colder regions pigmented CVS. like Blood Red sweet oranges develop excellent quality. Citrus CVS. cannot continue growth below 13°C and above 40°C, but they can endure a temperature around 0 to −2°C without injury depending upon the CV. and the duration of the cold period. Citrus fruits are usually damaged by frost; species appear to resist frost in the following descending order: mandarin, sour orange, sweet orange, chakotra, grapefruit, sweet lime, lemon, kaghzi lime, and citron. In hotter regions, the trees produce more growth and the fruit grows faster and ripens earlier, but in these regions the fruit frequently suffers from sunburn and is unmarketable. The maximum temperature that can be endured by citrus ranges from 46°–50°C. The citrus fruit most resistant to high temperature is grapefruit.

Time of blooming, maturity, and fruit quality are all influenced by temperature. Kinnow mandarin matures in February–March in the central regions, and as late as April in the northern regions of Pakistan. Humidity also affects the productivity and quality of citrus fruits. Some citrus cultivars, particularly sweet oranges, produce more in regions with higher humidity. Fruit drop is less and they tend to be smoother skinned, juicier, and better-looking in these regions than in drier regions.

13.1.3 Propagation and rootstocks

Citrus varieties are propagated by both sexual and asexual methods. Generally rootstocks are propagated sexually through seeds, and most commercial varieties are propagated by various asexual methods.

Raising of rootstocks. Rootstocks are generally tougher species and are more resistant to diseases and adverse soil and other environmental conditions than the varieties that are grafted onto them. With almost all citrus cultivars, the choice of rootstock is very important. They affect longevity, productivity, quality, and precocity of the scion trees.

Most citrus seeds are polyembryonic and produce seedlings true to type. Citrus rootstock species are therefore propagated through seeds in the nursery. The fruit for seed extraction is taken from healthy, vigorous trees by the first week of September. The fruit is cut into halves and twisted to squeeze out the contents. The seeds are then separated from the pulp and juice and washed to remove the gelatinous material. They are dried by placing them in partial shade. It is best to grow the seeds fresh as they cannot be stored for long. Seeds stored in charcoal under dry storage conditions lose nearly 50% of their viability after 45 days, therefore storage of rootstock seeds is not generally advisable; but if it is necessary to ship them far, storage in ground charcoal or packed in vacuum tins at a temperature of 3–10°C is recommended.

Seeds are sown either in September or in February–March. Sowing earlier or later than this affects germination adversely, so for early or late sowing the seed rate is increased. Around 70–80% of rough lemon seeds germinate. Sour orange, a commonly used rootstock in the heavy soils of NWFP, gives about 66% germination at its optimum sowing period, mid-September. Temperature is the main factor affecting germination, but maturity of seed, species, soil conditions, and method of sowing also affect it.

The seeds are sown on well–manured and well prepared raised seedbeds located in a warm place on good, well–drained soil. The seedlings should be protected from frost if sown in September and from heat if sown in February–March, with suitable shelters. These seedlings are ready for transplanting in 6–12 months. Transplanting of citrus seedlings should be done in September–October or in February–March. Planting is done in rows at alternate distances of 25 and 60 cm apart and 25 cm from plant to plant. At transplanting, only uniformly sized seedlings are transplanted while all the small, weak, diseased, and bench–rooted seedlings are eliminated. During transplanting the seedlings are kept in the shade in a cool place.

Asexual propagation. One year after transplanting, the rootstock seedlings are ready for budding. Budding is a method of asexual propagation in which only one bud is removed from the scion variety and is inserted by a special technique in the stock seedling. T–budding is the usual method of citrus asexual propagation. Budding may be done any time during spring (Feb–March) and autumn (Aug–Oct) when the bark slips freely or when both the stock and scion plants show active growth. The bud should be selected from a healthy and productive tree. Use only mature wood with white streaks on it and never get buds from a water sprout or from immature and angular wood. The buds will unite and show signs of growth after 3–4 weeks. When the bud has grown to 25–30 cm, the rootstock is cut back close above the bud union. All sprouts of the stock must be removed from time to time.

Cutting and layering. Planting of semi–hard and hard–wood stem cuttings is the commercial propagation method most used for sweet lime, while layering is commonly practiced in kaghzi lime. Both of these cultivars are also propagated through seeds.

Cuttings of 25–30 cm length are made in Jan–Feb and Sept–Oct with a round cut at the lower end and a slanting cut at the upper end. The upper end cut is made about 4 cm above the last bud. Before planting, the cuttings are callused for better results. Direct planting also gives above 60% success if the cuttings are properly cared for afterwards. Treatments of round ends with Seradix A or indolebutyric acid further enhances root initiation and increases the success ratio.

Grafting. Grafting of citrus can be successfully done even when it is late in the season for budding. Veneer grafting and T–grafting are commonly used for asexual propagation of grapefruit and kaghzi lime which have thorns in the leaf axils and cannot be easily budded. In grafting, unlike budding, 3–4 buds are kept on the scion wood, which is inserted in the bark of the stem of the rootstock seedling.

13.1.4 Cultural practices

Citrus orchards are planted during spring or autumn, the spring season being generally better. The orchards are usually planted in the square system at a planting distance of 7–8 m depending upon the variety and environmental conditions.

Cultivation and interculture. During the early years of orchard trees, only light cultivation is required to eradicate weeds. In later years, particularly at the bearing age, ploughing and hoeing are required more frequently to provide a more suitable environment for root growth. Observation indicates that orchards which are ploughed five to six times from May to September perform better than those which are intercropped and cannot be ploughed during this time. Keeping the orchard clean during winter also minimizes the danger of frost injury.

Growing of crops like wheat and cotton between the trees is commonly practiced in Pakistan, but these crops ultimately spoil the trees and shorten their productive life. The different irrigation requirements of wheat and heavy insect infestation on cotton damage the trees seriously. Growing of some intercrops during the unproductive stage of the trees is permissible with qualifications. Intercrops must not be very tall or exhaustive, and should not carry common diseases or insects. Early income is no doubt increased by intercropping, but sometimes irreparable losses are inflicted on the orchards.

Intercropping may prove helpful in adding humus to the soil where hot temperatures generally deplete it of organic matter. But this practice can be repeated only once in 3–5 years. For this purpose, leguminous crops are better. In Pakistan, berseem fodder is a common winter intercrop. According to recent observations, this practice has proved devastating and is a major cause of the decline of many orchards. The fodder needs frequent irrigation during winter, while citrus fruits are very sensitive to excessive irrigation. For citrus fruit, it is always wise to manage clean cultivation except for casual intercropping of leguminous crops for green manuring.

Irrigation. Citrus trees are evergreen and need appropriate irrigation all year round. Successful citrus culture thus requires a permanent source of water. However, over–irrigation is generally more harmful for citrus fruits than under–irrigation, and this principle should guide the irrigation schedule for citrus orchards. The frequency and amount of irrigation depend upon climatic conditions, soil type, kind and age of trees, number of trees per hectare, and rootstock used. Citrus orchards should be given restricted watering during winter, with one heavy irrigation about two weeks before blooming. This stimulates fruit bud differentiation and results in heavy bloom. During blooming, irrigation is generally not recommended, but the author has observed that in sweet lime, an irrigation during the peak bloom produces good results. After fruit set, it is critical to provide regular irrigation. Failure to provide fortnightly irrigation during April–July will enhance fruit drop, restrict fruit development, and result in low yield and small fruit. During winter, monthly irrigation is enough to maintain the health of the trees.

Irrigation may be supplied through the basin system or modified–basin system. The **basin** and **modified–basin systems** restrict the supply of water to the spread of the trees by constructing basins. This is good for younger orchards, but trees over 10 years old attain a size where flood irrigation performs better. In the basin system, a straight channel connects the basin of each tree in that line, while in the modified–basin system a water channel is constructed in the middle of each alternate row and from there it is connected with the basin of each tree. Both of the systems use less water than flood irrigation. The modified system further helps to avoid the spread of disease carried by water from one plant to other. Also, if fertilizer is applied, each tree will have its own share within the bounds of its basin. If water is abundant, flood irrigation can be practiced. For flood irrigation, levelling of the fields is very important. Several other irrigation systems are also practiced in various countries; among the most important are drip irrigation, furrow irrigation, alternate irrigation, and sprinkler irrigation.

Irrigation must be given to citrus trees before evident wilting. The need for irrigation is determined by various methods:

1. BY ESTIMATION OF FRUIT GROWTH RATE: 5–10 fruit are tagged on a branch and their circumferences measured daily between 7 and 9 A.M. Soon after irrigation they gain size rapidly, then slow down, and finally stop. As soon as the rate of size increase slows down, the trees should immediately be irrigated.

2. BY OBSERVING SYMPTOMS OF WILTING: When young leaves start curling and growth of young shoots almost ceases the orchard should be irrigated, although it is a little late.

3. WITH A TENSIOMETER: A **tensiometer** is an instrument which measures the amount of water in the soil. It consists of a long porcelain tube with a gauge fixed on the top. Tensiometers are put under the tree within the drip line. The tube is filled with water, and as the soil is depleted of water, the water in the tube moves out into the soil and a vacuum is created in the tube. This suction influences the gauge and the needle moves upward from 0, the saturation point. At the 40 mark, the trees should be irrigated. When fruit are in their early stage (first 8 weeks), the trees are irrigated at the 25–30 mark.

Fertilizer. Commercial citrus orchards are seldom profitable without annual fertilization. Like other plants, citrus needs 16 essential elements for satisfactory growth. Hydrogen, oxygen, and carbon are available from water and air, while the other 13 are mainly absorbed by the roots from the soil. Among the major elements absorbed from the soil are nitrogen, phosphorus, potassium, magnesium, and calcium. N, P, and K are repeatedly applied in the form of fertilizer, while the rest are taken up from the soil resources. In some soils, although sufficient quantities of elements are present they are not available because of the soil reaction state. If such soil conditions cannot be corrected, foliar sprays of these elements are used. Foliar spray, particularly of trace elements like zinc, manganese, iron, copper, and boron is frequently applied to orchard trees grown in alkaline soil conditions to overcome the effects of malnutrition. Much research has been done to determine the elements and quantities needed, the proper time and method of application, and their effects on growth and yield (see Chapter 8).

For fertilizer requirements, no general rule can be recommended, as it depends on soil type, climate, cultivar, and rootstock effect. That is why experiments are conducted locally in order to formulate recommendations. At present no appropriate recommendations are available here in Pakistan. They are available for California (Embleton et al. 1978), Brazil (Rodriguez and Gallo 1984), Spain (Puiggros et al. 1969), Sierra Leone (Diest 1980), and Florida (de Gues 1973). These all differ from each other and none of them would be entirely reliable under our conditions. In some recent experiments it was observed that although the experimental trees were within the nutritional standards recommended for California orchards, fertilizer application increased vegetative growth and improved yield (Akhtar 1991).

Early experiments conducted on sweet oranges and grapefruit in the Punjab had found that continuous use of chemical nitrogenous fertilizers like ammonium sulphate and sodium nitrate enhanced yield over control but also induced mottling. This situation was remedied by the addition of FYM (Anon. 1936; Singh et al. 1945; Ali 1962). A common finding of all these experiments is that the use of FYM as a part of the fertilizer programme for citrus is immensely important. Besides increasing yield, it noticeably improves fruit quality. The results of experiments on three–year–old Valencia late orange budded on sour orange stock at Peshawar (Said and Inayatullah 1962) indicated a good response to nitrogen and FYM, while application of phosphorus and potash alone did not affect the yield of sweet oranges.

Fertilization recommendations. For citrus orchards in Pakistan, nitrogen is required in fairly large amounts (1–1.5 kg per tree); also, in most of our orchards, response to phosphorus and potassium is obtained only in combination with N application. Of the various types of fertilizers, FYM is considered best for orchards. If FYM is available, chemical fertilizers should be used as a supplement; it is recommended that half the nutrient requirements be supplied by FYM and half by chemical fertilizers. In general, it is recommended that FYM always be used in orchard management. If it is not available in sufficient quantity, green manuring may be substituted. Organic matter keeps the soil soft and aerated and enhances the absorption efficiency of the relatively few root hairs of citrus. The development of mycorrhiza further improves the conditions for growth of citrus trees, as the fungus helps to increase the absorption of water and minerals by the roots.

Of NPK, the most frequently applied fertilizer is N. It is absorbed mostly in nitrate form. Since nitrate is negatively charged, it cannot stay long on the soil particles and excess nitrate is rapidly leached. Phosphorus and potassium move more slowly and stay in the rhizosphere for a longer time. Therefore, for orchard trees it is recommended that the total dose of N be applied in two equal parts or three if the trees are weak, while P and K be applied only once a year.

In applying fertilizer to citrus fruits, well–rotted FYM should be applied during November and December so that it decomposes by the onset of active growth and bloom in the spring. The first dose of nitrogenous and other fertilizers is given in February, and the other half in April when fruit setting is completed. August–September applications produce excellent results, enhancing blooming, fruit set, and yield. If the trees are weak, a third dose can be applied 10 days before blooming in February. This schedule has helped trees to keep their foliage, which usually drops excessively during winter, healthy year–round. Excessive loss of foliage during the winter reduces yield, whereas if foliage remains intact as a result of fall fertilization, yields are increased.

Pruning. Pruning, the judicious removal of any vegetative part, is an important cultural operation for fruiting trees. Pruning prolongs the bearing age of the tree. An unpruned tree becomes very large, inhibiting light penetration. As a result, leaf sprout is decreased and photosynthetic activity remains low. Fruiting area is restricted to the periphery of the trees; and the inside of the tree remains unproductive and behaves as parasitic wood, consuming tree resources without producing anything.

Pruning is done both to develop a strong tree skeleton and to increase production of marketable fruit. Pruning in citrus is quite controversial. Some recommend heavy annual or biennial pruning, while others are in favor of light to moderate pruning, removing only diseased and dead wood, water sprouts, and other undesirable growth. This wide difference of opinion is mainly due to lack of sufficient experimental evidence under different growing conditions of citrus fruit in subtropical to tropical regions. In Pakistan, lack of sufficient and proper pruning results in shorter tree life. In contrast, in California, where regular pruning is practiced, citrus trees remain productive even at ages of 75–100 years.

Pruning of young trees is called **training**. Its main objective is to provide a strong skeleton for the tree. For a young citrus tree to have a good framework, no branch should be allowed on the main stem within 60–75 cm of the ground. For the formation of a good framework, two to four evenly spaced main limbs should be selected to provide balance, symmetry, and maximum strength.

At the onset of bearing the objectives of pruning change, and more importance is placed on producing maximum fruit of good quality by maintaining a balance between fruiting and vegetative wood. In some countries, mechanical pruning is done to keep the trees within bounds, and any branch extending beyond that limit is pruned both from the sides and top. In Pakistan, however, no systematic pruning is followed except for removal of dead, diseased, and unwanted twigs like water sprouts, or branches which cross each other and spoil the skeleton of the tree. It is recommended that light pruning be done to allow sunlight to penetrate the tree and keep its inside branches productive.

Pruning of lemon is essential to overcome its habit of apical dominance and to induce fruiting. New shoots that grow from old pruned wood tend not to bloom until their second year. Different varieties, however, can behave differently and succulent shoots may bloom in the following spring. With lemon also, pruning has a depressing effect on yield during the year the tree is pruned, but in subsequent years, the yield improves.

Citrus trees are pruned during late winter or early spring when there is no fruit on the trees. With some early CVS., pruning during September–

October can also be done provided winter is not so severe as to kill the new shoots. With late varieties like Valencia Late and Kinnow, pruning is done after harvesting the crop in March–April. There will be some loss of the new crop which was already set, but this is compensated for by better quality of the remaining fruit that year, and heavy yield the following year. In Pakistan, there is at present not enough experimental evidence to recommend a specific pruning regime.

13.1.5 Physiological problems

Unfruitfulness. Sometimes citrus cultivars do not produce a commercial crop for years. This situation may be attributed to genetic or physiological causes like incompatibility, heterostyly, and ovule abortion. Physiological causes involve hormonal or nutritional imbalances which result in general weakness or excessive vegetative growth, preventing the trees from blooming and bearing fruit. General weakness may be remedied by an appropriate fertilizer programme, while excessive vegetative growth may be controlled by exposure of the trees to drought stress or by withholding nutrition. This may also be achieved by root pruning or ringing of the main branches or even the main stem. Climatic factors also affect tree physiology, and certain varieties refuse to produce in particular environments. For example, Washington Navel does not produce fruit in Faisalabad. Genetic and evolutionary causes of unfruitfulness can be evaluated by studying problem trees and then making decisions about planting pollinizer trees or changing the variety.

Alternate bearing. Alternate bearing in fruit trees is the habit of bearing heavily in one year called the **on–year** and very little or not at all in the second year called the **off–year**. This characteristic is manifested by several citrus cultivars like the mandarins Kinnow and Wilking, the sweet oranges Valencia Late and Washington Navel, and Marsh seedless grapefruit. So far there is no specific technique to overcome alternate bearing in citrus. However, many experts agree that it is due to nutritional imbalance in the affected trees. Therefore, strategies to overcome alternate bearing aim at improving the tree's nutritional status, both organic and inorganic. In other words, improving the C:N ratio seems to be the answer. Some measures suggested to overcome alternate bearing are: heavy manuring during the off–year, thinning of fruit during the on–year, delayed harvest during the off–year, and early harvesting during the on–year.

Fruit drop. Another serious problem in citrus orchards is fruit drop, which is sometimes so severe that nothing is left on the trees. Drop starts from blooming and continues until harvesting. Drop at different stages is categorized as flower drop, June drop, and pre–harvest drop.

1. FLOWER DROP. Citrus cultivars bloom heavily. An average tree of Pineapple sweet orange in Faisalabad bears around 50,000 flowers. It is evident that not all of them can mature to fruit. Commonly 95% will drop and the rest will set fruit, but with weak trees deficient in various nutrients, particularly N, sometimes over 99% of the flowers drop. This is either due to mutual competition of flowers or to defective development of their sexual parts and the resulting inability to set fruit. Such drop is relatively less in healthy and leafy trees. Weather conditions like hot and dry climate during blooming or heavy rains accompanied by wind storms also play a major role in excessively heavy flower drop.

2. JUNE DROP OR YOUNG FRUIT DROP. Drop of young fruit occurs in May–June. This is another natural load–shedding by the trees. This drop consists of poorly developing fruits or those which cannot survive dry conditions and high temperatures. Under these unfavorable conditions embryo abortion occurs, which stops seed development and synthesis of hormones and ultimately causes fruit drop. Young fruit drop is quite significant; 96% of the set fruit in Pineapple sweet orange, and 75% in Kinnow mandarin drops.

3. PRE–HARVEST DROP. This drop of fruit is during the mature stage and is of great concern to growers. It is a loss of full–grown fruit, and therefore a lot of research has been conducted to overcome this problem. The drop occurs when a premature abscission layer develops, which is presumed to be due to the failure of auxin synthesis within the tree. Pre–harvest drop in citrus is being controlled effectively by spraying growth hormones like gibberellic acid, 2,4–D (dichlorophenoxyacetic acid), 2,4,5–T (trichlorophenoxyacetic acid), and NAA (napthalene acetic acid).

Granulation. This condition is characterized by enlarged, hardened, and apparently dry juice vesicles. It may affect all or part of a fruit, usually towards the stem end. One–third, one–half, and in extreme cases a whole fruit may be granulated, e.g. in certain pummelo CVS. Such fruit do not show any external signs, but excessively affected fruit are heavier. Cell walls of granulated vesicles are thickened, sugar content decreases, and mineral content is higher than normal, but water content remains at normal levels.

The cause of granulation, also called ***riciness***, is not well understood. However, large fruit have more granulation than smaller ones, and granulation increases if harvesting is delayed. Frequent irrigation or water standing in the root zone for a long time may also increase the incidence of granulation. To overcome the problem, spot picking (picking of large fruit earlier), reduced irrigation, and spraying of 2,4–D are suggested. Some of the cultivars most affected are Mosambi, Valencia Late, Washington Navel, pummelo, and Feutrell's Early.

13.1.6 Diseases and insects

Both the foliage and fruit of citrus are quite vulnerable to a number of diseases and insects. Diseases and insects are discussed in detail in Chapter 11 and only the most serious problems of citrus are mentioned briefly here.

Citrus canker (*Xanthomonas compestris* P. cv. Citri). This is the only bacterial disease of citrus. The disease appears on leaves, branches, and fruit. Bacteria travel through the air or insects and attack the emerging young leaves. Small yellowish spots develop first on the lower and then on the upper surface of the leaves. These get bigger and become more numerous, then turn hard and brown, and become raised from the surface of the tissue. The diseased areas die and sometimes drop out, leaving holes in the leaves. Similar disease spots appear on twigs and fruit. Fruit drop is common. Kaghzi lime is the most susceptible cultivar, and the disease renders the fruit unfit for marketing. Other susceptible cultivars are grapefruit and lemons. Sweet lime and sweet oranges are less susceptible.

To minimize the incidence of citrus canker, select healthy nursery plants, prune away the severely attacked parts of the plants, and spray Bordeaux mixture during spring.

Citrus withertip. This disease is caused by *Colletotrichum gloeosporioides*. The pathogen attacks all the aerial parts of the plant—branches, leaves, and fruit. Withering of branches from tip towards base is the most characteristic symptom. Diseased twigs turn silvery gray and become distinct from the healthy green portion. Leaf fall is common and in severe attacks the trees are killed. The pathogen attacks only weak plants. It can be controlled by improving growth conditions in the orchards, or by spraying with copper-based fungicides.

Citrus psylla. This is a very destructive insect frequently found in citrus orchards. The adult is brown with antennae black at the apex. It is a sucking insect and attacks the trees during and after blooming. The fresh growth important for fruiting is its main target. Infestation is controlled by a prebloom prophylactic spray of insecticides during January–February.

Citrus leaf miner. This is a small silvery-white insect with black eyes and wings fringed with hairs. The larvae make silvery-white galleries on the young leaves and tender shoots. Infested leaves curl up and finally wither and dry up. Young nursery plants suffer the most, although full-grown trees are not immune. Spraying with insecticides during leaf emergence and pruning and burning of affected twigs in December are effective in controlling the leaf miner. Citrus hedges should be avoided in orchards.

13.2 Mango

13.2.1 Botany

Mango (*Mangifera indica* L.) belongs to the family Anacardiaceae.The genus *Mangifera* consists of 40 species, of which only a few produce edible fruit. These are *M. odorata* Griff; *M. foetida* Lour, *M. sylvatica* Roxb., *M. lagenifera* Griff, *M. langar* Miq, and *M. similis*. The fruit of the species other than *indica* is inferior. All species are found in the tropical regions of southern Asia.

The cultivated mango grows to an impressive–sized tree. Under favorable conditions it may attain a height of 25 meters. The trees are mostly evergreen and erect; only a few cultivars have prostrate growing trees. The tree is very long–lived, and some original trees in Lakh Bagh (Durbanga, India) planted by Akbar the Moghal Emperor are still growing after 300 years. The trees do not grow continuously but have a unique habit of producing periodic flushes of various colored leaves when young. The colors of young leaves are specific (reddish, wine–colored, or pale) and may be used as a dependable character for variety identification. In mango–growing regions of Pakistan, flushes appear from February–October.

On a bearing tree, shoots of the first flush are apt to be from terminal buds on non–flowering shoots from the preceding year. The second flush is from lateral buds on shoots with a terminal inflorescence that is setting little or no fruit. A third flush may grow from lateral buds on shoots that have ripened a crop, and a fourth from lateral buds on shoots greatly weakened by a crop in the current or preceding year (Chandler 1958).

Leaves of mango are arranged spirally on a shoot and a general phylotaxy 3/8 is observed. The leaves tend to stay on a growth flush for little more than a year, they fall usually in waves during winter or the dry season.

Flowers. Flower buds differentiate in September and October. Blooming starts by the end of February and continues until mid–April. Some early varieties like Early Gold bloom continuously from December through mid–March. The flowers are borne terminally on panicles 30–50 cm long which carry usually 500–1500 flowers; in some cases they have up to 7000 flowers. The flowers are small, yellowish or pinkish in color, and predominantly male (67–90%) or hermaphrodite (10–30%) (personal observation on Dusehri, Samar Bahisht, Langra, and Seedling 1990). Staminate flowers contain only stamens. In mango there is only one fertile stamen, plus 5–6 abortive stamens called **staminodes**. The fertile stamen is longer, with a dark–red anther filled with pollen. The perfect flowers have a small ovary, resting on a disc. The style is lateral and curved upward; the stigma is simple, small, and terminal. There are 4–5 sepals and 5 petals, which spread free from the disc and are twice as long as the sepals.

The inflorescence is borne terminally on flushes. The frequency of inflorescence depends upon the age of the flushes. The following data is from Khan and Singh (1946):

April flush	37% bear inflorescence
May flush	29% "
June flush	20% "
July flush	12% "
August flush	7% "

Fruit. The mango fruit is a large drupe. The skin is the epicarp, the pulp or flesh is the mesocarp, and the stone is the endocarp. The seed is inside the hard endocarp.

The size of different varieties of mangoes is extremely variable. Some are as small as plums, while others weigh as much as 2 kg. The skin is smooth and thick, and commonly yellow or greenish when mature. Some varieties have a crimson red blush on a yellow skin. Mango fruit, unlike citrus, develops rapidly after fruit set and is ready for harvesting within 13–20 weeks, depending upon variety and climate. When a fruit matures and starts ripening, it abscises from the tree and drops. For marketing, the fruit should be picked just prior to this stage.

Distribution and importance. Mangoes are an important tree fruit of the tropical and subtropical regions of the world. Mangoes are grown on a large area in Pakistan, India, Java, the Philippines, the West Indies, and in Hawaii and Florida in the USA. Pakistan produces excellent mangoes on an area of 82,700 ha, with a total production of 766,000 metric tons annually. Mangoes stand second in area and production, after citrus fruits. They are mainly grown in Punjab and Sindh (more in Punjab), while only a small area in Balochistan in Qalat Division and a still smaller area in NWFP in Peshawar and D.I. Khan Districts is under mango orchards (Table 13.1).

Mangoes are well established as an item of international trade. Because of their superb quality, Pakistani mangoes sell at good prices in Europe, Canada, the Gulf, and Far Eastern countries. Pakistan exported fresh mangoes worth Rs. 69 million during 1989 (Asian Development Bank 1990).

13.2.2 Soil and climate

Mangoes can be grown on a wide range of soil types, but deep, well–drained sandy loam soils give the best results. The subsoil should be free of hardpan, sticky clay, and waterlogged conditions. Mangoes are grown on soils with pH ranging from 5.5–8.7.

The mango is a fruit of the tropical regions. Excessive rains, however, may tend to produce continuous growth flushes, while for induction of fruit a drought period or low temperature stress, as are found in subtropical regions, is important. This is probably why the per–hectare yield in Punjab is better than in Sindh. Freedom from rains, cloudiness, and frost during flowering is particularly important. During ripening, dry hot conditions are helpful in improving the quality of mangoes. Heavy frost during winter may kill even big trees.

13.2.3 Propagation and cultural practices

Mango is propagated both sexually and asexually. Sexually propagated mangoes are called seedling or *desi* mangoes. Sexual propagation is now restricted to raising rootstocks, which are not true to type. In some tropical countries mango seeds are polyembryonic and produce several true–to–type seedlings of which only one is zygotic. At early stages of growth it is not possible to identify which one is the zygotic seedling.

To raise rootstock, fresh mango stones are sown during July–August in lines either directly in beds in the ground, or in a heavily manured brick–lined bed from where the seedlings are shifted to beds in August–September when their new leaves turn green. These seedlings remain there for a year. When they attain a suitable thickness and height they are propagated asexually by side or veneer grafting either in the beds or shifted to suitable sized earthen pots and grafted by **inarching** (approach grafting).

Layout. Mangoes are usually planted in the square system with fillers in the centre of the squares. Planting distance depends upon soil, variety, climate, and system of orchard management. Under good management the planting distance for Dusehri and Anwar Ratol should be 11 m, while for Sindhri, Langra, and Samar Bahisht the recommended planting distance is 15 × 15 m. Transplanting is done in well prepared soil and pre–dug pits during March and September every year.

Irrigation. In arid and semi–arid regions, irrigation is essential for a good harvest. In mango–growing tracts in Pakistan, fortnightly irrigation after fruit set until the monsoon rains tends to prevent fruit drop. Such a schedule also helps to improve the size and quality of fruit. A heavy irrigation just before blooming promotes vigorous bloom. Withholding irrigation during fruit–bud differentiation is helpful.

Manuring. Addition of the correct amount of fertilizer at the right time is essential for commercial production. A liberal addition of FYM results in excellent performance in mango orchards. In the Punjab, trees eight years old or more are provided with 100 kg of dry well–rotted FYM or 1.5 kg of N and 500 g of K per tree. In the Multan region, fertilizer is applied half after fruit harvest and the other half in February. When FYM is used, it is all added in August–September. In Faisalabad Division, due to early onset of winter, the preferred times for fertilization are in February and April. In Sindh, Jagirdar and Maniyar (1967) recommended 100 kg of FYM to a bearing tree, or 2 kg of N per tree in three doses: 500 g in January, 500 g in March, and 1000 g per tree in August after harvesting. For young trees, both in Punjab and Sindh, only FYM is recommended, at the rate of 5–10 kg per plant.

Pruning. Mango has no specific pruning schedule, except when trees are young to give them a good shape. With mature trees, some people prune away some branches during the on-year to avoid blooming, and to get new shoots which can bloom in the coming year. In Laos, deblossoming is practiced to reduce the number of fruit and to ensure a better crop in the succeeding year.

13.2.4 Special problems

Malformation. Malformation is a serious threat to mango culture in Pakistan. It affects both the vegetative parts and the inflorescence. Malformed branches show stunted growth, with small and bunchy leaves; this disorder is thus called '**bunchy top**'. Malformed inflorescences grow unabated beyond the blooming season and sometimes continue growing even until November. Most of the flowers on malformed inflorescences are male and fruit is hardly ever obtained from them. This problem is present in almost every orchard in the country, with variable intensity. The cause is not yet known, but viruses, fungi, mites, and deficiency of elements have all been considered as possible causes. Good management reduces the problem to some extent.

Alternate bearing. Most of the commercial varieties of mango tend to produce crops in an alternate or biennial pattern. The factors affecting it are environmental, physiological, and genetic. Environmental factors like frost, heavy rains, hail, diseases, and insects at blooming cause heavy damage, which results in a light crop on the tree that year. Thus food accumulates in the tree, which may induce a bumper crop the next year. This heavy crop exhausts the tree, and it not able to recoup by the following year. As a result, it again produces a light crop and a rhythm may be established in otherwise regularly bearing trees. Alternate bearing may be physiological, due to an imbalance in the carbohydrate–nitrogen ratio, deficiency of certain minerals, more shoot growth in one year and less in the next, old age of the tree, and low intensity of blooming. It is also presumed to be a genetic disorder because some varieties like Totapari, Romani, Fazli, Neelum, and Kelepad bear regularly, while all commercial varieties of Pakistan have the biennial bearing habit.

There is no established control measure, but improved cultural practices including deblossoming, ringing and girdling, and control of pests and diseases are helpful in lessening the problem.

13.2.5 Insects and diseases

There are many insects and diseases which attack the fruit, inflorescence, twigs, and roots of mango. The most serious are discussed here.

Mango hopper. Also called *am ka tela*, this is a jumping insect which creates a noise on the leaves when the population is great. It is a very serious threat to mango; the damage is so great that if the problem is not noticed in time the whole orchard may be barren. The insects suck the sap from the leaves and panicles as its attack coincides with the time of blooming. It is a more serious threat with thick plantation. Thinning of the plantation for good light also reduces the insect population. To save the fruit, spraying insecticides even during the first 15 days of blooming is useful. Pre–blossoming spray from mid–December to late January is done for complete control of the insect.

Mango mealybug. This is another serious pest which causes heavy losses to the tree and the crop. The nymphs of the female insects suck sap from the panicles and leaves. They crawl up the tree during February and feed there until May; then as the temperature rises they climb down and find shelter in cracks in the soil and die there. The female is filled with eggs which hatch next year and again cause damage. Control of the mealybug is possible only through a series of systematic efforts like destruction of eggs during November–December, by putting slippery and sticky bands on the main stem to stop their crawling on the tree, and dusting and spraying of insecticides during various stages of insect activity.

Anthracnose. This is a very common and widespread fungal disease of mangoes, most common on Samar Bahisht. It causes a lot of damage to the twigs, leaves, panicles, and fruit. Severity increases with excessive rains and humidity, with shedding of leaves and flowers, and young fruit drop. Fruit cannot be shipped, as it rots soon after picking. The disease can be controlled by timely spraying of copper–based fungicides.

Powdery mildew. This disease has emerged as a serious threat to mangoes during recent years. It spreads rapidly during warm, humid weather, and prevalence of such weather during blooming causes heavy crop loss. It attacks the inflorescence, which looks purplish with a sprinkling of white fungus. The panicles dry out, and the flowers drop completely without setting any crop. The disease can be controlled by a spray of Bordeaux mixture before or during blooming.

13.3 Guava

Guava is indigenous to tropical America, Peru, Mexico, and Cuba, where it grows wild as bushes. Now it is grown all over the world for its high dietary value and good taste. It is a very rich source of vitamin C.

In Pakistan, it is grown on 46,200 hectares with a total annual production of 347,300 tonnes. Thus it has attained the status of fourth most important fruit of the country and third most important of Punjab. It is grown on a large scale in Sheikhupura, Gujranwala, and Lahore Districts, and on a smaller scale throughout the plains of Punjab. In Sindh, excellent pear-shaped guavas with a smaller seed core are grown in Larkana, Dadu, Shikarpur, and Hyderabad Districts. In NWFP, Mardan and Hazara Districts are famous for excellent guavas.

13.3.1 Botany

Guava (*Psidium guajava*) grown in Pakistan and India belongs to the family Myrtaceae. Some other edible species of this genus are grown in various parts of the world, like *P. cattleianum* (Strawberry guava), *P. araca* (Feijoa), and *P. guineense* (Brazilian guava).

The guava tree is naturally spreading and bushy. Tree height is generally 4–5 m, but well–trained older trees may reach a height of 9 m. Bark of the trunk tends to be bright, smooth, and scaly. The leaves are light–green, 7–15 cm long, oblong, and finely pubescent below. The young twigs have opposite leaves, are 4–angled, and bear flowers and fruit. Flower buds are mixed. The flowers are complete, and are borne singly or in groups of 2–3 in the axils of the leaves. A tuft of stamens is characteristic of the genus. The fruit is a berry with a large seedy core; it may be round or pyriform, red, yellow, or white–fleshed. Skin may be smooth, or ridgy and waxy. The fruit has a dominant musky flavor with mild acidity. Guava gives two crops a year.

13.3.2 Soil and climate

Guavas can be grown on a wide variety of soils, from heavy clay to light sandy soil, and with a range of pH 4.5 (acidic) to 8.5 (alkaline). It is tolerant to wet and saline conditions. On good soil and with proper care the trees are highly productive.

Guavas are grown in tropical and subtropical regions of the world. They produce a high yield of good quality fruit in climates where there is a distinct winter. Guava is cold–sensitive and cannot resist long periods of frost. A mean summer temperature of 20°C is a minimum requirement for commercial guava production; temperature lower than this may lead to loss of crop. The optimum temperature lies from 23–28°C. Guava does not tolerate the high temperatures of the desert regions. A mild tropical climate induces fruiting throughout the year, while under subtropical conditions two distinct crops are obtained. Rains and high humidity during ripening cause damage to the skin of the fruit.

13.3.3 Propagation and planting

Guavas are generally propagated by seeds. Selected fruits are soaked in water for several days until the seeds separate and settle to the bottom. The seeds thus obtained germinate better by having the testa somewhat soft. These seeds are planted in beds.

Some good varieties may be propagated vegetatively through various budding and grafting techniques. Soft–wood stem cuttings are also successfully used for asexual propagation. For budding and grafting, seedlings are raised as rootstock and side–grafted or budded during March or August. In the region of Hyderabad, Sindh, layering is very successful.

Plants grown in the nursery are later shifted to orchards and planted in the square system. Guava trees are also raised as fillers in mango orchards. If managed well, the trees start bearing two years after planting.

13.3.4 Cultural practices

Guava in a commercial orchard requires careful irrigation. When the trees are young, irrigation frequency is high. Bearing guavas are irrigated heavily during fruiting to induce good development. Withholding irrigation during spring reduces bloom, and prolonging the drought conditions until summer will inhibit fruiting, either totally or so that only a few small early–ripening fruit are obtained. Thus some growers resume irrigating their orchards during late summer to induce growth and bloom and produce a heavy winter crop which is normally free of fruit fly attack.

Farmyard manure or chemical fertilizers are also used to increase yield and quality. About 40 kg of well–rotted FYM in winter or 500 g N in two doses with 4 kg of single superphosphate and 1 kg of potassium sulphate per tree during July–August for the winter crop produce excellent results. Spray of urea has also produced good results.

Guava trees are pruned frequently, but only to remove the suckers which sprout all the year round, particularly during spring and monsoon. Undesired limbs or dry wood are also removed. These days, in the guava–growing regions of Punjab summer fruit is also commonly removed in order to get a good winter crop.

13.3.5 Problems

Fruit fly. The fruit fly is very destructive to guava, particularly the summer crop, and failure to control it has forced growers to stop taking a summer crop and concentrate on getting a good winter crop. The female insect punctures the fruit to lay eggs. The larvae hatch after a few days and start eating the flesh of the fruit, which drops and quickly rots on the ground. To control the fruit fly, all fallen fruit should be buried or the trees sprayed at least twice with insecticides during the early days of fruit set.

Dieback. This problem has been observed recently, particularly in those orchards where spring and early summer irrigation is stopped to discourage the summer crop in the hope of a good winter crop. Even when irrigation is stopped, a few flowers appear and set fruit. To get rid of these young fruit the branches are beaten manually, but in this process a large number of leaves are also lost along with the fruit. Because of lack of available moisture and foliage, these trees cannot transpire sufficiently and are severely heat stressed. The high internal temperature affects the metabolic system of the trees adversely, and as a result they start dying. So far no other reason for this serious problem has been proposed.

13.4 Date palm

13.4.1 Botany

The cultivated palm (*Phoenix dactylifera* L.) belongs to the family Palmae. Closely related species are *P. canariensis*, *P. sylvestris*, and *P. humilis*. These are all dioecious but are fruitful with each other's pollen. The date palm is a feather palm and has 4–6 m long pinnate leaves called fronds with a distinct conical mid–rib which tapers along the length of the frond. The trunk is vertical and columnar, having almost the same girth (100–150 cm) all the way up. It is covered with the bases of pruned fronds. In time these wither away, making that part of the trunk smoother. Being a monocot, the tree has adventitious roots of roughly uniform thickness along their length.

The cultivated date palm is dioecious with independent male and female trees. Both male and female inflorescences are enveloped in spathes, which arise from the leaf axil in whorls and on maturity split longitudinally. Male spathes are shorter and wider than the female. Male spathes enclose 10,000–15,000 flowers against 2000–3000 in female spathes. Male flowers are sweet scented, and densely arranged on strands of the inflorescence, with 3 sepals, 3 distinct petals, and 6 stamens. Female flowers are bead–like structures individually spaced on the inflorescences. Each flower has 3 indistinct sepals and 3 petals. There are 3 carpels with a hook–like stigma. Only one of these develops into a fruit; the other two universally abort during development.

The size of the fruit is extremely variable, ranging from 2–60 g in weight, 18–110 mm in length, and 8–32 mm in width. The fruit has a single seed which also varies in weight from .5–4 g, in length from 12–36 mm, and in breadth from 6–13 mm. The seed is ventrally grooved, a characteristic of genus *Phoenix*, and has a hard endosperm composed of cellulose.

Distribution and history. The date palm has been cultivated since 3000 B.C. Almost all palms are indigenous to hot climates. Date palms are commercially grown in Asia and Africa; some important date–growing countries are Iraq, Iran, Pakistan, Saudi Arabia, Morocco, Algeria, Tunis, and Egypt. Dates are also grown in some regions of Australia, the USA, and Mexico.

In Pakistan, dates are the third most important fruit and are grown on an area of 41,800 ha, with a total production of 284,000 metric tons per annum. Punjab has the highest acreage, followed by Sindh, Balochistan, and NWFP (Table 13.1).

Varieties. There are several good cultivars grown in Pakistan; some of the more common and promising of them are:

Punjab	Hillawi, Khardrawi, Zaidi
Sindh	Aseel
NWFP	Dhaki
Balochistan	Berni, Begum Jangi

13.4.2 Climate and soil

Climate. The five most important factors of climate are temperature, rainfall, humidity, light, and wind. These all affect growth and productivity; in general, date palm cultivation requires high temperature, much sunlight, low humidity, low rainfall, and the absence of high winds.

Date palms tolerate quite low temperatures but cease growth activity below 10°C. They are very resistant to heat and may tolerate temperatures up to 58°C for a short while. In date–growing regions a temperature between 45–50°C generally prevails during the ripening period. Date palms bloom only when the temperature rises above 18°C, and fruit setting and development take place only if the temperature is above 25°C. The earlier dates are produced where the temperature first rises above 18°C; in regions where this temperature is attained later, fruit production is later. In general, a high minimum temperature is more important in determining the maturation timing of dates than a high maximum temperature.

For proper ripening and quality, the interaction of temperature and humidity is extremely important. Hot, dry regions produce hard and dry rather than syrupy dates. High humidity may delay ripening and cause other problems like black nose. Rains during pollination and ripening are devastating. In Punjab, monsoon rains are a major obstacle to date culture, as the time of ripening just coincides with the rainy season.

Soil. In principle, the date palm requires light, deep, well–drained soil. Rocky, calcareous, and compact soils are not fit for date culture. In deserts, where there is a shortage of water, a hardpan at a depth of 3–5 m will benefit date growing. Dates may be grown on soils affected by salinity, a problem of arid regions which are the home of the date palm. But highly saline soils are unfit for date cultivation.

13.4.3 Propagation

Dates are propagated both sexually and asexually. Sexual propagation is through seeds and does not produce true–to–type fruit. Also, as with other tree fruits, sexually reproduced trees take a longer time to bloom than asexually reproduced ones.

Asexual propagation is done through suckers, also known as offshoots, or axillary buds. They generally emerge from the underground stem or sometimes from the aerial parts. The rooted suckers are carefully removed from the tree for further propagation. Suckers weighing 18–20 kg or about 3–5 years old propagate successfully in orchards. The suckers are removed during February–March and August–September for planting in orchards. If care is observed in removing them, and during and after planting, the success rate is over 80%. Orchards are planted in the square system at a planting distance of 7 m. Male trees (3%) are also planted at various locations in the orchard for pollination.

Pollination. All commercial date–palm orchards are artificially pollinated. Several methods, both manual and mechanical, are practiced in various date–growing countries. Among them are mechanical date pollinators, spray of pollen in nutrient solutions, and hand pollination. In Pakistan, pollination is usually done by hand. Flowering strands are separated from the mature, ready–to–open male inflorescence in groups of twos and threes. They are warmed in sunlight for a short time so that when shaken they will shed their pollen. A man carrying a bag of male sprigs climbs the tree and pollinates the opened female spathes by shaking the male sprigs onto the female inflorescence. The operation is quite laborious and has to be repeated four or five times because the female spathes do not all open at the same time. If it rains immediately after pollination, the operation has to be repeated in order to ensure good fruit set.

13.4.4 Cultural practices

The date palm is quite a hardy tree and can be grown even under difficult conditions. But for commercial growing, where a high yield of good quality is desired, they require congenial growing conditions. Additional fertilization is also necessary: 1 kg of N for a full–grown tree. This is better if supplied through FYM, either alone or in combination with ammonium sulphate.

Irrigation. The water requirement of the date palm varies with climatic conditions, water table, type of soil, and the age of the trees. The date palm requires more water during fruit development than at other times of the year. Newly planted orchards require daily irrigation for a month at least, while established orchards require irrigation weekly during fruit development and once a month during the rest of the year.

Cultivation. Deep cultivation around the trees will damage their root system, and should be avoided. On heavy clay soil, cultivation is done mainly to facilitate water penetration. Shallow ploughing or harrowing is helpful to keep weeds under control and roots proliferating.

Pruning. No systematic pruning is needed except removal of the dropped leaves. In some vigorous trees or cultivars some green fronds are also removed to overcome the humid micro-environment around the fruiting bunches. This is especially true in Deglet Noor. Also, the removal of spines from the bases of fronds facilitates pollination and minimizes puncturing of growing fruit. Under certain conditions, some fruiting stands or parts of them are also pruned to improve the size of the remaining fruit. Removal of bunches, however, is not practiced.

13.5 Banana

Bananas are among the oldest fruits, and are a staple food in some tropical regions. The banana belongs to the genus *Musa* of the monocot family Musaceae. Two table species are *M. cavendishii* (dwarf banana), and *M. sapientum* (tall banana); *M. paradisiaca* is used for cooking. Some important banana-growing countries are Mexico, Jamaica, Guatemala, Brazil, Malaysia, Costa Rica, and India. Many varieties are grown in different countries, but in Pakistan some of the important ones are Basrai or Mirpuri, Sonkel, Safri, and Chini Champa.

In Pakistan, experimental banana cultivation was started after Independence, and soon after its success in Sindh the banana emerged as an important fruit crop. It is now grown on 23,500 ha with a total annual production of 209,800 metric tons. It is mostly cultivated in Hyderabad, Khairpur, and Karachi Divisions of Sindh. It is also produced on comparatively small acreages in some parts of NWFP and Punjab (Table 13.1).

Botany. The banana, unlike other fruit trees, has a rhizomelike underground stem, a perennial organ which sends out a pseudostem, a tubelike structure composed of circular and tightly packed non-woody leaf sheaths. The outermost layer consists of the oldest leaf. The inflorescence emerges erect from the centre of the physiologically mature pseudostem but later bends downward and hangs like a pendulum. The epigynous flowers covered in red bracts are arranged on the peduncle in spirals in sets of twos, called

hands. All the hands on an inflorescence are collectively called a **bunch**. Each hand consists of 12–20 fruits called **fingers**. The inflorescence terminates in a reddish ball, the **heart**, which contains only the male or unfertilized flowers. When male flowers start appearing in the terminal part of the inflorescence, removal of the heart results in better development of fruit. The fruit is a triploid seedless berry. The exocarp (skin) separates easily, and the mesocarp and endocarp are the edible part of the fruit.

Climate. Banana is a typical tropical fruit, grown only in the frost-free, humid, and hot regions of the world. Dry weather and strong winds are equally destructive to it. The province of Sindh provides quite suitable conditions for banana growing, while the dry hot summer and frosty winter conditions of Punjab, NWFP, and Balochistan are generally unfavorable except in some localities along the rivers or lakes.

13.5.1 Propagation

Cultivated commercial varieties are seedless, hence vegetative propagation is the rule. Bananas are grown for several years on the same site; the crops grown after the first year of plantation are called **ratoons**. The underground rhizome stem sends out several suckers, a few of which are used for ratoon crops. The others, along with the roots, are shifted for further propagation. One-meter tall, narrow-leaved suckers known as **sword suckers** are usually preferred for propagation, while broad-leaved or **water suckers** are rejected, as they produce late and inferior quality fruit. New plantation is done preferably during February–March, or also in the monsoon months. Planting distance ranges from 2–3 m, depending upon the species, variety, and climatic factors. The pseudostems bear only once (monocarpic) after 9–12 months of plantation and then are chopped away.

13.5.2 Cultural practices

Banana is a surface feeder and therefore prefers clean cultivation on fertile, well-drained soils. It is a fast-growing and exhaustive crop, and copious fertilizer and manure application is essential for good yield and quality. Jagirdar (1961) reported that 250 kg N, 26 kg P, and 500 kg K are removed by one growth cycle of banana to produce 72 tons of fruit/ha. For early vegetative growth, N is important; P helps in fruit formation; and K helps in increasing the yield and quality of fruit.

Banana plantation requires abundant available moisture in the root zone because the plants have a large leaf area and transpiration losses. In Pakistan, where rains are insufficient in the banana-growing tracts, liberal irrigation is supplied. Drought stress at any stage of growth is harmful.

Banana plants are severely pruned; the bearing stem is completely removed after the fruit is harvested. During the cropping season, the numerous suckers which sprout from the underground rhizome are removed, except for one or two suckers which may be allowed to grow to get a ratoon crop. Unchecked sucker growth lowers the yield and quality of fruit. Bananas are harvested before complete ripening, while still at a slightly angular stage; the exact stage of ripening depends upon the marketing centre. After harvesting, the bananas are cured to stimulate ripening.

13.6 Loquat

This fruit is grown in the cooler subtropical regions of the world. In Pakistan, it is found in sub–mountainous districts of Punjab like Lahore, Gujranwala, Jhelum, and Rawalpindi; and in Mardan District in NWFP. Loquats arrive in the market when citrus fruits are finished and mangoes are yet to ripen. Although filled with large seeds which constitute 2/3 of the fruit, they bring a good income because of the timing of their ripening and marketing.

Botany. Loquat (*Eriobotrya japonica*) belongs to the family Rosaceae, to which apples and the stone fruits belong. The tree is evergreen, unlike most other members of this family. The leaves are long, leathery, and symmetrical. The simple flower buds bloom into branched panicles at the end of new shoots during November–December. A flower has 5 sepals, 5 petals, 5 carpels each with 2 ovules, and 20 stamens. The pale–yellow fruit is a pome derived from the inferior ovary and torus.

13.6.1 Climate and soil

Loquat performs well in the mild climate of subtropical regions with an average annual rainfall of 50–100 mm. Early onset of hot summer winds causes sunburn and poor development, and faulty ripening of the fruit. Under wet conditions, the fruit remains insipid. It requires bright sunshine and a mild environment for proper development of size, sweetness, and flavor. The trees are fairly tolerant to drought and frost, but hot desiccating winds can easily kill them.

The loquat is successfully grown on a wide range of soils from light sandy loam to heavy soil, but the ideal is a well–drained light loam. Subsoil containing gravel and a high percentage of calcium is unfit for loquat production. Ample water supply and effective drainage are important for commercial cultivation of the fruit.

13.6.2 Propagation and cultural practices

Loquats are commonly propagated through seeds, although the seedlings take a long time to bear and are not true to type. The seeds are planted during the fruit ripening period. Propagation may also be by cuttings, layering, budding, and grafting. Propagation by cuttings is difficult and the success ratio is quite low, while layering is easy and satisfactory. Shield budding using loquat or quince seedlings as rootstocks is preferred, as it is easier and the success rate is higher. Asexual propagation is carried out during Feb–March and July–Aug.

Since loquats are generally propagated through seeds, named varieties are not common; but a few such varieties are Golden Yellow, Thames Pride, Tanaka, and Improved Golden Yellow.

Cultural practices. Orchards are planted in the square system during Feb–March and Sept–Oct at a planting distance of 5–8 m depending upon the rootstock. The young saplings are staked to avoid breakage by wind.

For a good crop loquat trees require copious irrigation, but they cannot tolerate standing water. Trees may do well with even a little water, except during the fruiting season when enough moisture is required for good fruit development.

Manuring is important in commercial loquat orchards. To get a good crop, 50 kg of well-rotted FYM mixed with 2 kg of bonemeal and 5 kg of wood ash should be added to each bearing tree every year. Green manure can also be ploughed in during the rainy season. Loquat trees bear heavily and regularly; therefore fully-bearing plantations seven years old or more should be manured twice a year so that the tree does not exhaust itself.

Pruning, as with other evergreen fruits, is restricted to the removal of dead, diseased, or otherwise unwanted wood. Thinning of fruit may be practiced on heavily bearing trees to get large sized fruits.

13.7 Ber

Ber or jujube is a very common fruit grown in the warm subtropical regions of Pakistan. It is mostly grown as windbreak or border trees, and in the warm areas of Punjab and northern Sindh some orchards are planted.

Botany. Jujube belongs to the genus *Zizyphus* of family Rhamnaceae. Several species of *Zizyphus* bear edible fruit, but only two, *Z. jujuba* and *Z. mauritiana*, are commercially important. Ber grown in Pakistan appears to be *Z. mauritiana*, although it is frequently misidentified as *Z. jujuba*. Several seedling and grafted varieties are available in the country, including Umran No. 9, Umran No. 13, Kernal Local, and Gohr.

Salient characteristics of the tree, foliage, flowers, and fruit of *Z. mauritiana* are: the tree is spreading and recumbent, with vine-like branches. The leaves are dark and densely tomentose (hairy) on the underside, and are shed during summer or after crop harvesting. Flowers are borne in autumn on fresh growth, and the fruit is a drupe ripening in winter or early spring. It is generally juicy with two or three seeds in a hard endocarp.

Soil and climate. Jujube can be grown on any soil, even on moderately alkaline ones, but flourishes best on deep, well–drained soil. It is extremely drought–resistant and can also tolerate waterlogged conditions. It is quite hardy and can tolerate both extremely hot and dry summer conditions up to a temperature of 45–50°C and *short* periods of frost; it cannot tolerate extended or severe frost.

Propagation. Ber trees are raised from seeds and budded by the ring method in June or shield–budded in March–April and Aug–Sept. Budding is generally carried out *in situ*, since transplanting success is poor because of its deep tap root. Two or three seeds are planted at a site, which germinate in three weeks. They are shield–budded at the age of 18 months. The budded plants start bearing during the second or third year.

The planting distance for ber generally ranges from 12–13 m apart in an orchard, but if the trees are planted as windbreak, the distance between trees is reduced to 8–10 m. The trees grow vigorously and the long slender branches quickly fill the space between the trees. These branches may also easily break under the weight of fruit, therefore pruning is highly desirable. Ber trees bear on current growth, so if they are not pruned regularly every 2–3 years, the heavy yield produced will be of inferior quality, and frequent breakage of limbs will spoil the main framework of the trees.

Irrigation. Ber is drought–resistant, but irrigation during fruit development improves fruit yield and quality. For orchard trees, application of 3–4 baskets of manure during the rainy season will have a good effect on tree health and productivity. Fruit flies are the most serious pest which attack the fruit during the early stages; therefore timely spray of an appropriate insecticide is very important.

13.8 Custard apple

The custard apple (*Annona squamosa*) and some other similar fruits like cherimoya (*Annona cherimola*), ramphal (*Annona reticulata*), and soursop (*Annona muricata*), belong to the family Annonaceae. Custard apple is also called sweetsop or sugar apple and locally *sharifa* or *sitaphal*. Custard apples are generally used fresh as table fruit, and the pulp can also be mixed with milk or ice cream. From the seeds an oil is extracted which is used in paint and also has high insecticidal value. The oil contains traces of hydrocyanic acid, which is also present in the leaves, bark, roots, and seeds. Custard apple is distributed throughout the tropics and warmer subtropics. In Pakistan, its cultivation is restricted to Sindh.

Botany. The custard apple is a large tree about 5 m tall with light–green, elongated pinnate leaves, relatively small green flowers, and pale–green, irregular knobby fruit. The tree is semi–deciduous and sheds some old leaves during winter. The tree blooms twice a year. The fruit of the March bloom is available in July, and from the July bloom in December.

Soil and climate. The custard apple is quite tolerant to various soil conditions. Its roots are shallow and it does not require deep soil, but the drainage must be good. It prefers sandy soils, but is also grown on heavy soils as well as on almost pure sands. Custard apple is not very particular about climate, except that during flowering it prefers dry, hot conditions, which promote fruit set and yield.

Propagation. The custard apple is mostly propagated through seeds which are planted *in situ* because the seedlings do not stand transplanting well. In some countries, T–budding is successfully done using cherimoya or its own seedlings as rootstock. Seedlings or budded saplings are planted in the orchards during February–March and in September. Planting distance varies from 7–8 m in the square system, depending upon soil fertility.

Fertilization and irrigation. Maintenance of soil fertility by an appropriate programme of manuring is important for commercial orchards. The trees are drought–tolerant, but in Mirpurkhas weekly irrigation produces a better yield than fortnightly irrigation. One or two irrigations during ripening are particularly important for enhancing the fruit yield and quality. Light pruning of old wood during late winter promotes new branches for better fruit bearing. The fruit is harvested while still firm but turning from green to creamy yellow and showing signs of cracking between the segments. Keeping the fruit for 2–3 days in straw is enough for complete ripening.

13.9 Papaya

The papaya (*Carica papaya*) belongs to the family Caricaceae. It grows in almost all the tropical regions of the world. In Pakistan, it is grown in Karachi and Thatta Districts on an area of about 500 hectares.

Botany. Papaya plants are dioecious and start bearing within 10–12 months, while blooming starts 4–6 months after seeding. The plants are straight, usually with a single stem 1–3 m tall with an umbrella of leaves on the top. The stem is fragile and hollow, and easily damaged by strong winds. The 5–7 leaves are dentately lobed on long pedicels. The leaves remain only on the top of the fruiting area while the older ones abscise.

Male and female flowers appear simultaneously on separate plants. The female flower is large, yellow, and has 5 large petals surrounding a large ovary. The stigma is also 5–lobed and each lobe is further branched. Female flowers are borne singly in groups of about 3, almost sessile at the base or axil of the leaf stalk. The male flowers are much smaller and are borne in the middle or at the end of a meter–long inflorescence stalk. The 5 petals terminate in a tubular structure or corolla. The sepals are inconspicuous as in the female flowers. Pollination is by insects and air currents.

The fruit is an oblong, pyriform, or nearly cylindrical, large and fleshy berry. The outer color may be green, yellow, or orange–yellow when the fruit is ripe. There are numerous, small, black seeds in the cavity of the fruit.

Culture. Papaya is planted at a distance of 1.5 m on fertile or heavily–manured, light, well–drained soils. Close planting is important because sexual propagation results in a ratio of around 50:50 male and female plants, which can only be identified on blooming. At that point most of the males are uprooted, leaving only a few for pollination.

Papaya grows in tropical regions with a warm, humid climate. Mature trees can endure light frost. Papaya grows and bears well in regions where the summer temperature ranges from 38–40°C and the winter temperature does not frequently go below 5°C, and the average annual humidity is 70%. The largest yield is taken from young trees two to three years old; trees become uneconomical after five years. Mature fruits are harvested when still hard but turning yellow, particularly for distant markets.

During summer papaya trees are irrigated weekly, and in winter fortnightly. The plants must be irrigated regularly, as lack of moisture results in stunted growth and poor fruiting.

13.10 Chiku

Botany. The chiku (*Achras zapota*), also known as sapodilla, sapota, or zapota, belongs to the family Sapotaceae. It is a fruit of tropical regions, but temperatures above 44°C are injurious. In Pakistan, it is grown on a small acreage in coastal districts of Sindh. The tree is evergreen with dense dark–green foliage. It blooms twice a year, once in Mar–Apr and again in Aug–Sept. Flowers are borne singly in the axil of the young rosette of leaves or bracts in a clustered configuration. The spring bloom and crop are both generally larger than those of autumn. The fruit is a round, oval, or conical berry. The skin color is brownish and the flesh is soft, granular, and very sweet. There are generally 1–4 hard black seeds.

Culture. Chiku is grown in a variety of soils, but deep, sandy loams are best. It is tolerant of alkaline soils and brackish water. For commercial production, well–prepared fertile soils are selected and abundant FYM added. Either seedlings or plants propagated by asexual methods of *gooti* or grafted on *khirni* (*Mimusops hexandra*) are planted at a distance of 10 m in the square system. Weekly irrigations, particularly during summer, the fruiting season, are advised. Delay in irrigation during any stage of fruit development results in loss in yield. The fruit are carefully picked when they reach full size, and later ripen and become soft in crates after 4–6 days.

13.11 Apple

The apple is one of the most important tree fruits of the world. In Pakistan its cultivation is limited and is restricted to the northern hilly tracts of Punjab and NWFP, and the Quetta region of Balochistan.

Botany. The apple belongs to the family Rosaceae and the sub–family Pomoideae. Other sub–family members are pear and quince. The botanical name(s) of apple are used inconsistently. Some reported names are *Pyrus malus* L, *Malus sylvestris*, and *Malus malus*. The most common name is *Pyrus malus* L. The genus *Pyrus* is further subdivided into two sub–genera, *Malus* and *Pyrophorum*. *Malus* is the sub–genus of apples and *Pyrophorum* of pears. The *Malus* genus has no grit cells, styles are united at the base, the calyx tube is open, the flowers are colored, and the fruit globular. The flower is perfect with 5 sepals, 5 petals, 5 carpels with 2 ovules each, and 15–20 stamens. The tree is upright spreading, deciduous, and carries mixed buds which produce flowers and mostly spur–bearing leaves .

Varieties. Some 5000–6000 varieties of apples are described in the literature; however, only a few are important for any particular region. To select a variety, important characteristics are studied under local conditions. A selected variety should be: an annual bearer, good to very good in taste quality, attractive in appearance, relatively pest–resistant, productive, and hardy. Good storage and handling quality further increase the value of the variety. Some important varieties grown in various apple–growing regions of Pakistan are: in the Quetta region, Kashmiri, Kandhari, Amri, Kulu, and Qalat Special; in the Murree Hills, Kapai Red Beauty of Bath, Delicious, Kashmir Amri, Banki, Golden Delicious, Red Delicious, and Sky Spur are recommended. In addition, some low–chilling new varieties have been introduced like Tropical Beauty, Enna, and Ein Sheimer. Enna is reported to perform well at lower altitudes.

13.11.1 Climate and soil

Apples need a cooler climate than most fruits because apple buds have the longest rest period and hence require more chilling than those of other deciduous fruits. Apples can endure quite low temperatures, but temperatures of −30°C and rapid fluctuation in winter from relatively warm to extremely cold temperatures are harmful. Uniformly severe cold nights during the rest period are favorable. Spring frosts or freezes shortly before, during, or after bloom are destructive to apple production.

Apples thrive and produce the best in a relatively long, cool, and slow growing season, the type of climate which usually prevails at altitudes of 1700–2500 m. Apples grown at lower altitudes produce less yield of inferior quality. The quality of fruit grown at lower altitudes is particularly lowered by high summer temperatures. Therefore, suitable summer and winter temperatures are the main climatic considerations in selecting a place for apple cultivation. Localities having a mean summer temperature up to 27°C or preferably lower are considered suitable.

Rainfall is important, as in hilly areas there is seldom any other source of water. For apples, 750 mm rainfall well distributed throughout the year is ideal. In Punjab, the northern hilly tracts receive most of their rain during the monsoon, while the Quetta region usually faces a shortage of irrigation water. Late spring frosts and hail during the early stages of fruit development often damage apple crops in the Murree Hills. Continuous strong winds, particularly during blooming, are another hazard in apple culture.

Soil. Apple trees grow and bear fruit in a wide range of soils, but the most suitable appears to be deep, well–drained, fertile loam which permits free root development. Waterlogged soils or those where subsoil water rises into the root zone even for a short time during the growth period should be avoided. Hardpan soils are also undesirable, as are light, shallow, gravelly soils, which are not capable of retaining adequate moisture.

13.11.2 Propagation and rootstocks

Apple trees are budded on crab apple, on various East Malling types, and on seedlings of some commercial apple varieties. In the Murree Hills, crab apple is the only popularly used rootstock, though efforts are already underway to introduce some foreign rootstocks. Besides crab apple, some other rootstocks are used in Sariab, Quetta and Tarnab, Peshawar, including M–1, M–7, and M–9. M–1 is a very outstanding rootstock because of its good nursery performance; also, it is vigorous on dry soils. M–7 is rapidly growing, semi–dwarf, and precocious. M–9 is most suitable for heavy soils, and is dwarfing in nature.

Crab apple (*Pyrus baccata*) is propagated through seed and stooling. Trees grafted on this rootstock are vigorous. To raise crab apple seedlings, the seeds are extracted in May–June and stored in a cool, dry place until December, when they are either sown directly in the soil or stratified in boxes. In Murree, both direct sowing and stratification produce similar results. The seeds begin to germinate in March, and are later transplanted to other beds, where they are ready for budding by the next summer.

Stooling, also called mound–layering, is another method of raising rootstock. About two–year–old crab apple plants are cut back to ground level in the spring. Several shoots then emerge, and when they are 10–15 cm tall, moist and friable soil is hilled up around them up to half the height of the shoots. This process is repeated two or three times until July, when the final height of the hill thus built up is 15–20 cm and the shoots are 45 cm or more in length. Many of these shoots will produce roots by the following spring. Then the rooted shoots can be removed from the parent stool in early spring. The unrooted shoots and parent stool are kept exposed to light during spring so that they will develop new shoots, and then the parent is again hilled up to encourage root development in these shoots.

Propagation of scion varieties. Scion varieties are propagated by vegetative methods, generally grafting or budding. Cuttings and layerings are not used because scion varieties are difficult to establish on their own roots. Budding is done during the active growth stage, while grafting is done during dormancy in winter or early spring. Some important methods of grafting of apples are whip or tongue grafting and cleft grafting; for budding, the ring and T–budding techniques are often practiced.

13.11.3 Cultural practices

Planting plans. On sloping land, apple orchards are laid out on contours along the slopes. Contour planting has the advantage that soil and rain water are conserved, resulting in better tree growth and production over a longer period of time. Also, the terrace channels are convenient for irrigation. On level land, the square system is the main method of plantation. New plantation is done in early spring when the saplings are still dormant. The distance between two trees is 6–7 m. Wide spacing improves tree vigor and also promotes better fruit color due to prolonged exposure to sunlight.

Good soil management is critical to successful crop production. It ensures a good supply of organic matter, prevents soil erosion, preserves moisture, and supplies nutrients regularly to the plants. Soils are managed clean or with sod. In hilly areas, sod helps to overcome the problem of soil erosion and in maintenance of organic matter, but brings its own problems like competition, less nitrogen supply, and harbouring of insects and rodents. In the Murree Hills, the trees are hoed under their canopy during early summer to eradicate weeds and grass, and from the monsoon onward grass is allowed to grow, and mowed from time to time.

Manuring. Manuring is an important cultural operation which is regularly performed in commercial orchards. At least three major elements, nitrogen, phosphorus and potassium, are applied through fertilizers. Fertilization recommendations based on leaf analysis for apples are to maintain leaf N at 1.8–2.6%, P at 0.12–0.22%, K at 1.2–1.8%, Mg at 0.24–0.4%, Ca at 1.0–1.5%, S at 150–2300 ppm, Fe at 50–100 ppm, B at 20–40 ppm, Zn at 25–50 ppm, Cu at 5–12 ppm, and Mo at 0.5–1.0 ppm. The needs of NPK, however, change with variety, and with the location of the orchard.

Apples require relatively high K levels, while the need for P is quite low. Organic manure, either FYM or stable manure, is more suitable than chemical fertilizers. For a full–grown apple tree, 80–100 kg of well–rotted FYM is applied in November before the first snow. For a full–bearing tree, fertilizer doses of 1.5–2 kg of ammonium sulphate, 2 kg of potassium sulphate, and 3 kg of superphosphate are recommended. In general, N is applied in two doses: one soon after snow clearance, and the other after fruit set. The fertilizer is spread under the canopy.

Irrigation. Irrigation during blooming helps to increase fruit set and subsequent fruit development. In the Murree Hills, apple orchards are mainly dependant on natural rainfall except in the months of May–June. In these months irrigation is supplied from springs. In Quetta, there is no rainfall during summer and only 250 mm during winter. Water for irrigation is supplied through *karez* and springs. When the fruit is almost mature, light and infrequent irrigation is advised. This helps the fruit attain timely ripening and full color development.

Pruning. The pruning of apple trees is often ignored by orchardists, mainly because they are unaware of the objectives of this practice. Both young and bearing trees are pruned to increase orchard output, mostly during the dormant season. Young trees are pruned to train them to a desirable shape so that the limbs constituting the main framework are strong enough to bear heavy crops of fruit. Such trees are easily managed for spraying, thinning, and picking of fruit.

Full–bearing trees are pruned to remove old or undesirable wood and to maintain the tree in a vigorous condition for regular and efficient production of high quality fruit. The tops of mature trees are kept low and spread within bounds to facilitate harvesting and management. The kind and amount of pruning of a tree will depend upon its age, original framework, growth characteristics, and fruiting habit of the variety.

13.11.4 Problems

Fruit setting and pollination. Apple flowers are hermaphroditic, but many varieties are self–unfruitful because of self–incompatibility. Some varieties are partially self–incompatible, some slightly so, and some almost completely. A moderate degree of self–incompatibility is desirable because it results in moderate fruit setting which minimizes the need for fruit thinning and reduces the incidence of alternate bearing. Cross–pollination by insects and wind generally produces better results than self–pollination. For cross–pollination the honeybee is the most important insect; on a quiet sunny day a bee visits an average of over 5000 flowers. Prevalence of temperatures lower than 8°C, rains, or winds are unfavorable conditions for the flight of honeybees. Some self–pollinated or self–fruitful varieties are Baldwin, Grimes Golden, Rome Beauty, and Cox's Orange Pippin. Partially self–unfruitful varieties are Golden Delicious, Jonathan, and Red Astrakhan. Completely self–unfruitful varieties are McIntosh, Gravenstein, Red Delicious, and Stayman Winesap. The problem of pollination and fruit setting can be overcome by looking for pollenizers and keeping beehives in the orchards.

Hailstorms. Hail is a particularly serious problem in the apple–growing sectors of the Murree Hills. In some years the storms are so severe that the whole crop is destroyed. Hail injures spurs, knocks down flowers, and damages young as well as half–grown fruit. Hailstorms usually occur during late spring and early summer when the fruit is still developing. They are a natural calamity, but the extent of losses may be reduced by covering the trees with nylon nets. This practice is recommended, but doing it on a large scale is expensive and impractical.

Insect pests. Apples are attacked by several insects and diseases, a few of which are very serious. The **codling moth** is the most destructive of the insect pests, and is common both in Murree Hills orchards and in Quetta. It damages the inside of the fruit without apparent signs. The larvae burrow through the skin or enter the fruit through the calyx tube. They eat the immature seeds and the flesh around the core. For pupation, they come out and find shelter in the tree or beneath the trees in fallen leaves. Another generation of insects appears and starts the cycle again. Control may be attained either by systematic spraying of insecticides, or by wrapping the main stem with burlap to trap the pupating larvae and later destroying them.

13.12 Pear

Botany. The pear belongs to the genus *Pyrus* and sub–genus *Pyrophorum* of the family Rosaceae. There are around 30 species of pear in the genus *Pyrus*; the most commonly cultivated species is *Pyrus communis* L.

The genus is characterized by white flowers with the style free at the base. The fruit have a closed calyx tube and contain grit cells in the flesh, clustered around the carpels and in the core. The tree of *P. communis* is strong, upright, and deciduous. The leaves are mostly oblong to ovate, hard in texture, bright–green, and serrated. The flowers are white and borne on spurs in clusters of 4–12 with leaves. Each flower is hermaphroditic and has 5 persistent sepals, 5 petals, 5 carpels, 15–20 stamens, and an inferior ovary. The fruit is a pome which, like the apple, is a false fruit since it develops from the torus and floral tube instead of the ovary. Pears are universally self–incompatible and hence produce more fruit when they are cross–pollinated. Many of the good cultivars are parthenocarpic.

Varieties. Important varieties grown in Pakistan are Clapp's Favourite, Hardy, Bose, Bartlett, Comice, LeConte, Kieffer, and Samar Kandi.

Climate and soil. Pears are an important fruit crop in the temperate regions of the world. Pear buds require about 900–1000 hours of chilling below 5°C to open in the spring. In Pakistan, the cold winters needed to meet the chilling requirements of most pear varieties are easily found above 1000 meters in the colder areas of Mardan, Hazara, Quetta, the Peshawar region, and the Murree Hills.

Summer and winter temperatures are important determinants of the time of flowering, ripening, and quality of pears. Pear buds and wood are less resistant to cold than those of apple, but the blossoms and young fruit seem more resistant. High summer temperatures are withstood better by pear than apple, and flavor of the fruit is not impaired by high summer temperatures to the extent that it is with apples.

The pear tree is tolerant of a wide range of soil conditions. It can withstand both excessive moisture and drought and is very tolerant of poorly aerated soils. It can grow even in shallow hardpan soils, and can also resist a fair amount of alkali. For a good yield, deep, well–drained, not–too–heavy, fertile soil, free of wet conditions and excessive alkali is the most suitable.

Rootstock and propagation. Like many fruit species, pears are propagated by vegetative methods like cleft and whip grafting during the dormant season. For the production of standard pear trees, seedlings of different varieties of common pear and Japanese pear are used; for producing dwarf trees, quince is the universal rootstock. In Pakistan, seed of *batangi*, a wild cultivar of common pear, is more commonly used.

Cultural practices. Pear orchards in Pakistan are planted on terraces in the hills, and in the square system at a planting distance of 6–8 m in the plains of Peshawar, Hazara, and Mardan. Planting time is late winter or early spring when the plants are still dormant. In young pear orchards moisture should be available throughout the growing season to attain good growth. Bearing trees grown in the Murree Hills are irrigated during May and June from perennial springs. In the Peshawar region, weekly irrigation is given from flowering to fruit ripening; suspension of irrigation at any stage adversely affects fruit quality and yield.

Young trees require a sufficient amount of minerals for growth, and bearing trees require additional amounts for fruit production. Fertilizer and manuring recommendations vary with tree age, type of soil, and other aspects of the growing environment. If animal dung is available, it should be preferred, as it supplies all the essential elements to the trees. About 80 kg of well–rotted manure is applied to a full–grown tree just before the first snowfall in November. If chemical fertilizers are also available, both chemical fertilizer and FYM should be applied in equal proportions. In many pear–growing regions of the world, the greatest response has been obtained with nitrogenous fertilizers to produce good growth and fruit. In the Peshawar region the recommended practice is to supply 2–3 kg of ammonium sulphate per tree. This is to be applied in three equal doses: the first 2–3 weeks before bud opening, the second dose when the fruit is of almond size, and the third application a month after the second.

Pruning is done essentially to maintain the framework and regulate production of the trees. Pruning for desired results is related to several factors like soil management, climate, age and tree vigor, and variety. Therefore, no one method can be universally recommended. Only intelligent pruning carried out keeping in view the above–mentioned factors will have the desired specific influences on normal tree growth, consistently good cropping, fruit quality, and economical management of the trees. It has been observed in a number of experiments that a pruned but unfertilized pear tree produced a better crop than a fertilized but unpruned tree.

Pruning of a bearing tree should accomplish the following objectives.

- Vigorous wood growth should be encouraged.
- Weak wood and sprouts should be removed.
- Fruiting wood should be evenly distributed.
- The tree center should be open to allow light penetration.
- Tree height should allow convenient spraying, thinning, and harvesting.
- Pruning should help to regulate bearing, size, and quality of fruit.
- Interfering branches, dead wood, and diseased parts should be removed.

13.13 Plums

Botany. Plums belong to the group of deciduous fruits commonly known as stone or drupe fruits. They are true fruits and are characterized by having a distinct three–layered pericarp— an exocarp, mesocarp, and a stony endocarp which encloses the seed. The other members of the group are apricots, peaches, almonds, and cherries. Plums belong to the genus *Prunus*, sub–genus *Prunophora* (plum and apricot) of the family Rosaceae. There are about 15 cultivated species of plums, of which *Prunus domestica* (European plum) and *P. salicina* (Japanese plum) are the most common. Important varieties produced in various regions of Pakistan are:
European plums: Fezle Manani, Grand Duke, Gauzales, and Formusa;
Japanese plums: Burbank, Wickson, and Methley.

Trees of *P. domestica* plum are medium–sized with a round top reaching a height of 10 m or more. The leaves are broad, coarsely serrate, usually pubescent beneath, and appear with the bloom or after it. Flowers are complete and perigynous, with 5 pure–white petals borne in pairs in lateral spurs or shoots. Fruit color may be green, yellow, red, or purplish–blue, and the size and shape is variable. Varieties of *domestica* are self–fruitful or partially self–incompatible, and a few are self–sterile.

Climate and soil. Plums are mostly grown in temperate regions, while some low–chilling varieties may be grown in the milder parts of subtropical regions. In Pakistan, most of the area under plums is in the milder regions of NWFP, and there is some acreage in the Murree Hills of Rawalpindi District in Punjab, and in Quetta and Qalat divisions of Balochistan.

Chilling requirements of domestic plum are 800–900 hours below 5°C. The Japanese plum requires less chilling than European plums and hence blooms earlier. As a result, it is more often damaged by spring frosts. In the Murree Hills, planting Japanese plums on northern slopes delays blooming and helps the trees escape frost damage. In relatively warmer climates, European plums develop better flavor than the Japanese varieties. The best flavor in European plums develops in places where the temperature seldom goes above 38°C. The fruit will remain acidic if the lower temperature falls below 15°C.

Soil requirements depend upon the type of rootstock. When budded on Myrobalan stock, the plum prefers heavy soil, well–drained and fertile; however, it can also tolerate poorly–aerated soils. Plums budded on peach stock may be grown on relatively light, shallow soil; but if apricot stock is used, alkaline soils may also be suitable. Almond stock for plums may allow plums to grow successfully in warm, dry soils. In general, most Japanese varieties will do better in lighter soils, while many European varieties prefer loam and clay loam soils.

Propagation and rootstocks. Plums are budded on plum, peach, almond, and apricot rootstocks. Of these, plum is the most commonly used worldwide. In Pakistan, plum is popularly used as rootstock in the Murree Hills, peach in NWFP, and almond in Balochistan. The stones of all the above mentioned rootstocks are collected during the fruiting season and stored in a cool, dry place. The stones are then stratified in December–January and sown in lines during February in well–prepared seedbeds. Some vigorous seedlings may be ring–budded in May, others shield–budded in July–August, and still others cleft–grafted during late winter when the seedlings are still in dormancy.

Orchard layout. Plums are planted in squares or in a quincunx layout. In Balochistan, the hexagonal system is also used. Planting distance generally varies from 5–7 m depending upon the variety, soil, climate, rootstock, and pruning system. Cultural practices are recommended according to local conditions and varietal characteristics.

Irrigation. Irrigation is essential because of scanty rainfall in many plum–growing areas of Pakistan. Frequency of irrigation is adjusted depending on prevailing local conditions.

Fertilization. Plums respond fairly rapidly to nitrogenous fertilizers. Bearing trees need liberal supplies of N in early spring to help in fruit setting. Besides this, organic matter, P, and K are also applied to commercial plum orchards.

Pruning. Plums are usually trained to vase form. Mature *domestica* trees are not pruned much because they bear on spurs; Japanese plums, on the other hand, bear fruit on shoots and hence are regularly pruned. Pruning of *domestica* plum varies with varieties, but necessary pruning is carried out to maintain the skeleton.

Fruit thinning is a common commercial practice, as under favorable conditions fruit set is very heavy. Thinning increases fruit size and leads to a larger proportion of marketable crop. Manual as well as chemical thinning is done after late spring frost and natural June drop.

13.14 Apricot

The cultivated apricot (*Prunus armeniaca*) of the family Rosaceae is a stone or drupe fruit. In Pakistan, it is cultivated from 300–2000 m. Varieties grown at the higher elevations are usually superior in quality. In Punjab, apricot cultivation is common in the Soan Valley and the Murree Hills. Haripur Hazara, Peshawar, Mardan, Kohat, and Malakand Divisions are important in NWFP, while in Balochistan it is the Quetta valley. Excellent apricots are produced in the Northern Areas. Important commercial varieties grown in various parts of Pakistan are Red Flesh Early, Old Cap, Charmaghzi, Moorpark, Nuri, and Shakarpara.

Soil, climatic requirements, and cultural practices. These are almost the same as for plums. Apricots are propagated by grafting during dormancy, or budding during the months of May–June on seedling rootstocks like Hari, a wild apricot, peach, or almonds.

Training and pruning. Apricot trees are trained to a modified–leader system. With bearing trees, the side branches on the main branches should be tipped if they are longer than 25 cm because they bear flower buds all along their length. Long side branches on the inside of the trees are cut back to one–third of their length. Vigorous but unfruitful lateral branches are completely removed. Renewal wood develops at the base in both cases and the fruit is, therefore, carried close to the main limbs and leaders.

Fruit thinning is necessary as apricot trees may set over 40% of their flowers. In thinning, small and weaker fruits are removed and clusters are reduced soon after the stones are completely hardened.

13.15 Peach

The peach (*Prunus persica*) is the second most important stone fruit (after plums) in Pakistan. It belongs to the sub–genus *Amygdalus* of the Rosaceae family, as does almond. The cultivated peach tree is generally vigorously growing and spreading in nature. The leaves are broadly lanceolate and coarsely serrated. The simple flowering buds have a dense wooly appearance. They are located laterally in leaf axils, and borne on one–year–old branches singly or in pairs with a leaf bud in between. They open before the leaf buds in the early spring. The flower is hermaphroditic and perigynous, the ovary being slightly lower than the 5 pinkish petals. The fruit is a drupe which encloses a bony endocarp with a ridged surface. The fruit is fuzzy, either flat or round, beaked, and sutured. It may be red, yellow, or white–fleshed, and may be sweet, sour, or bitter in taste.

Cling and freestone are two important cultivated types of peaches produced worldwide. In Pakistan, peaches are grown in the same regions of NWFP and Balochistan as plums.

Varieties. There are both dwarf and tall–growing varieties. In the Peshawar region about four dozen varieties have been tested. A few of them are recommended: Jubilee, Golden, Robin, Wiggin, and 6–A. The last, being a late variety, is very important. In Quetta, Golden Early, Shah Pasand, and Shireen are important.

Soil and climate. Peaches can grow on a wide range of soils, depending upon the rootstock used. With light sandy loam soil, they can be budded on peach seedlings; if the soil is heavy and moist, plum rootstock is most suitable. For almond rootstock, dry, light warm soil is preferable. In general, rich, well–drained, loamy soils with deep subsoil are most suitable. The peach tree is not tolerant of wet conditions.

Peach has widely variable chilling requirements—from 200 hours below 5°C for low–chilling varieties to more than 1000 hours. Peach orchards are located at from 600–1500 m elevation. Peaches and plums are grown in almost similar environmental conditions, except that some low–chilling peach cultivars can also be grown in warmer places. For good flavor and better quality peaches, mild summer temperatures of 27–30°C are required.

Propagation and rootstock. Peach seedlings are the most commonly used rootstock for peaches, while almond (in Quetta) and plum are also sometimes used. At Tarnab, Yellow Local peach gives better performance as rootstock for peach than apricot and plum. Common methods of budding peach are ring and shield, which are done during summer, and cleft–grafting, which is done during late winter.

Vegetatively produced plants of commercial varieties are planted during early spring before sprouting. The square layout system is used, with a planting distance of 7–8 m depending upon soil type, rootstock, and variety.

Cultural practices. The soil or orchard floor is kept covered or clean–cultivated depending upon the site and other local conditions. In the Murree Hills, at lower heights where peaches can be cultivated, orchards are kept clean during the dry period (April–June), while weeds and grass are allowed to grow during the rainy season (July–August) to conserve the soil. In Peshawar, where the finest peaches are grown, non–bearing orchards are intercropped with vegetables and berseem, while bearing orchards are kept clean.

Peach orchards must be regularly irrigated to obtain optimum yield; adequate water is particularly necessary during the pit–hardening and fruit maturing stages. A deficiency or negligence in irrigation during the growing season will result in reduced fruit size. Excessive irrigation, on the other hand, may cause early yellowing and falling of leaves.

Peaches bear only on previous–year growth, therefore for regular fruit production, vigorous shoot growth must be encouraged during the growing season. It is induced with heavy N–fertilization in addition to proper irrigation and pruning. Peaches respond well to N and K under most conditions and have higher requirements of these two elements than even apples. An average–sized, bearing peach tree is supplied with 60–80 kg FYM annually depending upon the type of soil, crop yield, and vigor of the tree. The manure is applied during late autumn when the trees are dormant. For chemical fertilizers, in the Peshawar region the general recommendation is 2–4 kg ammonium sulphate per tree. This fertilizer should be applied in two equal doses: first in early spring and then after fruit–setting in April.

Pruning. Peach trees are trained according to the central–leader or modified–central–leader system, which gives them a strong frame. Annual pruning of bearing trees is essential for peaches, as it is related to their typical bearing habit. Peaches bear laterally on previous–year growth. Fruit buds may be located in the center or on the basal portion of the shoots in different varieties. The position of the buds affects the severity of pruning. If the buds are towards the base, pruning of one–year–old shoots should be more severe than when they are located in the center. Pruning is also necessary to get new shoots on this wood for the following year's crop.

Fruit thinning is carried out to get more fruit of commercial size. The necessity for thinning can be minimized by proper pruning.

Problems. Some viral and fungal diseases can eliminate a whole plantation. The viruses causing peach yellow and mosaic are particularly dangerous. Peach leaf curl is a serious fungal disease frequently found in peach orchards. Gummosis sometimes constitutes a serious threat to peach plantations.

13.16 Almond

The almond (*Prunus dulcis*, formerly known as *P. amygdalus*), family Rosaceae, is a stone fruit important for its nutritious and highly valued seeds. Botanically the almond is close to the peach. Worldwide almost all important varieties of almond are self–sterile.

Varieties. Sweet almonds are classified into three types. 1) Hard–shelled varieties are Peerless and Drake. (2) Soft–shelled types are Jordonalo and Neplus Ultra. (3) A paper–shelled variety grown in Balochistan and the Soan Valley is Nonpareil. Most varieties of almond bear regularly.

In Pakistan, about 7200 hectares of almonds are cultivated in small areas in Qalat and Quetta divisions of Balochistan, and in small areas of Swat and the tribal areas (FATA). Recently almond cultivation has been introduced in the Soan valley of Punjab. Almonds are grown in slightly warmer environments than peaches, as their wood and flower buds are less resistant to cold weather during dormancy, and they become still more tender on blooming and at the start of growth. Almonds bloom early and hence frequently fall victim to spring frosts.

Propagation. Almonds are propagated vegetatively. The best rootstocks are seedlings of either bitter or sweet almond. Peach is also commonly used, and sometimes apricot seedlings. Grafting is done during the dormant season or budding during summer.

Cultural practices. Since almost all varieties of almond are self–sterile, planting pollenizer varieties and keeping hives of honeybees in the orchards are very important to ensure cross–pollination. Since almonds bloom quite early in the season, when the activities of the honeybee are at a low point, and since almond pollen is not carried by the wind, close planting of the pollenizers together with the commercial varieties is helpful to promote pollination and fruit–setting. After pollination, it takes 6–8 months for the nuts to ripen, depending upon the summer temperature and the variety. In hot summers they ripen earlier than in mild summers. On full maturity the exocarp and mesocarp which form the hull of the fruit split open, and the endocarp is released and dropped on the ground.

The tree is deep–rooted, and so needs deep, well–drained, moderately fertile soil. Sites with poor drainage or waterlogging are not suitable for almonds. In Balochistan, farmers generally believe that almonds can be grown without irrigation. Even though almonds are drought–resistant, casual irrigation during summer brings better results by reducing fruit drop and producing large–sized kernels. Manuring on time is also required.

During their non–bearing age, almonds are pruned for training, but during bearing, unlike peaches, they are pruned very little. Almonds bear on short spurs which remain productive for five years, and also on one–year–old branches. Thus only a little pruning on alternate years is recommended. In contrast to all other stone fruits, thinning is not done for this fruit.

13.17 Persimmon

Botany. The persimmon (*Diospyros kaki*) belongs to the family Ebenaceae. *D. kaki* trees are deciduous and medium–sized (8–12 m tall). The leaves are ovate, elliptic to oblong ovate, acuminate apex, pubescent beneath, and glabrous and shining above. Persimmon trees are monoecious, dioecious, or polygamous (with hermaphrodite, pistillate, and staminate flowers). The varieties cultivated in Pakistan like Haychia and Fuyu usually carry only pistillate flowers. Flowers are borne in the leaf axils of the current season's growth. Female flowers are large and solitary, while male flowers are smaller and grouped in clusters of 3 or 4. Male flowers have 16–24 stamens in two rows and the calyx and corolla are 4–lobed. The pistillate flowers have a leaf–like, 4–lobed calyx, 4–lobed light–yellow corolla, 8 abortive stamens, and a globose 8–celled ovary. Pollination, which takes place through insects, may or may not be necessary. Pollination affects the flesh color of the fruit and reduces astringency, though most of the cultivated varieties are parthenocarpic. The fruit is a berry which varies in size, shape, color, and quality. The fruit of some varieties is oblong or conical, while that of others is flat like a tomato. Fruit color ranges from orange–yellow to deep orange. The fruit is sweet and soft only when fully ripe.

In Pakistan, the persimmon is gaining popularity and is cultivated in cool subtropical regions like Chakwal, Jhelum, Rawalpindi, and Taxila in the Punjab; and Mardan, Hazara, and Peshawar Divisions of NWFP. Some varieties grown in Pakistan are Hachiya, Fuyu, Aman Kaki, and Marko Italy. These are seedless.

Soil and climate. Persimmons flourish well on sandy loam soils with good drainage and sufficient organic matter. The persimmon prefers a mild climate. It flowers late in the spring and ripens in early to late October, depending upon temperature conditions. It has moderate chilling requirements of 400–500 hours and is fairly tolerant to 0°C. It is sensitive to high temperatures, and in places where summer temperatures rise above 38°C fruit yield and quality is considerably reduced.

Propagation and rootstocks. Seedlings of *D. lotus* (*amlok*) or *D. kaki* are commonly used as rootstock; *D. kaki* is considered to be superior. Persimmons are propagated from seed, and also by ring and T–budding in spring and summer, and cleft and tongue–grafting in late winter.

Culture. Persimmon seedlings, or budded/grafted plants are transplanted during late winter when they are still dormant. The planting distance between plants and rows is 7–8 m. Transplanting success is not very high, therefore considerable post–transplanting care must be taken.

Where there is not much rainfall, persimmons require frequent irrigation during summer. In winter, there is little need of irrigation. Irrigation should be restricted during flowering and fruit setting.

Use of nitrogenous fertilizers is necessary for good health and fruit production, but an excessive dose may cause the tree to grow vigorously. Excessive shedding of young fruit is not uncommon in such trees. For a bearing tree, 40 kg of well–rotted FYM is added during the autumn. Artificial fertilizers are applied during the early spring.

Persimmons are trained to a modified–leader form, and four or five angled shoots are developed. During the bearing age, practically no pruning is done except for annual removal of dead, broken, diseased, and interfering branches. Wood which has borne fruit should also be removed while the sterile wood on main limbs sends out new shoots to bear fruits. Fruit thinning is practiced with some heavily bearing trees. If there is enough fruit drop at an early stage, thinning may not be needed.

Unripe persimmons are highly astringent, and are not edible in this condition. Some treatments to induce ripening are storing the fruit in airtight containers, inducing mechanical bruises, treating with ethanol, or storing in airtight rooms containing CO_2. These procedures enhance ripening by turning the excessively soluble tannins to an insoluble form.

13.18 Grapes

The European grape (*Vitis vinifera*) is the most important species of the family Vitaceae. Two other commercial species are *V. labrusca* (American grape) and *V. rotundifolia* (Muscadine grape). Worldwide, grapes are grown for eating, wine, juice, and raisins. European grapes are grown in mild subtropical conditions, while both of the other species do well under temperate conditions. *Vinifera* grapes have a thick skin which firmly adheres to the sweet pulp. American and Muscadine grapes have a thin skin which separates easily from the relatively acid pulp. Muscadine grapes have small bunches, and large seeds. Their berries have a strong musky flavor and shedding is frequent.

Botany. Grapes are produced on vines which are trained on wires running on both sides of the plant. The buds develop in the leaf axils. Leaf buds produce only leafy shoots, while fruit buds produce shoots which also carry one or two flower clusters opposite the leaves. Tendrils also develop on these shoots opposite some of the leaves. *Labrusca* grapes develop tendrils opposite each lateral leaf, which support the shoots by coiling around the wires. Inflorescences are produced laterally on the third or fourth node of one–year–old shoots called canes. The flowers are usually complete and perfect, but male or female flowers are also found in some varieties. Each flower has 5 green sepals, 5 greenish white petals, 5 stamens, and 2 carpels each containing 2 ovules. Blooming is followed by vegetative growth after dormancy is over. Blooming is more rapid at temperatures between 18° and 21°C and is retarded if the temperature reaches 37°C. Pollination, which is carried out by insects, takes place even in seedless cultivars. The fruit is a berry of variable size in each variety and is borne in clusters.

In Pakistan, only European grapes are cultivated for eating. Over 70% of the grapes are grown in Balochistan, while there is some acreage in NWFP. Some important varieties grown are:

Seedless: Sunda Khani, Sra Kishmish, and Askari
Seeded: Haita, Black Prince, Sahihi, Hussaini, and Tandon

Soil and climate. Grapes are well adapted to a great variety of soils, ranging from gravelly sands to heavy clays, from shallow to deep, and from low to high fertility level. However, poorly drained or highly alkaline soils should be avoided. For good production, fertile soils with good structure which permit sufficient root growth should be selected.

Grapevines are deciduous and bear fruit under mild climatic conditions. They require a winter which is cold enough to fulfill their chilling requirements, but it should not freeze often. European grapes grow and bear well under a long, warm to hot, dry summer. This type of climate is found in Quetta Division of Balochistan. The failure of grape cultivation in Punjab is largely due to the onset of monsoon weather during the ripening period. Only early varieties can be successful under these conditions.

Propagation. Grapes are most commonly propagated from hard-wood stem cuttings. Cuttings 30–40 cm long are made in late winter when vines are pruned. The cuttings are prepared by making a round cut at the proximal end and a slanting cut at the distal end. For better rooting the cuttings are callused by burying them upside down in moist soil in shade, but they can also be planted directly in well-prepared beds. Other methods of propagation are layering and grafting.

Culture. Grapevines are planted in early spring at 3–3.5 m distance depending upon the fertility of the soil, climate, variety, and the pruning system. In Balochistan, grapes are planted in trenches prepared at a distance of 3–4 m apart. Normally manures are not frequently applied, but in commercial orchards addition of about 5 kg of FYM per bearing vine is desirable for the maintenance of health and yield.

Irrigation during the warm and dry season is required for satisfactory berry development. Irrigation during flowering and ripening is not advised as it may delay ripening, reduces berry size, and increases shedding.

Training and pruning. Pruning, which is done in the late winter, is the most important operation in grape culture. Probably no other fruit plant is so severely pruned. Pruning gives the vines proper shape and a size which facilitates insect and disease control. It also keeps the fruiting wood near the main stem and well distributed all over the vine. Vines are trained to several systems, three of which—the head system, cane system, and cordon system—are very important.

HEAD SYSTEM: The vines are trained to one straight stem for the first two years to a height of 1 m. At the head, short branches (arms) are allowed to develop, on which every year at the time of pruning 8–12 short spurs are left (headed–back, one–year–old shoots). This system does not need any supporting structure, but good results can only be obtained in those varieties which bear on the lowest second or third bud of the cane.

CANE SYSTEM: In varieties where the basal two or three buds do not produce a crop, it is necessary to leave 4–6 fruit canes, each 30–100 cm long. To support the long canes and to keep the fruit above them, trellises are necessary.

CORDON SYSTEM: In this system the main trunk is 2.5–3.5 m long and is trained horizontally on wires. The fruiting spurs are left all along the length of the trunk. This system is good for vigorous varieties of grapes which produce large bunches and are grown in rich soils.

Table 13.2 Production (thousand metric tons) and area (thousand ha) of deciduous fruits. (Figures are 5–year averages. Figures in parentheses indicate area.)

Year	Punjab	Sindh	NWFP	Balochistan	Pakistan
			APRICOTS		
1975–80	0.3	-	6.2	23.3	29.8
	(0.1)	-	(0.6)	(2.1)	(2.8)
1980–85	0.3	-	8.1	34.6	43.0
	(0.1)	-	(0.9)	(2.8)*	(3.8)+
1985–90	0.5	-	14.0	51.3	65.8
	(0.1)	-	(1.3)	(4.2)	(5.6)
			PEACHES		
1975–80	1.5	-	1.6	6.8	9.9
	(0.3)	-	(0.2)	(0.7)	(1.2)
1980–85	0.5	-	2.1	8.3	10.9
	(0.1)	-	(0.3)	(0.8)	(1.2)
1985–90	0.5	-	3.3	12.1	15.9
	(0.1)	-	(0.5)	(1.3)	(1.8)
			PEARS		
1975–80	8.1	-	22.5	0.8	31.4
	(1.2)	-	(1.8)	(0.1)	(3.1)
1980–85	3.0	-	29.8	0.8	33.6

Year	Punjab	Sindh	NWFP	Balochistan	Pakistan
	(0.5)	–	(2.4)	(0.1)	(3.0)
1985–90	2.4	–	30.3	1.0	33.7
	(0.3)	–	(2.4)	(0.1)	(2.8)
			PLUMS		
1975–80	0.2	–	21.3	9.0	30.5
	(a)	–	(1.8)	(0.8)	(2.5)
1980–85	0.8	–	28.5	11.4	40.7
	(0.1)	–	(2.7)	(0.9)	(3.7)
1985–90	1.3	–	30.4	15.2	46.9
	(0.2)	–	(2.8)	(1.3)	(4.3)
			ALMONDS		
1975–80	0.1	–	0.3	22.7	23.1
	(a)	–	(0.2)	(6.1)	(6.3)
1980–85	0.1	–	0.6	25.0	25.7+
	(a)	–	(0.2)	(6.4)	(6.6)
1985–90	(b)	–	0.8	29.1	29.9
	(a)	–	(0.3)	(7.0)	(7.3)
			APPLES		
1975–80	3.1	0.2	39.7	41.4	84.4
	(0.5)	(0.1)	(3.4)	(5.5)	(9.5)
1980–85	3.6	0.2	53.0	67.4	124.2
	(0.5)	(0.1)	(4.5)	(7.8)	(12.9)
1985–90	8.3	0.2	92.2	103.5	204.2
	(1.2)	(0.1)	(6.6)	(11.9)	(19.8)
			GRAPES		
1975–80	0.3#	(b)	0.2	28.5	28.9
	(a)	(a)	(0.1)	(2.4)	(2.5)
1980–85	0.3	(b)	0.1	26.0	26.4
	(0.1)	(a)	(a)	(2.6)	(2.7)
1985–90	0.2	(b)	0.3	30.1	30.5+
	(a)	(a)	(0.1)	(3.0)	(3.0)

Source: Data from *Agricultural Statistics of Pakistan, 1989–90*:92–101.
* *ASP, 1989–90* has 2.3, which is in error.
\# *ASP, 1989-90* has 0.2, a typographical error.
\+ Pakistan totals differ by .1 from those in *ASP* due to rounding differences.
(a) Less than 50 hectares. (b) Less than 50 tonnes.

PRACTICAL EXERCISES

1. Determine the harvesting time for apples, mangoes, stone fruits, and various citrus fruits by using several different maturity indices.

2. Practice sexual and asexual propagation of various fruits.
3. Practice training and pruning various deciduous fruits.
4. Practice laying out an orchard in the square and the hexagonal systems.
5. Study flower and fruit morphology of some important fruits of your region by examining the flowers and fruits.
6. Examine buds of various fruits to determine their structure.

QUESTIONS

1. Name some of the most important fruits of Pakistan, indicating their total acreage, production, and the regions where they are produced.
2. Discuss the botanical characteristics of citrus, mango, date palm, papaya, apple, peach, persimmon, and grape.
3. Name some commercial species of citrus fruit and comment on their economic importance.
4. What cultural practices are important in evergreen and deciduous fruits?
5. What is alternate bearing? In what fruits it is more pronounced? Suggest some measures to overcome the problem.
6. Describe the problems of mango malformation in detail.
7. Mention and discuss some salient problems of citrus, guava, apple, peach, almond, and grape.
8. Describe the cultivation of two evergreen and two deciduous fruits.
9. By what methods are various tree fruits propagated?
10. How are date palm, banana, custard apple, pear, and persimmon propagated?
11. Name several different commercial species of plum and grape. Describe briefly the characteristics of each species.
12. What makes a persimmon astringent? How is this astringency reduced?
13. What are the characteristics and environmental requirements of *Zizyphus jujuba* and *Zizyphus mauritiana*.
14. List at least four tree fruits which bear twice a year and also state the conditions favorable for their good bloom.

15. Discuss and differentiate pollination and pruning requirements for various stone fruits.

REFERENCES

Agricultural Research Institute, Sariab. Annual research report of Project Director of Horticulture. Sariab, Quetta: Agri. Res. Inst., Sariab; 1990.

Agricultural Research Institute, Tarnab. Annual research report of Project Director, Horticulture. Tarnab, Peshawar: Agri. Res. Inst., Tarnab; 1990.

Akhtar, N. Effect of time of fertilizer application on winter injury and yield in some commercial cvs. of citrus. Faisalabad: Deptt. of Hort., Univ. of Agri.; 1991. M.Sc. thesis.

Ali, S. Some experiments on manuring of citrus in the Punjab. Punjab Fruit J. 26–7(90–99):108–119; 1962.

Anonymous. A summary of the important results arrived from agricultural stations and research officers in the Punjab. Lahore: Deptt. of Agri. of Punjab; 1936:p. 137–138.

Asian Development Bank. Pakistan fruit and vegetable export marketing studies. Technical report. Asian Development Bank, 1990.

Ayub Agricultural Research Institute. Annual research report of Director of Horticulture. Faisalabad: Ayub Agri. Res. Inst., 1990.

Chandler, W. H. Evergreen orchards. Philadelphia, PA: Lea and Febiger; 1958.

Chaudhary, M. I. Studies on the physiological aspects of erratic bearing in sweet lime. Faisalabad: Deptt. of Hort., Univ. of Agri., 1988. Ph.D thesis.

Chaudhary, M. I., Malik, M. N. A study on the blooming and cropping behaviour of some commercial citrus cultivars of the Punjab. Pak. J. Agri. Res. 26:278–283; 1989.

de Gues, J. D. Fertilizer guide for the tropics. In: Samson, J. A. Tropical fruits. London, New York: Longman, 1973:p. 102.

Diest, A. V. Citrus: cultivation measures. In: Samson, J. A. Tropical fruits. London, New York: Longman; 1980:p. 103.

Embleton, T. W., Jones, W. W., Pallares, C., Platt, A. G. Effect of fertilization of citrus on fruit quality and ground water nitrate pollination potential. Proc. Int'l. Soc. Citriculture; 1978:p. 280–285.

Government of Pakistan. Agricultural statistics of Pakistan, 1989–90. Islamabad: GOP, Ministry of Food, Agriculture and Co-operatives; 1991.

Hayes, W. B. Other sub–tropical fruits: the jujube. Fruit growing in India. Allahabad: Kitabistan; 1957:p. 394–398.

Hodgson, R. W. Taxonomy and nomenclature in citrus. In: Proc. of the 2nd Conf. of the International Organization of Citrus Viruses, held in Italy, 1961:p. 1–7.

Jagirdar, S. A. P. Banana culture in Pakistan. Punjab Fruit J. 14(84–85):11–25; 1961.

Jagirdar, S. A. P.; Maniyar, A. K. What is wrong with citrus in Sindh. Punjab Fruit J. (Citrus Number) 16–17(90–99):56–73; 1962.

Jagirdar, S. A. P.; Maniyar, A. K. Mango culture in Sindh. Punjab Fruit J. 13(82–83):221–243; 1967.

Khan, M. D.; Singh, H. The floral count and fruit setting studies in some North Indian mangoes. Punjab Fruit J. 37:35–42; 1946.

Puiggros, T.; Franciosi, R.; Morin, C. A preliminary nutritional study of Washington Navel oranges in the central coast of Peru. Proc. Int'l. Soc. Citri. 3:1613–1617; 1969.

Rodriguez, C.; Gallo, J. R. Levantamento de estado nutrieional de pomares Citricos de Sao Paulo: Pela analise foliar, *Bragantia campinas* (sp). Proc. Int'l. Soc. Citr. 1:151–154; 1984.

Said, M.; Inayatullah. Sweet orange cultivation in Peshawar and Dera Ismail Khan. Punjab Fruit J. 16–7(90–99):81–83; 1962.

Singh, L.; Singh, B.; Khan, A. A. Fertilizer experiments with sweet oranges growing on rough lemon. Punjab Fruit J. 9(33–34):113–114; 1945.

Swingle, W. T. Botany of citrus and its wild relatives of the orange sub–family. In: Reuther, W. ed. The citrus industry. Vol. 1. Berkeley, CA: University of California Press; 1943: p. 129–172.

Winkler, A. J.; Cook, J. A.; Kliewer, W. M.; Lider, L. A. General viticulture. Berkeley: University of California Press; 1974.

Wardowski, W. F.; Nagy, S.; Grierson, W. Preface of fresh citrus fruits. Westport, Connecticut: Avi Publishing Co.; 1986.

Zielinski, Q. B. Modern systematic pomology. Dubuque, Iowa: W. M. C. Brown; 1955.

Outline of Chapter 14

14. VEGETABLE CROPS

Abdul Fatah Baloch[1]

LEARNING OBJECTIVES

After studying this chapter, you should be able to:

- Discuss the effect of specific plant nutrients on the growth and yield of different vegetables under various agro–climatic conditions in Pakistan
- Determine the most economical fertilizer levels for maximal yields of specific crops
- Discuss the cultivation methods, climatic requirements, and planting times of various tuber crops
- State the comparative advantages and disadvantages of different methods of sowing and planting root crops
- Evaluate the bulb crop cultivars used in various parts of Pakistan
- Compare the planting times, seed rates, yields, and food value of various cole crops
- Discuss the planting times and methods, and cultural practices appropriate for various cucurbit crops
- Understand the effects of different NPK levels on the growth and quality of okra

Vegetable crops can be divided into the following 10 groups according to their morphology and habits of growth.

1. Tuber crops: potato, sweet potato
2. Root crops: radish, carrot, turnip, beet root
3. Bulb crops: onion, garlic
4. Leafy vegetables: spinach, fenugreek, amaranth, lettuce, purslane
5. Solanaceous fruits: tomato, chili, brinjal
6. Cole crops: cabbage, cauliflower, kohlrabi

---

[1]Dr. Baloch is Associate Professor of Horticulture, Sindh Agriculture University, Tandojam, Sindh.

7. Cucurbits: cucumber, muskmelon, watermelon, Indian squash, bitter gourd, sponge gourd, bottle gourd, pumpkin or red gourd
8. Peas
9. Beans: cowpea, cluster beans, hyacinth bean
10. Okra

14.1 Tuber crops

14.1.1 Potato

The potato (*Solanum tuberosum* L.) belongs to the family Solanaceae. There are many related species, but only the potato and a few others bear tubers.

Potatoes are an important staple food crop. Annual world production of potatoes surpasses that of all other vegetables and ranks with wheat and rice. Introduced into Europe in the 16th century by Spanish explorers, it is cultivated throughout the world including the Indo–Pakistan subcontinent. Potatoes are used alone and mixed with meat, fish, chicken, and vegetables, and in big cities and towns, potato chips are prepared and sold commercially. The potato is rich in starch and contains protein, minerals, and a fair amount of vitamins, particularly vitamin C.

Climate. The potato has a wide range of seasonal adaptability. It is a cool–season crop and is moderately frost–tolerant. Temperature during the growing season has long been recognized as one of the most important factors influencing yield. Young plants grow best at a temperature of 24°C; later growth is favoured at 18°C. Tuber production reaches a maximum at 20°C, decreases with rise in temperature, and at about 30°C tuber production stops entirely. Short days are beneficial for tuber production.

Soil. The potato develops best on deep, fertile, sandy to clay loams with good water retention capacity. Because the potato has a relatively weak, shallow root system, impermeable layers in the soil limit rooting depth, which restricts water availability to the plant in dry periods. Thus soil compaction can greatly reduce potato yields. Aeration of the soil has a great effect on the set and development of tubers.

Manures and fertilizers. Potato has a high nutrient requirement; a yield of 25 tonnes of tubers/ha depletes the soil of 119 kg of N, 50 kg of P, and 225 kg of K. In light soils and places where organic manures are not easily available, green manuring is beneficial. Well–rotted FYM at 30 t/ha should be incorporated into the soil three to four weeks before planting. Research on fertilizer requirements conducted in the Punjab and Sindh suggests that 100 kg N, 80–100 kg of P_2O_5, and 50 kg of K_2O per hectare are sufficient to obtain good yields. Shah and Ismail (1983), and Hussain (1968) have reported that better yields of potatoes can be obtained by application of 125–220 kg N, 33–75 kg P_2O_5, and 75–111 kg K_2O. They also found that the cost:benefit ratio ranged between 1:5.10 and 1:11.59.

Spacing. In hilly areas, potatoes are spaced at 60 × 30 cm and the tubers are planted in furrows. In the plains, however, they are planted on ridges spaced 30 cm apart; the distance between ridges is kept at 45–60 cm.

Seed rate. Potatoes are propagated by tubers, planted either whole or cut into pieces. To obtain maximum yields, healthy, disease-free tubers, free from mixture of other varieties, should be used. Seed rate depends on tuber size; 800–1000 kg/ha is generally recommended. Choudhry et al. (1990) conducted a trial comparing improved seed of the Swat and Kaghan varieties bought from the Kalam Integrated Development Project with seed from the local market. Potatoes grown from the improved seed gave better soil coverage, had a lower incidence of virus attack and higher tuber weight, and gave higher yields than those from the locally-purchased unimproved seed.

Irrigation. Potatoes need frequent irrigation. The first irrigation should be given immediately after sowing, and thereafter at one-week intervals. After tuber formation, the frequency of irrigation is decreased. Care must be taken while irrigating that the water does not reach the top of the ridge. Irrigation should be stopped a few days before harvesting.

Harvesting. The size of the tuber increases until the vines become dry. In developed countries, potatoes are harvested by mechanical harvesters. In Pakistan, they are manually harvested by digging up the ridges with a spade. Care must be taken that tubers are not injured during the process. After harvesting, the crop should be kept in the shade.

Varieties. The varieties of potato commonly grown in Pakistan are: Prima, Alpha, Vekaro, Alladin, Baroka, Cardinal, Desiree, Ultimas, Multa, and Lala Faisal.

Yield. The average yield of potato is 20–25 tonnes/ha, however it depends upon the variety, cultural practices, and location. Hanan et al. (1987) reported that yield varies with location and varieties planted. They found Thalassa a good yielder in Peshawar conditions (Table 14.1). Deho et al. (1983) found that Multa gave better yields than other varieties in Tandojam conditions (Table 14.2), and Jagirdar and Jandan (1972) found that Desiree gave higher yields than other varieties in Mirpurkhas (Sindh) conditions (Table 14.3).

Table 14.1 Total yield of various French potato cultivars in Peshawar region during spring and autumn crop

Cultivar	Yield, spring crop (t/ha)	Yield, autumn crop (t/ha)
Appalo	24.47	24.00
Claudia	30.41	25.00
Claustra	17.41	22.00

Cultivar	Yield, spring crop (t/ha)	Yield, autumn crop (t/ha)
Cardinal	30.97	21.00
Eurika	23.30	13.00
Lamia	23.55	22.00
Lola	30.29	15.00
Sahel	28.16	18.00
Thalassa	33.93	23.00

Source: Hanan et al. (1987).

Table 14.2 Total yield of potato varieties under Tandojam conditions

Variety	Yield (t/ha)*
Mirka	12.55 d
Atleet	10.20 e
Patrones	17.88 b
Spartan	09.36 f
Multa	22.92 a
Bintje	13.20 c

* Values followed by different letters are significantly different from each other. Source: Deho, et al. (1983).

Table 14.3 Total yield of potato varieties in Mirpurkhas conditions

Variety	Total yield (maunds/acre)
Desiree	200.3
Bintje	176.0
Patrones	168.0
Atleet	168.0
Ultimus	162.8
Multa	145.4
Dijkhuis 52–211	139.3
Radosa	138.7
Mentore	137.7
Furore	130.8
Ostara	130.7
C.B.52–105–9	125.0
Spartan	122.6
Gineke	118.8
Arka	110.8
Alpha	108.5
Eigenhiemer	105.2
Sientje	100.9

Source: Jagirdar and Jandan (1972).

14.1.2 Sweet potato

Sweet potato (*Ipomoea batatas*) belongs to the family Convolvulaceae. It is native to tropical America where it is called *camote* or *kumara*, and now it is an important tropical and subtropical crop worldwide. The leaves of the sweet potato are quite nutritious and are eaten as a leafy vegetable.

The major uses of the sweet potato are for human food and for the manufacture of starch and alcohol. It contains 16% starch and 4% sugar, vitamins A and C, and is a source of iron.

Climate. The sweet potato is a tropical and subtropical crop. It requires warm weather and a growing season of four months and is quite susceptible to frost. More drought–resistant than many other vegetables, it will produce an adequate crop without irrigation in semiarid conditions. Annual rainfall of 102 cm (40") or irrigation is needed for good yields.

Soil. Well–drained sand, sandy loam, and loamy sand, with a clay or clay loam subsoil are considered ideal for sweet potatoes. Optimum pH ranges from 5.8–6.7. Heavy and very rich soils result in too much vegetative growth and poor root shape.

Manures and fertilizers. Sweet potatoes can be grown in soils of varying fertility; however, heavy application of manures promotes excessive growth. Sweet potatoes are seldom fertilized and are often neglected in tropical regions.

Spacing. Selected tubers are planted 5–6 cm deep in well–prepared nursery beds at spacings of 30 cm in rows 45 cm apart. The sprouts are cut and planted for further growth in another nursery. Ultimately, when the nursery vines reach a sufficient length, cuttings are made and planted at about 60 cm between rows and 30 cm within the row. The length of the cutting depends on the internode length since each cutting should have at least four nodes. Cuttings from the upper portion of the vines are preferred, as they produce more tuberous roots. The general practice is to bury the two middle nodes and expose the two extreme ones. The soil around the cuttings is firmly compacted, and planting should be followed immediately by irrigation. During the monsoon, cuttings should be planted when it is drizzling.

Seed rate. Sweet potatoes are grown from sprouts or draws produced from tubers, and also from cuttings. To get good yields, it is advisable to plant selected tubers in the nursery and then propagate them by sprouts or vine cuttings. About 50,000–60,000 sets are required to plant one hectare.

Planting times. Planting time varies considerably in different parts of Pakistan. In hilly areas, vine cuttings are planted during June and July. Where irrigation facilities are available earlier, planting is done in April–May, and gives high yields. In the plains, planting is done during Oct–Nov. In some areas, both summer and winter crops are grown. Cuttings for summer are planted in May and June and for winter in Oct–Nov.

Irrigation. The plants can tolerate considerable dry weather. Water is normally best applied after the plants are set and before they nearly cover the ground. Frequent light irrigations are preferable to heavy applications at long intervals, but continuous moisture near the surface promotes shallow roots. The newly–planted crop shouid be irrigated on the fourth day after planting, and again on the eighth and ninth days. Thereafter, 10–12 irrigations are given at 10–day intervals.

Varieties. Two varieties of sweet potato are commonly grown in Pakistan: (a) red skin with white flesh, and (b) white skin with white flesh. The variety with red skin and white flesh is sweet and is eaten raw as well as cooked or roasted. The white–fleshed variety which produces long tubers is cultivated commercially.

Harvesting. Usually the crop takes 90–100 days to mature. When the rop matures, its leaves turn pale and later become slightly yellow. The nlarging tubers cause the soil to crack noticeably. If cut and exposed to the r, the surface of a mature tuber dries up quickly, while an immature tuber remains moist and turns dark in colour.

Yield. Yield varies with variety and region. The average yield is about ,000–12,000 kg/ha. If good management practices are followed, yield can e increased to 18,000–20,000 kg/ha.

14.2 Root crops

nportant commercial crops belonging to this group are radish, turnip, rrot, and beet. A number of other root vegetables, which are commercially s important in Pakistan, also fall in this group. They are parsnip, rutabaga, sify, chervil, skirret, and celeriac. All these crops thrive well in a cool nate. However, a number of varieties of radish, turnip, and carrot have n developed which grow well in a comparatively warm climate. All the crops, though belonging to a number of different families, have nearly same cultural requirements.

14.2.1 Radish

The radish (*Rapahnus sativus* L.) belongs to the family Cruciferae. It is related to cabbage and mustard, but belongs to a different genus. The fleshy edible portion of the root develops from both the primary root and the hypocotyl. It is both annual and biennial. The radish probably originated in central and western China and the Indo–Pakistan subcontinent.

Radishes are a favourite crop for kitchen gardens because they are easily grown and are ready for use in three to six weeks after sowing. Radishes are very useful as intercrops or companion plantings between rows or plants of slower growth. They are grown for their fleshy edible roots, which are eaten raw, as salad, or cooked as a vegetable. They have a cooling effect, prevent constipation, increase appetite, and are very tasty when roots and leaves are cooked together. Radishes are recommended to patients suffering from piles, liver trouble, enlarged spleen, and jaundice. Radish leaves are cooked as leafy vegetables in different forms and are very rich in minerals and vitamins A and C.

Climate. The radish is a cool–season crop, is adversely affected by hot, dry weather, and remains in prime condition only for a few days. Radishes are heat tolerant, but to develop its flavour, texture and size, a cool season with temperatures from 10–15°C is required, although varieties differ in temperature sensitivity. To be mild, tender, and attractive, radishes must be grown rapidly. They are both a spring and fall crop.

Soil. Radishes can be grown on all soil types, but they prefer soil that is light, rich, moist, and friable. Radishes can be grown on soils that are not suitable for other root crops. They can also be grown well on fairly acidic soils. For early crops, sandy or sandy loam soils are best, and for summer radishes a cool, moist soil gives better results.

Manures and fertilizers. Fertile soil supplemented by well–rotted manure or fertilizer encourages quick growth. Fresh manure should not be used just before seeding. A yield of 20 t/ha removes 120 kg of N, 65 kg of P, and 100 kg of K. Shaikh et al. (1990a) observed that increasing N with $K_2 0$ (160+200 kg/ha) progressively increased root yield (194.71 quintals/ha.)

Seed rate and spacing. To sow one hectare, 10–12 kg of seed is required, depending on the variety. For commercial planting, graded seed should be used so that plants come up more uniformly. Seeds are planted on ridges 30 cm wide, with 5–8 cm between plants.

Sowing time. For a succession of crisp, tender roots, several plantings should be made at intervals of about 10 days. By the proper selection of varieties, radishes may be had throughout the year. In Pakistan, seeds are sown from August onwards. Tropical types sown later than November bolt early. Often another spring or summer crop of radishes is taken. In places where the climate is mild, radishes can be grown all year round.

Irrigation. Frequent irrigations should be given at 10–day intervals.

Varieties. Radish roots vary greatly in size, shape, and external colour, as well as the length of time they remain edible. Varieties can be divided broadly into two groups: European or temperate types, and Asiatic or tropical and subtropical types. The temperate varieties are small and mild, and are mostly raised for salad.

CLASSIFICATION OF RADISH VARIETIES BY SEASON

A. EARLY OR SPRING TYPES: comparatively mild, quick–maturing, reach edible maturity in 24–30 days.

1. round, globe, or turnip-shaped
2. half-long, olive, or oval-shaped
3. long

B. SUMMER TYPES: larger and keep better than type A; mature in 30–48 days.
   1. round
   2. long

C. WINTER TYPES: quite pungent, large; mature in 50–60 days or more, planted so as to reach a desirable size in autumn. Some common varieties are Desi White, Shae-mai, Mino Early, and Mirpurkhas Selection.

Bhatti et al. (1983) found that Mino Ochall and Shamura were the best yielders. Yousaf and Shafi (1971) observed that among different varieties, Burpee Red Giant, White Icicle, Burpee White, Sparkler, Scarlet Globe, and French Breakfast were desirable for root characters.

Harvesting. Harvesting starts when the roots are of edible or marketable size and are still tender. Delay in harvesting, particularly of the temperate types, may make the roots pithy and unsuitable for the market. Radishes are pulled by hand, tied in bunches of 6–12 with their tops in the field, and washed to remove soil and give them a fresh appearance.

Yield. European varieties yield about 6000–8000 kg/ha. Asiatic types yield from 20,000–30,000 kg/ha.

14.2.2 Carrot

Carrot (*Daucus carota* L.) belongs to the family Umbelliferae. The species is probably native to Asia, and is now under cultivation in many countries. The carrot is an important vegetable because of its large yield per unit area throughout the world and its increasing importance as human food. It is orange-yellow in colour, which adds attractiveness to foods on a plate, and makes it rich in carotene, a precursor of vitamin A. It contains appreciable quantities of thiamine and riboflavin, and is also high in sugar. It is eaten raw as well as cooked in curries and is used for pickles and sweetmeats. Black carrots are used for the preparation of a beverage called *kanji*, a good appetizer in the early summer.

Climate. The carrot is a cool-season crop, though some tropical types can tolerate quite high temperatures. Colour development and root growth are affected by temperature. Carrots grown at 10–15°C develop poor colour. Those grown at 15–20°C will develop a good colour, while at 20–25°C the colour will be less bright. The highest temperatures produce the shortest roots, while the longest roots are produced in the range 10–15°C. The effect of temperature varies with soil conditions and variety.

Soil. Carrots prefer a loose–textured, deep, fertile, well–drained, loamy soil. The land chosen should be capable of being plowed 20–30 cm deep. Carrots do not grow well on highly acidic soil. The maximum yield is expected around pH 6.5, and yields are extremely low at pH 2.5 or less.

For an early crop, sandy loam is preferred; but for large yields, silt or silt loam is better. Market carrots are mostly grown on sandy loam and other fairly light soil types. Fairly rapid growth is desired, as slow–growing roots may be inferior in flavour, smoothness, and texture. Carrots grown for canning or dehydration may be sown on heavier soils since shape is less important than in roots intended for the fresh market. Heavier soil often provides greater fertility and water–holding capacity and higher yields are obtained.

Manures and fertilizers. It is estimated that a yield of about 275 kg/ha removes 125 kg of K, 40 kg of N, and 22.5 kg of P from the soil. Well–rotted FYM should be applied at the rate of about 30 tonnes/ha and should be supplemented by NPK according to the fertility level of the soil. Bangulzai (1990) observed that increasing the rate of N and K_20 progressively increased carrot yield. Shaikh et al. (1990b) found that increasing N with K–fertilizers resulted in increased yield. Leghari (1989) found that application of 100 kg/ha N, 100 kg/ha P_20_5, and 125 kg/ha K_20 gave a higher yield.

Seed rate. With local varieties, the recommended seed rate ranges from 10–15 kg/ha, whereas with imported varieties, it is about 10 kg/ha. More seed is needed for early planting than for in–time sowing. The seeds are planted on ridges 60 cm apart, and the field is irrigated immediately after planting.

Sowing. Local varieties are planted in September and October, whereas imported varieties are planted in November. Late planting can be done even in December. If the crop is planted thickly, it is thinned about three to four weeks after sowing, and a distance of 4–6 cm between plants is maintained. Plant spacing is very important for the development of proper roots.

Irrigation. Insufficient moisture will result in low yields.

Varieties. Carrot varieties can be grouped into two types: temperate and tropical. The temperate types have a biennial life cycle, whereas the tropical types are annual. Some of the common varieties grown are Red Long Lord and Mirpurkhas Selection.

Harvesting. Early carrots for the market are pulled when they are partially developed. They are normally dug out with a spade or a *khurpa* when the soil is fairly moist. The roots are partially graded, and all split and branched roots are left in the field. Roots are trimmed and washed before being sent to market.

Yield. The tropical types usually give high yields. The yield varies from 20,000–30,000 kg/ha.

14.2.3 Turnip

The turnip (*Brassica rapa* L.) belongs to the family Cruciferae. It has also been classified as species *compestris* and var. Rapa. The turnip is probably of European or Asiatic origin. In hot climates, it is grown as a winter vegetable; and in areas where summers are cool it is grown as a summer crop. Both its enlarged root and its foliage are used as salad, cooked, and pickled. The young tops are very rich in minerals and vitamins, particularly for their content of calcium and iron and vitamins A, B, and C.

Climate. Turnips are a cool–weather crop. They also grow in warm weather, but the best quality is produced under cool–weather conditions. Other climatic requirements are the same as for radishes.

Soil. Turnips can be grown on all soil types. The best crops are produced on fertile, deep, and moist sandy loam soils. Thorough preparation by deep ploughing is necessary. To secure high yields and quality, the soil must be fertile and there must be a constant supply of moisture until the roots have attained marketable size.

Manures and fertilizers. FYM at 20–25 t/ha should be applied before the land is prepared. This should be supplemented by 35 kg of N, and 50 kg each of P and K before sowing the seed. Another dose of 35 kg N should be top–dressed when the roots just start forming.

Spacing. Turnip seed is generally planted in rows 30-45 cm apart. Seed drills are used, and the seed is covered about 2–4 cm deep. The seed rate is about 3–4 kg/ha.

Sowing time. The turnip is grown almost entirely from seed. Since it does not thrive in hot weather, in the plains the tropical types should be sown from August to September and the temperate types from September to December. In the hills, they are grown from March to May.

Irrigation. The first irrigation is given immediately after sowing, while subsequent ones are at intervals of 7–8 days. In warm weather this should be reduced to 4–5 days. Irrigation frequency should be increased during the period of active root development.

Varieties. There are generally two classes of turnips: (1) European or temperate, (2) Asiatic or tropical. The temperate varieties are sweeter and more palatable, whereas the tropical types are more pungent and better for pickles, and are earlier and more heat–resistant. Temperate varieties include Purple Top, White Globe, Snowball, and Golden Ball. Tropical types are red, white, or dull–coloured. Desi Red is one such variety.

Harvesting. Turnips are pulled when the roots reach marketable size, and the leaves are removed from the base. For use as greens, the plants are pulled or cut at the soil surface and the foliage is cooked. Young turnips are sometimes bunched, but most of them are topped and packed.

Yield. Yields vary from 20,000–40,000 kg/ha.

14.2.4 Beet

The beet (*Beta vulgaris* L.) is a member of the Chenopodiaceae ('goose-foot') family. The garden beet, stock beet or sugar beet, and chard are important members. Beets probably originated in the Mediterranean region or in Central Europe. Beets are grown for the table, as a truck crop, and also for canning. They are eaten raw or cooked with other vegetables and with meat; and are rich in minerals, vitamin C, and sugar. Good quality beets have dark internal colour throughout the root, but under unfavourable conditions may have alternate white and coloured layers. Good colour is not necessarily related to high sugar content, but poor colour seems regularly correlated with low sugar content.

Climate. Beets are grown as a cool-weather crop. They can also be grown in warm weather, but the best quality beets are produced in cool weather. The crop is grown in September and October. Temperature affects maturity, and at temperatures below 10°C the crop sets seeds before reaching marketable size.

Soil. Beets can be grown in all soil types but do best on moist, well-drained, deep, friable loam. Alluvial soils such as silt loam are excellent for commercial farms. Sandy loams are also good for the development of beet roots. Heavy soils are not desirable since root development is impaired and unsymmetrical shapes may result. Beets are very sensitive to acidic conditions, and if the soil is very acidic lime should be applied to bring soil pH to around 6.0–6.5.

Manures and fertilizers. Well-rotted FYM at 50–60 t/ha should be incorporated during soil preparation, or preferably applied to the previous crop. Application of fertilizers at high rates—300 kg ammonium sulphate, 625 kg superphosphate, and 125 kg muriate of potash—may benefit the crop. Half of the ammonium sulphate should be supplied as top-dressing in split doses, with one heavy dose when the plants are about 10 cm tall.

One tonne of beet roots removes 2.25 kg of N, 4.5 kg of P, and 4.5 kg of K. Thus, for an average soil the manurial recommendation is 60–70 kg/ha N, 100–230 kg/ha P, and 60–70 kg/ha K. On sandy soils, the quantities of N and K should be increased. Beets have a high boron requirement; boron deficiency results in internal root symptoms like black spot or dry rot.

Beets must make rapid and continuous growth to develop high quality; therefore a good supply of available N, P, and K is necessary. On sandy and sandy loam soils, manure adds both fertilizer and humus. Well-rotted manure can be applied to good advantage shortly before seeding. Green crops are useful to improve soil texture. Heavy soils can be made more friable, and sandy types will hold more water if organic matter is incorporated.

Seed rate. About 5–6 kg/ha of seeds are required. The rate depends on several factors such as size of planting, percentage of germination, methods of seeding, and market. Beet seed may retain its viability for five to six years, but as a rule it is not wise to sow seed that is more than two years old.

Irrigation. The row or furrow method of supplying water is used in irrigated areas, and irrigation at 10–day intervals should be given.

Varieties. Beet varieties are classified by shape. They are flat, short, top–shaped, deep–oblate to round, globular to oval, half–long, and long.

Harvesting. Bunch or body beets are harvested at a diameter of 3–4 cm. In some areas, the beets are pulled by hand and injured or dead leaves removed before bunching. They are tied together in bunches of four to six with the tops and washed to remove any adhering soil.

Yield. About 20,000–25,000 kg/ha.

14.3 Bulb crops

14.3.1 Onion

The onion (*Allium cepa* L.) belongs to the family Amaryllidaceae. The onion originated in Afghanistan, the area of Tajikistan and Uzbekistan, western Tien Shan, and India. Western Asia and the areas around the Mediterranean Sea are secondary centres of development. Onions have been cultivated and used as a food throughout history. They are consumed in their early green stage and also after the bulbs mature. They are used by rich and poor alike, and are often called the 'poor man's food'. The onion has its own distinctive flavour and is used in soups, meat dishes, salads, and sandwiches, and is cooked alone as a vegetable. Its pungency is due to the presence of a volatile oil (allyl propyl disulphide). The mature bulbs contain some starch, appreciable quantities of sugars, some protein, and vitamins A, B, and C.

Climate. The onion is a cool–season crop. Onions are affected by the length of the day, and daylength requirements of different varieties may differ. Therefore only acclimatised varieties should be grown. A relatively high temperature and a long photoperiod are essential for bulb formation in most of the commercial varieties grown in Pakistan. However, for seed–stalk development, temperature is more important than daylength.

Soil. Onions are grown on nearly all types of soils, but they grow more successfully on fertile soils which are rich in humus and well–drained. Onions are sensitive to high acidity and produce their maximum yield over a fairly narrow range of soil reaction; optimum pH ranges from 5.8–6.5.

Manures and fertilizers. Fertilizer requirements vary with soil type, previous fertilizer application, purpose for which the crop is grown, and many other factors. A yield of 30 tonnes of onions is reported to remove 15 kg of N, 42 kg P, and 130 kg of K (Choudhry 1979). The onion, with its limited root system, responds to good fertilizer practices on almost any soil. Nutrients are usually applied in the form of commercial fertilizers.

FYM at the rate of 40–50 t/ha should be added several weeks before planting. NPK fertilizer at the rate of 120–60–60 kg/ha is recommended. Half the dose of N is to be top-dressed after about a month; the rest of the fertilizer can be added at the time of transplanting. As Pakistani soils are generally rich in K, the dose of this element can safely be reduced. Baloch et al. (1991) conducted fertilization experiments on onions and found that application of 125 kg N with 75 kg K_20 gave the highest yield (Table 14.4).

Table 14.4 Onion yield as affected by different nitrogen and potassium fertilizer combination levels

Treatment: N + K_2O (kg/ha)	Yield per hectare (tonnes)
75 + 75	13.67
75 + 100	13.93
100 + 75	17.19
100 + 100	16.93
125 + 75	**22.66**
125 + 100	22.40

Source: Baloch et al. (1991).

Mangrio et al. (1987) reported that application of 100 kg P_20_5/ha gave a higher yield than 70 kg P_20_5/ha, and that K_20 fertilizer did not show any significant effect on bulb yield of onions. Shaikh et al. (1987b) found that application of 90 kg N/ha increased the yield of bulbs. Rehman et al. (1978) reported that a higher rate of N with an optimum dose of K increases onion yield. The highest yields were obtained from plants receiving N at 56, P_2O_5 at 67, and K_2O at 67 kg/ha.

Spacing. Onion seedlings are transplanted onto ridges, which should be made about 60 cm apart. Seedlings are transplanted to both sides of the ridges at a distance of 10–15 cm between plants. Khushk et al. (1990) report that spacing of 20 cm between rows and 10 cm between plants gave good yields. Mangrio and Baloch (1985) suggest that a plant population ranging from 50–62 plants/m^2 appears to give the maximum yield. Mangrio and Bozdar (1985) observe that with increasing plant density the number of onion bulbs increases but the vertical and horizontal dimensions and weight of each bulb decrease. Rashid and Rashid (1978) reported that bulb size and weight increased with increasing inter and intra-row spacings, but total bulb yield was higher with closer spacing.

Seed rate. The seed rate depends upon the purpose for which planting is done and the distance intended between rows. In nursery, seed is sown at the rate of about 8–10 kg/ha. For direct sowing, about 25 kg/ha of seed are used, and when the plants are 6–8 weeks old, they are thinned to a proper distance. When bulbs are sown, about 1000–1200 kg/ha are needed. They are dibbled 15 cm apart in rows 45 cm apart.

Irrigation. The correct timing of irrigation is important. Soil, both surface and subsoil, current weather conditions, and age of the crop need to be considered in deciding when to irrigate and how much water to use. A seedbed is usually irrigated immediately before transplanting. The moisture content of the soil should be kept optimum. Irrigation after a dry spell may cause the outer scales to split. In such situations, the farmer must decide what to do based on the condition of the crop. If it has reached marketable size, it should be harvested; but if it is not ready to market, then irrigation is the solution. Irrigation is stopped when the tops mature and start falling.

Varieties. The commercial varieties of onion grown in Pakistan are Faisalabad Early, Pindi Red, Ghotki, Local White, Phulkara, Thanobola Khan, Red Nasik, and Desi Red.

Planting time. Planting time varies with locality, type of onion, and method of propagation. For early planting, seedlings are raised between June and September and transplanted from July to October. In the plains of Pakistan, for the winter crop, seeds are sown from the middle of October to the end of November. In December and early January, when the seedlings reach the age of 8–10 weeks, they are transplanted. For spring onions, seeds are sown earlier. In the hills, the seeds are sown from March to June.

Harvesting. Onions may be harvested either as green bunch onions or as mature ripe bulbs. Green bunch onions are harvested when they are about the diameter of a lead pencil and a small bulb has formed. A well–matured, ripe bulb is firm and has the shape and colour typical of the variety. Maturity is indicated by the tops dropping just above the bulb while the leaves are still green. Mature bulbs can be pulled out easily by hand if the soil is not hard or compact; otherwise they can be dug out with a shovel.

Yield. A yield of 25–30 t/ha has been recorded.

14.3.2 Garlic

Garlic (*Allium sativum* L.) belongs to the family Alliaceae. It has compound bulbs containing 10–16 cloves enclosed by a thin membranous sheath.

Garlic is native to southern Europe and western Asia; it is an ancient crop highly valued as a seasoning. It is eaten daily by thousands of people in food products like soups and pickles, and is used as a spice or condiment throughout Pakistan. Garlic salt is also made from pulverized dehydrated garlic. It has a higher nutritive value than other bulb crops and its preparations are administered as cures for some lingering stomach diseases, sore eyes, and earache. It reduces the level of cholesterol in the blood, and is recommended for heart patients. Also, it contains a considerable amount of calcium, phosphorus, and vitamin C.

Climate. Generally, garlic also grows well in places where onions are grown; but it performs better where summer and winter temperatures are moderate. In Pakistan, it is usually grown as a cool–season crop. Long days are considered better for the growth of quality cloves.

Soil. Soil requirements are the same as for onions except that garlic needs a richer soil. It thrives on fertile, well–drained, loamy soils — friable sandy loam well supplied with organic matter. Heavy clay soils may result in poor bulb shape and make harvesting difficult.

Manures and fertilizers. The cultivation, fertilization, manuring, and general care of garlic are about the same as for onions. The main fertilizer requirement is for N, which should be about 125–150 kg/ha, applied in two doses for best results. Korejo (1983) reported that application of 60 kg/ha of N is an appropriate dose for getting good yields of garlic.

Spacing and seed rate. Garlic is propagated by planting good cloves which are detached individually from the mother bulb. Cloves are planted singly, dibbled 7.5 cm apart and 2.5 cm deep in rows spaced 15–20 cm apart. About 400–500 kg/ha is sufficient. The cloves must not be planted so deep that the soil checks their swelling nor so shallow that they get washed out.

Planting time. The crop is planted from Sept–Nov in the plains.

Irrigation. Frequent irrigation is required for good development of the cloves. In order to get good–sized cloves, earthing–up is also necessary. Irrigation should be stopped when the leaves start withering.

Varieties. Faisalabad White, Lahsan Gulabi, and Lahsan Purple are commonly grown. Lashari (1984) found that the purple variety was superior to all varieties in per hectare yield.

Harvesting. Garlic is ready for harvesting when the tops turn brownish or yellowish and dry up. The plants are pulled by hand after loosening the soil, and placed in the field to dry for a week or more. After this, the tops and roots are removed by hand with sickles. The top is cut about 2.5 cm above the bulb and the roots are trimmed close to the bulb. When they are completely dry the bulbs are packed.

Yield. The yield varies from 4,000–10,000 kg/ha depending on the soil, variety, and care taken.

14.4 Leafy vegetables (potherbs or greens)

Leafy vegetables are generally grown for their tender, edible foliage. The importance of green leafy vegetables in the diet has long been recognized because they are very rich in minerals and vitamins A and C. They also supply needed roughage in the daily diet. Nutritionists recommend daily consumption of at least 116 g of leafy vegetables.

A variety of greens are cultivated in Pakistan. The leaves of certain plants which grow wild are also used as vegetables. Some vegetables like cabbage, Chinese cabbage, and kale, usually considered in other groups, are in fact green leafy vegetables. The leaves of some shrubs and trees are also used as potherbs or greens. The leafy vegetables are grown throughout the year, and their culture is relatively simple. Spinach, lettuce, fenugreek, and mustard are suitable for growing during winter. Others like amaranth and purslane can be grown during the summer.

14.4.1 Spinach

Spinach (*Spinacia oleracea* L.) belongs to the family Chenopodiaceae. It was first cultivated by the Arabs, and probably originated in southwest Asia. The Moors (Berbers and Muslims of Arab descent) took it to Spain from where it spread throughout the world. Spinach is an annual dioecious plant. It grows rapidly, especially in spells of dry weather with bright sunshine, and has a tendency to go to seed quickly.

The leaves are rounded, broad, and fairly thick, with a large amount of parenchymatous tissue. The plant has an upright growth habit, its height reaching about a metre at flowering, depending on the variety. The edible portion of the winter type is the leaf, and both the leaves and shoots of the summer types are eaten.

Spinach is rich in vitamin A, contains appreciable ascorbic acid and riboflavin, and has a small quantity of thiamine. It also contains iron and calcium, but the calcium is unavailable because it unites with oxalic acid to form calcium oxalate.

Climate. Spinach thrives during relatively cool weather and withstands freezing better than most vegetable crops; it does not do well during hot weather. It is sown several times a year. In regions with a mild winter, the crop is grown mostly during winter and early spring, while in other regions it is grown both as a spring and a fall crop. In the plains, the main crop is sown from September to November, and in the hills from March to May.

Soil. Spinach can be grown on all type of soils, with clay loam soils considered especially suitable. It also tolerates slightly alkaline soils. Deep cultivation and thorough preparation of the soil are essential. Good drainage is important for healthy leaves. Weed–infested soils should be avoided.

Manures and fertilizers. Spinach should be fertilized with 20–30 t/ha of FYM, well mixed with the soil one month before sowing. For leafy growth, N–fertilizer is the first requirement. A top–dressing of about 20 kg/ha of N after every cutting gives a quick growth of tender, green, succulent leaves.

Spacing and seed rate. Generally spinach is sown by broadcasting in thoroughly prepared soil. The seed should be evenly sown at about 20–25 kg/ha. It is, however, better to sow the seed in rows about 15–20 cm apart. This facilitates weeding in the earlier stages of growth. At some places it is grown on ridges 30–45 cm apart. For early planting, ridges are made; normal planting is done on flat beds. Good quality seed produces tender, high–quality leaves, good and quick rejuvenation after each cutting, and late appearance of the seed stalk.

Irrigation. Abundant moisture is needed to produce dense growth. Besides irrigation immediately after sowing, irrigation should be continued at 8–10 day intervals.

Varieties. Local or desi *palak*, English *palak*, and Kandiari varieties are commercially grown in Pakistan.

Harvesting. The leaves are picked or cut from the base when they are 15–30 cm long. The crop is ready for a first cutting within six to eight weeks. The larger leaves are selected as they develop, emphasizing tenderness and succulence. The leaves should be plucked so that the plant is not injured. Three to four cuttings are taken.

Yield. The average yield during the best season is about 8,000–10,000 kg/ha of green leaves.

14.4.2 Fenugreek

Fenugreek (*Trigonella foenumgraecum* L.) belongs to the family Leguminoseae. Fenugreek is considered to be native to eastern Europe and Ethiopia. The green leaves are used as a vegetable and are quite rich in protein, minerals, and vitamin C. The seeds and dry leaves are used for flavouring stews and curries. It also has some medicinal value and is used to prevent constipation, relieve indigestion, stimulate the spleen and liver, and as a diuretic. Fenugreek is grown all over Pakistan for its green leaves and seeds. Generally two varieties, local and Kasuri, are cultivated.

Climate. Fenugreek is a cool–season crop; it is fairly frost–tolerant and moderately heat–tolerant, but good quality leaves are produced at low temperatures.

Soil, manures, and fertilizers. Soil requirements are about the same as for spinach. About 8–10 tonnes of FYM should be mixed thoroughly with the soil one month prior to sowing. A top–dressing of N at the rate of 50 kg/ha should be given after every cutting.

Spacing and seed rate. About 20 kg/ha of local and 15 kg/ha of Kasuri fenugreek seed are required. The seed is sown in rows 20–25 cm apart. It must be free of contaminating seeds of weeds or other varieties.

Irrigation. Frequent irrigation is necessary for quick growth. Ample moisture in the soil keeps the plants succulent and makes more cuttings possible. During summer the crop should be irrigated every week, whereas in winter irrigation should be given every fortnight.

Harvesting. The crop takes about six to eight weeks from sowing to the first cutting, which is made when the plants are up to 18 cm tall. The plants are cut 2 cm above the ground, leaving the slabs, which produce new stalks which are then cut in turn.

Yield. Yield varies according to the variety. Local fenugreek may give a yield of 7000–8000 kg/ha, whereas Kasuri fenugreek gives a higher yield of about 9000–10,000 kg/ha.

14.4.3 Amaranth

Amaranth (*Amaranthus tricolor* L.), local name *chaulai,* belongs to the family Amaranthaceae. There are about six different cultivated species: *A. blitum* and *A. tricolor*, which are known as *Amaranth* and *badi chaulai,* respectively; *A. caudatus* which is grown for seed; and *A. mangostanus*, *A. lividus*, and *A. dubius* which are grown on a smaller scale for green leaves.

The amaranth is considered to be native to the Indo–Pakistan subcontinent and occurs wild or as a weed in fields. It reaches the market when other greens are scarce. The tender leaves and shoots are a good source of proteins, iron and other minerals, and vitamins A, B, and C. This is the most common leafy vegetable grown during summer in Pakistan. Amaranth *chaulai* is suitable for sowing in spring and gives green leaves throughout the summer. The plants are dwarf with small leaves, and rejuvenate quickly after each cutting. The plants of *badi chaulai* are tall with thick stems and large leaves. This type can be grown in the summer as well as in the rainy season.

Climate. Amaranth is warm–season crop and grows best in the rainy season. It is also grown on medium elevations.

Soil, manures, and fertilizers. A moderately fertile and well–drained soil is preferable for amaranth. Manure and fertilizer requirements are the same as for other leafy vegetables.

Spacing and seed rate. Spacing at about 20–25 cm is the same as for fenugreek. About 2–3 kg/ha good quality seed is required.

Irrigation. Irrigation is the same as for other leafy vegetables.

Varieties. Green and red–leaved varieties are commercially grown in various parts of the country; the green types are more common than the red.

Harvesting. Cutting of the leaves starts when the plants are 25–30 cm tall. Only the fully–grown side leaves are removed. The tops of the plants may also be cut, leaving the lower leaves to produce new shoots in their axils. Picking is continued once a week.

Yield. The average yield is about 6000–7000 kg/ha.

14.4.4 Lettuce

Lettuce (*Lactuca sativa* L.) belongs to the family Compositae. Originating around Asia Minor, Iran, and Turkey, lettuce has been used as a salad crop for more than 2500 years. In Pakistan, cultivation of lettuce is limited to the large cities. Rich in minerals and vitamins, it is commonly used as a salad.

Climate. Lettuce prefers a relatively cool growing season with average monthly temperatures of 12–15°C. At high temperatures it bolts to seed, the leaves become bitter, and the development of tip burn and rot is accelerated. Head lettuce rots in hot humid weather. Lettuce seed does not germinate properly when the soil temperature is above 30°C.

Soil, manures, and fertilizers. The crop can be grown successfully on many soils, but sandy loams and loams are best. The plants must grow rapidly and without check; hence heavy manuring is needed. Usually 10–15 tonnes of FYM and fertilizers to supply 100 kg/ha of N and 60 kg/ha each of P and K are applied.

Spacing and seed rate. Lettuce is grown from seed; the seedlings are transplanted when they are four weeks old. They are planted at a distance of 25–30 cm in rows 45–60 cm apart. About 2–3 kg/ha of seed is required.

Irrigation. Water should be supplied to keep soil moisture conditions uniform. An irrigation frequency of 5–7 days may be enough. Drainage is as important as irrigation since an excess of soil moisture may cause rotting.

Varieties. The seed of various types of lettuce are usually imported from the USA and Europe. Lettuces fall into head and leafy varieties. Leafy varieties are (i) crisp with crumpled leaves, or (ii) buttery with soft smooth leaves. Memon (1991) observed that the varieties Red Sails, Salad Bibb, and Butter Crunch give yields of 6121, 5493, and 5388 kg/ha, respectively.

Harvesting. The crop is gathered as soon as plants attain an acceptable size, or alternate plants are removed for use while thinning. The leaves should not be allowed to become tough. Harvesting should be done in the morning and produce must be sent to market immediately.

Yield. Average yields of head lettuce range from 9000–10,000 kg/ha; the leafy varieties give relatively higher yields.

14.4.5 Purslane

Purslane (*Portulaca oleracea* L.) belongs to the family Portulaceae. *Portulaca* includes 100 or more herbaceous species, found throughout the warmer parts of the world. The plant grows wild in most parts of the tropics, sometimes as a weed, but the cultivated types differ greatly from their wild relatives in flavour as well as size.

Purslane is grown widely in arid regions for its succulent leaves. The improved types have broad leaves and erect shoots under thick planting. Generally sown in early spring, purslane will grow in the hottest climates, and often provides the only green vegetable after others have succumbed to heat. Unimproved types are often creepers with small succulent, fleshy leaves and tender stems which can be cooked separately as well as with meat, gram, or pulse. Purslane is a source of vitamin C. Calcium oxalate crystals are commonly laid down in the leaves and stems.

The succulent branches are cut when they reach 15–20 cm in length. The crop is ready for the first cutting about a month after sowing. The fruit, a small, almost spherical capsule which splits transversely when mature, contains small black or dark-brown seeds with an oily feeling skin.

Climate. Purslane is a hot-weather crop which requires a warm climate for good leafy growth.

Soil, manures, and fertilizers. The plant grows wild as a weed and is also cultivated on moderately fertile soils. The land is ploughed thoroughly two to four times before the seeds are sown. A basal dressing of 12–15 t/ha of bulky manure is given.

Seed rate and sowing. In the plains, March to June is the common sowing season, while in the hills April–September is normal. Seed at the rate of 2.5–4 kg/ha is mixed with sand, then broadcast in beds, and raked to cover. The beds are then watered. Sowing is done fortnightly.

Irrigation. Irrigation is light; except for a few weekly irrigations at the beginning, watering is done only if necessary.

Varieties. There are no identified commercial varieties of purslane. However, some local flowering varieties are under cultivation in Hyderabad.

Harvesting. The crop is pulled about eight weeks after sowing. The leaves are bundled, packed, and sent to the market immediately.

Yield. The yield is about 7500 kg/ha.

14.5 Solanaceous fruits

14.5.1 Tomato

The tomato (*Lycopersicon esculentum*) belongs to the Solanaceae (nightshade) family. It is native to tropical America where its indigenous name was *tomati*. From Mexico the tomato was taken to Europe and then to Asia. An important vegetable crop, it is grown in most home gardens and by market gardeners and truck farmers. It is also produced by forcing in greenhouses. It can be eaten either fresh or processed into many different products. It plays a vital role in maintaining health and vigour. Conn and Stumph (1970) reported that tomatoes are very helpful in healing wounds because of the antibiotic properties found in the ripe fruit. It is a good source of vitamins A, B, and C.

Climate. The tomato is a warm–season crop which needs a relatively long season to produce profitable yields. It requires 80–120 days from seeding to bearing, depending on type. Low temperatures without actual freezing inhibit fruiting. During heat and drought immature tomato blossoms drop rapidly due to increased transpiration. Fruit setting may not occur if the temperature is too high or too low at pollination. Comparatively cool nights with temperatures of 15–20°C and warm day temperatures between 25 and 30°C are best for optimum yields. Most tomatoes do not produce fruit when night and day temperatures exceed 25° and 35°C, respectively. The most difficult period for growing tomatoes in the plains is from mid–May to June when the day and night temperatures are very high, and during the monsoon season when there is rapid development of insects, diseases, and weeds.

Soil. Tomatoes can be grown on almost all soil types from sandy to heavy clay, at a pH range between 5.5 and 7.5. For an early crop, sandy loam is considered best. Loam, clay loam, and silt loam containing adequate organic matter give the highest yields.

Manures and fertilizers. The tomato plant requires large quantities of readily available plant foods. The use of manures and commercial fertilizers varies with locality and soil. A yield of 16,000 kg of tomatoes removes 50 kg of N, 16 kg of P, and 65 kg of K from the soil. For a good yield, 20–25 t/ha of well–rotted FYM should be incorporated in the soil at the time of preparation. About 80 kg/ha P and 40 kg/ha K should be applied at the time of transplanting. 100 kg/ha N should be applied in three splits.

Baloch et al. (1988) found that application of 100 kg/ha of K_2O resulted in the maximum yield. Rahim and Amjad (1985) conducted trials of fertilizer application to different varieties and found that fertilizer treatment of 101 kg/ha of N, and 67 kg/ha of P_2O_5 gave the highest yield. The Pakistan Agricultural Research Council (PARC) has recommended broadcast application of fertilizer before transplanting of 100 kg/ha N, 150 kg/ha P_2O_5, and 60 kg/ha K_2O. For optimum yields, the full quantity of P_2O_5, and half of the N and the K_2O should be applied before transplanting the seedlings; the remaining half of the N and K_2O should be given 6–8 weeks after transplanting. Ginai (1970) observed that application of 56–57 kg/ha N, 112–168 kg/ha P_2O_5, and 112–168 kg/ha K_2O resulted in higher yields.

Spacing and seed rate. High quality, true–to–type seed should be used in order to obtain high yields and a uniform crop. About 500–600 g/ha of seed is enough. Channels should be prepared 1.0–1.5 m apart for transplanting the seedlings. Tomato seedlings are transplanted on both the sides of the channels at a plant–to–plant distance of 60 cm.

Time and method of sowing. The time of sowing is largely determined by when it is desired to transplant the seedlings to the field. Plants are usually started in specially prepared nursery seedbeds, several weeks ahead of transplanting time. In the plains, the times for seeding for raising nursery and for transplanting seedlings to the field are as follows.

Time of seeding for nursery	Time of transplanting
Seed planted in July/August	August/September
Seed planted in September	October
Seed planted in November	February/March

In the high hills, where tomato is grown as a summer crop, seed is planted for raising nursery in March–April and the seedlings are usually transplanted in May–June.

Irrigation. The water requirement of tomatoes depends on the soil and climatic conditions of the particular area. During the warmer season, irrigation may be needed as often as every 5–6 days on sandy soils, and at 10–12 day intervals on heavier soils.

When the temperature is very high, weekly irrigation should be given. During cold periods, fortnightly irrigation is needed to ensure fruit setting and maturation. A period of drought followed by sudden heavy watering during fruit development may cause the fruit to crack. The tomato plant is susceptible to 'wet feet', therefore production areas should be well drained.

Varieties. Some local varieties grown are Lyallpur Selection I and Peshawar Long. All the cultivars commercially grown in the country have been introduced from abroad, mainly from Europe and America. The principal cultivars grown are T–10, Money Maker, Roma, Red Top, and Marglobe. In choosing cultivars for tomato production, the following points should be kept in mind: (1) whether the crop is grown for home consumption or for general marketing; (2) length of the growing season; (3) yield; (4) susceptibility to diseases; and (5) ability to withstand handling during transportation.

Harvesting. The fruit can be picked every three days. Fruit should be picked when they have just started changing colour. The stage of maturity at which tomatoes are picked depends on the purpose for which they are grown and the distance over which they are to be transported. Immature green, mature green, turning half–ripe or pink, and ripe are some of the recognized stages.

Yield. The yield varies from 20,000–24,000 kg/ha.

14.5.2 Chili

Chilies (*Capsicum annum* L., *Capsicum frutescens* L.) belong to the family Solanaceae. Cultivated chilies are of American origin, and have been discovered in the prehistoric remains of Peru. Columbus carried seeds from America to Spain in 1493, and the cultivation of chilies spread rapidly to other parts of Europe. They were probably introduced to the Indo–Pakistan subcontinent by the Portuguese in the middle of the 17th century, where some types established themselves so quickly that some botanists believed that chilies had their origin in the East. However, the word *chili* strongly suggests its importation from South America. In Pakistan, chilies are cultivated on a large scale as a cash crop.

Chilies are rich in vitamins, especially vitamins A and C, and also contain appreciable amounts of calcium, phosphorus, and iron. The hot types are a source of the digestive stimulant capsaicin.

Climate. Chilies thrive best in warm climates with a long, warm growing season where there is little danger of injury due to frost. The crop takes 120–150 days to mature.

Chilies have a wide range of cultivation, being grown under both tropical and subtropical conditions. The plant grows over a wide range of elevation, from sea level to 2000 m. It thrives in areas having moderate (60–120 cm) rainfall. Heavier rainfall during the growing season is not favourable as it leads to rotting; however, the crop can even be grown under such conditions if it is planted after the heavy rains.

The crop is irrigated in low–rainfall areas and when grown as a hot–weather crop. In general, chilies require a temperature of 20–25°C. Unfavourable temperature and water supply are the main causes of bud, blossom, and fruit drops.

Soil. Chilies grow best on well–drained silt or clay loam. Waterlogged and alkali soil are not suitable.

Manures and fertilizer. Fertilizer regimes for chilies are similar to those for tomatoes. Khan et al. (1982) observed that application of N, alone or in combination with P, resulted in increased yield of peppers. Baloch et al. (1989) observed that high doses of P_20_5 and K_20 significantly increased the dry fruit yield of chilies. They further found that application of 12.5 kg each of P_20_5 and K_20, along with the application of a basic dose of 12.5 g N per plant, seems to be a better manurial programme for getting high yields from chilies.

Spacing and seed rate. About 1.50–2.0 kg seed is required to plant one hectare, depending on the varieties. The seedlings are transplanted when they are about 4–8 weeks old. They are planted 30 cm apart on the sides of ridges, which should be prepared at a distance of 60 cm apart.

Sowing time. Chili seeds are sown in nursery beds in February, March, and April for a summer crop, in August for an autumn–winter crop, and in November for the spring–summer crop. In the hills, seeds are sown in March–April.

Irrigation. Irrigation should be done in the same way as for tomatoes.

Varieties. The varieties of chilies fall into three groups: (1) long pungent types, including pickling types; (2) bell–shaped, mild thick–fleshed types used as vegetables, commonly known as sweet peppers or *Simla ki mirch*; and (3) small, round varieties shaped like jujubes (*ber*). Common varieties grown are:

Group 1	Talhari and Ghotki
Group 2	Sanam and Gola Peshawari
Group 3	Lawangi, Tatapuri, and Burewala

Harvesting. The stage of maturity at which chilies are picked depends on the type and the purpose for which they are grown. Chilies which are used as vegetables are generally picked while they are fully grown but still green. Those used for pickles are picked either green or ripe; and chilies used for drying are picked when red and fully ripe.

Yield. Normally, the yield of fresh green chilies is three to four times that of dry chilies. Yields of 1500–2500 kg/ha have been obtained.

14.5.3 Brinjal or eggplant

The brinjal (*Solanum melongena* L.) belongs to the family Solanaceae (nightshade). It is one of the most common vegetable crops grown in Pakistan, and is probably native to India. It has been under cultivation in the subcontinent since ancient times and is available in the market year–round. Brinjal is considered a staple vegetable. It contains vitamins A and B, and its nutritive value varies in different varieties. The brinjal has Ayurvedic medicinal properties, and white brinjal is said to be good for diabetic patients.

Climate. Brinjal is a tender plant that requires a long, warm growing season for successful production. It is killed by severe frost, and a long spell of chilly frost–free weather may also damage the crop. The late varieties, however, withstand mild frost. A daily mean temperature of 13–21°C is most favourable for successful production.

Soil. Fertile and well–drained soils are desirable for brinjal cultivation. It can be grown on different kinds of soil, but does best on silt loams or clay loams. With a short growing season, a sandy or sandy loam soil is preferable to a heavier one since lighter soils warm up earlier in the spring.

Manures and fertilizers. If the soil is not naturally fertile manures and fertilizers should be added. Land preparation depends on the type of soil and the previous crop. FYM at the rate of 20–25 t/ha should be incorporated at the time of soil preparation. This should be supplemented with about 100 kg/ha N, 50 kg/ha P, and 50 kg/ha K.

Seed rate, sowing, and timing. About 1–2 kg of seed is sufficient to plant a hectare. The seeds are sown in raised nursery beds 6–12 mm deep in rows 5 cm apart. The bottom few centimetres of the nursery bed should have small chips and sand to allow good drainage. When they are four to six weeks old, the seedlings are transplanted, at a distance of 60 × 45 cm apart.

The main seasons for sowing brinjals are:

1. Autumn–winter crop: Seed is sown in June and seedlings transplanted in July.
2. Spring–winter crop: Seed is sown in early November and the seedlings transplanted during January and early February. The nursery bed should be protected from frost.
3. Rainy–season crop: Seed is sown in March and seedlings transplanted in April.

To get brinjals round the year, sowing is done in September or early October and the seedlings are transplanted in November. Special care should be taken to protect this crop from frost. In the hilly areas, however, the seed is sown in April and the seedlings are transplanted in May.

Irrigation. Transplanting is done in standing water. The first irrigation should be given four days after transplanting. In summer, weekly irrigation is necessary and should be given well in time. In winter, however, fortnightly irrigation is preferable.

Harvesting. The brinjal can be eaten from the time it is one–third grown until it is fully ripe. The fruit is harvested before it is fully ripe, but it should be well–coloured. When fully ripened on the plant, the fruit becomes greenish bronze, the flesh becomes dry and tough, and the seed gets hard. Harvesting is done by cutting with a knife or plucking by hand.

Varieties. The commercial varieties under cultivation are of two types, classified according to the shape of the fruit. One group includes round, medium or large size fruits, and the second includes the long, thin fruits. Common local varieties are Neelum, White Egg, Mirpurkhas Selection, Multan Selection, and Sarhandi Long. Common imported varieties are Black Beauty, Black King, and Pusa Purple Long.

Yield. The average yield varies from 22,000–25,000 kg/ha.

14.6 Cole crops

All cole crops are hardy and thrive in cool weather. The group includes cabbage, cauliflower, knol–khol, broccoli, Brussels sprouts, and Chinese cabbage. All crops of this group have developed from the sea or wild cliff cabbage (*Brassica oleracea*) known as coleworts, from which the name *cole* derives. Cultural requirements for all the crops in this group are similar.

14.6.1 Cauliflower

The cauliflower (*Brassica oleracea* L. var. Botrytis) belongs to the family Cruciferae. Cauliflower (Latin *caulis* 'cabbage' and *floris* 'flower') probably originated in Cyprus. It is grown for its tender, white head or curd. It is used in curries and soups, and for pickles. It is rich in minerals, carbohydrates, and vitamins A and C.

Climate. Cauliflower is a delicate crop and can be damaged by freezing weather near harvesting. Moderately cool weather and a uniformly good supply of soil moisture are essential. Cauliflower produces the best curds in a cool, moist climate, but is damaged by snow. In the plains it is available from September to May. The plants may fail to form desirable heads in dry, hot weather, which causes the heads to develop prematurely and bolt or button. However, there are varieties which can withstand quite high temperatures. The optimum monthly temperature for curd formation is 15–20°C, with an average maximum of 25°C and average minimum of 8°C. The early varieties require higher temperatures and longer daylengths. A late type cauliflower is commonly called heading broccoli.

Soil, manures, and fertilizers. Cauliflower can be grown on any good, well–drained soil, but a fairly deep loamy soil is desired. Sandy loams and silt loams are best. Cauliflower is sensitive to high acidity and grows best on a neutral to slightly acid soil, pH 6.0–6.8. A liberal quantity of manure is highly desirable, and well–rotted manure can be used with great freedom as a fertilizer for cauliflower. Cauliflower removes large quantities of major nutrients from the soil. A yield of 50 t/ha removes approximately 220 kg of N, 85 kg of P, and 270 kg of K. About 15–20 t/ha of FYM with 120 kg/ha N, 80 kg/ha P, and 40 kg/ha K should be applied just before transplanting. Bukhari and Shahani (1983) suggested that application of 112 kg/ha N in combination with FYM was the best treatment for obtaining a high yield. Rizvi and Jagirdar (1967) also concluded that nitrogenous fertilizer in addition to organic manures helped produce a good yield.

Seed rate, transplanting, and spacing. For an early crop, 600–750 g and for a late crop 350–400 g of seeds are required for one hectare. In the seedbed, the seedlings often suffer from damping off. Poor seed is a frequent cause of failure. Four to six–week–old seedlings are transplanted into a well prepared field. The planting distance depends on soil fertility, season, variety, and market demand. For early varieties, 45 cm and for late varieties 60 cm spacing is maintained both between plants and between rows. In some markets, small to medium–sized curds are preferred to large ones. When growing cauliflower for such markets, closer spacing should be used.

Irrigation. Cauliflower is a shallow–rooted crop and most of its roots are within 45–60 cm of the surface. After transplanting, an early crop will probably need irrigation twice a week and a late crop once a week. Watering immediately after planting is usually helpful, especially in warm, sunny weather when the plants soon droop. When the heads begin to form, weekly irrigation is wise.

Varieties. Some of the common varieties grown in different parts of Pakistan are Agati, Faisalabad, Local Early, Cheen-ka-Moti, Sadiqabad, and Mirpurkhas-Ka-Moti. Apart from these cultivars, some foreign varieties are also imported for cultivation.

Harvesting. The heads are harvested when they attain proper size and condition, i.e. before they become ripe, discoloured, or are otherwise blemished. It is better to cut a little early than too late. The heads do not all develop uniformly, and so a number of cuttings are necessary. Medium-sized heads (15–20 cm diameter) are in greatest demand. The curds are harvested retaining part of the stems so that the stub thus left protects the curds during transport.

Yield. Yield varies with variety and planting date. An average of 20–30 t/ha is common. Early crops give low and mid-season crops higher yields. Miano et al. (1990) conducted varietal trials and found that different varieties of cauliflower give highly significant variation in yield per hectare.

14.6.2 Cabbage

Cabbage (*Brassica oleracea* L. var. Capitata) belongs to the family Cruciferae. Possibly originating in Europe (Greece or Italy) about 2000–2500 B.C., it has long been an important vegetable in the USA and Europe, but its introduction to the Indo–Pakistan subcontinent is relatively recent. Cabbage is a biennial crop, grown for its enlarged, edible, terminal buds. Most commercial varieties produce more or less round heads, with slightly tightened, folded leaves. Cabbage is rich in minerals and in vitamins A, B1, B2, and C. It has a cooling effect, increases appetite, aids digestion, helps prevent constipation, and is good for diabetics. Cabbage is used for cooking, salad, and pickling.

Climate. A cool-season crop, cabbage grows well in cool, moist climates, and tolerates frost and extreme cold better than cauliflower. In the plains, cabbage can be grown from October to January. In the hills, it is grown as a spring and early summer crop.

Soil. Cabbage can be grown in all types of soil, but sandy loam gives the heaviest yields. It does poorly in highly acidic soils, and should not be grown in soils with pH lower than 5.5 or higher than 6.5.

Manures and fertilizers. Cabbage is a shallow-rooted crop and is a heavy feeder. A 50 t/ha yield removes 220 kg of N, 100 kg of P, and 220 kg of K. For a good crop, 15–20 tonnes of FYM before transplanting, and about 160 kg/ha N, 120 kg/ha P, and 60 kg/ha K should be applied.

Seed rate, transplanting, and spacing. About 1–1.5 kg seed yields enough seedlings for one hectare. The method of raising seedlings and transplanting is the same as for cauliflower. Four to six–week–old seedlings are transplanted to ridges. The planting distance depends on soil fertility, season, variety, and market demand. Relatively wide spacing (60 × 60 cm) promotes earliness and larger heads. Yield per hectare and number of heads harvested are usually increased by close spacing (45 × 45 cm).

Irrigation. Cabbage transplants are generally irrigated. The first irrigation is done soon after the seedlings are set, unless adequate moisture is already present in the soil. However, a light watering usually helps the plants to re–establish themselves rapidly. About three more irrigations are given to late cabbage and two to early cabbage. It is irrigated heavily when the heads are well developed and quite firm. Additional irrigation may be needed on light, sandy soils.

Varieties. One variety currently cultivated on a large scale is Golden Acre. Other varieties are Copenhagen Market, and Drumhead Early and Late. All these varieties are imported.

Harvesting. Cabbage is harvested when the heads are of suitable size, and firm but tender. Harvesting early cabbage usually begins when the heads are large and firm enough for use. Later cabbage is not harvested until the heads are full and hard, or until the green cover leaves begin to curl back slightly, exposing the whiter leaves. Heads burst if harvested beyond this stage. Late cabbage is harvested by grasping the head in one hand, bending it slightly, and cutting it off with a knife.

Yield. The average yield varies between 20–25 t/ha for the early crop and 30–35 t/ha for the late crop.

14.6.3 Kohlrabi

Kohlrabi or knol–khol (*Brassica caulorapa* L.) belongs to the family Cruciferae and is closely related to cabbage. The word *kohlrabi* is German, *kohl* meaning 'cabbage' and *rabi* meaning 'turnip'. It resembles an above–ground turnip, with the edible fleshy enlargement of the stem developing entirely above ground. Kohlrabi is an excellent vegetable if it is used before it becomes tough and fibrous. It is high in minerals and vitamin A and C. The leaves may be used for greens and the heads can be cooked in various ways. One way is to boil them, add salt and vinegar, and serve as a salad.

Climate. Knol–khol is a cool–season crop which grows well in a relatively cool moist climate. It can withstand extreme cold and frost better than other cool–season crops.

Soil, manures, and fertilizers. A rich soil will produce excellent knol–khol and a liberal dressing of manure is desirable. If manure is not available, green manure crops and commercial fertilizer may be used. A yield of

20 t/ha removes about 100 kg N, 85 kg of P, and 170 kg of K. Manurial and fertilizer requirements are the same as for cabbage and cauliflower.

Seed rate, transplanting, and spacing. About 1–1.5 kg/ha of seed is sown in the nursery. In the plains, sowing is usually in August, September, and October for the early, main, and late varieties, respectively. Seeds are originally sown in nursery beds, and the seedlings are ready for transplanting in three to five weeks depending on weather conditions. The usual spacing in the field is about 30 cm between rows and 20 cm between plants.

Irrigation. Irrigation recommendations are the same as for cabbage and cauliflower.

Varieties. Local and two exotic varieties, White Vienna and Purple Vienna, are under cultivation.

Harvesting. Knol-khol should be harvested when the swollen stem reaches a diameter of 5–7 cm and before it becomes tough and woody. The plants are harvested by pulling them out.

Yield. The average yield is 25–30 t/ha.

14.7 Cucurbits

Cucurbits are the largest group of summer vegetable crops grown in the tropics and subtropics. They include cucumbers, melons, gourds, pumpkins, squashes, and a large number of (mostly) vine crops. Most of them are monoecious, with only a few dioecious. A number of hermaphrodite and andromonoecious cultivars of some crops are also available. They are all summer crops and are susceptible to frost. The cultural requirements of all crops in this group are more or less the same.

14.7.1 Cucumber

The cucumber (*Cucumis sativus* L.) belongs to the family Cucurbitaceae. The plants are ancient crops, probably native to Asia and Africa. Cucumber is cultivated on a large scale in the Indo–Pakistan subcontinent. The fruit contains vitamins B and C. It is eaten raw with salt and as a salad with onion and tomato. It has a cooling effect, prevents constipation, and is helpful in checking jaundice. Cucumber seeds are used in Ayurvedic preparations; it is believed that the oil extracted from cucumber seeds is good for the brain and the body.

Climate. These plants require high temperatures and are planted either in the spring or summer. They are very sensitive to cold and are killed by even a light frost. There is no germination below 11°C, and at 18°C the germination rate is only about 68 percent. High germination rates and quick germination are obtained at 25–30°C. Sowing for the spring–summer crop may be done early in January–March or for the rainy-season crop from late June to July depending on local conditions.

Soil. Cucumbers can be planted on a variety of soils ranging from sandy to clay loam. For an early crop, sandy to sandy loam soils should be selected if possible, and for high yields, loam, silt loam, and clay loam. A pH of 5.5–6.7 is preferable, and the soil should be well drained.

Manures and fertilizers. A basal dose of 35–45 t/ha of FYM should be worked into the soil before sowing. In addition, organic fertilizers at the rate of 50 kg/ha ammonium sulphate, 100 kg/ha superphosphate, and about 55 kg/ha of potassium sulphate should be applied at final field preparation. Additionally, ammonium sulphate can also be used as a top–dressing at the rate of 40–60 kg/ha in two doses, the first when the plants start to develop runners and the second when fruiting has started.

Seed rate and sowing. About 2.5–3 kg/ha of seed is sufficient. Beds about 1.5 m wide are made in the field. Seeds are sown on both sides of the raised beds at the rate of 4–6 seeds per hill, the hills being .5 m apart. In sowing, the seeds are not bunched but scattered over an area of 15–20 cm to give the seedlings space for development. Just before the seedlings crowd, and about three weeks after sowing when the danger of beetle infestation is almost over, the plants are thinned to two plants per hill. The seedlings may have to be staked in areas with high rainfall.

Irrigation. Irrigation should be given weekly; the duration can be increased or decreased depending on the temperature. Care should be taken that irrigation water does not reach the foliage or the fruit of the plant.

Varieties. Cucumber plants generally bear male and female flowers separately on the same plant. Usually there are more male flowers than female flowers which bear fruit. An ideal variety has a high female:male flower ratio. Normally male flowers appear first and female later. The lower the main stem node on which the female flower appears, the earlier is the variety. The female:male flower ratio increases with the age of the plant.

Varieties of cucumbers are generally classified into four groups:

EUROPEAN–AMERICAN: Some of the English cultivated varieties
WEST ASIATIC: A very xerophytic species
CHINESE: Very large, long fruits with semi–glossy rind
INDO–PAK–JAPANESE: Himalayan and hermaphrodite types

A large number of varieties are grown in Pakistan, ranging from small pickling cucumbers to very large and thick varieties. Some of them are:

1. JAPANESE LONG GREEN: Long, green fruits, 30–35 cm in length.
2. STRAIGHT EIGHT: White spined, medium–long, thick, straight, green fruit.

3. BALAM KHIRA: Net–like structure, pale–green and dull–red colour, medium–sized fruit
4. LOCAL OR DESI VARIETIES: Sialkot Local, Commander, PARC–I, and Beittalpha.

Miano et al. (1991) investigated exotic cucumber varieties and found that spring Swallow F1 gave better results for yield in Sindh conditions.

Harvesting. The crop matures 60–75 days after sowing. The fruits should be harvested as soon as they are ready. If they are allowed to mature, the growth of new fruits is retarded and yields are decreased considerably. Picking should be done every third or fourth day.

Yield. The average yield is 12,000–14,000 kg/ha.

14.7.2 Muskmelon

The muskmelon (*Cucumis melo* L.) belongs to the family Cucurbitaceae. It is probably native to the hot valleys of Iran and northwest India. Muskmelon is an important crop, occasionally cooked as a vegetable in its green stage, but normally eaten ripe as a fruit. It is very wholesome and nutritious, prevents constipation, and is a good source of vitamins A, B, and C.

Climate. Muskmelons require more heat than cucumbers and do not withstand temperatures below 5°C. Optimum germination is at temperatures of 27–29°C. Hot dry air and sunshine during fruit ripening are needed for the development of high sugar content and good taste.

Soil. For an early crop, sandy loam soil is considered best, though the crop can be cultivated successfully on well–drained loam to clay loam soils. The soil should be cultivated deeply and thoroughly, especially in areas where the crop is not irrigated.

Manures and fertilizers. Rapid growth is required early in the season, so fields manured heavily for the preceding crop are excellent. Otherwise, FYM is added at the rate of 50 t/ha. About 110 kg/ha of superphosphate, 110 kg/ha of muriate of potash, and 125 kg/ha of ammonium sulphate should be incorporated at the time of final soil preparation. A side dressing of ammonium sulphate at 120–125 kg/ha may be given after the fruits set. Fertilizer should be applied according to the fertility status of the soil.

Spacing and seed rate. The land is ploughed deep (25–30 cm) and heavily irrigated. When the land is ready for planting, the crop is drilled at a distance of 2 m. Where irrigation water is applied, raised beds 2 m wide are prepared and seeds are planted 50–60 cm apart on both sides of the beds. About 4–5 kg seed is enough for planting one hectare.

Irrigation. Crops sown on river beds do not need irrigation. The plants get sufficient moisture through capillary action from the subsoil water. Crops sown elsewhere do need irrigation. The crop requires an abundance of moisture during the early stage of vine growth. Once the melons are fully grown, excessive watering should be avoided. Generally, five to six waterings are required.

Varieties. Local cultivars include Jahangir, Agaita Sufaida, T–96, Ravi, Chichawatni, and Jasalmari (Sanghari). Imported cultivars include Honeydew, Tam Dew, Canara, and NV–990. Mahmood et al. (1988) conducted a study on exotic and local cultivars of muskmelon. They reported that local cvs. T–96 and Chichawatni were better than the exotic cultivars. Hussain et al. (1986) also reported that cvs. T–96 and Jahangir were higher yielders than the exotic cultivars Duce, Top Mark, and Compo. Similar results were also reported by Malik (1977). Sarwar et al. (1975) conducted a study on seven muskmelon cvs. They found that cv. T–96 matured earliest and outyielded all the others.

Harvesting. Picking of the fruits at the right stage, especially with imported varieties, is crucial. The following indications can help in deciding when the fruit is ready to be picked.

- Softening of rind. This can be observed or felt with the thumb.
- The base of the pedicel changes from green to waxy.
- Ripened fruits emit a musky odour.
- Parting of the fruit from the stem leaving a circular depression, or full-slip stage, when the fruit detaches spontaneously from the vine.

Yield. The average yield is 10,000–15,000 kg/ha.

14.7.3 Watermelon

The watermelon (*Citrullus lanatus* Thanb. Mavs.) belongs to the family Cucurbitaceae. Watermelon is a monoecious, cross-pollinated crop. Native to Africa, it is one of the most important vegetable crops grown all over Pakistan. When ripe, the fruit is sweet; very occasionally is is cooked as a vegetable when immature. Watermelon is used as dessert and a thirst quencher in very dry parts of the world. It is relished by both people and livestock, particularly goats, as a source of water. It is diuretic and slightly purgative, but is comparatively poor in nutritive value. The roasted seeds are popular as a food, especially in west Africa and southern China, and they contain a semi-drying oil. Cultivars with very small fruits and hard, white flesh are used to make preserves and pickles.

Climate. Watermelon is a warm-weather crop, requiring a long, warm growing season. It needs higher temperatures and a longer growing season than muskmelon for optimum development, and is much less affected in quality by atmospheric humidity.

Soil. Fertile, well-drained, sandy loam soils are preferred. The optimum pH range is 5.5–7.0, but it can tolerate soils as acid as pH 5.

Manures, fertilizers, and irrigation. Manurial and irrigation practices are similar to those followed for muskmelon.

Seed rate and planting time. Seed is required at the rate of 2.5–4 kg/ha. In arid regions, watermelon is planted in July–August; in irrigated tracts it is planted in February–March and June–July.

Varieties. Sugar Baby is the commercial variety which fetches the highest prices because of its black rind, pink flesh, and sweet taste. Hussain and Zafar (1976) compared five watermelon varieties with the standard cv. Sugar Baby and reported that Sugar Baby gave the highest yield.

Harvesting. Watermelon needs to be at the proper stage of maturity when harvested. Size of the fruit and rind colour are not good indicators of the proper stage. Change of colour of the portion of the fruit which rests on the ground from white to creamy yellow indicates maturity. A metallic sound when the melon is tapped signals immaturity, whereas a heavy dull sound indicates ripeness. Dying of the tendril at the stem end and a smooth appearance and total absence of hair on the stem which attaches the melon to the vine are also signs of maturity.

Yield. Yields average 12,000–15,000 kg/ha.

14.7.4 Indian squash

Indian squash (*Citrullus vulgaris* L.) belongs to the family Cucurbitaceae, and is native to India. Also called squash melon or round melon, it is an important summer vegetable of Pakistan. It is harvested when the seeds are still soft and the fruit is tender and cooked in curry alone or mixed with other vegetables or pulses. It has cooling effects and contains vitamin A.

Climate. Indian squash cannot tolerate low temperature or even light frost. The crop performs well in a dry, warm climate. The seeds germinate best at temperatures of 27°C.

Soil, manures, and fertilizers. Sandy loam to medium loam with average fertility and good drainage is considered the best soil. For an early crop, sandy loam soils are better. Besides a basal dose of about 20–30 t/ha of FYM, 50 kg/ha N, 100 kg/ha P_20_5, and 40 kg/ha K_20 are essential, in addition to top-dressing with 50 kg/ha N, about 40–50 days after sowing.

Seed rate, sowing, and planting time. Seed is required at the rate of 2.5–5 kg/ha. Raised beds are made about 1–1.5 m wide with irrigation furrows between them, and 3–4 seeds per hill are sown about 1 cm deep. Seeds germinate in 6–15 days on sandy soils. Seeds are also broadcast by hand and the surface is planked slightly. The field is then made into plots of convenient size. Seeds are sown from the end of January to mid-February in the plains, and from May to July on hills near the edges of each bed.

Irrigation. Irrigation should be done as soon as most of the seeds have germinated. After that, the crop is irrigated once every four to five days.

Varieties. There are two types, one with dark-green and the other with light-green fruit.

Harvesting. To prevent damage to the plant, the fruit should not be pulled off the vines. Fruit should be picked when they are of normal size, tender, hairy, and soft-seeded. The crop is ready 45–50 days after sowing.

Yield. 7,000–10,000 kg/ha.

14.7.5 Bitter gourd or balsam pear

Bitter gourd (*Momordica charantia* L.) belongs to the family Cucurbitaceae. There are 60 species, of which only two, *charantia* and *cochinchinensis*, are important as vegetables. The bitter gourd is believed to have originated in the tropics and is widely distributed in China, Malaysia, Indo–Pakistan, and tropical Africa.

Bitter gourd is one of the most popular cucurbitaceous vegetables in Pakistan. The fruit is cooked in many ways—fried, boiled, and cooked in curries. The bitter taste is liked by some and is perceived to indicate medicinal properties. The fruit is said to be vermicidal and a cure for stomach disorders. Extracts are used to cure rheumatism and disorders of the liver and spleen, and to control diabetes. The bitter gourd has high nutritive value, ranking first among the cucurbits in iron and vitamin C content. The seeds yield a clear brown oil which is edible, though not popular.

Climate. The bitter gourd is a summer crop and requires warm weather, but can tolerate temperatures as low as 18–10° C.

Soil, manures, and fertilizers. Fertile sandy loam to medium loam are considered the best soils for this crop. Well–rotted FYM at 40–50 t/ha is applied one month before sowing. In addition to this, 50 kg/ha N is given at the time of flowering.

Seed rate and sowing. To plant one hectare, 5–6 kg seeds are needed. The seeds are sown on raised beds 1.5–2 m wide. Three to four seeds are sown in hills which are 20–30 cm apart. After complete germination, one to two healthy plants are left and the extra ones are pulled out.

Planting time. In the plains, the summer crop is sown during January–March, and the rainy season crop from June to July. In the hills, the seed is sown from March to June.

Irrigation. Irrigation is usually done at seven–day intervals, depending upon the temperature.

Varieties. Local Selection, Faisalabad Long, Ghotki, Lucknow, and M–70 are some common varieties.

Harvesting. Harvesting is done when the fruit is still young and tender. On ripening, the fruit changes colour from green to yellow and orange. Picking should be done every other day. If the fruit is not picked when tender and is allowed to ripen on the vines, bearing decreases.

Yield. The yield varies from 8000–12,000 kg/ha.

14.7.6 Sponge gourd and ribbed gourd or ridged gourd

There are eight species in the genus *Luffa* but only two, sponge gourd (*Luffa cylindrica* Roem) and ridge gourd (*Luffa acutangula* Roxb.) are important as vegetable crops.

The sponge gourd and ridged gourd are common vegetables throughout Pakistan. There is little difference in the nutritive value of these two species. The fruit of both is edible when young. The dry, fibrous, inner portion of the mature fruit is used for bathing, cleaning utensils, making shoe soles, and also as filters in factories. Some industrial oil is also extracted from its seeds.

Climate. These summer crops require warm weather and like high humidity; they have a wide range of adaptability. The smooth sponge gourd has more tolerance for cold than the ridged gourd.

Soil, manure, and fertilizers. A well–drained, fertile, sandy to medium loam is considered the best soil for these crops. The soil should be well prepared. Fertilizer regimes are the same as for other cucurbits.

Seed rate and sowing. To plant one hectare, 4–6 kg of seed is required. Seeds are sown on raised beds 1–1.5 m wide. The beds are 1 m apart, with furrows between them for irrigation. Seeds are sown on both sides of these beds at a distance of 15–25 cm. To get a good germination percentage, the seeds should be soaked for six to eight hours before sowing. Two to three seeds are planted per hill, which are later thinned to one or two plants per hill. If support is provided for climbing, yield is considerably increased.

Planting time. The early crop is sown from February to March, and the late crop from June to July.

Irrigation. Irrigation is done at weekly intervals. The frequency of irrigation, however, also depends upon the planting time.

Varieties. Two common varieties are generally cultivated. The ridge gourd has longitudinal ribs and is light–green in colour. Its seeds are dark–black and rough textured. The other variety is called *ghia tori* and has a smooth, dark–green skin. Its seeds are either white or black.

Yield. Yield averages 10,000–15,000 kg/ha.

14.7.7 Bottle gourd

The bottle gourd (*Lagenaria sicerarta* Molina) belongs to the family Cucurbitaceae. This crop is native to tropical Africa and Asia, and wild species are found in India.

The bottle–like shape of the fruit and its use as a container for wines and spirits in the past gave it the common name of bottle gourd. It is extensively grown in Pakistan. In the green stage, the fruit are used as vegetables and also for sweetmeats. The hard shell of the fruit has many uses—as water jugs, domestic utensils, for making musical instruments, and floats for fishing nets. The pulp, young stems, and leaves have medicinal value.

Climate. The bottle gourd is a summer crop; it can withstand cold somewhat better than other cucurbits, but it cannot tolerate frost. Most varieties are season–specific, and so different varieties are recommended for summer and for the rainy season.

Soil. Soil requirements are the same as for the other cucurbits. Neutral soil is best, but satisfactory yields are obtained at pH 6.0–7.0.

Manures and fertilizers. The bottle gourd is a shallow–rooted crop and responds well to top–dressing of fertilizers. It does not grow well on poorly manured soils. The general practice is to supply about 30–50 t/ha of FYM at land preparation along with 60 kg/ha of P and 50 kg/ha of N.

Seed rate and sowing. About 4–5 kg/ha of seed is required. Sowing recommendations are the same as for watermelon. The seeds are planted in February–March and June–July. In *kacha* lands the crop is planted in December and January. The rainy season crop is usually staked, often trained on a bower made of bamboo and sticks. The summer crop is not staked.

Irrigation. Irrigation requirements are the same as for other cucurbits.

Varieties. These vary from very long and slender to thick and round. Each locality has local varieties. Some summer varieties are Summer Prolific Long, Summer Prolific Round, and Water Flower Gourd. Some local varieties grown are Sialkot Round and Louki.

Harvesting. The fruit should be harvested when it is still tender, separating it from the plant by cutting the stem with a knife.

Yield. The average yield per hectare is 10,000–15,000 kg.

14.7.8 Pumpkin

Pumpkin (*Cucurbita moschata* Poir) belongs to the Cucurbitaceae family. Probably native to tropical America, it grows best in the wet tropics, and is more widespread in the tropics than many other species. The plant is an annual, having a climbing or trailing habit. Its large fruit is used both in the immature and the mature stages and is extensively cultivated. The fruit is a source of vitamins A, B, and C. The fully–mature fruit cooks well, and keeps well for months if properly stored in a ventilated place. The delicate shoots and leaves are used to a limited extent as a cooked vegetable.

Climate. Pumpkin does best in hot, humid climates, but can withstand a certain degree of cold. It can be cultivated in areas with low temperature and high humidity. This crop also can tolerate shade to some extent.

Soil. A well–drained and fertile soil is desirable. On lighter soils, the fruit matures a bit earlier. The plant can be grown on a medium acid soil but performs best at pH 6.0–6.5.

Manures and fertilizers. FYM at the rate of 20–25 t/ha is added at soil preparation. 40–80–40 kg/ha NPK is also applied as a basal dose. Then at 40–45 days after sowing, 40 kg/ha N is applied as a top-dressing.

Seed rate and sowing. To plant one hectare, 5–7 kg seeds are needed. The plants grow quickly and spread widely; hence they should be allowed enough space. For an irrigated crop, the seeds are sown in well-prepared beds about 3 m wide at a distance of about 75 cm to 1 m apart in hills. Three or four seeds are sown per hill about 2–3 cm deep. When the crop is raised without irrigation, beds are not made and sowing is done by drilling the seeds in rows 1.5 m apart. The plants are thinned later to a distance of 1 m, keeping one or occasionally two plants per hill.

Sowing time. In the plains, sowing is done during January–March and June–July. In the hills, March–June is the common period for sowing.

Irrigation. The crop should be irrigated at four to five-day intervals.

Varieties. Only local varieties are commercially grown.

Harvesting. The crop is harvested depending upon market demand. The fruit is cut with a piece of peduncle attached, which facilitates handling.

Yield. The yield is about 15,000–25,000 kg/ha.

14.8 Peas

Peas (*Pisum sativum* L.) belong to the Leguminoseae family. There are two sub-species of garden pea: *Pisum sativum* sub-species *hartense,* and the field pea *P. sativum* sub-species *arvense*. Both these sub-species have edible pods. Peas are thought to have originated in Ethiopia and parts of Europe and Asia. Peas are an excellent human food, either eaten as a vegetable or in soup. Cooked green peas are a rich source of protein, a good source of vitamins A, B, and C, and contain a high proportion of minerals.

Climate. Peas are a cool-season crop and grow best in cool weather with ample moisture. The plants will tolerate considerable cold and light frost, but the flowers and green pods are often damaged by heavy frost. Yields are usually higher and the quality better when the crop matures under cool conditions. They do not do well in dry, hot weather. Hot, dry weather interferes with the seed set and lowers the quality of pods.

Soil. Peas are grown on a great many kinds of soils, from light sandy loams to heavy clays. Ample drainage is essential for good pea production. For a very early crop, a sandy loam is preferred. For large yields, where earliness is not as important, a well-drained clay loam or silt loam is better. The garden pea prefers a slightly acid soil, but will not tolerate excess acidity. The most favourable pH is between 6.0 and 7.5; if pH falls below 5.5, peas will not do well.

Manures and fertilizers. Well–rotted FYM at the rate of 20–25 t/ha should be added and worked into the soil. Superphosphate at the rate of 350–400 kg/ha, and 100 kg/ha of potassium sulphate or muriate of potash should be applied at final land preparation before sowing. A dressing of nitrogen as ammonium sulphate at the rate of 200–300 kg/ha at the time of pod formation will be beneficial. Bhatti (1985) states that a fertilizer dose of 45 kg/ha N, 90 kg/ha P_20_5, and 90 kg/ha K_20 produced the best yield of peas. Ahmed and Shafi (1978) reported that increased NPK levels resulted in increased yield per hectare. Bukhari and Rizvi (1973) recommended 80–80–40 lbs/acre (90–90–45 kg/ha) of NPK for peas in Sindh .

Seed rate and spacing. The seed rate for an early crop is 100–120 kg/ha per hectare, whereas for a normal or late planting it is 80–90 kg/ha. The minimum temperature for germination is about 10°C and the optimum about 22°C. Seeds should be dibbled about 30 cm apart on both sides of ridges spaced 1 m apart. For normal and late planting, the width of the furrows can be increased to 1.25 m. Peas can also be planted on levelled flat beds drilled in rows, with 45–60 cm between rows.

Sowing time. In plains areas peas are sown from October to December; whereas in hilly areas they are planted after mid–March to the end of May.

Irrigation. For proper germination, a pre–sowing irrigation is advisable, and if the soil is dry at the time of sowing, a light irrigation after sowing may be needed for proper germination. Thereafter, during the dry weather, light irrigation may be applied at intervals of 10–15 days. One to two irrigations at the time of flowering and fruit–setting are essential.

Varieties. Two imported varieties, Kalvex and Blue Bantam, are grown on a large scale in Hyderabad Division.

Harvesting. Peas lose quality rapidly after maturity and must be picked at the proper stage and used promptly. Harvesting starts six to eight weeks after seeding or about 21 days after blooming. Green peas to be used fresh are hand harvested; three to four pickings are usually necessary. The pods are picked when fully green and well developed and before the peas harden. Uniformity is an important criterion for canning and dehydration.

Pea quality is determined by many factors, but high quality is generally defined as tenderness and high sugar content. During maturity the sugar content decreases rapidly. There is a rapid increase in content of starch and other polysaccharides and insoluble nitrogen compounds, mainly proteins.

Yield. Early varieties usually yield about 3000–4000 kg/ha of green pods. Mid–season and late varieties give higher yields, about 6000–7000 kg/ha. Shaikh (1980) claims that sowing peas in mid–season gives a higher yield.

14.9 Beans

There are at least 18 types of cultivated beans, belonging to eight genera of the family Leguminoseae. They vary greatly in flavour, season, and other characteristics. Beans may be used as green vegetables or dry as pulses, depending on the harvest stage. Cowpeas, cluster beans, and hyacinth beans are the most important species used as vegetables and have somewhat fleshier–walled pods with less fibre at the green stage. Dry beans are an important source of food, and are usually classed as a field crop. All beans except broad beans are susceptible to frost and are grown as a summer crop.

14.9.1 Cowpea

Cowpeas (*Vigna sinensis* Savi) belong to the family Leguminoseae. They have their origin in Central Africa, the Mediterranean regions, and South Asia including India. A leguminous plant, the cowpea fixes atmospheric nitrogen and is thus a soil–improving crop, often cultivated as an intercrop. Cowpeas are considered to be hot, dry, diuretic, and difficult to digest. However, they possess high nutritive value, being rich in protein, minerals, and carbohydrates. Numerous dishes can be prepared from this vegetable. Both the green pods and the dried seeds are eaten.

Climate. Cowpeas are a warm–season crop and cannot withstand cold weather. They can be grown both in spring and rainy seasons in the plains, but cannot tolerate heavy rainfall. Different varieties respond differently to temperature and daylength, and distinct varieties are recommended for spring and rainy seasons.

Soil, manures, and fertilizers. Cowpeas can be grown in all kinds of soil, provided that there is no drainage problem and the soil is rich in organic matter. Sandy to sandy loam soils are best. FYM at 20–25 t/ha is applied as a basal dressing. In addition, a mixture of 60–80 kg/ha N, 40–60 kg/ha P_2O_5, and 40 kg/ha K_20 should be thoroughly worked into the soil before sowing.

Seed rate and spacing. A seed rate of 12–18 kg/ha is sufficient, depending upon the spacing. Sowing is done directly in the field. March and September sowing is done by dibbling in rows 40 cm apart with 20 cm between plants. For June sowing, making the rows about 60 cm apart and keeping plants about 30 cm apart is recommended to facilitate cultural operations and give the best crop.

Irrigation. For an independent crop, a total of four irrigations are needed, and for an intercrop two to three irrigations are recommended. As the plants are sensitive to excess water, the irrigation should be very light.

Varieties. Local, Pusa Barsati, and Mirpurkhas are some widely–grown varieties.

Harvesting. Cowpeas are harvested for two purposes. (1) Green pods which are mature but not ripe are harvested for fresh consumption. (2) Pods are harvested when fully ripe and dry. The dried seeds are separated and sold as dried beans and for seed.

Yield. The average yields of grain and of green pods are about 800–1200 kg/ha and 5000–6000 kg/ha, respectively.

14.9.2 Cluster bean

The cluster bean (*Syamopsis tetragonoloba* L., Taub.) belongs to the family Leguminoseae. In the plains it is generally planted twice a year, in spring and in the rainy season. It is thought to have originated in India or Africa.

The cluster bean is gradually gaining in importance. Its tender pods are used as a green vegetable, and are available in the market from April to November. The green pods are rich in protein, vitamins A and C, and iron. It can also be used as a green manure crop, and some varieties are used for manufacture of gum. The seeds are an article of export trade.

Climate. This plant is very hardy and relatively resistant to drought. In regions with high summer temperatures and low rainfall it has produced excellent crops where no other vegetable crop could be grown successfully.

Soil, manure, and fertilizers. The cluster bean is adapted to all types of soil but does best in well–drained sandy loam. At field preparation, 10–15 t/ha of FYM are incorporated into the soil. Then immediately before sowing, 20–30 kg N, 60 kg P_20_5 and 40 kg K_20 per hectare are applied.

Seed rate and spacing. A spacing of 45–60 cm between rows and 20–30 cm within the row is recommended. If the cluster bean is sown alone, 10–15 kg/ha seed is required, but if it is sown mixed with other crops or as a border crop around the main crop less seed will be needed.

Irrigation. Cluster bean requires very little irrigation, but timely watering can increase yield. With an early–sown crop, the field is irrigated when necessary. There is no need to irrigate the rainy–season crop.

Varieties. There are several varieties, distinguished by their height and the size and shape of the pods. The local varieties are classified as *kachhan* and *kunchi* according to the shape of their pods. The former has long, straight pods, and the latter medium–sized, curved pods. Varieties can also be classified according to (1) branching type [a. single–stem, b. multiple–stem]; (2) presence of hair [a. hairless (glossy), b. hairy (glabrous)]; (3) the kind of fruit.

Harvesting. Picking of the green pods continues over a considerable time because they continue developing as the plant grows. All of the three to six pods in the clusters which develop in leaf axils usually mature at the same time. Green pods intended to be eaten as vegetables are harvested when they attain a marketable size and are still tender.

Yield. The yield of green pods varies from 5000–6000 kg/ha, and the yield of seeds is about 1000 kg/ha.

14.9.3 Hyacinth bean

The hyacinth bean (*Dolichos lablab* Roxb. and L.) belongs to the family Leguminoseae. The name *Dolichos* is derived from the Greek word meaning 'long' and *lablab* is from an Arabic word which describes the dull rattle of the seeds in the dry pod. It originated in the Indo–Pakistan subcontinent and has spread to other parts of the world. It is highly drought–resistant, spreading in nature, and is often grown along with castor, which provides support for climbing. This plant is used both as green pods and dry beans. Its nutritive value is greater than that of the French bean.

Climate. This plant requires a fairly cool climate but cannot withstand a severe frost. The minimum temperature at which the seeds germinate is 5°C and the optimum temperature for germination about 22°C. The hyacinth bean can be sown either as an early–spring or an autumn crop.

Soil, manures, and fertilizers. It grows well on all types of soil, from sandy loam to clay loam. The fertilization regime for hyacinth beans is the same as that for cowpeas.

Seed rate and spacing. The same seed rates apply as for cowpeas. The plant is of trailing nature, and the crop spreads as a vine. It requires support for climbing, and appropriate spacing depends on the particular variety.

Irrigation. Depending upon the severity of the temperature, three to four irrigations are required.

Varieties. There are two broad groups — the bushy field varieties and the twining–pole garden varieties.

Harvesting. Harvesting is done in the same manner as for cowpeas.

Yield. A yield of 3000–5000 kg/ha of green pods is normal.

14.10 Okra or lady's finger

Okra (*Abelmoschus esculentus* L. Moench) belongs to the family Malvaceae. Okra originated in tropical Africa, was grown in the Mediterranean region, and wild forms are also found in India. Okra is a popular summer crop; the young tender pods are cooked in curries, stewed, and used in soups. It is a good source of vitamins A, B, and C, and is also rich in protein, minerals, and iodine. When ripe, the black or brown white–eyed seeds are sometimes roasted and used as a substitute for coffee. The stem of the okra plant provides fibre which is used in the paper industry.

Climate. Okra is very sensitive to frost and will not thrive when there is continuous cold. It grows best in hot weather, and needs a long season with warm nights. Okra seeds will not germinate below 20°C.

Soil. Any good soil will produce a satisfactory crop of okra if other conditions are favourable. Okra thrives on all kinds of soils, but it grows best in a well-manured, friable medium loam soil. It is only slightly tolerant to acidity, and the optimum pH ranges from 6–6.8.

Manures and fertilizers. About 25–30 t/ha of well-rotted FYM should be applied one month before sowing and mixed well with the soil. Before sowing, 350 kg/ha of superphosphate, 125 kg/ha of muriate of potash, and 300 kg/ha of ammonium sulphate should be drilled in the rows. A month after sowing 300 kg/ha of ammonium sulphate should be top-dressed.

Sowing method and spacing. Okra is a deep-rooted crop, therefore thorough land preparation is necessary. The seed coat is hard and does not absorb water readily. Soaking in water for about 24 hours (or until it splits) just before sowing increases germination rate and percentage. The early crop is planted on ridges, whereas normal planting is done in flat beds. Ridges 60 cm wide are prepared and the seeds are drilled on both sides of the ridges, keeping the row-to-row distance about 45–60 cm and the plant-to-plant distance about 20–30 cm.

Seed rate and sowing time. Okra is propagated by seeds at the rate of 18–20 kg/ha. There are two sowing dates or growing seasons of okra: (1) February–March, (2) June–July. Rizvi and Jagirdar (1976) reported that row-to-row spacing of 1.5 feet gave a better yield than 2 or 2.5 feet.

Irrigation. The crop should be irrigated every fifth or sixth day during spring and summer when needed. In all, six to eight irrigations are required.

Varieties. The following local cultivars are grown commercially: Faisalabadi-I, Faisalabad-II, Mirpurkhas-I, Mirpurkhas-II, Tarnab-13, PARC Green, and Spiny Local. In addition, some exotic cultivars are also grown; they are Emerald, Clemson, Perkins Dwarf, Pusa Green, and Pusa Sawani.

Harvesting. Okra is picked every second or third day starting when the first pods are formed. The best time to pick the pods is 6–7 days after the flowers open, depending on the variety and the season. Plant growth and yield are both adversely affected if the pods are not harvested when young.

Yield. The average yield is about 8000–10,000 kg/ha of green pods. Baloch et al. (1990) conducted experiments to assess varietal performance. They reported that the Emerald cultivar gave a higher yield than Clemson, Perkins Dwarf, Pusa Sawani, and Pusa Green. The yields of these cultivars are given in Table 14.5 for comparison.

Table 14.5 Yields of five okra cultivars in Tandojam conditions

Variety	Pod yield (kg/ha)
Emerald	7,912.5
Pusa Sawani	4,237.5
Clemson	4,062.5
Pusa Green	3,942.0
Perkins Dwarf	3,895.0

Source: Baloch et al. (1990).

Sadiq et al. (1988) performed experiments on five okra cultivars under Peshawar conditions. Maximum pod weights found are given in Table 14.6.

Table 14.6 Maximum pod weights for five okra cultivars, Peshawar conditions

Cultivar	Pod weight (g)
Clemson	718.32
T–13	694.33
Pusa Green	661.70
Perkins Dwarf	660.82
Pusa Sawani	625.92

Source: Sadiq et al. (1988).

Sheikh et al. (1987) conducted yield experiments on six okra cultivars under Tandojam conditions. The results are summarized in Table 14.7.

Table 14.7 Average yields of six okra cultivars at Tandojam

Cultivar	Average pod yield (t/ha)
Ahmedabadi	27.385
Faisalabadi–I	21.025
Mirpurkhas–II	17.275
Green Pusa	17.162
Faisalabad–II	16.880
Mirpurkhas–I	15.775

Source: Sheikh et al. (1987).

Summary and recommendations. The production of fresh fruits and vegetables is an important and growing industry in Pakistan. Production units vary from the small kitchen garden to large areas grown exclusively for city markets. Since the yield of high-quality fruits and vegetables is greatly dependent upon using good seed, proper varieties, and good cultural practices, and on employing economic disease and pest control, the dissemination to growers of the material presented in this chapter becomes an important role for researchers and extension workers in horticulture.

STUDY QUESTIONS

1. Describe the importance and nutritional value of vegetables in the daily diet.
2. Give the fertilizer requirements of vegetable crops.
3. (a) Name the solanaceous vegetables. (b) Discuss the nutritive value of each of them.
4. Discuss the climate and soil requirements for various tuber crops.
5. Describe the irrigation and nutritional requirements of various root crops.
6. (a) How are the important bulb crops planted? (b) Discuss the medicinal values of bulb crops.
7. Discuss the seed rates, recommended planting practices, and post-planting care of several important leafy vegetable crops.
8. Give recommended planting times and spacings for cole crops.
9. Discuss the best practices for land preparation, planting, manuring, and irrigation of various cucurbits.
10. (a) Describe the recommended planting, seed rate, and irrigation practices for peas. (b) Discuss the food value of peas.
11. Summarize the importance, seed rate, and yield per hectare of beans.
12. Describe the planting method, seed rate, and varieties of okra.
13. What is right and what is wrong about the statement: "Plants make food, and plants absorb food from the soil"?
14. Differentiate between cabbage, cauliflower, and kohlrabi (knol–khol).
15. List the members of the family Brassica, giving their local, English, and botanical names.
16. The climate of Pakistan is suitable for raising high quality *Brassica* vegetables. Indicate the best regions for growing cabbage, cauliflower, and kohlrabi.
17. Describe and compare the three main types of onions.

18. The climate of Balochistan, NWFP, and the Pothwar Plateau is particularly favourable for high yields of potatoes. Explain why this is so.
19. Name three common characteristics of root vegetable crops.
20. List some vegetables containing carotene.
21. Why are tomato yields relatively low in Pakistan?
22. What are the irrigation and fertilizer requirements of cucumber?
23. Discuss the appropriate time period for harvesting watermelons and muskmelons.
24. How do pumpkins and squashes differ from each other?
25. Why do beans and peas require less nitrogen fertilizer than most other vegetable crops?
26. Beans are usually extremely vegetative and unfruitful when grown in highly fertile soil. Explain why this is so.
27. What types of beans are grown in your local area? Discuss the type of market for which they are grown, for example the long-distance market, the local market, or the cannery.
28. How would you obtain reliable information on varieties of horticultural crops suitable for your location?
29. Why are leafy vegetables important for the human body?

REFERENCES

Ahmed, S.; Shafi, M. Effect of different levels of N,P and K on the yield of peas (*Pisum sativum*). Pakistan J. Agri. Res. 13(1):417; 1978.

Baloch, A. F.; Baloch, M. A.; Deho, N. A. Effect of phosphorus and potassium on the yield of red chilies (*Capsicum frutescens* L.). Pak. J. AAEVS. 5(1–2):74; 1989.

Baloch, A. F.; Qayyum, S. M.; Baloch, M. A. Growth and yield performance of okra (*Abelmoschus esculentus* L.) cultivars. Gomal Univ. J. Res. 10(2):91; 1990.

Baloch, A. F.; Qayyum, S. M.; Mangrio, H. K.; Hassan, A. H.; Newton, Z. M. Agronomists study effect of potassium on the growth and quality of fruit (Tomato production). Pak. Agri. 10(2):48; 1988.

Baloch, M. A.; Baloch, A. F.; Baloch, G.; Ansari, A. H.; Qayyum, S. M. Growth and yield response of onion to different nitrogen and potassium fertilizer combination levels. Sarhad J. Agric. 4(4):63; 1991.

Bangulzai, N. A. Effect of different levels of nitrogen (N) and potassium (K_2O) on the growth and yield of carrot (*Daucus carota* L.). Tandojam: Sindh Agriculture University; 1990. M.Sc. Thesis.

Bhatti, M. H.; Inayatullah, H.; Khan, S.; Shakoor, A. An evaluation of exotic and local cultivars of radish (*Raphanus sativus*). Pak. J. Agri. Res. 4(1):17; 1983.

Bhatti, M. S. Effect of different doses of NPK fertilizer on the growth and production of peas (*Pisum sativum*) variety Blue Bantam. Tandojam: Sindh Agriculture University; 1985. M.Sc. Thesis.

Bukhari, S. J. I.; Rizvi, I. A. S. A pamphlet on cultivation of peas. Mirpurkhas, Sindh: Cooperative Fruit Development Corporation, Government Fruit Research Station; 1973.

Bukhari, S. J. I.; Shahani, N. M. To study the effect of application of nitrogen in the form of farmyard manure and urea on the growth and yield of cauliflower (*Brassica oleracea* L.) Var. Botrytis. Sindh J. Agric. Res. 3(6):1–69; 1983.

Choudhry, B. Vegetables. 6th ed. New Delhi, India: National Book Trust; 1979.

Choudhry, J. A.; Mughal, S. M..; Khan, N. A. Impact of improved seed sources on growth, characteristics and yield of potato. Sarhad J. of Agric. 6(3):249; 1990.

Conn, E. E.; Stumph, P. K. Outlines of biochemistry. 3rd ed. New York: John Wiley and Sons; 1970.

Deho, N. A.; Gondal, N. A.; Arain, A. L.; Nenwani, S. S. Comparative performance of imported potato varieties for their yield performance under local conditions. Sindh J. Agric. Res. 3(6):1:69; 1983.

George, W. W.; McCollum, J. P. Vegetable crops. 3rd ed. Danville, Illinois: Interstate; 1980.

Ginai, M. A. Vegetable culture. Lahore: Punjab Department of Agriculture Publication 31; 1970.

Gondal, N. A.; Arain, A. L.; Deho, N. A.; Aslam, M. A varietal trial on okra (*Hibiscus esculentus*) Pak. Agri. 7(4):55; 1985.

Hanan, A.; Ali, N.; Inayatullah, H. Comparative performance of the French potato cultivars in Peshawar valley. Sarhad J. Agric. 3(4):463; 1987.

Hussain, A. Effect of various inorganic fertilizers on the yield and quality of potato grown in the autumn. Pak. J. Sci. 20(5–6):271; 1968.

Hussain, A.; Saeed, M. Our vegetables. Lahore: Punjab Department of Agri. W. Pak. 71; 1967.

Hussain, A.; Zafar, A. M. Comparative studies in watermelon varieties. Lyallpur J. Agric. Res. 14(3):165–169; 1976.

Hussain, S. I.; Mahmood, K.; Malik, M. N. Comparative morphology and qualitative evaluation of some exotic and local muskmelon (*Cucumis melo* L.) cultivars. Pak. J. Agri. Res. 7(193–197); 1986.

Jagirdar, S. A. P.; Jandan, D. M. Evaluation of exotic potato varieties for adaptability and yield performance in Sindh. Agric. Pak. 23(2–3):109; 1972.

Katyal, S. L. Vegetable growing in India. 2nd ed. Calcutta: Oxford and IBH Publishing Co.; 1988.

Khan, S.; Inayatullah, H.; Ishtiaq, M. Effect of nitrogen alone and in combination with phosphorus and potassium on yield of red peppers. Pak. J. Agri. Res. 3(2):21; 1982.

Khoso, A. W. Growing vegetables in Sindh. Tandojam: Department of Agronomy, Sindh Agriculture University; 1988.

Khushk, A. M.; Miano, N. M.; Ansari, A. H.; Mari, M. I. Influence of inter and intra row spacing on the yield and yield component of onion (*Allium cepa* L.). Sarhad J. of Agric. 6(2):147; 1990.

Korejo, N. A. Effect of NPK fertilizers application on garlic. Tandojam: Sindh Agriculture University; 1983. M.Sc. Thesis.

Lashari, M. R. R. Varietal trial on garlic (*Allium sativum* L.) Tandojam: Sindh Agriculture University; 1984. M.Sc. Thesis.

Leghari, M. A. Influence of phosphorus and potassium with uniform dose of nitrogen on the growth and yield of carrot (*Daucus carota*). Tandojam, Sindh: Sindh Agriculture University; 1989. M.Sc. Thesis.

Mahmood, K. K.; Zahur, A.; Hussain, S. I.; Mahmood, T. Comparative studies on the evaluation of some foreign and local muskmelon cultivars. Pak. J. Agri. Res. 9(4); 1988.

Malik, M. N. Comparison of some foreign varieties with local varieties of muskmelon at Faisalabad. Annual Res. Report. 1976–77. Faisalabad: University of Agriculture; 1977.

Mangrio, H. K.; Baloch, A. F. Plant population studies on onion (*Allium cepa* L.). Pak. J. AAEVS. 1(2–4):81; 1985.

Mangrio, H. K.; Baloch, A. F.; Baloch, M. A. The size and yield of onion bulbs in response to application of phosphorus (P_20_5) and (K_20). Sarhad J. Agric. 3(4):477; 1987.

Mangrio, H. K.; Bozdar, A. A. The effect of plant density on the bulb size and yield of onion. Pak. J. AAEVS. 1(1):21; 1985.

Memon, A. W. G. Varietal trial on lettuce (*Lactuca sativa* L.). Tandojam: Sindh Agricultue University; 1991. M.Sc. Thesis.

Miano, N. M.; Memon, G. H.; Razzak, A. Performance of eight exotic varieties of cauliflower under Sindh conditions. Pak. Agri. 12(1):29; 1990.

Miano, N. M.; Memon, G. H.; Ghilzai, A. N.; Khushik, A. M. Varietal trial on cucumber (*Cucumis sativus* L.). Sindh J. of Res. 8(1):30; 1991.

Pakistan Agricultural Research Council. Current agro–technology for tomato production. Crop Production Bulletin No. 5. Islamabad: PARC; 1981.

Rahim, A.; Amjad, M. Response of various tomato cultivars to different levels of fertilizers. J. Agric. Res. 23(2):93; 1985.

Rashid, M. A.; Rashid, M. M. The effect of spacing on the yield of onion. Hort. Abstr. 48(6):481; 1978.

Rehman, S.; Talukder, M. R.; Mian, A. M. Effect of nitrogen, phosphorus and potash on the bulb size and yield of onion. Hort. Abstr. 48(6):481; 1978.

Rizvi, I. A. S.; Jagirdar, S. A. P. The effect of manures and ammonium sulphate on cauliflower. Agric. Pak. 18(3):293; 1967.

Rizvi, I. A. S.; Jagirdar, S. A. P. Effect of irrigation interval, spacings and planting methods on the yield of lady's finger. Hort. Abstr. 46(2):6802; 1976.

Sadiq, W. M.; Amin, N.U.; Shahzoor. Performance of okra cultivars under soil and climatic condition of Peshawar. Sarhad J. Agric. 4(5):633; 1988.

Sarwar, I.; Shafi, M.; Asi, A. R. Evaluation of some exotic varieties of muskmelon. Lyallpur. J. Agric. Res. 13:465–472; 1975.

Shah, M. A.; Ismail, M. Effect of different levels of NPK on potato yield. Pak. J. Agri. Res. 4(2):71; 1983.

Shaikh, M. A.; Ansari, A. H.; Qayyum, S. M.; Baloch, N. A. Okra cultivars. Tandojam experts' study of vegetative behaviour in Okra. Pak. Agri. 9(1):40; 1987–a.

Shaikh, M. A.; Ansari, A. H.; Qayyum, S. M.; Baloch, M. B.; Qayyum, S. A.; Hassan, A. M. Technical report on effect of nitrogen doses on the growth and crop yield of onion. Pak. Agri. 9(11):35; 1987–b.

Shaikh, M. A.; Baloch, Q. B.; Baloch, D. M. Influence of nitrogen and potassium fertilizer levels on the productivity of *Raphanus sativus* L. Pak. Phyton. 2:81; 1990–a.

Shaikh, M. A.; Baloch, Q. B.; Baloch, D. M. Effect of different levels of nitrogen and potassium fertilizer on the growth and yield of carrot (*Daucus carota* L.). Pak. Phyton. 2:85; 1990–b.

Shaikh, M. R. Effect of different sowing dates on the growth and production of peas (*Pisum sativum*). Tandojam: Sindh Agriculture University; 1980. M.Sc. Thesis.

Shanmugavelu, K. G. Production technology of vegetable crops. Bombay: Oxford IBH Publishing Co.; 1989.

Splittstoessor, W. E. Vegetable growing handbook. 2nd ed. Westport, Connecticut: AVI Publishing; 1984.

Thompson, H. C.; Kelly, W. C. Vegetable crops. 5th ed. London: McGraw–Hill Book Co.; 1957.

Yousaf, M.; Shafi, M. Evaluation of some exotic varieties of radish. Hort. Abstr. 41(2):476; 1971.

Outline of Chapter 15

15. FLORICULTURE AND LANDSCAPE GARDENING

Daud Ahmad Khan[1]

OBJECTIVES

A person who studies this chapter should be able to:

- Discuss the elements of design as they relate to gardens
- Recommend the most appropriate species of plants for each part of a garden plot in any specified part of the country

An intimate relationship exists between man and plants. Growing plants of diverse nature around the dwellings of mankind is an age-old practice, a manifestation of man's desire to live with nature. In today's stressful life, plants are becoming more highly valued for their positive influence on human behaviour. Plants provide a variety of colour, form, texture, and pattern in the landscape. Besides their intangible aesthetic values, trees furnish a close contact with nature and enhance our surroundings. They contribute to the beauty of buildings by softening architectural lines and emphasizing structural details.

In ancient times groves of trees having religious or other significance were preserved and maintained as a tradition. Gardens for private use were established by emperors, kings, and wealthy people. The Hanging Gardens of Babylon, once considered to be one of the seven wonders of the world, were established by Nebuchadnezzar II in the sixth century B.C. Gardens were planted in ancient Egypt and China. In the Indo–Pakistan subcontinent, Persian influence brought by the Mughal emperors determined the form of intimate enclosed gardens arranged in unified compositions of flowers, fruit trees, water, and shade. The most notable examples of great royal gardens on a formal plan occur at the Taj Mahal in Agra (India), Nishat Bagh in Kashmir, and Shalimar Bagh and Jehangir's tomb in Lahore.

---

[1]Dr. Daud Ahmad Khan is former Professor and Chairman, Dept. of Horticulture, University of Agriculture, Faisalabad and former Consultant Horticulture, PARC, Islamabad.

15.1 Garden design

Design is an orderly arrangement of living and inanimate objects to achieve a pleasing objective. To make a garden requires the use of creative instinct and imagination. A garden established in an orderly manner is a pleasure to everyone. The discussion of design here follows Janick (1979:563–64).

15.1.1 Elements of design

Colour. There is tremendous variation in the colouration of different plant organs—flowers, fruit, foliage, stem, and bark. Indeed, the most attractive colours are those of natural flowers, fruits, and foliage. Each colour presents an assortment of hues in all intensities. The most predominant and indispensable is green, in all possible tones and values. Green is pervasive, pleasing, and restful to the eyes and the mind. Red, orange, and yellow suggest warmth, while blue, purple, and green are regarded as cool colours. Another perception is that the warm colours seem to advance towards the observer and the cool colours seem to remain in the background. The pure colours of the spectrum—red, blue, and yellow—look closer to the observer than when tinted towards neutral grey. Therefore, these perceptual sensations associated with different colours give life to a landscape and create the effect of advance or recession (Janick 1979:563–64).

Texture. In design, texture means the *visual* effect of *tactile* surface quality. Plant materials like leaf, stem, and bark vary in texture; some are smooth, soft, and fine—like flower petals; others are rough, hard, and rigid—like the bark of an acacia tree. The different textures of plant materials and species can be arranged to produce a visual effect of contrast. Therefore textural differences among plants should be one of the considerations in the choice of species for planting in a landscape.

Form. The terms *form* and *habit* are often used while talking about the appearance of plants. **Habit** can be described as the direction of growth, and **form** as the shape of a three-dimensional object. With trees, the *habit* aspect is generally most prominent, while with shrubs *form* is most salient. In deciding on the visual suitability of plants with particular habits, other visual qualities like colour should also influence plant choice.

There is a field of psychotherapy using plants. Designers and plant therapists believe that certain combinations of objects with distinctive forms and habits produce effects of stress, repose, or balance.

Line. Line guides the eye from one part of a design to another; it defines shape and structure, and can evoke emotional responses. For these reasons straight and curved lines are combined in designs. Triangles, rectangles, circles, and modifications of these and other geometric forms are used in designing.

15.1.2 Principles of design

Balance. Balance refers to the arrangement of objects around a vertical axis. Balance implies equilibrium, which may be real or illusory. The exact matching of species and positions of the plants on each side of a dominant centerpiece like a building creates balance and symmetry. The formal landscapes in front of the Taj Mahal at Agra and Jehangir's Tomb at Lahore are superb examples of balance. In informal designing, balance is not achieved merely by exact matching of species and positions of plants on each side of an axis, but by an intelligent use of colour, texture, and form.

Rhythm. Rhythm refers to the occurence of some object, sound, or event at regular intervals of space or time. Very common examples of rhythm are the heartbeat, a drum beat, or repetition of a motif in a design. In the context of landscape designing it would be the repetition of a plant, tree, or flower bed at equal intervals in a garden.

Emphasis. As in a painting, in a garden there is usually some special object to be focused on and highlighted. This aspect of design is achieved by careful placement of dominant plants or other object(s) which give emphasis to the design. A large tree, colourful shrubs or flowers, or features like a pool or terrace can be used to create accent.

Harmony. Harmony results when all the materials in the design combine to produce a pleasing effect, and depends on good use of scale and combination of plant materials. The landscape designer should aim at unity in the overall design, and coherence of its various parts and elements.

15.1.3 Types of design

Selection of plants and styles of plantation have changed with changing life styles throughout history. The criteria for plant selection and planting designs change according to the cultural traditions and heritage of the people, based on their varying perceptions of beauty. Primarily, there are two styles of gardening: FORMAL and INFORMAL. Both styles are of ancient origin.

Formal design. This style of gardening, also referred to as the Muslim style, is commonly practiced in the Middle East and Indo–Pakistan. This style is very simple and uncomplicated. It is based on geometric symmetry on both sides of a chosen axis. The illumination contrasts between day and night, and between areas of shade and light are taken into account. Trees are planted to provide a maximum of shade in order to moderate the unkind effects of the climate. Summer's high temperatures and dry air necessitate the construction of water reservoirs, humidification through fountains and waterfalls, and efficient irrigation systems. Plants are used to delineate and emphasize formal patterns by the recurrence of one species at regular intervals. Cypress avenues are usually planted to direct attention to a vista beyond the garden or to spotlight some feature in the garden. Topiary work is another characteristic of this garden design.

Informal design. The informal design of gardening is generally referred to as Japanese garden design. It is a naturalistic design in which balance and harmony is achieved through curved, non–geometric lines without repetition of any form, and many artistic techniques, avoiding symmetrical planning. The visual effect of one plant is considered in relation to another plant. A major consideration is given to the contrasts between the forms of the various plants and the habit or line effects. Short–lived seasonal effects of strong visual contrast are especially sought after. Trees and ground flora are planted for spring and autumn effects. Mountain and river landscapes are simulated by mounds, streams, and certain plants.

15 1.4 Landscape architecture

Landscape architecture is a branch of design dealing with integrating man–made structures with their natural surroundings. It is concerned with the arrangement of space for utility and beauty using natural as well as structural materials to integrate man, building, and site both functionally and aesthetically. It is primarily a fine art whose most important function is to create and preserve beauty where people live and work and to promote the comfort, convenience, and health of urban populations.

Rock gardening. Rock gardening is a branch of landscape horticulture involving growing ornamental plants in rocky areas like a natural formation. There are many types of rock gardens: wall, naturalistic, moraine, scree, and alpine meadow gardens. In a **wall garden**, plants are planted in a wall constructed with flat rocks without cement or mortar. The rocks are placed in horizontal layers with alternate joints and holes. The holes are filled with good soil and plants are inserted. The wall is built against a terrace with a slight backward slope. A **naturalistic rock garden** is constructed to look like a real mountain site. A **moraine garden** mimics an alpine setting watered by melting snow. A **scree garden** is like a moraine garden except that water is supplied by over–the–top sprinkling.

Water gardening. This branch of landscape horticulture deals with growing ornamental plants in natural or artificial ponds and pools. Apart from aesthetics, water gardens have cooling and refreshing effects on the surroundings. In this type of gardening, hoeing, irrigation, and weeding are not required. The ponds or pools are located in open places away from trees, preferably at the foot of a rock garden, as a lawn feature, or in a paved courtyard. The pool should receive at least four hours of sunshine. Plants like lotus, water lily, umbrella plant, calla lily, watercress, tradescantia, and ferns are suitable for growing in and around such gardens.

15.2 Lawns

A lawn is often the heart of a garden. It is a source of continuing pleasure; the green of the grass is soothing to the eyes and consoling to the mind. The open expanse of a lawn imparts dignity to a site and provides the foreground to any architectural or landscape work. Utilitarian sites like recreation grounds and playing fields are all planted and maintained as lawns.

15.2.1 Varieties

The grasses used for turf or lawns vary in different regions according to the soil and climatic conditions. In Pakistan, which mostly has a subtropical climate, Bermuda grass (*Cynodon dactylon*) is generally used for lawns. This grass is a native of the Indo–Pakistan subcontinent. It grows well in summer, relishing high temperatures and full sunshine, but does not thrive in shade. Bermuda grass is very resistant to drought and high temperatures but has a tendency to frost damage and winterkill. It has all the characteristics of an excellent turf; it spreads by stolons which root readily from each joint and quickly produce a thick turf. It develops quickly and stands regular mowing as well as being resistant to drought and diseases. One drawback is that its bright green colour fades during winter in open and exposed lawns. Lawns in protected places may retain their colour during winter. Some of the important varieties are Tifway, Tifgreen, Tifton, Tifdwarf, and Dhaka.

15.2.2 Establishment of new lawns

A good lawn is a blessing; it is a source of joy and visual pleasure. Three things are necessary for the establishment of a new lawn. It should be planted on a piece of good, fertile, well–drained land receiving a maximum of sunshine. Seed or turf of a superior variety should be selected; and cultural practices should be appropriate and timely.

Preparation of land. The land to be planted as lawn should be thoroughly dug out or deeply ploughed many times. All debris, rubbish, and stones should be removed and the land levelled. It should then be irrigated to check the finer levelling and minor depressions and irregularities corrected. Often the land is ploughed again and well–rotted FYM thoroughly mixed into the soil at the rate of 1 t/100 sq m. Before grass is planted, 10–12.5 kg of superphosphate and 5 kg of ammonium sulphate/10 sq m are broadcast and mixed properly, taking care that the final levelling is not disturbed.

One should remember that a lawn remains for many years, while a crop is sown every season. Therefore, any mistake committed during the establishment of a lawn will be difficult to correct later. Levelling is the most important step in planting a lawn.

Time of planting. Since Bermuda grass is a summer–growing grass, it should be planted so that it is well established before the onset of winter. The best planting time is in July at the beginning of the rainy season. Where irrigation water is ample, planting may be done in March–April. There should be no shortage of water during the following hot and dry months.

Planting. There are many ways to plant lawn grasses: seeding, sprigging, plugging, or sodding. **Sprigging** is the most economical and commonly used method for turfing in Pakistan. Sprigs of well–grown grass are taken from any established lawn or grass nursery. Clusters of 3–5 sprigs are planted at a distance of 10–15 cm each way. The basal portions of the sprigs are buried and the upper portions left above ground. The soil around the transplanted sprigs is then patted firm. Grass from 1 sq m is enough to plant about 15 sq m of lawn. Irrigation is given immediately after planting.

The method of **sodding** or turfing is costly and is employed to achieve immediate results. Blocks of suitable sizes are cut from grass–nursery lawns and removed along with about 3.5–5 cm of soil. These blocks are set together on the prepared soil in the same fashion as flooring is done with tiles. Arrangements should be made to keep the blocks pressed down so that they do not float when they are irrigated.

For seeding, the bed is prepared properly as for any crop. In an ideal seedbed, when the soil has settled down it does not show deep footprints when walked on and the surface is crumbly. The best time for seeding is early summer so that the lawn is ready before the onset of winter.

Seed should be selected with care keeping in view the environmental conditions and the purpose of the lawn. For 1000 sq m of lawn, 2–3 kg of seed is sufficient. The seed is sown after the plot has been levelled and finely prepared. At sowing the field should be slightly moist—neither dry nor wet. To achieve even seed distribution, the plot is divided into 2–3 m widths and the seed divided into the same number of lots as there are strips. Each lot of seed is mixed with ample coarse sand and broadcast evenly in the plot. Half the seed is sown while moving in an east–west direction and the remaining half while moving north–south so that no gaps are left. Each plot is sown in the same way and the seed is covered with a .5 cm layer of fine leaf mold, or the surface is raked lightly. After sowing is completed, a light roller is run over the field and the plots are irrigated, preferably by light sprinklers. Care is taken to see that the seed is not disturbed. The field must be kept moist until the seed germinates in two to three weeks.

15.2.3 Maintenance

Rolling. After the first irrigation, when the soil is still moist, the lawn is rolled with a medium–weight roller. Rolling is carried out after each irrigation until the lawn is established.

Mowing. The first mowing is delayed until the swards have grown 10–15 cm high. The first cutting is preferably done with a sickle; thereafter it can be done with a lawn mower. After the first cutting, the lawn is mowed when the swards are about 5 cm tall. Close mowing is not recommended; the mower should be set at 12 mm and should have a grass catcher so that the cut grass does not fall on the lawn.

Weeding. Because of the use of FYM and canal water for irrigation, seeds of many pernicious weeds find their way into lawns and grow at the expense of lawn grass. Weeding should, therefore, be carried out very steadfastly and carefully. There are two approaches to keeping a lawn weed–free: (a) weed inhibiting practices, and (b) weed eradication practices.

WEED INHIBITING PRACTICES: (1) Application of fertilizers instead of compost, since compost contains a lot of weed seeds. If compost is used, it should be treated so that all weed seeds are killed. (2) Occasional spiking and raking so that weeds are uprooted and raked out. (3) Controlling earthworms and other pests and diseases.

WEED ERADICATION PRACTICES: This approach demands a combination of cultural practices and application of agro–chemicals. Many weed–specific chemicals (weedicides/herbicides) are available in the market. These should be used with utmost care under the direction of experts.

Irrigation. Bermuda is a fast–growing grass and needs frequent watering for a healthy and beautiful appearance. If irrigation is delayed the grass becomes coarse and looks shabby. During summer, when growth is faster, irrigation is needed more frequently (every 5–7 days); in cold weather, when growth is almost stopped, the intervals are longer. At each irrigation the lawn should be thoroughly soaked so that the entire root zone is properly wetted. Light irrigations restrict root growth to near the soil surface. Lawn sprinklers are very useful for this purpose. They have the advantage of also washing the grass and making the lawn look fresh and clean.

Fertilization. A well–fed lawn is greener and more even than a poorly–fed one, and it also has fewer weeds since the vigorous turf chokes out most of them. Nutrients are continuously being removed in grass clippings; therefore the soil has to be replenished every season by adding 2.5 kg of ammonium sulphate for each 100 sq m of lawn. Irrigation is applied immediately afterwards. The best times to apply fertilizer are as new growth commences, in the beginning of spring, and then again after the rainy season.

Aeration. Unlike agricultural crops, lawns are not hoed. Aeration, however, must be provided to the roots once a year. In the fall, after a close mowing, lawns are rolled with peg/prick rollers. The pegs or pricks penetrate the soil to a depth of about 1 cm. Superphosphate is also applied afterwards so that it lodges in the peg holes. Aeration is followed by irrigation.

Thatching. Bermuda grass spreads along the surface by means of stolons. The stolons make roots at each joint so that after a few months the whole lawn surface is covered by a thick mat of stolons and looks coarse and whitish. To remedy this situation, lawns are scraped once a year. The whole surface is scraped to remove the thatch. The best time for scraping is at the beginning of the rainy season. After scraping, the soil is lightly loosened with rakes. FYM at the rate of 500 kg/sq m and 2.5 kg/sq m of ammonium sulphate are added and properly mixed. The ground is then irrigated. During this operation care is taken that the level of the lawn is not disturbed.

15.3 Ground covers

Places which are rocky, steep, or otherwise unsuitable for turfing are planted with vegetation which will cover the ground and make the place more beautiful. Such vegetation is known as ground cover. These plants usually grow into dense mats, mounds or bushes, checking the growth of weeds or covering bare patches of soil, or filling gaps in shrubbery and perennial borders. They beautify difficult sites where it would be impossible to mow a lawn, such as steep banks, dry corners, shady places, or long areas.

Plant selection depends upon the site, soil, and climatic conditions. A variety of ground cover plants are available, ranging from low-growing, mat-forming plants to creepers and small bushes. Many flowering perennials make excellent ground covers. Some popular ground cover plants are: gazania, mesembryanthemum, juniper, arctotis, tree chrysanthemum, ivy, dichondra, chlorophytum, verbena, hypericum, and sedum.

15.4 Annual flowers

Annuals are flowering plants whose seeds are sown every year, produce blossoms the same season, and are killed by extreme weather. Tender perennials, like snapdragons and pansies, are also treated as annuals. According to their flowering season, they are classified as summer or winter annuals.

These plants are grown for their aesthetic value, diversity of colour, form, size, growth habit, and seasonality; they give a quick and inexpensive display of colour. In a garden, annuals are used for several purposes: to provide a mass of colour in front of hedges; in individual beds, borders, pots and planters; to fill shrubbery beds in the early stages; and to cover up beds of bulbs after they have bloomed. Smaller-growing annuals are planted for low edging around beds and walks, and tall-growing annuals are planted to provide quick-growing screens to hide unsightly scenes. Annuals with showy and long–keeping flowers are cultivated for cut flowers, while others like helichrysum and acroclinium are particularly suited for drying.

Success in growing annual flowers depends on starting with good seed, sowing or transplanting at the proper time, soil fertility and preparation, proper distancing, and appropriate after–care.

Soil preparation. Soil preparation is started well ahead of planting time. The soil is dug to a depth of 20–30 cm and turned over completely. All stones, pebbles, pieces of wood, and rhizomes or roots which might sprout again should be removed, but all leaves, grass, stems, and dead roots are buried. Re–spading is done three or four times at weekly intervals. All weeds should be pulled before they set seed. The soil should again be spaded just before planting.

Fertilizers and FYM are added and mixed thoroughly. A 5–6 cm layer of well–rotted FYM or leaf mold, and 1–2 kg of ammonium sulfate suffice for each 10 sq m. In heavy, clayey soils a 2–3 cm layer of unwashed sand should be added. Finally, the soil surface is raked smooth so that it is ready for planting.

Seed procurement. Seed should be fresh, as it loses its viability with age and improper storage. If purchased much before it is actually required, it should be kept in a cool and dry place until it is planted. F_1 hybrid seed is expensive, but superior in quality to F_2; however, it has to be purchased fresh every year. Seed should be clean and free of weed seeds. Samples including shrivelled and underdeveloped seeds should be rejected.

Sowing time. Time of sowing depends upon germination, growth, and flowering responses to environmental conditions, which vary from place to place. In the ecological regions where there is a short growing season, as in the hilly areas, all annuals are grown during summer when the danger of frost is over; while in places with longer growing seasons, i.e. the plains, annuals are grown at different times of the year. Seeds of most winter annuals are sown in September–October and transplanted in November–December, while summer annuals are sown in March–April and transplanted in April–May. Still another group called *monsoon annuals*, e.g. cock's comb and celosia, are sown in July–August and transplanted in August–September. Seeds of salvia, dahlias, hollyhocks, balsam, and marigolds are also sown in August–September.

Seed sowing. For preparing nurseries, the seedbed should be finely pulverized but also firm. A 1–2 cm layer of well–decayed leaf mold is spread on the surface. Shallow furrows are made at 5–7 cm distances. Seeds are sown singly and then covered with finely–sieved leaf mold. Planting depth should be about twice the diameter or thickness of the seed. The seedbed is watered thoroughly with a fine mist nozzle. Water should not run off.

In order to prevent the seedbed from drying, it can be covered with newspaper, which is removed after the seedlings emerge. To avoid rain damage, seeds may be sown in 5–7 cm deep wooden flats, which can be moved indoors as required. Seeds of plants susceptible to damping off are treated with fungicides.

The seedlings of larkspur, poppies, and sweet peas are difficult to transplant; therefore these are either sown direct or sown in peat pots and then transplanted along with the pots. Morning glory, spiderplant, cornflowers, and nasturtium reseed freely; thus there is no need for re-sowing if they are to be grown again in the same beds.

Germination time. Germination of seed depends upon many factors like internal physiological conditions of the seed, moisture, temperature, and light. The average number of days taken from sowing to germination varies from species to species. Some germination times are given here.

5 days	Ageratum, cornflower, cosmos, dahlia, four-o-clock, marigold, morning glory, pink, stock, strawflower, zinnia
8 days	China aster, calliopsis, nasturtium
10 days	Baby's breath, balsam, calendula, cock's comb, coleus, forget-me-not, pansy, petunia, phlox, portulacca, scabiosa, spider plant
15 days	Globe amaranth, impatiens, salpiglossis, scarlet sage, snapdragon, summer cyprus, vinca
20 days	Candytuft, gaillardia, larkspur, lupine, rudbeckia, verbena

Transplanting. Small, young seedlings stand transplanting better than larger, older ones. Seedlings of summer annuals may be ready to transplant 2–4 weeks after sowing. Those of winter annuals are transplanted 2–6 weeks after sowing. The plants are lifted carefully so that the roots are not injured. Seedlings are separated carefully, allowing as much soil as possible to remain with the roots. The soil should be neither too dry nor too wet.

Evening is the best time for transplanting; uprooted plants are kept in the shade so that the roots do not dry. The seedlings are planted immediately after removal from the nursery, in holes which are made wide enough to accommodate the root system undisturbed. After transplanting, the soil around each seedling should be firmed and watered thoroughly.

Plant spacing. Plant-to-plant distances depend on the size (spread) of the plant, climatic conditions, and the fertility of the soil. However, the following plant spacings are recommended for various common plants; these distances may be adjusted according to local conditions.

20–30 cm	Larkspur, lupine, pansy, phlox, sweet pea, poppy, snapdragon, stock, calendula

25–30 cm	Candytuft, nasturtium, pink, scarlet sage, zinnia, baby's breath, china aster, cock's comb, coleus, cosmos, forget–me–not, gaillardia, globe amaranth, impatiens, portulacca, salpiglossis, strawflower, sweet alyssum, vinca
30–35 cm	Calliopsis, marigold, rudbeckia, balsam, cornflower, dahlia, four–o–clock, petunia, scabiosa, spider plant, sunflower

Thinning. If seeds are sown direct in the flower bed, they are thinned after developing two true leaves, taking care that the remaining plants are at proper distances according to their size and growth habit.

Pinching. The growing tips of the seedlings of many plants like ageratum, snapdragon, spiderplant, and verbena are pinched off in order to encourage branching and to keep them short and compact.

Training and supporting. Vines like sweet peas, cyprus vine, clitoria and weak–stemmed, heavy–flowered plants like carnations, snapdragons, and salpiglossis need staking or use of special plastic netting.

Watering. Flower beds are moistened thoroughly when they are moderately dry. Irrigation at noon time during summer in hot climates is not a good practice; it is better to irrigate in the evening. Sprinkling water on the foliage and flowers should be avoided.

Hoeing. Hoeing loosens the soil, allowing better respiration of roots, and eradicates weeds. Hoeing is done when the soil has moderately dried. Hoeing should not be deep, otherwise the finer roots will be cut. It should be finely done, so that it can help to conserve moisture effectively as well as provide aeration to the roots. After hoeing, all weeds should be removed and the bed firmed. Weeding must be completed before seed formation.

Suitability for different purposes. The size of a plant, its growing habit, and certain other foliar and floral characteristics make it suitable for specific purposes. Flowering annual plants are classified according to the purposes for which they are most often used.

Edging	Ageratum, candytuft, pansy, portulacca, sweet alyssum
Bedding	China aster, balsam, calendula, candytuft, cock's comb, cornflower, cosmos, dahlia, impatiens, marigold, nasturtium, pansy, petunia, phlox, portulacca, salvia, snapdragon, stock, strawflower, verbena, zinnia
Borders	Gypsophilla, gaillardia, globe amaranth, lupine, pink, poppy, salvia, sweet alyssum
Screen	Cosmos, larkspur, morning glory, summer cypress, sweet peas, clitoria, hollyhock, sunflower

Cut flowers China aster, ageratum, gypsophilla, calendula, cornflower, cosmos, dahlia, gallardia, globe amaranth, larkspur, marigold, pansy, snapdragon, stock, strawflower, sunflower, verbena, zinnia

The common and botanical names of selected flowering annuals are given in Table 15.1.

Table 15.1 Selected flowering annuals

Common name	Botanical name
	WINTER FLOWERS
Acroclinium	*Helipterum roseum*
Alyssum	*Alyssum maritimum*
Antirrhinum (snapdragon)	*Antirrhinum majus*
Arctotis (African daisy)	*Arctotis stoechadifolia*
Aster	*Callistephus chinensis*
Brachycome	*Brachycome iberidifolia*
Buttercup	*Ranunculus acris*
Calendula	*Calendula officinalis*
Candytuft	*Iberis amara*
Carnation	*Dianthus caryophyllus*
Cineraria	*Senecio cruentus*
Clarkia	*Clarkia elegans*
Cornflower	*Centaurea cyanus*
Cosmos	*Cosmos bipinnatus*
Chrysanthemum	*Chrysanthemum morifolium*
Delphinium	*Delphinium ajacis*
Dianthus	*Dianthus chinensis*
Dimorphotheca (Marigold)	*Dimorphotheca aurantiaca*
English daisy	*Bellis perenne*
Hollyhock	*Althea rosea*
Iceplant	*Mesembryanthemum, Cryophytum crystallinum*
Lineria	*Lineria maroccana*
Linum	*Linum grandiflorum*
Nasturtium	*Tropaeolum majus*
Nemesia	*Nemesia strumosa*
Painted daisy	*Chrysanthemum carinatum*
Pansy	*Viola tricolor*
Petunia	*Petunia hybrida*
Phlox	*Phlox drummondii*
Poppy (Shirley)	*Papaver rhoeas*

Common name	Botanical name
Salpiglossis	*Salpiglossis sinuata*
Salvia, scarlet sage	*Salvia splendens*
Schizanthus	*Schizanthus wisetonensis*
Statice	*Limonium bonduellii*
Stock	*Matthiola incana*
Strawflower	*Helichrysum bracteatum*
Sweet pea	*Lathyrus odoratus*
Sweet sultan	*Centaurea moschata*
Sweet William	*Dianthus barbatus*
Verbena	*Verbena hybrida*
Wallflower	*Cheiranthus cheiri*
SUMMER FLOWERS	
Amaranthus (tassel flower)	*Amaranthus caudatus*
Balsam	*Impatiens holsti*
Cosmos	*Cosmos sulphureus*
Celosia	*Celosia plumosa*
Cockscomb	*Celosia argentea cristata*
Gaillardia	*Gaillardia amblyodon*
Gomphrena (globe amaranth)	*Gomphrena globosa*
Sunflower	*Helianthus annuus*
Kochia	*Kochia scoparia*
Marigold (African)	*Tagetes erecta*
Marigold (French)	*Tagetes patula*
Morning glory	*Ipomoea purpurea*
Morning glory (dwarf)	*Convolvulus tricolor*
Portulaca (moss rose)	*Portulaca grandiflora*
Zinnia	*Zinnia elegans*

15.5 Perennials

Herbaceous plants which live for more than two growing seasons are known as **perennials**. There is a wide range of ornamental plants in this group, which differ in size, growth habit, flower shape, size, colour, flowering season, foliage, and hardiness. They are indispensable for landscape planting. Once planted they remain for several seasons, and can usually be multiplied easily.

Because of their ease of planting, wide range of growth habits, and beauty, perennials are suitable for many purposes: ground cover and grass substitutes, shade gardens, wild gardens, naturalization, cut flowers, screen and background planting, herbaceous borders, and landscape structure.

15.5.1 Selection of perennials

Perennials remain in the soil for several years; one must, therefore, be very careful in selecting them. With the tremendous diversity of perennial species, many factors influence their selection. Major considerations are: size, form, and growth habit; seasonality of bloom, bloom colour, bloom duration, and recurrence of blooming; foliage characteristics like leaf size, form, texture, colour, seasonality of colour, and hardiness; purposes of planting; permanence and hardiness; and problems relating to maintenance and care of the plants.

Besides the above characteristics, the plants selected should grow well in the soil and climatic conditions of the locale selected and tolerate drought, heat, cold, and diseases. They should be true perennials and not biennials, should bear beautiful blooms, and have pleasing shapes. They should have beautiful foliage and long, strong stems, be able to stand without support, and be easy to maintain. They should have a long blooming season and should not be spreading and weedy.

Growth habits. Plant height and form are the major criteria for selecting plants for particular situations. For plant height, plants are classified as: small (less than 30 cm), small–medium (30–60 cm), medium (60–90 cm), tall (90–120 cm), and very tall (more than 120 cm). Plant form refers to the appearance of the entire plant within its extremities. The following forms are recognized: compact (*Bellis perennis*, anchusa); spreading-rooting (phlox, vinca, ranunculus); underground stems suckering (coreopsis, aster); erect (*lilium*, salvia, delphinium); sprawling (alyssum, gaillardia).

Blooming habit. This characteristic includes many factors such as blooming singly or in clusters, in raceme or in cymes, terminally or laterally, size, shape, colour, seasonality, recurrence and periodicity of blooming, and fragrance. An attractive display of flowers, well-distributed over the bed, should be available throughout the year, particularly from early spring to late autumn. In certain places winter blooming is also possible.

Foliage characteristics. Good foliage throughout the season is essential to the appearance of perennial borders. It will frequently soften many otherwise harsh combinations. Selected plants should have beautiful foliage and retain it for long periods. Flowers, though, should not be hidden by too-large leaves.

Purpose. As described earlier, perennials are planted to achieve certain purposes; while making a selection of plants they should be objectively examined for their suitability for the specific purpose required.

Permanence. Perennials last for several seasons, but not forever. Their longevity or permanence depends upon many factors—mainly hardiness to vagaries of weather, resistance to pests, adaptability to varying soil conditions, and ability to survive under unfavourable circumstances. The more permanent and hardy the selected plants, the less will be the upkeep, care, and replacement.

Ease of maintenance and care. The performance of a plant depends on the environment in which it is grown. Each plant has its optimum growing conditions. Some plants grow better in relatively low temperatures, while others prefer warm temperatures. Similarly, different plants prefer different types of soils: light, heavy, sandy, calcareous. They also differ in sunlight or shade growing conditions, and water requirements. Therefore in choosing landscape plants, great emphasis should be put on their suitability for the environmental conditions in the locality where they are to be planted.

For proper maintenance and care of perennial plants, the following tasks may need to be done.

- Hoeing, weeding, and forking over the soil if the plants are panned down by rain.
- Irrigation when needed, recognizing that over-irrigation is more harmful than under-irrigation.
- Staking of tall plants and those bearing heavy flowers.
- Removal of old and unsightly leaves and flowers to improve the appearance of plants and encourage further flowering.
- Feeding with appropriate fertilizers at proper times.
- Lifting and dividing every three to four years: autumn is the best time to divide perennials.
- Protection against pests, diseases, and inclement weather conditions.

15.5.2 Planning a border

Designing a border is a creative act, like painting a landscape. Apart from esthetics, one should have a thorough knowledge of the characteristics of perennial plants. One has to take into account the size of the border, bloom succession, colour combinations, plant combinations, and plant size. Inspiration for planning comes from knowledge, observation, and experiment.

The dimensions of the border are important. Narrow, 1 m-wide borders will have space for only two heights. Borders 150–180 cm wide may have space for three or four different heights. For maximum effect, the border should be 150–180 cm wide. The length of the border has a close relationship with its width; this is a matter of scale and proportion. A very wide but short border appears unbalanced and odd. A narrow border 1–2 m wide could be about 15 m long, and a wider one could be as long as 25 m.

Plans should always be prepared on paper and revised as often as necessary. The planner should see that the border has an even distribution of blooms over its entire length and breadth all year round. Concentration of blooms at one place is bad planning; therefore plants blooming at the same time should be distributed throughout the border. Similarly, assignment of plants according to their sizes has to be done thoughtfully. Taller plants should be planted in the back row and the smaller plants in the front row. Plants of medium height should be adjusted according to their heights in the middle rows. The plants in the first row should have good foliage. The border should slope from back to front throughout its length.

Colour combination is very important in any artistic work. Without proper blending of colours the desired effect of a border is not achieved. However, satisfactory effects can be realized by merely having an adequate amount of healthy, vigorous foliage to act as background for the blooms and a buffer between any chance discordant colours. For narrow borders it is advisable to use single colours; warm colours in varying hues are appropriate here. Deep-blue is effectively contrasted with yellow. Similarly, white combines favourably with blue.

15.5.3 Preparation of beds and borders

Preparation of beds and borders begins with good soil preparation. Since soil preparation is less frequent for perennials, the initial mixing of the proper amounts of organic manure and fertilizers is very important. There should be at least 30 cm or more of this material for successful growth and sustained development of perennial plants.

The common and botanical names of selected perennial flowers are given in Table 15.2.

Table 15.2 Selected perennial flowers

Common name	Botanical name	Family
Periwinkle	*Vinca minor*	Apocynaceae
Anchusa	*Anchusa barrelieri*	Boraginaceae
Forget-me-not	*Myosotis scorpioides*	Boraginaceae
Bellflower	*Campanula carpatica*	Campanulaceae
Baby's breath	*Gypsophila paniculata*	Caryophyllaceae
Carnation	*Dianthus caryophyllus*	Caryophyllaceae
Pink (Maiden)	*Dianthus deltoides*	Caryophyllaceae
Sweet William	*Dianthus barbatus*	Caryophyllaceae
Aster	*Callistephus alpinus*	Compositae
Aster	*Callistephus subcaerulens*	Compositae
Blanket flower	*Gaillardia aristata*	Compositae
Chrysanthemum	*Chrysanthemum morifolium*	Compositae
Coneflower	*Rudbeckia speciosa*	Compositae

Common name	Botanical name	Family
Cornflower	*Centaurea montana*	Compositae
Dahlia	*Dahlia coccinea*	Compositae
English daisy	*Bellis perennis*	Compositae
Goldenrod	*Solidago canadensis*	Compositae
Painted daisy	*Pyrethrum roseum*	Compositae
Sunflower	*Helianthus orgyalis*	Compositae
Tickseed	*Coreopsis lanceolata*	Compositae
Candytuft	*Iberis sempervirens*	Cruciferaceae
Goldentuft	*Alyssum saxatile*	Cruciferaceae
Wallflower	*Cheiranthus allionii*	Cruciferaceae
Scabiosa	*Scabiosa columbaria*	Dipsaceae
Iris	*Iris* sp.	Iridaceae
Iris (bearded)	*Iris germanica*	Iridaceae
Iris (beardless)	*Iris pseudacorus*	Iridaceae
Iris (crested)	*Iris tectorum*	Iridaceae
Iris (English)	*Iris xiphioides*	Iridaceae
Salvia	*Salvia azurea*	Labiateae
Day lily (tawny)	*Hemerocallis fulva*	Liliaceae
Day lily (lemon)	*Hemerocallis flava*	Liliaceae
Lilium	*Lilium philippinense*	Liliaceae
Lilium	*Lilium formosum*	Liliaceae
Lilium	*Lilium regale*	Liliaceae
Lobelia	*Lobelia cardinalis*	Lobeliaceae
Hollyhock	*Althea rosea*	Malvaceae
Four o'clock	*Mirabilis jalapa*	Nyctaginaceae
Iceland poppy	*Papaver nudicale*	Papaveraceae
Oriental poppy	*Papaver orientale*	Papaveraceae
Statice	*Limonium latifolia*	Plumbaginaceae
Primrose	*Primula polyantha*	Primulaceae
Buttercup	*Ranunculus acris*	Ranunculaceae
Larkspur	*Delphinium grandiflorum*	Ranunculaceae
Monk's hood	*Aconitum fischeri*	Ranunculaceae
Windflower	*Anemone coronaria*	Ranunculaceae
Windflower	*Anemone japonica*	Ranunculaceae
Foxglove	*Digitalis purpurea*	Scrophulariaceae
Verbena	*Verbena tenera*	Verbenaceae
Verbena	*Verbena canadensis*	Verbenaceae
Violet	*Viola papilionacea*	Violaceae

15.6 Flowering bulbs

Flowering bulbous plants include all the plants having bulbs, corms, or tubers—underground thickened roots or stems. These structures maintain life during the dormant period or during adverse conditions, and also store food for rapid above–ground growth. These plants are very popular for the exquisite beauty of their flowers, attractive colours, unique flower shapes, plant forms, ease of culture and transportation, and longevity of the flowers. They can be grown in beds, in gaps between shrubs, and in shady places and pots.

True bulbs are of two types: **tunicated** as narcissus and **scaly** as lilium. The structure consists of a very shortened stem with fleshy leaves which act as storage for the food materials. A **corm** is also a shortened stem but it is solid, and nodes and internodes are visible on the structure; a typical example is gladiolus. **Tubers** are thickened roots like that of dahlia. Canna and ranunculus have typical fleshy rhizomes and fleshy roots, respectively.

Bulbs fall into three classes based on their adaptibility: hardy, half–hardy, and tender. **Hardy** bulbs can be left in the ground for several years, whereas **half–hardy** bulbs can be left in the ground during winter in regions with mild climates but require protection in colder climates. **Tender bulbs**, however, must be taken indoors in winter even in mild climates. Planting of hardy bulbs should be done in autumn. Anemone and perishable bulbs like freesias should be planted in September. Small bulbs like crocus, scilla, and muscari are best planted during October or early November. All others are planted during November–December, depending upon the species and when flowers are desired (Laurie and Ries 1950:298).

A medium consisting of four parts rich soil, three parts coarse sand, four parts well–rotted compost, and one part bonemeal is considered good for growing bulbs in containers. Most bulbs are planted to a depth of about three times their own diameter or thickness, with a minimum of 5 cm. The amaryllis bulb is planted so that its neck protrudes above ground.

Cultural operations include all those practices which are necessary for perennials. Tender bulbs are, however, dug out after their foliage has dried. These bulbs are stored until needed for planting next season.

Some of the important bulbs are: narcissus, tulip, hyacinth, crocus, gladiolus, amaryllis, tuberose, canna, dahlia, caladium, lilium, and freesia.

15.7 Roses

Roses are woody perennials including shrubs, bushes of various sizes, ramblers, and climbers, as well as very small plants known as miniatures. There is a tremendous diversity of growth habit, flower form, and colour among roses. The rose is the most popular of the flowers because of its beauty and fragrance and is rightly called the queen of flowers. No garden is considered complete without roses.

Classification. Modern cultivars of roses have resulted from a long process of hybridization. Consequently, rose classification is quite complex. A botanical or exhaustive professional classification of roses is not needed here, and a brief, practically-oriented list of the main groups and their important attributes will suffice.

Bedding roses are smaller than shrubs. They include the following groups.

HYBRID TEAS are bedding roses with large, shapely, and often fragrant blooms. Usually solitary blooms are borne at the terminals of branches produced during the current season. Blue Moon, Chicago Peace, Ena Harkness, Fragrant Cloud, Mischief, Peace, and Super Star belong to this class.

FLORIBUNDAS are also good bedding roses. Their blooms are smaller and are borne terminally in clusters on current-season branches. Daily Sketch, Fashion, Iceberg, Seven Seas, Spartan, and Tip Top belong to this group.

GRANDIFLORA roses have the characteristics of both the groups described above. The blooms are shapely, large, and borne in clusters of a smaller number of flowers. These are also referred to as the floribunda—hybrid tea type. An example is Queen Elizabeth.

POLYANTHA POMPON are bushy bedding roses which flower freely and bear smaller blooms. The plants are of restricted growth. Golden Salmon, Paul Crampel, and Gloria Mundi are examples.

MINIATURE or fairy roses bear small flowers, and are small and compact. They flower like the hybrid teas and floribundas. Examples are Bambino, Gold Coin, Mimi, Baby Pinnochio, Beauty Secret, Starina, and Sweet Fairy.

Ramblers flower only during summer on the preceding year's canes. The canes produce only once, and canes which have flowered are cut out in the pruning season. **Climbing roses** are of rather indeterminate growth, possess relatively large flowers and stiff wood, and do not renew themselves from the base every year. **Shrub roses** are of erect growing habit, smaller than small trees, and bigger than the groups already described.

Soils. Roses prefer porous and well-drained, medium-heavy clay loam soils, containing 20–30% clay. They do not like heavier soils.

Planting. Fall or late winter is the best time for planting. Potted plants may be planted in early spring. For planting, dig a hole large enough to permit the roots to be spread out. Before planting, trim all injured roots and cut back the tops to 15–20 cm. The plants are set out so that the stock–scion union is level with the soil surface. Saplings raised from cuttings should be set at the same level as those grown in the nursery. The soil is packed thoroughly around the roots, and a mound of soil is made around the branches to prevent rapid evaporation and withering before the roots become established. Newly set out plants are irrigated immediately after planting. Bedding roses are planted at 1–1.5 m distances.

Pruning. Pruning is a most important, yearly operation. All suckers, weak branches, and dry, diseased, and otherwise undesirable growth is removed. Healthy and vigorous branches are left on the plants and cut, leaving only 15–20 cm from the base. Early spring, before the start of new growth, is the best time for pruning.

Other cultural operations are similar to those for perennials.

15.8 Trees, shrubs, and vines

Trees have always been valued as symbols of life, strength, and beauty. The demand for trees has greatly increased because of better understanding of their benefits and beauty, rapid urbanization, and improvement in people's financial and educational status. Trees not only contribute to our enjoyment, consolation, and well–being, but also are useful in enhancing the beauty and dignity of the environment. Trees in landscaping are often used to form vistas, frame views, create accents and emphasis, and reduce the boredom of brick and mortar. They provide shade and enjoyable fragrance. Because of all these attributes, landscape trees add to the value of real estate.

Plants influence our physical environment by purifying the air and reducing noise and erosion. Plants moderate the effect of hot and harsh summer climates. They provide shade, reduce temperatures, increase humidity, and break the force of strong winds. They can increase fog precipitation and snow deposition.

Trees in an urban setting. Beautification is the primary aim of planting trees in towns and cities, but besides aesthetics, plants serve useful purposes. Shrubs can screen headlight glare from oncoming traffic and reduce glare and reflection from the sun or artificial lights. Plants can be used to direct foot and vehicular traffic, and can serve as road dividers or barriers to out–of–control vehicles.

Trees to be planted in cities should be hardy enough to withstand the stresses of the urban environment—gas, smoke, dust, bad drainage, poor, hard and impervious soil, heat, light, and radiation. They often must contend not only with impoverished dry soil, but also with intense summer heat, multiplied by reflection from building walls and pavement. City air is also highly polluted with toxic gases which cause the rapid defoliation of all but a few highly resistant species.

Trees should be selected carefully for adaptation to the harsh conditions found in cities. A smooth and waxy leaf surface is more easily washed by rains or winds. Smoke and soot do not adhere to a smooth surface as much as to rough, pubescent surfaces. Trees that renew their leaves frequently can withstand the bad effects of gas and soot better than trees that keep their leaves for a long time. Shallow–rooted species are better for planting along roads. Trees observed to be resistant to pollution are *Ailanthus*, *Gladitschia*, *Albizzia julibrissim*, *Alstonia scholaris*, and magnolia.

Plants should be matched to the places intended to be planted. For example, the shapes and sizes of the following plants are suitable for planting in the types of places indicated.

Eucalyptus or *Albizzia*	highways
Poinciana, *Washingtonia*, *Ficus*	wide city roads
Tecoma stans, *Schinus terebinthifolia*	narrow streets
Sterculia	small gardens, where space is limited

During the planning process one should avoid planting trees or shrubs near electricity, water, gas, and drainage systems along roads and streets, or in houses. The planting hole should be made sufficiently deep and wide, and all the old soil should be replaced by fresh fertile soil.

Trees newly planted in the city should be guarded from compaction and physical damage by pedestrians by putting metal grills or metal cylinders around them. They should be properly cared for, which includes training, pruning, fertilization, irrigation, and protection against pests, diseases, and vandalism. Whenever necessary, replacements should be made promptly with the same type and size of plant.

Trees in home gardens. Trees are very important in the home garden. Besides their aesthetic value, they play an important role in moderating the micro–climate of the house. They protect the house and its dwellers from extremes of weather, and can shield it from storm winds which can otherwise damage the property. Trees filter and purify the air from dust, and provide shade and protection against ultraviolet radiation.

Species selected for the home should be adapted to the particular climatic zone. Plants of colder areas like maples, pines, platanus, and oaks cannot grow in tropical areas; similarly, plants of warmer areas, like cassias, delonix, magnolia, and palms do not grow in colder areas. Another consideration is the size of the home ground and the amount of space available. In a small area only smaller trees can be grown, whereas if the area is big enough a number of large and small trees can be accommodated and the choice is broadened. The shape and size of the trees at maturity should be in a proper relationship to the size of the house.

A third consideration is the functional usefulness of the trees. The number of species which can withstand the polluted atmosphere, poor soil, and limited soil water penetration in towns and cities is limited. Trees for seashore homes should be able to tolerate salt–laden winds. Poplars, willows, silver maple, and mulberry invade and clog sewer lines. Deep–rooted trees are less competitive with lawns and flower beds than shallow–rooted ones. Species with broad leaves require more leaf disposal than those with tiny leaves. Trees for shade should have thick foliage and spreading habit. Trees used for windbreaks and screens should be evergreen, retaining branches close to the base throughout their lives. They should be deep–rooted, with tough wood and the flexibility to withstand strong winds.

The names and height characteristics of several trees, shrubs, and vines are presented in Table 15.3.

Table 15.3 Selected trees, shrubs, and vines

Common name	Botanical name	Ht (ft)	Family
	Big, umbrella shaped, evergreen		
Aam	*Mangifera indica*	50–60	Anacardiaceae
Kanak champa	*Pterospermum acerifolium*	60–70	Sterculiaceae
Andrikni	*Bischofia javanica*	50–70	Euphorbiaceae
Bargad	*Ficus bengalensis*	50–60	Moraceae
Pilkhan	*Ficus infectoria*	30–35	Moraceae
Peepal	*Ficus religiosa*	40–60	Moraceae
Peepli	*Ficus retusa*	30–35	Moraceae
Jaman	*Eugenia jambolana*	50–60	Myrtaceae
Chinese elm	*Ulmus parvifolia*	40–60	Ulmaceae
	Big, umbrella shaped, deciduous		
Kangar	*Pistacia integgerima*	50–60	Anacardiaceae
Tallow tree	*Sapium sebiferum*	40–60	Euphorbiaceae
Sufaid shreen	*Albizzia procera*	60–80	Leguminoseae
Simbal	*Salmelia malabarica*	80–100	Malvaceae
Tun	*Cedrela toona*	50–60	Meliaceae
Neem	*Azadirachta indica*	50–60	Meliaceae
	Big, columnar, evergreen		
Alstonia	*Alstonia scholaris*	60–80	Apocynaceae
Arjan	*Terminalia arjuna*	80–100	Combretaceae
Eucalyptus	*Eucalyptus citriodora*	80–90	Eucalyptus
Cheel	*Pinus helepensis*	40–45	Pinaceae
Cheel	*Pinus roxburghii*	60–80	Pinaceae
Silver oak	*Grevillea robusta*	25–30	Proteaceae
	Big, columnar, deciduous		
Sheesham	*Dalbergia sissoo*	60–80	Leguminosae
Coral tree	*Erytherina suberosa*	50–60	Leguminosae
Seedless ash	*Fraxinus pennsylvanica*	50–60	Oleaceae
Chinar	*Platanus orientalis*	80–100	Platanaceae

Common name	Botanical name	Ht (ft)	Family
Poplar	*Populus euamericana*	50–60	Salicaceae
Kar	*Celtis australis*	40–50	Urticaceae
	Small, umbrella shaped, evergreen		
Maple	*Acer oblongum*	40–50	Aceraceae
Kali mirch	*Schinus molley*	30–40	Anacardiaceae
Putanjan	*Putranjiva roxburghii*	35–40	Euphorbia
Kachnar	*Bauhinia purpurea*	20–25	Leguminosae
Amaltas	*Cassia fistula*	30–40	Leguminosae
Sukh chaen	*Pongamia glabra*	25–30	Leguminosae
Magnolia	*Magnolia grandiflora*	20–25	Magnoliaceae
Bottle brush	*Callistemon lanceolatus*	20–25	Myrtaceae
Willow, Laila	*Salix tetrasperma*	50–60	Salicaceae
	Small, umbrella shaped, deciduous		
Beeri patta	*Heterophragma adeno-phyllum*	30–40	Bignoniaceae
Persimmon	*Diospyros kaki*	25–30	Ebenaceae
Silk tree	*Albizzia julibrissin*	30–40	Leguminosae
Kachnar	*Bauhinia alba*	30–40	Leguminosae
Kachnar	*Bauhinia variegata*	20–25	Leguminosae
Gul mohar	*Poinciana regia*	30–40	Leguminosae
Bakain	*Melia azaderch*	25–30	Meliaceae
Toot	*Morus alba*	30–40	Moraceae
Samandar phal	*Barringtonia acutangula*	20–25	Myrtaceae
Gul fanoos	*Lagerstroemia flosreginae*	30–35	Lythraceae
Aaru	*Prunus persica*	20–25	Rosaceae
Weeping willow	*Salix babylonica*	30–35	Salicaceae
	Small, columnar, evergreen		
Saroo	*Cupressus sempervirens*	35–50	Pinaceae
Jacaranda	*Jacaranda mimosifolia*	20–30	Leguminoseae
Black locust	*Robinia pseudoacacia*	25–30	Leguminoseae
	Small, columnar, deciduous		
Nashpati	*Pyrus communis*	20–25	Rosaceae
Poplar	*Populus nigra*	25–30	Salicaceae
	Tall shrubs		
Frangipani	*Plumeria obtusa*	10–12	Apocynaceae
Thevitia	*Thevitia nerifolia*	12–15	Apocynaceae
Tecoma	*Tecoma stans*	10–15	Bignoniaceae
Morpankh	*Thuja orientalis*	15–20	Pinaceae
Barbados pride	*Poinciana pulcherrima*	12–15	Euphorbiaceae
Cassia	*Cassia glauca*	8–12	Leguminoseae
Pahari gurhal	*Hibiscus mutabilis*	10–12	Malvaceae

Common name	Botanical name	Ht (ft)	Family
Gul fanoos	*Lagerstroemia speciosa*	15–20	Lythraceae
Haar–singhar	*Nyctanthes arbortristis*	12–15	Oleaceae
Anar	*Punica granatum*		Punicaceae
Hamelia	*Hamelia patens*	15–20	Rubiaceae
Khatta	*Citrus aurantium*	15–20	Rutaceae
Kamni	*Murraya exotica*	15–20	Rutaceae
	Small shrubs		
Eranthemum	*Eranthemum nervosum*	5–8	Acanthaceae
Kaner	*Nerium odorum*	8–10	Apocynaceae
Chandni	*Tabernae montana coroneria*	5–8	Apocynaceae
Poinsettia	*Poinsettia pulcherrima*	5–8	Euphorbiaceae
Buddleia	*Buddleia asiatica*	8–10	Loganiaceae
Buddleia	*Buddleia hybrida*	8–10	Loganiaceae
Gurhal	*Hibiscus rosa sinensis*	8–10	Malvaceae
Pahari gurhal	*Hibiscus syriacus*	8–12	Malvaceae
Chambeli	*Jasminum grandiflorum*	15–20	Oleaceae
Peeli chambeli	*Jasminum humile*	10–15	Oleaceae
Motia	*Jasminum sambac*	3–5	Oleaceae
Gardenia	*Gardenia florida*	8–10	Rubiaceae
Raat ki rani	*Cestrum nocturnum*	5–8	Solanaceae
	Palms		
Coconut palm	*Cocos nucifera*		Palmaceae
Kentia	*Howea forsteriana*		Palmaceae
Chinese palm	*Livistona chinensis*		Palmaceae
Common date palm	*Phoenix dactylifera*		Palmaceae
Sabal palm	*Rhapis sabal*		Palmaceae
Royal palm	*Roystonea regia*		Palmaceae
Washington fan palm	*Washingtonia filfera*		Palmaceae
Cycas	*Cycas revoluta*		Cycadaceae
	Creepers		
Bignonia	*Bignonia venusta*		Bignoniaceae
Tecoma	*Tecoma grandiflora*		Bignoniaceae
Honeysuckle	*Lonicera japonica*		Caprifoliaceae
Rangoon creeper	*Quisqualis indica*		Combretaceae
Railway creeper	*Ipomoea carnea*		Convolvulaceae
Porana	*Porana paniculata*		Convolvulaceae
Banisteria	*Banisteria laurifolia*		Malpighiaceae
Bougainvillea	*Bougainvillea glabra*		Nyctaginaceae

Common name	Botanical name	Ht (ft)	Family
Bougainvillea	*Bougainvillea spectabilis*		Nyctaginaceae
Chambeli	*Jasminum officinale*		Oleaceae
Antigonon	*Antigonon leptopus*		Polygonaceae

15.9 House plants

A large number of plant species, selected from diverse ecological regions, are grown in containers for decoration inside and around buildings. These house plants must grow under non–natural conditions, usually confined to pots of varying sizes. The most popular ones are those which remain healthy and grow slowly. Plants that double in size in a few months outgrow their usefulness as decorative plants. Plants which grow well under shady or semi–shady conditions keep well indoors and are referred to as indoor plants. Indoor decoration with plants is called interior landscaping or interior plant-scaping. The same principles are applied as for outdoor landscaping.

Today's life style, particularly in urban areas, is characterized by intense competition, causing stresses, strains, depression, and hypertension. Plants can serve as a source of beauty and serenity, providing relaxation and helping to enlarge the imagination and revive the spirit.

15.9.1 Causes of plant failures in the house

Plants grown inside are as though kept under house arrest. The environment inside a house is quite different from the natural habitat from which they are selected. The house environment has abnormal temperatures, restricted air circulation, unusual light conditions, low humidity, cooking gases, dust, and other pollutants, water of different composition, and artificial means of irrigation. All these conditions are not favourable for normal plant growth.

When selecting house plants, one should look for those which are relatively tolerant to an environment with limited fresh air, relatively dry air, narrower differences in day and night temperatures, limited space for the root system, limited lateral and vertical space for branching, over or under–watering abuse, unnatural light regimes, and unnatural water composition.

15.9.2 Growing requirements

Light. Light is essential for the growth and development of plants and many vital processes within plants. Three aspects of light are important — intensity, duration and colour. Each plant species has a minimum light requirement below which it cannot survive, and the selection of plants for the house must be based on this factor.

Temperature. All biochemical reactions responsible for the life of plants are influenced by temperature. Tropical or subtropical plants need relatively high temperatures for normal growth. Most of them grow satisfactorily at temperatures between 18 and 24°C, preferably with a 5–10°C drop at night. They are usually injured at temperatures below 4.5–7°C.

Moisture. Water is a major component of the plant's body, and judicious application of water is one of the key factors in the maintenance of house plants. The amount and the intervals between irrigations cannot be predetermined since they depend upon the environmental conditions. Overwatering is the major abuse with house plants, and must be avoided at all costs. The error, if any, should be on the dry side rather than the wet. When irrigation is done, the growing medium should be thoroughly wetted; excess water should drain off through the drainage hole.

Growing media. Not all house plants grow equally well in the same type of growing medium. An ideal growing medium is one which contains all the nutrients in available form, is porous and well-drained while retaining moisture well, and is light in weight. It should be low in soluble salts, with a suitable pH and sufficient cation exchange capacity. It should be uniform, disease and pest-free, and compatible with the natural habitat of the species.

Most house plants, except cacti, prefer a highly organic, acid growing medium (pH 5.8–6.5). A growing medium consisting of good garden soil, leaf compost, and sand in equal proportions by volume is considered good. Addition of wood shavings makes it airier. In order to prepare a nutritionally balanced medium, the following fertilizers are added to each cubic meter of the above mixture.

- 3 kg calcium carbonate to raise pH to a range of 5.8–6.2
- 3 kg single superphosphate to promote initial root growth and supply the plant with required phosphorus
- a micro-element mix composed of 30 g each of copper sulphate, zinc sulphate, manganese sulphate, and iron chelate

Sphagnum peat, a natural organic compost, and synthetic materials like vermiculite, perlite, or polystyrene beads used in highly industrialized countries to make the medium more porous and well-drained, are not available in Pakistan. Sand and tree bark perform the same function just as well.

Nutrition. Excesses or deficiencies of any of the essential elements cause problems for healthy plant growth. Therefore a well-managed fertilizer programme should provide balanced nutrition at all times. N, P, and K are needed in larger quantities than the remaining 13 essential elements. Indoor plants should not be allowed to grow large or fast. If overgrown, they lose their usefulness as indoor plants. The fertilization programme should be planned carefully so that the plant remains healthy and maintains a suitable shape and size. The following schedule of fertilizer application has proved satisfactory under our conditions in Pakistan.

1. No feeding to the newly potted plant for 2–3 months.
2. Monthly application of a mixture of potassium nitrate, calcium phosphate, and ammonium sulphate, in a ratio 1:1:2, in water solution at the rate of 2 g/litre of water.
3. Fortnightly application of nitrogenous fertilizer at the rate of 7 g of ammonium sulphate per litre of water.

In some countries slow release fertilizers are available which supply nutrients to the plants slowly for up to six months or more.

Potting and repotting. Pots or containers should be selected carefully. Their size should be suitable for the size of the plant. A small pot cannot hold enough growing medium to support a large plant. The container should be durable, light in weight, attractive in appearance, and easy to handle. Dark-coloured containers placed in the sun build up heat in the growing medium, which is harmful to the plant. The container should have a sufficient number of drainage holes which are covered with crocks, taking care that they are not clogged.

The best potting time is just before the normal active period of growth of each kind of plant. Use pots of proper size, neither too big nor too small, and cover the drainage hole with crocks. Prune the roots and branches in order to balance the two and minimize the transpiration surface. Irrigate immediately after potting

Care and maintenance—summary. Some common guidelines for the care and maintenance of house plants are summarized here. Use a soil medium compatible with the natural habitat of the species, and fertilize plants at proper intervals. Provide sufficient light but avoid direct sunlight. Ensure fresh air but avoid drafts. Water carefully, but do not over-water, avoiding splashing water on the leaves. Turn the containers slightly every third or fourth day for uniform growth, otherwise the plants will grow towards the light source. Do not let the leaves touch walls or furniture. Wipe smooth-leaved plants occasionally to keep them healthy and attractive; plants with hairy leaves should not be washed. Lightly prune plants occasionally to keep them trim. Pinching plants will make them bushier, and with flowering plants, disbudding produces larger blooms. Remove dead, off-colour, and spotted leaves as they appear. Every one or two years repot plants as they outgrow their pots, discarding the old soil and pruning the roots slightly. If a plant appears to be unwell, diagnose the problem in consultation with a plant protection expert and take appropriate measures. Acclimatize plants before moving them from one location to another.

A list of common house plants is presented in Table 15.4.

Table 15.4 Selected house plants

Common names	Botanical names	Family
Shrimp plant	*Beloperone guttata*	Acanthaceae
Bloodleaf plant	*Iresine lindeni*	Amaranthaceae
Chinese evergreen	*Aglaonema crispa*	Araceae
Cheese plant	*Monstera deliciosa*	Araceae
Dumb cane	*Dieffenbachia picta*	Araceae
Pothos	*Scindapsus aureus*	Araceae
Philodendron	*Philodendron melanochryson*	Araceae
Peace lily	*Spathiphyllum blandum*	Araceae
Arrowhead vine	*Syngonium podophyllum*	Araceae
English ivy	*Hedera helix*	Araliaceae
Umbrella tree	*Schefflera arboricola*	Araliaceae
Araucaria	*Araucaria heterophylla*	Araucariaceae
Busy lizzie	*Impatiens walleriana*	Balsaminaceae
Wax begonia	*Begonia semperflorens*	Begoniaceae
Begonia	*Begonia elatior*	Begoniaceae
Begonia	*Begonia rex* (hybrid)	Begoniaceae
Ananas	*Ananas comosa*	Bromeliaceae
Orange star	*Guzmania lingulata*	Bromeliaceae
Wandering jew	*Tradescantia fluminensis*	Commelinaceae
Good–luck plant	*Kalanchoe daigremontiana*	Crassulaceae
Flaming Katy	*Kalanchoe blossfeldiana*	Crassulaceae
Cycas	*Cycas revoluta*	Cycadaceae
Croton	*Codiaeum variegatum*	Euphorbiaceae
Crown–of–thorns	*Euphorbia splendens*	Euphorbiaceae
Poinsettia	*Euphorbia pulcherrima*	Euphorbiaceae
Geranium	*Pelargonium grandiflorum*	Geraniaceae
Asparagus	*Asparagus plumosus*	Liliaceae
Asparagus	*Asparagus sprengeri*	Liliaceae
Aspidistra	*Aspidistra elatior*	Liliaceae
Spider plant	*Chlorophytum comosum*	Liliaceae
Cordyline (ti plant)	*Cordyline terminalis*	Liliaceae
Dracaena	*Dracaena deremensis*	Liliaceae
Ruscus	*Ruscus hypoglossum*	Liliaceae
Sansevieria	*Sansevieria trifasciata*	Liliaceae
Calathea	*Calathea zebrina*	Marantaceae
Maranta	*Maranta leuconeura*	Marantaceae
Weeping fig	*Ficus benjamina*	Moraceae
Rubber plant	*Ficus elastica*	Moraceae
Common palm	*Livistona chinensis*	Palmaceae

Common names	Botanical names	Family
Ruscus	*Ruscus hypoglossum*	Liliaceae
Sansevieria	*Sansevieria trifasciata*	Liliaceae
Calathea	*Calathea zebrina*	Marantaceae
Maranta	*Maranta leuconeura*	Marantaceae
Weeping fig	*Ficus benjamina*	Moraceae
Rubber plant	*Ficus elastica*	Moraceae
Kentia palm	*Howeia belmoreana*	Palmaceae
Common palm	*Livistona chinensis*	Palmaceae
Peperomia	*Peperomia obtusifolia*	Piperaceae
Maidenhair fern	*Adiantum*	Polypodiaceae
Pomegranate	*Punica granatum*	Punicaceae
Ixora	*Ixora chinensis*	Rubiaceae
Aluminum plant	*Pilea cadierei*	Urticaceae
Artillery plant	*Pilea microphylla*	Urticaceae
Clerodendrum	*Clerodendron thomsonae*	Verbenaceae

QUESTIONS

1. What is meant by landscape architecture?
2. What considerations are important in planning the landscape in the premises of a house?
3. Name the elements of design.
4. What is the significance of colour in a design?
5. What is meant by scale and proportion?
6. List 10 common house plants suitable for your region.
7. Describe three indoor plants suitable for growing in your drawing room.
8. What are some drawbacks to growing plants in the house?
9. Why should feeding of house plants be kept at the maintenance level?
10. Give examples of (a) flowering indoor plants, (b) indoor trees, (c) indoor vines, (d) indoor foliage plants, (e) indoor succulents.
11. Name the most suitable grass for establishing a lawn in your region.
12. What is the most economical method of installing a lawn?
13. When is the best time for planting a lawn?
14. What are the disadvantages of close mowing?

15. What cultural practices limit the rooting depth of lawn grass?
16. Write a short note on the maintenance of a lawn.
17. What is the difference between a perennial and a biennial?
18. Name the major criteria for selecting perennials.
19. Describe the elements important for planning a perennial border.
20. Suggest perennials for planting in a wide border at your campus.
21. What are the major considerations in selecting the trees and shrubs to plant in a landscape?
22. What are best times for planting trees?
23. List 10 flowering trees with their flower colour and flowering times.
24. Describe each of the following: (a) Hybrid Tea rose, (b) Miniature rose, (c) Pompon rose.
25. How would you differentiate between (a) a floribunda and a grandiflora rose, and (b) a rambler and a climber rose?
26. What is the best time for pruning roses?
27. Write a short note on the feeding of roses.
28. What is a wall garden?
29. List some plants suitable for growing in rock and water gardens.

REFERENCES

Anonymous. Growing better bulbs. Your Garden 23(20):3–7; 1970.

Cooper, E. L. Agriscience, fundamentals and applications. Albany, New York: Delmar Publishers; 1990:p. 303–346.

Flemmer, W. III. Shade and ornamental trees in colour. New York: Grosset and Dunlap; 1965.

Graf, A. B. Exotica. East Rutherford, New Jersey: Roehars; 1976.

Grindal, E. W. Every day gardening in India. Bombay: D.B. Taraporevala Sons and Co; 1939.

Hackett, B. Planting design. London: E. and F. N. Spon Ltd.; 1979.

"Hakea". Year–round lawn care. Your Garden 23(8):3–9; 1970.

Harris, R. W. Arboriculture—care of trees, shrubs, and vines in the landscape. Englewood Cliffs, New Jersey: Prentice Hall; 1983.

Hollis, L. ed. The rose annual. St. Albans, Hertfordshire, UK: The Royal National Rose Society; 1969.

Hollis, L. Roses. London: The Hamlyn Publishing Group Ltd.; 1974.

Janick, J. Horticultural science. San Francisco: W. H. Freeman; 1979.

Laurie, A.; Ries, V. H. Floriculture, fundamentals and practices. New York: McGraw–Hill Book Company; 1950.

Little, B. ed. Allen Seale's complete guide to Australian gardening. Port Melbourne, Victoria, Australia: Treasure Press; 1985.

Manaker, G. H. Interior plantscapes, installation, maintenance and management. Englewood Cliffs, New Jersey: Prentice Hall; 1981.

Manning, L. The how and why of better gardening. Toronto, New York, London: D. Van Nostrand; 1955.

Mathias, M. E. ed. Flowering plants in the landscape. Berkeley, CA: University of California Press; 1982.

Perry, F. Gardening in color. London: The Hamlyn Publishing Group, Ltd.; 1963.

Pool, M. J. ed. 20th century decorating architecture and gardens. London: Weidenfeld and Nicolson; 1980.

Proctor, R. Perennials. London: Cassel Publishers, Ltd.; 1990.

Redwood, R. Ground cover plants. Your Garden 23(7):3; 1970.

Shepherd, H. W. Woody plants for landscape use in California. Cir. 109. Berkeley, CA: California Agricultural Extension Service, College of Agriculture, University of California, Berkeley.

Snyder, R. ed. The complete book for gardeners. Princeton, New Jersey: D. Van Nostrand; 1964.

Spickernell, C. ed. Nature's bounty. Karachi: Pakistan Herald Press; 1970.

Squire, D.; McHoy, P. eds. The book of house plants. 59 Grosvenor Street London: Octopus Books Ltd.; 1978.

Wyman, D. Shrubs and vines for American gardens. New York: Macmillan; London: Collier–Macmillan.

Younger, M. Spring bulbs. Your Garden 22(2):3, 34–35; 1969.

ABBREVIATIONS AND ACRONYMS

2,4–D 2,4–dichlorophenoxyacetic acid

2,4,5–TP 2,4,5–trichlorophenoxypropionic acid

2–AB 2–aminobutane

4–CPA 4–chlorophenoxyacetic acid

A ampere(s)

AARI Ayub Agricultural Research Institute (Faisalabad)

ADBP Agricultural Development Bank of Pakistan

ADP adenosine diphosphate, a product obtained from hydrolysis of ATP

AECL Atomic Energy Centre, Lahore

AFC antifungal complex

a.i. active ingredient

AMSL Agriculture Marketing and Storage Limited

ATP adenosine triphosphate

CA controlled atmosphere

cc cubic centimetre(s)

CCC Cycocel

cd candela: unit of luminous intensity

cm centimetre(s)/centimeter(s)

CoA Co-enzyme A

DNA deoxyribonucleic acid

E Einstein: the energy in 1 mole, or Avogadro's number (6.02×10^{23}), of photons

EDTA Fe–ethylene diamine tetra-acetic acid

FAO Food and Agriculture Organization of the United Nations

ft foot (feet)

ft–c foot–candle: the illuminance from a standard candle received on a one-foot-square surface one foot from the candle

g gram(s)

Gy Gray(s): unit of absorbed radiation dose (J/kg = $m^2 s^{-2}$)

h hour(s)

ha hectare(s)

IAA indoleacetic acid

IBA indolebutyric acid

in inch(es)

IPM integrated pest management

K Kelvin

kg kilogram(s)

kGy kilogray (SEE **Gy.**)

km kilometre(s)/kilometer(s)

KOCN potassium cyanate

l litre(s)/liter(s)

lm lumen: replaces foot–candle as unit of luminous flux; definition and formula: cd · sr

lx lux: one lumen per square meter

m metre(s)/meter(s)

MA modified atmosphere

MH maleic hydrazide

ml millilitre(s)/milliliter(s)

mm millimetre(s)/millimeter(s)

MMT million metric tonnes

mol mole(s)

mRNA messenger ribonucleic acid

NAA naphthaleneacetic acid

NADP nicotinamide adenine dinucleotide phosphate

NBF National Book Foundation

nm nanometre(s)/nanometer(s)

NWFP AU Northwest Frontier Province Agricultural University

PAR photosynthetically active radiation

PARC Pakistan Agricultural Research Council

PMA phenylmercuric acetate

ppm parts per million

PPP pentose phosphate pathway

RH relative humidity

RNA ribonucleic acid

RQ respiratory quotient

s second(s)

SAU Sindh Agriculture University

SI Système International d'Unités. Modernized system of metric units. It is constructed from seven basic units for independent quantities, plus units for plane and solid angles. This system is gradually replacing the centimeter–gram–second (CGS) and meter–kilogram–second (MKS) systems.

SOPP sodium orthophenyl phenate

sr steradian(s): unit of solid angle

t tonne(s)

TBZ thiabendazole

TCA tricarboxylic acid

TIPAN Transformation and Integration of the Provincial Agricultural Network

TSS total soluble solids

TTC triphenyl tetrazolium chloride

UAF University of Agriculture, Faisalabad

UGC University Grants Commission

USAID United States Agency for International Development

yd yard(s)

GLOSSARY

absolute humidity. The weight of water vapor per unit *volume* of air, expressed as grams per cubic meter.

absolute zero. The temperature at which matter contains no heat energy. It is equivalent to −273°C or −460.4°F.

achene. A small, one-seeded, indehiscent fruit developing from a single carpel, e.g. sunflower, safflower, buttercup, or sycamore. Generally achenes are found in aggregate form as in strawberries.

albedo. (1) Whitish mesocarp in citrus fruit below colored skin; (2) ratio of the intensity of light reflected from a surface to the intensity of light received.

albumen. Starchy, nutritive material accompanying the embryo in the seed.

allogamous. Structurally adapted for cross-pollination (flowers).

allogamy. (1) Cross-pollination. The transference of pollen grains from the anther of one flower to the stigma of another flower situated on another plant of a different variety or cultivar of a related species. (2) Fusion of male and female gametes derived from genetically dissimilar individuals of the same species.

amyloplasts. Sites of starch grain development; also known as **plastids**.

androecium. Male reproductive component of the flower.

anemophilous. Pollinated by wind-borne pollen (flowers).

anemophily. Pollination by wind.

angiosperms. Flower bearing plants with enclosed seeds—the largest, most widely spread, and most advanced group of plants.

annual. A plant that flowers, sets seeds, and dies within a single growing season.

anther. Pollen-producing element of the androecium.

anther sacs. Chambers or loculi of anther containing fine powdery or granular mass of pollen grains.

antipodal cells. Haploid cells, usually three in number, found in the embryo sac at the end opposite the micropyle.

anti–auxins. Compounds which competitively inhibit the action of auxins. Examples are coumarin, parascorbic acid (a naturally occurring compound), and phenylbutyric acid (a synthetic compound).

apical dominance. Dominant growth of central trunk; this occurs when hormones produced in the stem apices of branches travel down the stem and inhibit or reduce branching and growth from lateral buds.

apocarpous. Describes a gynoecium in which the carpels are free.

apomictic species. Species which produce seed from vegetative cells and not through sexual means. Polyembryonic species are an example.

areole. A modified cushion–like side shoot carrying spines or hair, unique to the cactus family. The areole gives rise to offsets and flowers.

aril. An appendage or an outer covering of a seed, arising from or near the hilum funiculus.

asexual propagation. Reproduction without union of gametes from two parents.

autogamy. Self–pollination, i.e. pollination occurring within a hermaphrodite or bisexual or perfect flower, or between two such flowers situated on the same plant, or of the same cultivar. This leads to fusion of male and female gametes derived from genetically similar sources, usually the same individual. **Autogamous** describes plants in which self–fertilization is the rule.

auxin. An organic substance which promotes growth along the longitudinal axis. A growth hormone is an auxin, but not all auxins are hormones. Synthetic auxins or growth regulators have been referred to as *hormones*, which is a misuse of the term. The term *hormone* should be reserved for the natural growth regulators, which are present in any organism/plant.

auxin antagonists. Substances which are themselves inactive, such as trans–cinnamic acid and DCA (2,4–dichloroanisole), but counteract the activities of auxins.

auxin synergists. Compounds that increase the effectiveness of the auxins. They cannot promote growth without the presence of auxins. Examples are ascorbic acid (vitamin C) and TIBA (2,3,5–tri–iodobenzoic acid).

available water. Difference between the wilting point and field capacity.

axil. The angle between leafstalk and stem. Any new–growth flower bud arising from an axil is called axillary.

bicarpellary. Describing a compound ovary with two carpels.

biennial. Producing vegetative growth (rosette) in the first season, and flowers and seeds in the second season.

bilocular ovary. Describes a syncarpous ovary with two chambers or loculi.

bisexual. Having stamen and pistil in the same flower.

bolt. To send up a flowering stalk with extended internodes, as in cauliflower.

bract. A leaf–like organ subtending an inflorescence.

budding. Insertion of a single bud as scion into the rootstock.

bulb. A storage organ that usually grows underground and contains embryo leaves and sometimes flowers.

bunchy top. A disorder affecting the vegetative parts and inflorescence of mango. Malformed branches show stunted growth, with small and bunchy leaves.

calorie. The amount of heat required to raise 1 g (cm^3) of water 1°C.

cambium layer. Single layer of living cells which lies between the bark and the sap wood.

campanulate. Bell–shaped, with a broad tube as long as or longer than broad, and a flaring limb or lobes.

capillary water. Soil water which is held by surface tension (cohesion) and by adhesion to the soil particles. It occurs as a thin film around particles in the micropores.

carotenoids. Accessory yellow, orange, or red fat–soluble pigments found in all photosynthesizing cells. They gather light during photosynthesis. These pigments are found in some fruits and vegetables.

carpel. One of the foliar units of a compound ovary or pistil; a female or megasporophyll of an angiosperm flower.

carpophore A wiry structure by which the schizocarp is suspended or supported from a short central axis.

cation exchange capacity. The quantity of exchangeable cations expressed in milliequivalents per 100 g of soil.

cauliflorous. Describes a plant with flowers or inflorescences on the stem or trunk.

chalaza. The basal part of an ovule where it is attached to the funiculus.

chemically combined water. Water which is a chemical component of soil particles.

chiasmata. Sites at which chromosomes break and join during crossing over.

chilling injury. Plant tissue damage at the membrane level to fruits of tropical and subtropical origin exposed to temperatures below 10°C but above their freezing point.

chilling requirement. The number of hours of cold after which dormancy will break.

chlorophyll. A green pigment found in plants; necessary for photosynthesis.

chloroplast. Plant organelle containing chlorophyll and surrounded by two bi-layered membrânes. The chloroplasts are the photosynthetic apparatus of the cell. Thylakoid, grana, and stroma are parts of the chloroplast.

chromophore. The light-absorbing portion of a pigment.

chromoplasts. Bodies which develop from mature chloroplasts on the degradation of chlorophyll. They contain carotenoids, the yellow/red pigments in many fruits.

chromosome. A threadlike structure in the nucleus carrying the genetic information.

citric acid cycle. SEE **tricarboxylic acid (TCA) cycle**.

cleistogamy. The production of flowers that do not open to expose the reproductive organs.

climacteric. A period in the development of certain fruits during which a series of biochemical changes is initiated by the autocatalytic production of ethylene, marking the change from growth to senescence and involving an increase in respiration.

clone. A population propagated vegetatively from a single original stock. More precisely defined as a population of plants of identical genotypic composition traceable to a single ancestral zygote; or a colony of cells with identical genetic makeup produced through mitosis.

coherent. Having like parts in close contact but not fused.

complete flower. A flower in which all the four whorls i.e., calyx, corolla, androecium, and gynoecium, are present.

connective tissue. The filament or tissue connecting the two cells or thecae of an anther.

corm. An enlarged base of an herbaceous stem with nodes and internodes visible on the structure; a typical example is gladiolus.

cotyledons. The seed leaf or leaves in the embryo.

cristae. Sheetlike or tubelike extensions formed from the highly folded inner membrane of the mitochondrion.

critical light period. The dividing line between day lengths favourable to vegetative growth and those inducing flowering and seed formation.

crossing over. The reciprocal exchange of segments between non-sister chromatids of homologous chromosomes leading to redistribution or recombination of linked genes.

cross-pollination. SEE **allogamy**.

cultivar. (1) A variety which has originated by cultivation rather than in the wild. (2) A named group of plants within a cultivated species which maintain its identity when propagated by sexual or asexual means; horticultural variety.

cutting. A portion of stem, root, shoot, or leaf removed from a parent plant and treated so that it produces roots and grows into a new plant.

deciduous. Shedding leaves annually at the end of the active growth period and producing new foliage at the end of the rest period.

definitive nucleus. Diploid nucleus found in the centre of embryo sac after the fusion of two haploid polar nuclei.

dehorning. Severe cutting back of trees and shrubs to keep them from becoming too tall.

dew point. The temperature at which relative humidity reaches 100 percent.

dichogamous. Having stigmas and anthers (of a flower) mature at different times (*Dacus*, *Delphinium*, and *Lactuca*).

dicotyledons. Angiosperms having embryos with two cotyledons.

dioecious. Having staminate (male) and female (pistillate) flowers on different plants (as in date, spinach).

diploid. Having two sets of chromosomes in the somatic nuclei.

division. Separating vegetative parts like rhizomes, offsets, crown, runners, and suckers from the parent plant and establishing them as independent plants.

double fertilization. The process where two male gametes enter the embryo sac and both participate in fertilization, one with the egg and the other with the definitive nucleus.

draws. Sprouts produced from tubers.

egg apparatus. Structure consisting of an egg and two synergids.

embryo. Rudimentary plant within a seed, usually developing from a zygote.

endocarp. The inner layer of pericarp or fruit wall.

endosperm. The starch and oil–containing tissue of many seeds.

entomophily. Pollination by insects.

epicarp. Outer layer of the pericarp or fruit wall.

epicotyl. The part of the plant embryo axis situated above the cotyledons.

epidermis. Outermost cells of the primary plant body.

epigynous. Describing an arrangement of floral parts in which stamens, sepals, and petals are inserted above the ovary, giving an inferior ovary.

epinastic agent. Substances which bring about swelling of cells instead of normal cell–elongation, causing twisting and deformation of leaves and stems.

etiolated. Showing symptoms of being grown in darkness or in dim light—weakness, with yellow or pale–green leaves, tall lanky stems, and a sickly appearance.

evapo–transpirational potential. A measure of both evaporation from the soil and transpiration by plants.

evergreen. Having leaves which persist through winter.

exine. (1) Thick outer wall of the seed; (2) outer membranous coat of a pollen grain.

exocarp. SEE **epicarp**.

false fruit. A fruit that incorporates tissues other than those derived from the gynoecium.

fermentation. The process whereby anaerobic organisms or aerobic organisms under anaerobic conditions accumulate either lactate or ethanol; the second stage in respiration.

fertilization. An essential part of sexual reproduction involving the fusion of haploid male and female gametes to form a diploid zygote.

field capacity. Water left after the drainage of gravitational water (after 24–48 hours). When all free water drains from the soil, it is said to be at field capacity.

filament. Stalk of a stamen.

flavedo. Coloured epicarp in citrus fruit.

floriculture. Flower production.

florigen. Hormone which induces flowering.

forcing (vegetable). Production of vegetables out of their normal season of outdoor production.

free water. SEE **gravitational water**.

freeze. Any situation where the average temperature drops below 0°C. **Radiational freeze** is associated with calm conditions, radiational cooling, and temperature inversion. **Advective freezing** occurs as a result of the invasion of a large, cold air mass.

frets. Thylakoids extending from stack to stack to form intergranal lamella.

funicle. A slender stalk by which each ovule is attached to a placenta.

gametogenesis. Formation of male and female reproductive cells in the flower.

gamophyllous. Having leaves, or foliar units, connate by their edges.

generative nuclei. The two male gametes that are formed by division of the generative cell.

gibberellins. Substances possessing the same carbon skeleton as gibberellin A_3 (GA_3) or very closely related to it, and biologically active in stimulating cell division, or cell elongation in plants.

glycolysis. The conversion of glucose, glucose–1–P, glucose–6–P, or fructose to pyruvic acid during respiration.

grafting. A process by which a piece of scion carrying one or more buds is attached to a rootstock in such a way that the cambiums of both scion and rootstock come in firm contact, so that the new secondary tissue resulting from cambial cell division in the scion and rootstock is closely knitted.

grana. Arrays of stacked thylakoids, the enclosed basic unit structure.

granulation. A condition characterized by enlarged, hardened, and apparently dry juice vesicles; also called **riciness.**

gravitational water. Water held very loosely in the soil macropores which moves downward and is lost quickly from the top of well–drained soils (also called **free water**).

green manuring. The process of returning an entire plant to the soil.

growth regulator. An organic compound which in a small quantity (around 1 ppm) modifies a physiological process in plants. These substances stimulate, inhibit, or otherwise alter growth. They include auxins, anti–auxins, epinastic agents, gibberellins, kinins, and inhibitors.

gynoecium. The female element of a flower.

habit. The normal direction of growth of a plant species.

haploid. A nucleus or individual containing only one representative of each chromosome of the chromosome complement.

heading back. Removal of only terminal portions of shoots or branches.

heat of fusion. The amount of heat required to change 1 g of a substance at its melting point from the solid to the liquid state.

heat of vaporization. The amount of heat required to change 1 g of a substance at its boiling point from the liquid to the vapor state.

heat unit. The sum of the differences between the daily mean temperatures in a given region and a fixed base temperature (commonly the minimum temperature at which growth occurs).

heaving. Upward displacement of soil caused by alternate freezing and thawing.

hemicellulose. Any of a variety of polysaccharides found in plant cell walls often in close association with cellulose.

herbaceous. Having soft, non–woody growth.

herkogamous. Describes plants in which self–pollination is impossible because of the relative position of stigma and anthers.

hermaphrodite. Having stamens and pistil in the same flower, i.e. containing both male and female sexual parts.

heterostylous. Having stamens and styles of different lengths (e.g. linseed).

Hill Reaction. Oxygen evolution by chloroplasts in the presence of light, water, and an artificial electron acceptor (potassium ferricyanide).

hilum. The point where the funicle meets the ovule.

hispid. Bristly, bearing dense, erect, straight, harshly stiff hairs.

homogamy. Simultaneous maturation of androecium and gynoecium.

homostylous. Having both stamens and styles of the same length.

hormone. A substance produced in any part of an organism and transferred to another part where it influences a specific physiological process.

hydrophily. Pollination by pollen carried on water.

hydroponics. Techniques of growing plants in solutions rather than in soil.

hygroscopic water. Water held very tightly in the form of a thin layer around the soil particle; it is not normally available to plants.

hypertrophy. Abnormal overdevelopment due to an increase in cell size.

hypobaric storage. Low-pressure (partial vacuum) storage.

hypocotyl. Axis or stem of a seedling below the cotyledons.

hypogynous. Inserted below the gynoecium and free from it.

hypophysis. Basal part of suspensor giving rise to the apex of the radicle.

imbricate. Having overlapping, shinglelike leaves.

incomplete flower. Flower lacking one of the perianth whorls, as calyx or corolla, or both.

inferior ovary. SEE **epigynous**.

inflorescence. A flowering system consisting of more than one flower.

infructescence. A multiple or composite fruit developing from the entire inflorescence.

insolation. Radiation received from the sun.

integument. One or two envelopes that cover the ovule.

intine. (1) Thin inner wall of the seed. (2) a thin, delicate, cellulose layer internal to the exine in a pollen grain.

kinins. A group of substances of small molecular weight which possess hormonelike properties and stimulate cell division.

Krebs cycle. SEE **tricarboxylic acid (TCA) cycle**.

layering. Rooting of shoots, stem, or branches while they are still attached to the parent plant.

lysimeter. A large container of soil in which a crop is grown; water loss is determined by weight loss of the entire container.

megaspore. The larger of the two types of haploid spores formed after meiosis in heterosporous species.

megasporogenesis. Development which takes place in the female part of the ovule (ovary) to give rise to female reproductive cells.

mericarp. Any of the one-seeded portions that result when a compound fruit divides at maturity.

mesocarp. The middle layer of a pericarp.

metaxenia. An effect of pollen on maternal fruit tissue.

metric tonne. 1000 kilograms.

micropyle. The opening or 'mouth' of the ovule.

microspore. The smaller of the two types of spore formed after meiosis in heterosporous species.

microsporogenesis. Formation of microspores or pollen grains in the anther of a flower.

mitochondrion (pl. mitochondria). A double-membrane-bounded organelle within the cell which contains the respiratory enzymes of the TCA cycle and the respiratory electron-transport system which synthesizes ATP.

modified-basin system. Irrigation system characterized by restricting the supply of irrigation to the spread of the trees by constructing basins around each trunk.

monocotyledons. A subclass of the angiosperms containing all the flowering plants having embryos with one cotyledon.

monoecious. Bearing unisexual flowers, with flowers of both sexes on the same plant.

monothecous. Having a unilocular anther lobe.

multilocular. Describing an ovary with many loculi or chambers.

NADPH. Nicotamide adenine dinucleotide phosphate reduced, an important source as reducing agent.

nectaries. The nectar-secreting organs.

neuter. Having neither stamens nor pistils (flower).

nucellar seedlings. Seedlings produced from the nucellus cells. In polyembryony, all the seedlings except the one zygotic seedling.

oceanic effect. The moderating climatic effect of oceans on land.

offsets. Small new plants which grow out from the parent and can be separated as independent plants.

olericulture. Vegetable production; the study of vegetable production.

oosphere. A large non-motile gamete that is rich in nutrients and is normally designated as female.

oospore. A thick-walled zygote that is formed following the fertilization of an oosphere.

organoleptic properties. The properties of food which are exhibited to the human senses, e.g. appearance, texture, taste, and flavour.

ovary. The swollen basal part of the carpel in angiosperms which contains the ovules.

ovule. The female gamete and its protective and nutritive tissue that develops into the seed after fertilization.

ovum. SEE **oosphere**.

panicle. An indeterminate branch system whose primary axis bears branched secondary axis and pedicellate flowers.

parenchyma. Relatively unspecialized tissue usually composed of more or less isodiametric polyhedral cells with thin non-lignified cellulose cell walls and living protoplasts.

pectin. A structural polysaccharide found in plant cell walls.

peduncle. A common leafless axis bearing several flowers.

perennial. Plant which lives for three years or longer.

perfect flower. A flower which contains both androecium and gynoecium.

perianth. Collective term for the calyx and corolla.

pericarp. The wall of a fruit, derived from the maturing ovary wall.

perisperm. Nutritive tissue derived from the nucellus that is found in the seeds of certain plants in which endosperm does not completely replace the nucellus.

petaloid. Petallike; resembling a petal in colour and shape.

photoperiodism. The developmental responses of plants to the relative lengths of light and dark periods.

photophosphorylation. Process involving a series of electron-transport coupling factors (proteins) which transfer electrons along an electrochemical gradient, yielding ATP in the presence of light.

photosynthesis. A process of CO_2 fixation in the presence of light and its conversion into sugars by green plants.

phototropism. The tendency of plants to bend in the direction of more intense light when they are irradiated unequally from two sides.

phytochrome. A photoreceptive pigment involved in photomorphogenesis which is ubiquitously present in all plant parts.

phytohormone. A hormone or regulator specifically produced in a plant, which regulates plant physiological processes.

pigment. A light-absorbing molecule.

pinching. Removal of a plant part soft enough to be pinched by hand.

pistillate. Female: having one or more pistils and no functional stamens.

placenta. The tissue by which spores, sporangia, or ovules are attached to the maternal tissue.

placentation. The mode of arrangement of the ovules within an ovary.

plant growth regulators. Synthetic products which produce reactions almost identical to those caused by natural hormones.

plant hormones. Natural substances produced by plant tissues in small quantities, especially at the growing points, and transported to other parts of a plant where they are required to regulate its activities.

plastids. SEE **amyloplasts.**

plough sole. A zone of compaction beneath the soil surface caused by ploughing with a heavy plough repeatedly along the same path.

plumule. Portion of the plant–embryo axis lying within the two cotyledons; first shoot above the cotyledons of a young plant.

pollen grain. A microspore which has become modified by wall modification and division of the microspore nucleus into tube and generative nucleus.

pollen mother cell. A somatic cell that, after meiosis, forms a tetrad of pollen grains.

pollen sac. A chamber in which pollen grains are formed in angiosperms and gymnosperms.

pollen tube. An outgrowth of the intine of the pollen grain that, on germination, emerges through an aperture in the exine and grows towards the egg, carrying male gametes with it.

pollination. Transfer of pollen from the anther to the stigma.

polyadelphous. Having stamens united into three or more groups.

polyandrous. Having separate stamens freely inserted on the receptacle.

polycarpellary. Having more than two carpels in the gynoecium.

polyembryony. Production of multiple embryos within a single seed.

polygamous. Having monosexual and bisexual flowers on the same plant.

pomology. The study of fruit production.

precambium. Beginning of the vascular system.

proembryo. The young plant individual after fertilization but before tissue differentiation into embryo and suspensor tissue.

protoderm. Precursor of the epidermis.

protogynous. Refers to plants in which maturation of stigmas takes place ahead of anthers (magnolia and mango).

pruning. Management of plant structure and fruiting wood.

pseudogamy. Embryo sac formed from stimulus (but not fertilization) by male gamete (false fertilization). The offspring may be haploid or diploid, showing only maternal characters.

raceme. An indeterminate single axis bearing pedicellate flowers.

radicle. The part of the plant–embryo axis directed towards the pointed end; embryonic root of a germinating seed.

ratoon. A shoot growing from the roots of a plant that has been cut down. A ratoon crop is grown from the roots of a previous crop without reseeding (as in banana and sugarcane).

relative humidity. The amount of water vapor present in the air as a percent of the amount at saturation for a given temperature and pressure.

replum. Partition between two loculi of cruciferous fruits.

respiration. An active process of all living organisms absorbing O_2 and releasing CO_2. It is the oxidative degradation of organic compounds (carbohydrates, fats, organic acids, and proteins) to yield usable energy.

respiratory quotient (RQ). The ratio of CO_2 evolved to O_2 uptake during respiration. It is an indication of the kind of substrate oxidized.

rest period. The period during the year when plants or their seeds are relatively inactive, producing little or no growth.

rhizome. Swollen underground stem; an organ of food storage and propagation.

riciness. SEE **granulation**.

rootstock. Plant or plant part on which a scion variety is budded or grafted.

rotate. Wheel-shaped, with a very short tube and a broad limb at right angles to it.

runner. Long, horizontal, prostrate shoot rooting to form a new plant at its tip or elsewhere along its length.

saline soils. Soils which contain harmful quantities of chlorides, sulphates, and carbonates of sodium, potassium, calcium, and magnesium.

scion. A bud or piece of twig grafted or budded onto a rootstock; it provides top and fruit bearing surfaces and synthesizes food which is transported to other parts of the plant.

sclerenchyma. Strengthening tissue with thick, often lignified cell walls and usually lacking a living protoplast at maturity.

secondary nucleus. A nucleus formed by the fusion of polar nuclei which gives rise to the endosperm after fertilization (SEE **definitive nucleus**).

seed. The ripened ovule.

self-pollination. SEE **autogamy**.

septa. Partition in a fruit or any other organ.

sexual propagation. In plants, reproduction by seed.

slurry. Fresh dung and urine combined.

sodic soils. Soils which have high levels of alkalis, particularly sodium hydroxide (also called **alkali soils**).

spadix. A racemose inflorescence in which the flowers are sessile and borne on an enlarged fleshy axis.

spathaceous. Resembling or furnished with a spathe.

specific heat. The amount of heat required to change the temperature of 1 g of a substance by 1 °C.

specific humidity. The weight of water vapor per unit *weight* of air, expressed as grams per kilogram.

spike. A racemose inflorescence in which the flowers are sessile and borne on an elongated axis.

sport. SEE **strain**. A variant among a clonal population.

staminate. Furnished with stamens but no functional pistil.

staminode. An abortive, infertile stamen which does not produce pollen, as found in mango flowers.

stigma. The receptive tip of the carpel, which receives pollen at pollination and on which the pollen grain germinates.

stolons. Runners by which plants reproduce vegetatively by sending out roots at each joint.

stone fruit. A drupe or drupelet.

stooling. A vegetative method of raising rootstock; also called **mound–layering**.

strain. A variant among a clone.

stratification. Storing of seeds in a cold humid environment for some time to allow an after–ripening process that promotes germination.

stroma. Material surrounding the thylakoids and within which the grana and fretwork are imbedded.

style. The sterile portion of the carpel between the ovary and the stigma.

subulate. Awl–shaped, tapering from base to apex and sharp–pointed.

sucrose. A non–reducing disaccharide of glucose and fructose.

superior ovary. SEE hypogynous.

suspensor. A line of cells that differentiates from the proembryo by mitosis and anchors the embryo in the parental tissue.

syconus. A type of pseudocarp in which achenes are borne inside the receptacle.

syncarpous. Having carpels united.

synergids. Two haploid nuclei at the micropylar end of the embryo sac.

tapetum. Layer of food-containing cells which encloses groups of spore mother cells in vascular plants.

tegmen. Thin, whitish inner layer of seed coat.

tensiometer. An instrument to measure the water tension/potential in soil.

testa. Outer, brown, comparatively thick seed coat.

thalamus. The receptacle of a flower.

thinning out. Removal of certain shoots, canes, spurs, or branches completely and entirely from the base.

tissue culture. *In vitro* and aseptic cultivation of any plant part (cells or tissues) on a nutrient medium in an artificial environment.

tonne. Metric ton(ne) (= 1000 kg).

tonoplast. Semi-permeable membrane surrounding the vacuoles.

topiary. Art of growing decoratively pruned and shaped trees and hedges.

tricarboxylic acid (TCA) cycle. A cyclic process whereby acetate is oxidized to CO_2 and water, yielding reduced NAD and FAD.

tricarpellary. Having three carpels.

trilocular. Of an ovary containing three chambers.

truck gardening. The production of vegetables in relatively large quantities for commercial markets.

tube nucleus. A large nucleus formed within the pollen grain of angiosperms along with one or two smaller generative nuclei. SEE ALSO **vegetative nucleus**.

unilocular ovary. Ovary containing a single chamber.

vacuoles. Fluid cell reservoirs containing various solutes like sugars, salts, and amino and organic acids.

valvate. Meeting by the edges without overlapping.

vapor pressure. That part of the total atmospheric pressure which is due to water vapor, expressed in millimeters of mercury (mm Hg) or millibars (1 mm Hg = 1.32 millibars).

vegetable. The edible portion of an herbaceous plant used fresh or processed.

vegetative nucleus. One of two nuclei formed when the nucleus of a pollen grain divides for the first time mitotically.

vernalization. Exposure to the cold period during the winter.

viticulture. The cultivation of grapes.

wilting point. The soil–water level reached after the depletion of all available capillary water.

xanthophyll. The oxidation product of carotenoids found in chloroplasts.

xenia. The direct influence of pollen upon endosperm.

zoophily. Pollination by pollen carried on animals.

zygote. The cell formed following the fusion of male and female gametes.

INDEX